Fundamentals for the Assessment of Risks from Environmental Radiation

NATO Science Series

A Series presenting the results of activities sponsored by the NATO Science Committee. The Series is published by IOS Press and Kluwer Academic Publishers, in conjunction with the NATO Scientific Affairs Division.

A. Life Sciences	IOS Press
B. Physics	Kluwer Academic Publishers
C. Mathematical and Physical Sciences	Kluwer Academic Publishers
D. Behavioural and Social Sciences	Kluwer Academic Publishers
E. Applied Sciences	Kluwer Academic Publishers
F. Computer and Systems Sciences	IOS Press

1. Disarmament Technologies	Kluwer Academic Publishers
2. Environmental Security	Kluwer Academic Publishers
3. High Technology	Kluwer Academic Publishers
4. Science and Technology Policy	IOS Press
5. Computer Networking	IOS Press

NATO-PCO-DATA BASE

The NATO Science Series continues the series of books published formerly in the NATO ASI Series. An electronic index to the NATO ASI Series provides full bibliographical references (with keywords and/or abstracts) to more than 50000 contributions from international scientists published in all sections of the NATO ASI Series.
Access to the NATO-PCO-DATA BASE is possible via CD-ROM "NATO-PCO-DATA BASE" with user-friendly retrieval software in English, French and German (© WTV GmbH and DATAWARE Technologies Inc. 1989).

The CD-ROM of the NATO ASI Series can be ordered from: PCO, Overijse, Belgium.

Series 2: Environmental Security – Vol. 55

Fundamentals for the Assessment of Risks from Environmental Radiation

edited by

Christa Baumstark-Khan
DLR Institute of Aerospace Medicine,
Radiation Biology Section,
Cologne, Germany

Stanislav Kozubek
Institute of Biophysics,
Czech Academy of Sciences,
Brno, Czech Republic

and

Gerda Horneck
DLR Institute of Aerospace Medicine,
Radiation Biology Section,
Cologne, Germany

Springer-Science+Business Media, B.V.

Proceedings of the NATO Advanced Research Workshop on
Fundamentals for the Assessment of Risks from Environmental Radiation
Brno, Czech Republic
6–10 October 1997

A C.I.P. Catalogue record for this book is available from the Library of Congress.

ISBN 978-0-7923-5668-4 ISBN 978-94-011-4585-5 (eBook)
DOI 10.1007/978-94-011-4585-5

Printed on acid-free paper

CONTENTS

Risk Assesssment

DNA Injuries, Damage Induction and Removal

Chromosome Damage

Cellular Responses

Tissue and Organism Responses

Radiation Protection Concepts.

Appendix

PREFACE

Human health as well as aquatic and terrestrial ecosystems are threatened from increasing levels of environmental radiation of various sources, many of them of anthropogenic causality: large areas of the former Soviet Union suffer from radioactive pollution, in particular after the Chernobyl accident; the increase in the incidence of UVB radiation at the Earth's surface as a result of a progressive depletion of stratospheric ozone is a global problem that requires international concerted actions; in areas of former uranium mining the natural radiation level is substantially increased due to elevated radon levels; a growing portion of the population involved in air traffic is exposed to increased levels of natural radiation; and with the International Space Station an increasing number of astronauts will be exposed to the complex field of cosmic radiation. To estimate the corresponding risks, a better knowledge of the underlying radiobiological mechanisms at the molecular, cellular and system level is required.

This book is the result of a multidisciplinary effort to discuss the current state of knowledge of the fundamental processes that result from interactions of environmental radiation - ionizing as well as UV radiation - with living matter and the existing radiation protection concepts, and then to define future research work needed as fundamental information for the assessment of risks from increased levels of environmental radiation to human health and ecosystem balance. It comprises the key lectures and statements presented at the NATO Advanced Research Workshop. "Fundamentals for the Assessment of Risks from Environmental Radiation" held in Brno, Czech Republic, 6-10 October 1997, as well as the recommendations for future research needed elaborated by the task groups of the ARW. The following scientists have served on the Organizing Committee:

Gerda Horneck, German Aerospace Center, Institute of Aerospace Medicine, Köln, Germany

Stanislav Kozubek, Institute of Biophysics, Academy of Sciences, Brno, Czech Republic

Eugene A. Krasavin, Joint Institute of Nuclear research, Department of Radiation and Radiobiological Research, Dubna, Russia

Christa Baumstark-Khan, German Aerospace Center, Institute of Aerospace Medicine, Köln, Germany

The international nature of the ARW is reflected by the composition of the participants coming from many countries which include China, Czech Republic, France, Germany, Hungary, Israel, Japan, Netherlands, Russia, Slovak Republic, South Africa, Sweden, Turkey, Ukraine, United Kingdom, United States and Vietnam.

The committee is most grateful to the North Atlantic Treaty Organization, Science Committee, for the outstanding support provided for the ARW and for the production of this monograph. It also acknowledges substantial financial support provided by the German Aerospace Center DLR, the Institute of Biophysics at Brno, and the companies ENVI-2000 at Brno, Gigahertz-Optik at Puchheim, and SOLAR-LIGHT at Brno, as well as the sponsorship by The European Late Effects Project of the European Commission, EULEP, and the Gesellschaft für Biologische Strahlenforschung (GBS). The professional organization of all local affairs by S. Kozubek, L. Rýznar, V. Fuciková, M. Skalniková, and A. Lisková is highly appreciated. We are especially grateful to Lisa Steimel for her valuable assistance during the whole organization of the ARW as well as for secretarial assistance in preparation of this monograph.

This monograph is dedicated in commemoration to Professor Horst Bücker, former head of the biophysics department at DLR, and Dr. Tracy Yang, director of the radiobiology lab at NASA JSC, two pioneers in environmental radiation research. Horst Bücker decisively influenced the development of space radiation biology, especially with the introduction of the BIOSTACK concept for the study of the biological effects of single heavy ions of cosmic radiation. He was director of a NATO Advanced Study Institute on Terrestrial Space Radiation and its Biological Effects held in 1987 in Greece. Tracy Yang was a leading scientist in the study of the biological effects of accelerated heavy ions at the cellular level and a strong promoter of space radiation biology in manned space flight issues. With Horst Bücker and Tracy Yang the radiobiological community has lost two outstanding and brilliant members.

Christa Baumstark-Khan

Stanislav Kozubek

Gerda Horneck

ASSESSMENT OF RISKS FROM ENVIRONMENTAL RADIATION.

UV and Skin Cancer

R. M. TYRRELL

Department of Pharmacy and Pharmacology, University of Bath
Claverton Down, Bath BA2 7AY, UK

Abstract

Sunlight is a major cause of non-melanoma skin cancer in humans and is a critical factor in melanoma development. DNA damage caused by solar radiation, particularly the UV-B (290-320 nm) component, is a crucial early event in the genesis of human skin cancer. Such damage, if incorrectly repaired, will lead to mutations in oncogenes or tumor suppressor genes and may also be involved in later steps of the multi-stage carcinogenesis process. Cellular models developed for the purpose of skin cancer risk estimates have therefore focused on studies of the relative occurrence, repairability and importance of DNA lesions induced by UV-B radiations. However, UV radiation including the longer UV-A (320-380 nm) component of sunlight can also modify signal transduction pathways, cell to cell signaling, gene expression and other processes that can originate from non-nuclear events and contribute to the carcinogenesis process. These multiple facets of the interaction of UV with biological material must be considered in attempts to understand the changes that occur in cells and eventually, whole organs and which will influence the biological outcome and the carcinogenic process. The oxidizing component of solar radiation (particularly the UV-A band) can clearly influence and even be the primary cause of cancer development. Risk estimates must therefore involve the determination of biological action spectra throughout the solar UV region both in whole animal models and in cultured cell systems for a variety of critical end-points (mutation, gene activation, etc). Interactions between different wavelength regions should also be taken into account in the evaluation of the biological effects of sunlight.

1. Introduction

Sunlight has been established beyond any doubt as being the major cause of most human squamous cell carcinoma (SCC) and basal cell carcinoma (BCC) although the correlation of total exposure (as estimated from latitude) is weaker for BCC than for SCC. The UV-B (290-320 nm) component of sunlight appears to be the major one re-

C. Baumstark-Khan et al. (eds.),
Fundamentals for the Assessment of Risks from Environmental Radiation, 1-6.

sponsible for non-melanoma skin cancer. There is also now general agreement from analysis of epidemiological data that melanoma incidence in humans is related to sunlight exposure but the wavelength dependence is much less clear and there are indications that the UV-A (320-380 nm) region of sunlight may play an important role. This is one factor in the renewed interest in the determination of action spectra for systems that range in complexity from damage to crucial biomolecules, particularly to DNA, through to skin cancer induction in animal models. The expectation is that they will provide further clues as to the wavelength dependence of the carcinogenic process itself and aid in risk estimates and development of protective measures against skin cancer in humans. The emphasis in the first part of this short overview will therefore be on some factors to consider in the determination of action spectra and how they may be used. DNA damage following solar UV and its repair will not be considered here since these are dealt with in depth later in this NATO workshop. However, since alterations in gene expression are likely to be involved in the carcinogenic process and there is a substantial amount of recent information on activation of specific genes by UV of different wavelengths, some perspective on these studies will be presented. These are particularly pertinent given that relatively low doses of oxidizing UV-A radiation strongly activate the expression of a set of genes quite distinct from those induced by shorter wavelength UV-B radiations.

2. The Determination and Applications of Action Spectra

A biological action spectrum is a description of the relative effectiveness of different wavelengths within the spectral region of study to produce a given response. Three important applications of action spectra are

1. the identification of chromophores
2. the development of biological dosimetry procedures and
3. risk evaluation of polychromatic UV sources including sunlight.

The criteria and limitations involved in the first two applications are very relevant to a rational employment of spectra in risk evaluation.To identify chromophores several quite strict conditions need to be met if the action spectrum is to be expected to parallel the absorption spectrum. These have been carefully listed by Jagger [1] and can be summarized as follows:-

1. The action mechanism must be the same at all wavelengths.
2. The quantum yield must be the same at all wavelengths.
3. The absorption spectrum of the chromophore must be the same in vitro as in vivo.
4. Absorption of light by inactive chromophores and light scattering must be negligible.
5. Only a small fraction of the incident light must be absorbed by the sample in the wavelength range of interest.
6. Repriocity of the exposure time and fluence rate must hold.

It should be noted that for comparison with absorption spectra, action spectra need to be "quantum" corrected at each wavelength to take account of the decrease in energy per photon with increase in wavelength. Long before DNA was identified as the genetic material, nucleic acids could be identified as the critical chromophore for the bactericidal action of UV from an action spectrum published by Gates [2]. However, for the entire spectrum of interest, such evaluations are rarely possible because the absorbing chromophores and crucial lesions will often change as a function of wavelength (see point 1). For example, the chromophore/target for UV-A induced cytotoxicity to cultured cells is not identifiable from action spectra. The most important limitation practically is that for anything but dilute solutions of biomolecules or suitable concentrations of single cells, wavelength dependent absorption by other chromophores will interfere (see later).

Biological UV dosimeters may be used either to substitute for a physical (radiometric) or chemical (actionometric) dosimeter in certain field applications after calibration against an accurate standard or they may be intended to provide a quantitative estimate of the magnitude of a given biological effect to be expected from a variable polychromatic source such as sunlight. In the latter case, the ideal biological detector would show the same wavelength dependence as the biological effect of interest (both or neither should be quantum corrected for the comparison). There are good arguments to support the notion that the UV sensitive *B. subtilis* spore system can reliably and accurately monitor and integrate the DNA damaging capacity of sunlight [3, 4, 5]. The action spectra for the spore dosimeter is shown in Fig. 1 together with the action spectra for inactivation of colony-forming ability of cultured populations of matched (same biopsy) human keratinocytes and fibroblasts. The divergence of the spectra in the UV-A region is a further indication that the cytotoxicity caused by the longer wavelength radiations is not simply due to DNA damage. Indeed most spectra diverge in this region and this will limit the use of a dosimeter that monitors DNA damage if the UV-A component for the effect is significant.

In skin cancer risk evaluation, two types of action spectra have been exploited. Firstly spectra derived in simple, essentially transparent, systems as described above where the spectra are expected to represent primary events. Additionally, action spectra for skin cancer in animal model systems continue to be determined. Action spectra for direct effects on DNA were originally compiled by Setlow [6] and proposed as a model for UV/visible radiation induced skin cancer. However, to predict the action of a polychromatic source such as sunlight on humans, wavelength dependent tissue absorption of radiation needs to be taken into account and spectral data are essential. For example, to calculate the relative cytotoxic effectiveness of sunlight to cells at the basal layer of the epidermis, in addition to an action spectrum for cytotoxicity to epidermal keratinocytes (see Fig. 1), data for transmission through human skin must be available [7]. If the product of this data is then further adjusted to take account of solar spectral irradiance under defined conditions, a curve for the relative biological effectiveness for killing of keratinocytes at the basal layer of the human epidermis can be derived. [8]. From such data, it has been calculated that even in noon sunlight, 50 percent of the killing effect may be due to UV-A radiation. Again this underlines the possible crucial importance of the long wavelength region, particularly where non-DNA targets may be involved. The action spectrum for induction of squamous cell carcinoma in hairless mice determined in Utrecht and Philadelphia (SCUPm) has a shoulder in the UV-A region which is enhanced 10-fold when the spectrum is corrected for transmission through human skin relative to rodent [9]. Although this does alter factors used in the estimation of human cancer risks, the UV-B component remains dominant. However, action spectra determined for melanoma induction in a hybrid fish model show orders of magnitude higher levels of induction in the UV-A region than predicted from the DNA action spectra [10]. Further studies are clearly necessary in a mammalian model for melanoma (e.g. Monodelphus domestica) [11]. Nevertheless, it is worth pointing out that human epidemiological studies indicate that melanoma may be related to acute episodes of early-life exposure sufficient to cause severe sunburn. Such exposures will presumably have caused severe destruction of the basal layer of the epidermis so that the cytotoxic component of sunlight, to which UV-A may be a major contributor (see above), would be crucial.

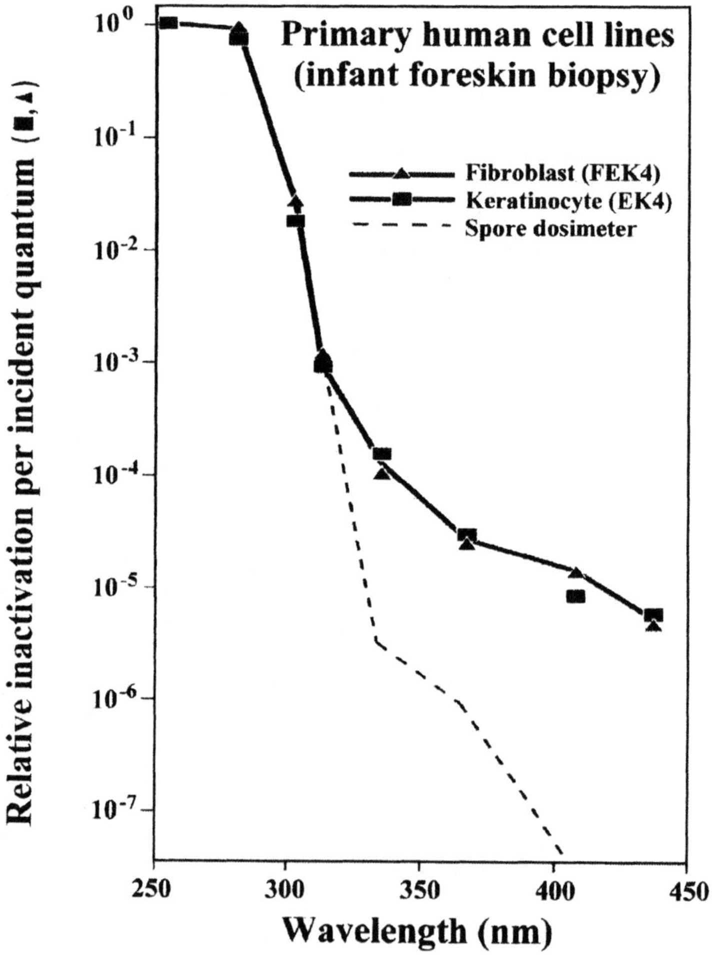

Figure 1. Quantum corrected action spectra for inactivation of human fibroblasts and epidermal keratinocytes derived from the same infant foreskin biopsy compared with the normalized quantum corrected action spectrum for the *B. Subtilis* UVSSP spore dosimeter (reproduced from [5] with permission from the publishers).

3. UV Activation of Genes

Given the potential importance of UV-A radiation and the likely importance of activation of gene expression in the carcinogenic process, the UV wavelength dependence for activation of pertinent genes merits consideration. UV-B effects result primarily from DNA damage whereas UV-A generates an oxidative stress that will cause a different spectrum of damage and effects. This is well illustrated by examples in the bacterium. *Escherichia coli*, where UV-C/UV-B radiation activates the SOS response linked to inducible DNA repair pathways [12], whereas UV-A activates the oxyR response linked to the induction of several antioxidant enzymes. UV-C/UV-B radiations induce many genes in eucaryotes but only in the case of the p53 protein have these been linked to DNA repair [for review see 13]. Furthermore, while UV activation of transcription factor complexes via defined signal transduction pathways has generated much recent interest, the high doses used cast doubt on the physiological relevance of many of the observations. Nevertheless, UV-B radiation does induce the expression of several genes in skin, at physiologically relevant doses.

For the most part, a quite different set of genes are activated by UV-A radiation and these include heme oxygenase 1, intercellular adhesion molecule 1, phospholipase A_2, the CL100 phosphatase and cyclooxygenase (our recent unpublished results). Collagenase and other metalloproteinases are induced by both UV-A and UV-B radiations. A pertinent point for this discussion is that activation of these genes does occur at physiologically relevant doses and at least for the case of heme oxygenase 1 (unpublished observations, this laboratory in collaboration with L. Applegate, and E. Frenck, Dermatology, CHUV) corresponding changes can be observed in human skin. Activation of the HO-1 gene underlies a widespread mammalian stress response [14, 15] and the transcriptional activation of this gene [16] is the largest that has been observed for gene activation by UV-A radiation or oxidants in general. Furthermore activation of this gene appears to be a crucial step in an adaptive response to oxidative membrane damage that involves increased ferritin levels [17]. Adaptation for protection against other types of damage is currently under investigation.

4. Summary

Action spectra, when appropriately used, can be of considerable value in evaluating human risk from UV. A knowledge of the wavelength dependence for non-melanoma and melanoma skin cancer is also crucial for the development of adequate protective measures such as the use of appropriately absorbing sunscreens. Recent information on melanoma incidence from human epidemiological as well as animal studies has emphasized the need to understand the primary events that occur in cells after UV-A as well as after UV-B radiation and reinforces the case for determining action spectra for additional parameters such as gene activation that do not necessarily involve DNA damage.

Acknowledgement. The work of the author is supported by the Association for International Cancer Research (UK), the Department of Health (UK) and the EU fourth framework program on Environment and Climate funded and administered by the Swiss Federal Office of Education and Science OFES (95.0509 and 95.0510).

5. References

1. Jagger, T., *Introduction to Research in Ultraviolet Photobiology*, Prentince-Hall, Englewood Cliffs NJ (1967).
2. Gates F. L. *J. Gen. Physiol.* **14**, 31-41 (1930).
3. Tyrrell R.M., Solar dosimetry with repair deficient bacterial spores: action spectra, photoproduct measurements and a comparison with other biological systems. *Photochem. Photobiol.* **27**, 571-579 (1978).
4. Munakata N. Biologically effective dose of solar ultraviolet radiation estimated by spore dosimetry in Tokyo since 1980. *Photochem. Photobiol.* **58**, 386-392 (1993).
5. Tyrrell, R.M. Biological dosimetry and action spectra. *J. Photochem. Photobiol B Biol.* **31**, 35-41 (1995).
6. Setlow R. B., The wavelengths in sunlight effective in producing skin cancer: a theoretical analysis. *Proc. Natl. Acad. Sci.* USA **71**, 3363 -3366 (1974).
7. W.A.G. Bruls, *PhD Thesis*, University of Utrecht (1984).
8. Tyrrell R. M. and Pidoux M., Action spectra for human skin cells: estimates of the relative cytotoxicity of the middle ultraviolet, near ultraviolet, and violet regions of sunlight on epidermal keratinocytes. *Cancer Res.* **47**, 1825-1829 (1987).
9. De Grujl F.R., Sterenborg H.J.C.M., Forbes P.D., Davies R.E., Cole C., Kelfkens G., Van Weelden H., Slaper H. and Van der Leun J.C., Wavelength dependence of skin cancer induction by ultraviolet irradiation of albino hairless mice. *Cancer Res.*, **53**, 53-60
10. Setlow R.B., Grist E., Thompson K. and Woodhead A.D., Wavelengths effective in induction of malignant melanoma. *Proc. Natl. Sci USA*, **90**, 6666-6670 (1993).
11. Ley R.D., Applegate L.A., Padilla R.S. and Stuart T.D., Ultraviolet radiation--induced malignant melanoma in Monodelphis domestica., *Photochem. Photobiol.* **50**, 1-5 (1989).
12. G.C. Walker in "*Escherichia coli* and *Salmonella typhimurium, Cellular and Molecular Biology* (F.C. Niehardt et al, eds.) pp 1346-1357, American Society of Microbiology, Washington DC (1987).
13. Tyrrell R. M., Activation of mammalian gene expression by the UV component of sunlight--from models to reality. *Bioessays* **18**, 139-148 (1996).
14. Keyse S.M. and Tyrrell R.M. Heme oxygenase is the major 32-kDa stress protein induced in human skin fibroblasts by UVA radiation, hydrogen peroxide, and sodium arsenite. *Proc. Natl. Acad. Sci. USA* **86**, 99-103 (1989).
15. Applegate L.A., Luscher P. and Tyrrell R.M., Induction of heme oxygenase: a general response to oxidant stress in cultured mammalian cells. *Cancer Res.* **51**, 974-978 (1991).
16. Keyse S.M., Applegate L.A., Tromvoukis Y. and Tyrrell R.M. Oxidant stress leads to transcriptional activation of the human heme oxygenase gene in cultured skin fibroblasts., *Mol. Cell. Biol.* **10**, 4967-4969 (1990).
17. Vile G.F., Basu-Modak S., Waltner C. andTyrrell R.M., Heme oxygenase 1 mediates an adaptive response to oxidative stress in human skin fibroblasts., *Proc. Natl. Acad. Sci. USA* **91**, 2607-2610 (1994).

ASSESSING RISK AT LOW DOSES.

Problems of Confounders and Statistical Power

R. E. SHORE

Department of Environmental Medicine, New York University Medical School
New York, NY 10016, U.S.A.

Abstract

Although a risk assessment based on extrapolation of risk from high-dose studies to the low-dose region gives the appearance of a relatively precise estimate of low-dose risk, in fact knowledge is required about the biological mechanisms involved before one can derive an appropriate extrapolation model.

To avoid this problem, one strategy is to attempt to assess risk from studies of populations with relatively low radiation exposures. This superficially appealing approach has distinct weaknesses which must be considered. First, the precision of the estimates (i.e., the narrowness of the confidence interval) is very limited with most low-dose data. Two corollaries to the lack of precision are that the sample size needed to detect an effect will be extremely large, and that the statistical power to detect an effect or a non-zero dose-response regression slope will often be limited. The lack of statistical power also means that if one should detect a "statistically significant" effect, the estimated magnitude of that effect is likely to be seriously biased on the high side.

Another problem with low-dose studies is that the magnitude of subtle biases or confounding in the data set may well be greater than the magnitude of the expected effect. This confounding can either mask a true effect or yield false-positive findings, so there is little certainty about the results.

In general, one can hope to obtain meaningful measures of risk from low-dose studies only if the number of persons studied is very large, as may be achieved by a pooling of studies. It is recommended that studies be developed studies that address gaps in our knowledge about the effects of ionizing radiation exposure, rather than focusing primarily on low-dose studies.

1. Sources of Uncertainty and Imprecision in Low-Dose Studies

It is generally agreed that existing epidemiologic data from the Japanese A-bomb survivors and those who received radiotherapy for a variety of medical conditions provide a

C. Baumstark-Khan et al. (eds.),
Fundamentals for the Assessment of Risks from Environmental Radiation, 7-14.

sound basis for estimating risks to populations whose radiation doses and dose-rates are relatively high. There is much less certainty about the magnitude of risks resulting from low doses and/or low dose-rates. It is intrinsically difficult to assess risk with accuracy (i.e., absence of bias) and precision when doses are below, say, 100 mSv or when they are delivered at rates of a few mSv per year.

The limitations on accuracy and precision occur for several reasons. First, human studies often have a fair degree of uncertainty in the dose estimates. In high-dose studies the dose uncertainties have a relatively small effect on the risk estimates (e.g., [1]), but in low-dose studies the ratio of the size of the dose uncertainties to the imputed doses may be relatively large and hence may obscure or distort associations.

A second source of imprecision is the heterogeneity in susceptibility to radiogenic cancer in exposed human populations. Genetic heterogeneity, about which knowledge is expanding rapidly with the discovery of various heritable tumor-suppressor gene mutations, is a major source of variability. There are also other sources of heterogeneity in radiosensitivity to cancer induction, such as gender [2], ethnicity [3], smoking habits [4], age-at-irradiation [2, 5], and age at risk [6].

Another source of imprecision and uncertainty is the fact that with a heterogeneous human population one is not dealing with a radiation exposure to organisms that exist in an otherwise controlled, uniform environment. Instead, humans have various amounts of many other exposures, some of which could be carcinogenic cofactors or protectors. The limited data on carcinogenic cofactors suggest that they can change not only the absolute probability of developing cancer but also the shape of the dose-response curve [7], which has important implications with regard to low-dose risk. A final category of uncertainty relates to variations in length and completeness of follow-up, accuracy of tumor diagnoses, and other methodological variations in epidemiologic studies.

Confounding variables that can bias study results (e.g., smoking history in studies of radon and lung cancer) are of special concern as a source of uncertainty and bias in low-dose studies. Typically, some of the potential confounding variables are measured either very crudely or not at all, so they are not controlled for adequately. Unlike high-dose studies, in a low-dose study the magnitude of confounding effects may be as large or larger than the exposure-caused effects and thereby lead to "false positive" or "false negative" results.

2. Statistical Power of Low-Dose Studies

Any study at low doses will have inadequate statistical power unless the sample size is extremely large [8, 9] due to the intrinsic problem of detecting a small effect, because the variance will be relatively much larger than the size of the effect to be detected, and in epidemiologic studies the variance may be further inflated by added uncertainties, as detailed above.

Moreover, the relationship between the size of the dose and the size of the sample required to detect its effect is nonlinear, even under the assumption of a linear dose-response association. To illustrate, if there is a linear relationship between radiation dose and lung cancer risk, and 500 subjects would need to be studied to detect an excess risk when the dose they received was 1 Gy, then about 29,000 subjects with doses of 100 mGy would need to be studied to detect an excess risk, or about 2,700,000 would be required if the dose were only 10 mGy. Put differently, a study with a mean dose of 10 mGy will require roughly 4,000-6,000 times as many subjects as a comparable population with a mean dose of 1 Gy. If the dose-response is sublinear, (i.e., quadratic with upward curvature) then the required sample sizes would increase even more yet at low doses.

What is the import of having low statistical power in a low-dose study? First, a null result is not very meaningful, since the probability of a null result was high. An implication of this is that if a low-dose study that lacks statistical power does not find a statistical excess of cancer, this null result is weak evidence at best as to whether there may be a dose threshold. Second, such a study does not have the capability of ruling out the low-to-moderate risks of interest, but only those that are unrealistically high, because low statistical power implies poor precision in the risk estimate. Hence, the confidence limits on the risk will have a high upper bound, so they are not very useful in quantifying risks. Third, when a statistically significant excess is seen in a study with low statistical power, it is likely that it is a "false positive" result due to chance. Positive results from a study with low statistical power therefore should be interpreted very cautiously. Finally, it is highly probable that the risk estimate for a statistically significant result in a study with low statistical power is seriously biased on the high side [8].

One important principle, motivated by the preceding discussion of the imprecision of low-dose epidemiologic studies and their potential for invalidity, is that one should not rely upon a single low-dose study, especially not a small one, as the primary basis for inferring there is no effect or for inferring a large effect. It is even less defensible to single out a particular *post hoc* subgroup in a study as demonstrating no effect (or a large effect), for this capitalizes on chance in such a way as to invalidate the statistical conclusion. Instead, one should seek to use a weight-of-evidence approach which incorporates all the pertinent data that are methodologically-sound [10].

3. Weaknesses of Certain Types of Studies

3.1. "ECOLOGIC" (AGGREGATE) STUDIES

In an "ecologic" (aggregate) study a correlation is obtained between average amounts of exposure in various geographic areas and the disease rates in those areas - as opposed to obtaining information on individuals' exposure and disease status. Examples are studies of cancer rates in various geographic areas as a function of levels of background exposure [11, 12] or proximity to nuclear plants [13, 14], of lung cancer rates in relation to estimated residential radon levels [15], and of birth defect rates in regions subjected to Chernobyl fallout [16, 17]. Ecologic studies are common because they are easy to perform since they often can use pre-existing data, and they appear to give very precise answers since they are typically based on large numbers of cases of the disease in question.

Unfortunately, ecologic studies have the potential for several serious statistical problems that are difficult to detect but can invalidate them. With the potential invalidity, such studies may be useful for hypothesis generation but cannot be viewed as providing definitive tests of hypotheses. Detailed technical discussions of the potential intrinsic weaknesses of ecologic studies can be found in [18-20]. Suffice it to say, briefly, that common epidemiologic problems (e.g., measurement error, nonlinear associations with age or duration of smoking) may have serious and uncontrollable effects in ecologic studies, and that efforts to control for bias and confounding through standardization or modeling may have unpredictable results in ecologic studies. Given these intrinsic problems with analyses of ecologic studies, one should be skeptical of ecologic studies of low-dose effects.

3.2. CASE-CONTROL STUDIES

Case-control studies, in which a series of cases of a disease from a defined population is compared, with respect to exposures of interest, to a matched or stratified random sample from the same population, can be effective ways to investigate relatively rare diseases and common exposures. However, when it comes to investigating the effects of low exposures and putatively small risks, modest biases can produce misleading results because the magnitude of the biases is likely to rival or exceed the magnitude of potential risks. Case-control studies are particularly subject to two kinds of bias. The first is selection bias, caused when the cases and controls that can be enlisted differ in the degree or ways to which they are unrepresentative of the population at risk. The second is the potential for recall bias when exposure information is based on self-reports. For example, diseased persons may have more motivation to report exposures than unaffected controls, so that the cases have more overreporting and/or less underreporting of exposure. An instructive example of this kind of bias was shown in Portsmouth Naval Shipyard studies in which an initial study based on self-report alleged an association between leukemia and radiation exposure at the Shipyard [21], but subsequent studies based on recorded exposures found no association [22, 23].

As another example, several case-control studies have shown apparent associations between self-reported exposures to medical or dental diagnostic radiation and various cancers, e.g., [24-27], but the two studies that have assessed the adequacy of anamnestic reporting of diagnostic radiation procedures found that there was 25-50% or more of inaccurate reporting of past exposures [28, 29]. Case-control studies of diagnostic irradiation that rely on comprehensive medical records (e.g. [30, 31]) are therefore more reliable because they do not suffer from recall bias.

3.3. COHORT STUDIES

In general, well-conducted cohort studies are the strongest type of epidemiologic study, short of randomized trials. A well-designed cohort study should permit one to obtain answers to the study questions with minimal bias in selection procedures, exposure assessment and ascertainment of confounding variables. Furthermore, it permits one to examine several disease endpoints of interest. Nevertheless, there are several potential weaknesses that cohort studies may have if they are small or poorly designed: too few cases of the disease in question (which yields low statistical power), different rates of follow-up and disease ascertainment (e.g., surveillance bias) in various dose groups, and self-selection factors in study participation.

4. How Large are Cancer Risks from Low-Dose and Protracted Radiation?

We have reasonably precise quantitative estimates of radiation-induced cancer risk for various organs based on a relatively large body of epidemiologic data [2, 32]. These estimates have been based primarily on studies with relatively high doses, because more precise estimates can be derived from high-dose studies than from low-dose studies. Nevertheless, the substantial low-dose data available suggest that the risks per unit dose at low doses and low dose rates are compatible with or less than those at higher doses. A major study of pooled data from >95,000 radiation workers with low, protracted doses showed that the risks of radiation-induced leukemia or other cancers are probably smaller than the risks predicted by linear extrapolation from the Japanese atomic bomb data [33]. Specifically, the Excess Relative Risks (ERRs) and 90% Confidence Intervals (CI) for leukemia were 2.6 Sv^{-1} (CI= 0.1, 7) in the worker cohort and 3.7 Sv^{-1} (CI= 2.0, 6.5) in the atomic bomb study. For all other cancers, the respective ERRs were -0.1 Sv^{-1} (CI= -0.4, 0.3) and 0.2 Sv^{-1} (CI= 0.05, 0.3). Of note was the fact that the confidence intervals in the worker study exclude risks that are more than twice the atomic bomb estimates [33]. The comparison of these two studies is reasonably persuasive evidence that a linear model provides appropriate, although perhaps conservative, estimates of the risks from low, protracted exposures. It is also strong evidence against the position some have taken that low-dose cancer risks are 5 or 10 times greater than would be predicted from the Japanese atomic bomb data, since the CIs in the worker study are incompatible with ERR estimates greater than twice the atomic bomb estimates of risk.

A more comprehensive look at studies with protracted or fractionated exposures is shown in Table 1 which presents a summary of available leukemia data from radiation studies that met certain criteria. A study had to have protracted or fractionated radiation with mean cumulative doses of less than 1 Sv, and it had to be fairly large (over 10 observed or expected leukemia deaths/cases). We performed a meta-analysis of these studies and derived a combined excess relative risk estimate of 230% Gy^{-1} (95% confidence interval = 25-780). In comparison, the most recent excess relative risk estimate for leukemia mortality from the Japanese atomic bomb study has been reported as about 460% Gy^{-1} (90% CI = 330-640) [42]. It can be seen that the two risks are statistically compatible, albeit the best estimate of the risk for the protracted-dose studies is two times lower than for the atomic bomb study with acute exposure. This suggests there may be less leukemia risk from protracted exposures than from acute exposures of equal magnitude.

5. Criteria For Defining Valuable Radiation Epidemiology Studies

The foregoing discussion has already identified several criteria having to do with technical characteristics of radiation studies that make them more, or less, valuable. These criteria include: having accurate individual doses, having a large enough irradiated population with a broad enough range of doses so that there is reasonable statistical power, having unbiased methods to select study subjects and to assess disease outcomes, and having sufficient information on confounding factors.

There are also substantive criteria that would make certain studies especially valuable. There are a number of gaps in our knowledge about radiation effects. Studies which would be informative with respect to these gaps should be given a high priority. The following is a brief list of some of the gaps in radiation epidemiology:

- Information on the degree to which children irradiated *in utero* are at high risk for childhood cancers other than leukemia.
- Information on the magnitude of adult cancer risk among those irradiated *in utero*.
- Information on intelligence deficits among those irradiated *in utero* during weeks 8-15 of gestation.
- Studies of iodine-131 effects among children that have an accurate dosimetry and an adequate sample size.
- Studies of the risk of early breast cancer (e.g., before age 45) among women who received radiation exposure to the breast in childhood or adolescence.
- Studies of possible synergisms between radiation exposure and having a high heritable cancer risk (e.g., a mutation in the *BRCA1*, *BRCA2*, or *ATM* gene).
- Genetic risks: Studies of the risk of heritable diseases or detrimental conditions in the F_1 generation as the result of gonadal exposure to parents. Re-assessments are currently ongoing with regard to genetic risks.
- Information on the effects of protracted or fractionated exposure, as compared to acute exposure. Current information suggests that fractionation effects to the breast and lung differ appreciably, and more needs to be learned about the mechanisms involved and about dose-protraction/fractionation effects in other organs.

TABLE 1. Risk Estimates of Radiation-Induced Leukemia, from Epidemiologic Studies where the Mean Bone-Marrow Dose was <1 Sv with Fractionated or Protracted Doses

Study Group (Reference)	*Mean Dose (mSv)*	*Observed / Expected*	*% Excess Relative Risk / Sv (95% CI)*	*Absolute Risk / 10^4 PY-Sv*
^{131}I for Hyperthyroidism [34]	~100	11/10.4	58 (-444, 838)	0.6
^{131}I for Hyperthyroidism [35]	60	34/36.2	-100 (-560,500)	-2.8
Diagnostic/ therapeutic ^{131}I [36]	14	195/179	643 (-410,1820)	12.2
Utah fallout study [37]	19	17/9.9	3800 (-320,11200)	--
Techa River releases [38]	370	27/12.3	324 (48, 600)	0.9
Chinese medical x-ray workers [39]	≈700	34/14.1	202 (43, 440)	0.9
US nuclear submariners [40]	~5	21/23.1	-178 (-840,730)	-1.3
Combined study: USA, UK, Canada [33]	40.2	119/109	218 (13, 570)	1.2
Savannah River workers [41]	46.2	25/19.6	591 (21, 1160)	--
COMBINED (Approximate)		**483/ 414**	**230 (25-780)**	**1.0**

a Only studies with at least 10 observed or expected leukemia cases were included.

In conclusion, it is recommended that resources for radiation epidemiology studies be devoted more to addressing the important substantive gaps in our knowledge and to tying together the rapidly developing fundamental biological knowledge about radiation effects with epidemiological studies in humans. More will be learned by focusing in these directions than by repetitively conducting low-dose studies with their inherent weaknesses.

Acknowledgment. Supported in part by center grants CA-16087 (Kaplan Cancer Center) from the US National Cancer Institute and ES-00260 from the US National Institute of Environmental Health Sciences.

6. References

1. Pierce, D.A., Preston, D., Stram, D. and Vaeth, M. (1991): Allowing for dose-estimation errors for the A-bomb survivor data. *J. Radiat. Res. (Japan)* **32 (Suppl)**, 108-121.
2. UNSCEAR (1994): *Sources and Effects of Ionizing Radiation (Annex A: Epidemiological studies of radiation carcinogenesis)*. In: United Nations Scientific Committee on the Effects of Atomic Radiation, 272 pp. New York: United Nations.
3. Shore, R., Albert, R., Reed, M., Harley, N. and Pasternack, B. (1984): Skin cancer incidence among children irradiated for ringworm of the scalp. *Radiat. Res.* **100**, 192-204.
4. Lubin, J., Boice, J., Edling, C., et al. (1995): Lung cancer in radon-exposed miners and estimation of risk from indoor exposure. *J. Nat. Cancer Inst.* **87**, 817-827.
5. Ron, E., Lubin, J., Shore, R., et al. (1995): Thyroid cancer after exposure to external radiation: a pooled analysis of seven studies. *Radiat. Res.* **141**, 259-277.
6. Kellerer, A.M. and Barclay, D. (1992): Age dependencies in the modelling of radiation carcinogenesis. *Radiat. Protect. Dosim.* **41**, 273-281.
7. Fry, R. and Ullrich, R. (1986): Combined effects of radiation and other agents. In: *Radiation Carcinogenesis*, eds. Upton, A., Albert, R., Burns, F. and Shore, R., pp. 437-454. New York: Elsevier.
8. Land, C.E. (1980): Estimating cancer risks from low doses of ionizing radiation. *Science* **209**, 1197-1203.
9. Shore, R.E. (1995): Epidemiological issues related to dose reconstruction. In: *Environmental Dose Reconstruction and Risk Implications*, ed. Till, J.E., pp. 245-260. Bethesda, MD: NCRP (National Council on Radiation Protection and Measurements).
10. Ashby, J., Doerrer, N., Flamm, W., et al. (1990): A scheme for classifying carcinogens. *Regul. Toxicol. Pharmacol.* **12**, 270-295.
11. Frigerio, N. and Stowe, R. (1983): Carcinogenic and genetic hazard from background radiation. In: *International Symposium on the Biological Effects of Low-level Radiation with special regard to Stochastic and Non-stochastic effects*, pp. 385-391. Vienna: IAEA.
12. Nambi, K.S. and Soman, S. (1987): Environmental radiation and cancer in India. *Health Phys.* **52**, 653-657.
13. Johnson, C.J. (1987): Cancer incidence patterns in the Denver metropolitan area in relation to the Rocky Flats plant. *Am. J. Epidemiol.* **126**, 153-155.
14. Jablon, S., Hrubec, Z. and Boice, J. (1991): Cancer in populations living near nuclear facilities. *J. Am. Med. Assoc.* **265**, 1403-1408.
15. Cohen, B.L. (1995): Test of the linear-no threshold theory of radiation carcinogenesis for inhaled radon decay products. *Health Phys.* **68**, 157-174.
16. Harjulehto, T., Rahola, T., Suomela, M., Arvela, H. and Saxen, L. (1991): Pregnancy outcome in Finland after the Chernobyl accident. *Biomed. Pharmacother.* **45**, 263-266.
17. Luning, G., Scheer, J., Schmidt, M. and Ziggel, H. (1989): Early infant mortality in West Germany before and after Chernobyl. *Lancet* **2**, 1081-1083.

18. Greenland, S. and Morgenstern, H. (1989): Ecological bias, confounding, and effect modification. *Int. J. Epidemiol.* **18**, 269-274.
19. Greenland, S. and Robins, J. (1994): Ecologic studies--biases, misconceptions, and counterexamples. *Am. J. Epidemiol.* **139**, 747-760.
20. Piantadosi, S. (1994): Invited commentary: Ecologic biases. *Am. J. Epidemiol.* **139**, 761-764.
21. Najarian, T. and Colton, T. (1978): Mortality from leukaemia and cancer in shipyard nuclear workers. *Lancet* **1**, 1018-1020.
22. Rinsky, R.A., Zumwalde, R., Waxweiler, R., et al. (1981): Cancer mortality at a naval nuclear shipyard. *Lancet* **1**, 231-235.
23. Stern, F., Waxweiler, R., Beaumont, J., et al. (1986): A case-control study of leukemia at a naval nuclear shipyard. *Am. J. Epidemiol.* **123**, 980-992.
24. Stewart, A., Pennybacker, W. and Barber, R. (1962): Adult leukaemias and diagnostic x rays. *Br. Med. J.* **2**, 882-890.
25. Gibson, R., Graham, S., Lilienfeld, A., et al. (1972): Irradiation in the epidemiology of leukemia among adults. *J. Nat. Cancer Inst.* **48**, 301-311.
26. Preston-Martin, S., Thomas, D., Yu, M. and Henderson, B. (1989): Diagnostic radiography as a risk factor for chronic myeloid and monocytic leukaemia (CML). *Br. J. Cancer* **59**, 639-644.
27. Preston-Martin, S., Henderson, B. and Bernstein, L. (1985): Medical and dental x-rays as risk factors for recently diagnosed tumors of the head. *Monogr. Nat. Cancer Inst.* **69**, 175-179.
28. Graham, S., Levin, M., Lilienfeld, A., et al. (1963): Methodological problems and design of the tristate leukemia survey. *Ann. N.Y. Acad. Sci.* **107**, 557-569.
29. Preston-Martin, S., Bernstein, L., Maldonado, A., Henderson, B. and White, S. (1985): A dental x-ray validation study: comparison of information from patient interviews and dental charts. *Am. J. Epidemiol.* **121**, 430-439.
30. Boice, J.D., Morin, M., Glass, A., et al. (1991): Diagnostic x-ray procedures and risk of leukemia, lymphoma, and multiple myeloma. *J. Am. Med. Assoc.* **265**, 1290-1294.
31. Inskip, P., Ekbom, A., Galanti, M., Grimelius, L. and Boice, J. (1995): Medical diagnostic x rays and thyroid cancer. *J. Nat. Cancer Inst.* **87**, 1613-1621.
32. ICRP (1991): 1990 Recommendations of the International Commission on Radiological Protection. *Ann. ICRP* **21 (Publication 60)**, 1-201.
33. Cardis, E., Gilbert, E., Carpenter, L., et al. (1995): Effects of low doses and low dose rates of external ionizing radiation: cancer mortality among nuclear industry workers in three countries. *Radiat. Res.* **142**, 117-132.
34. Saenger, E.L., Thomas, G. and Tompkins, E. (1968): Incidence of leukemia following treatment of hyperthyroidism. *J. Am. Med. Assoc.* **205**, 147-154.
35. Holm, L., Hall, P., Wiklund, K., et al. (1991): Cancer risk after iodine-131 therapy for hyperthyroidism. *J. Nat. Cancer Inst.* **83**, 1072-1077.
36. Holm, L. (1991): Cancer risks after diagnostic doses of ^{131}I with special reference to thyroid cancer. *Cancer Detect. Prev.* **15**, 27-30.
37. Stevens, W., Thomas, D., Lyon, J., et al. (1990): Leukemia in Utah and radioactive fallout from the Nevada test site. *J. Am. Med. Assoc.* **264**, 585-591.
38. Kossenko, M. (1996): Cancer mortality among Techa River residents and their offspring. *Health Phys.* **71**, 77-82.
39. Wang, J.X., Inskip, P., Boice, J., Jr. and Li, B. (1990): Cancer incidence among medical diagnostic X-ray workers in China, 1950 to 1985. *Int. J. Cancer* **45**, 889-895.
40. Charpentier, P., Ostfeld, A., Hadjimichael, O. and Hester, R. (1993): The mortality of US nuclear submariners, 1969-1982. *J. Occup. Med.* **35**, 501-509.
41. Cragle, D.L., Robertson-DeMers, K. and Watkins, J.P. (1997): Mortality among workers at a nuclear fuels production facility: the Savannah River Site, 1952-1986. *In Press* .
42. Pierce, D., Shimizu, Y., Preston, D., Vaeth, M. and Mabuchi, K. (1996): Studies of the mortality of atomic bomb survivors. Report 12, Part 1. Cancer: 1950-1990. *Radiat. Res.* **146**, 1-27.

RADIATION RISKS AT LOW RADIATION DOSES.

Lessons of Chernobyl, 1986-1996

V. K. IVANOV

Medical Radiological Research Center of RAMS
4 Korolev Str., Obninsk, 249020 Russia

Abstract

Few, if any, estimates of radiation risk for low doses of ionizing radiation (0.2-0.3 Gy) exist that are based on direct epidemiological studies. The prediction of the radiation induction of malignant tumors in this range of low doses is normally based on the extrapolation of observed risk coefficients from relatively high doses (> 1 Gy) to low doses. In this sense, the data accumulated since the Chernobyl accident are of singular value. Indeed, during the first 10 years of follow-up, large volumes of epidemiological data have been collected, characterizing the health status of hundreds of thousands of persons who received low doses.

This work focuses on the direct epidemiological assessment of the risk of radiation-induced solid cancer, leukemia and thyroid cancer in emergency workers (EW) after the Chernobyl accident. Analysis of radiation risks is performed on the basis of data of Russian National Medical and Dosimetric Registry (RNMDR) in which medical and dosimetric data on 168 thousand of emergency workers are kept.

1. Introduction

In 1986 immediately after the Chernobyl accident the Ministry of Public Health of the USSR adopted a large-scale program to establish in the country the All-Union Distributed Registry of persons exposed to radiation. Towards 1992 (by the time of collapse of the USSR) the data base of the Registry comprised medical and dosimetric information for 659,292 people including that for 284,919 emergency workers (EWs) (liquidators). Into establishment of the Registry all republics of the former Soviet Union as well as a wide range of scientific and practical institutions were involved [1, 2].

At present (in accordance with the Decree of the Government of Russia N 948 of 22.09.93) the National Radiation and Epidemiological Registry (NRER) operates in the country. The general customer of works on the Registry is Ministry on Emergency Situations. The leading organization is the Medical Radiological Research Center of

C. Baumstark-Khan et al. (eds.),
Fundamentals for the Assessment of Risks from Environmental Radiation, 15-24.

RAMS responsible for sampling primary medical and dosimetric data through 24 regional centers.

The NRER involves 3 main bases: the Registration List of persons exposed to radiation which is established on special dosimetric criteria from the Decree of the Government of Russia N 948); the Chernobyl Registry since 1992 the Russian National Medical and Dosimetric Registry (RNMDR); the Registry on Interdepartmental Expert Councils.

In this paper the radiation-epidemiological analysis of the Chernobyl Registry of Russia (RNMDR) [3-5] will be enlarged.

2. Methods

As of 1.1.1996 the RNMDR database comprises medical and dosimetric information for 168 thousand emergency workers. In 1986, 77.7 thousand people were involved in remediation works, 58.7 thousand - in 1987 and 31.6 thousand - in 1988-1990. As is known more than 200 thousand of emergency workers from Russia took part in remediation activities in the 30-km zone of the Chernobyl NPP in 1986-1990. Therefore, ten years after the disaster unaccounted emergency workers continue to be entered in RNMDR. At present, RNMDR has a rigid hierarchy (in line with the Decree of Russian Government): the national level of the registry is realized in Medical Radiological Research Center in Obninsk, and 20 regional centers of the Registry are responsible for collection of data of annual check-ups across Russia which are then provided to the national level.

From the standpoint of dosimetric information about the emergency workers cohort, the Registry includes only official doses of external irradiation. The accuracy of determination of external irradiation doses in emergency workers remains a very complicated issue. By our estimation, the degree of uncertainty in calculating individual doses may be as high as a factor of 2.5-3.0.

Of 168 thousand emergency workers registered in RNMDR to date, 119 thousand (71%) have individual doses of external exposure. As is seen from Table 1 the highest radiation exposures for emergency workers occurred in 1986: more than 4.5% of 46575 people received the officially reported doses exceeding the threshold value of 250 mGy.

TABLE 1. Dose distribution for liquidators by years of arrival in zone

Years of arrival	*Number of persons*	*Dose, mGy*					
		0-49	*50-99*	*100-149*	*150-199*	*200-249*	*>250*
1986	46575	18.2%	10.2%	10.1%	20.7%	36.3%	4.5%
1987	48077	24.0%	51.9%	9.7%	8.1%	5.8%	0.6%
1988-1990	24764	87.3%	9.7%	1.3%	0.7%	0.6%	0.4%
1986-1990	119416	34.5%	25.9%	8.1%	11.8%	17.5%	2.2%

TABLE 2. Age distribution for liquidators with an established external dose by years of arrival in zone

Years of arrival	*Age to 1st January 1996*								
	< 30	*30 - 34*	*35 - 39*	*40 - 44*	*45 - 49*	*50 - 54*	*55 - 59*	*60 - 64*	*> 65*
1986	4.7%	13.8%	17.9%	22.0%	26.9%	8.6%	4.8%	0.9%	0.4%
1987	1.8%	7.9%	19.5%	36.6%	27.6%	4.9%	1.2%	0.3%	0.1%
1988-1990	1.6%	2.8%	29.1%	42.2%	20.7%	2.6%	0.7%	0.2%	0.1%
1986-1990	2.9%	9.3%	20.8%	31.7%	25.9%	5.9%	2.6%	0.5%	0.3%

For prediction of distant stochastic radiation effects it is necessary to account for age distribution of subjects (Table 2). It should be underlined that the mean age of emergency workers at the time when they worked in the 30-km zone was 33.5 years, which means that the average life span at radiation risk (after exposure) exceeds 25 years. Today (by 1 January 1997) the majority of emergency workers (31.7%) lies in the age range of 40 to 44 years.

The primary goal of this study was to determine radiation risks of solid cancer, leukemia and thyroid cancer in the cohort of emergency workers. This was done based on the methods widely used in epidemiology. The derived coefficients of radiation risks were compared with the recommended values.

3. Results and Discussion

3.1. LEUKEMIA INCIDENCE

The present analysis considers 48 cases of leukemia in emergency workers verified by the Medical Radiological Research Center of RAMS and local health care establishments by 1 January 1994. Verification of leukemia is a complicated and lengthy procedure and, therefore, the study contains the analysis of incidence from 1986 to 1993 inclusive. By 1 January 1994 the RNMDR database contained medical and dosimetric information for 142 thousand emergency workers, among which 48 cases of leukemia were reported.

Our analysis of radiation risks accounts for all leukemia types (ICD-9 204.0-208.9). At the same time, it is common knowledge that chronic lymphocytic leukemia is not a radiation induced decease. The reason why all leukemia types were included in the study is because their number in 1986-1993 was few. Another important point to mention is comparison of observed and expected (including radiogenic) cases of leukemia. The expected number of all leukemia types was calculated using a multiplication prediction model in which coefficients were derived based on the Japanese cohort of atomic bomb survivors [6].

TABLE 3. SIR incidence of leukemia among EWs

Observation period years	*Observed number of cases*	*Expected number of cases*	*SIR*	*95% confidence interval*
1986-1989	14	12.3	113	62, 190
1990-1993	34	19.2	177	122, 247

Table 3 summarizes SIR (ratio of observed and expected incidence) estimates for leukemia in emergency workers for two time intervals: 1986-1989 and 1990-1993. It should be emphasized that in both cases SIR is more than 100%, i.e. incidence of emergency workers is higher that of the control reference group. At the same time, this difference from the reference with the confidence level of 95% is statistically significant only in 1990-1993 interval. It is also worth noting another point: the absence of radiation risk in 1986-1989 suggests that the assumptions in the prediction models regarding the latent period of 2-3 years in induction of radiogenic leukemia were correct.

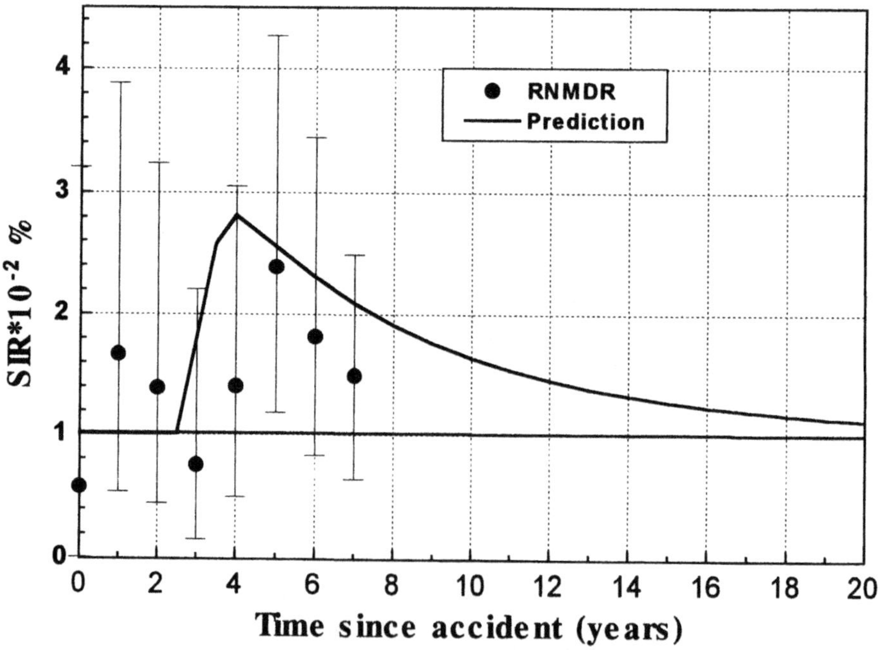

Figure 1. Actual data and prediction of the dynamics of leukemia incidence SIR in the emergency workers cohort.

As was mentioned above the study of leukemia (all types) in the Japanese cohort of atomic bomb survivors resulted in developing a dynamic model of radiation risk and we used this model for our predictions [6]. Besides the indicated prediction model, the main characteristics of the cohort of emergency workers were used: age and dose distribution and leukemia spontaneous incidence for Russia. Figure 1 shows data of the Chernobyl Registry on leukemia incidence in emergency workers and the corresponding prediction curve. What can be inferred from Figure 1? First, the prediction and actual data are in good agreement within error limits. Secondly, as follows from the prediction and can be clearly seen from the Registry data, the peak of radiogenic leukemia occurred 4-5 years after the accident with the attributable risk (AR) of 45-60% (AR=1-1/SIR). This suggests that each second leukemia detected in emergency workers today is radiation induced. Though the proportion of radiogenic leukemia in emergency workers is expected to decrease steadily with time, continuation of studies this area is one of the priority tasks of RNMDR.

3.2. THYROID CANCER INCIDENCE

By 1 January 1995 RNMDR included 47 thyroid cancers in emergency workers. These were diagnosed at different time since exposure: the time interval between entry into the 30-km zone and detection ranged from 1 to 8 years.

The histologic types of the tumor were: 42.8% of patients had follicular cancer, 33.3% - papillar cancer, in 14.3% cases the tumor was some kind of carcinoma. Table 8 presents main epidemiological data on incidence. Overall, 28 thyroid cancers were detected in the emergency workers of 1986, 15 - in emergency workers of 1987 and 4 - those of 1988-1990.

Table 4 shows standardized incidence ratios (SIR) for the observation period of 1986-1990 which is a latent period for thyroid cancer and those for the post-latent period of 1991-1994. It should be noted that during 1986-1990 SIR does not differ significantly from 100% for all groups of emergency workers of 1986, 1987, 1988-1990. At the same time, in the post-latent period SIR by far exceeds 100% (except 1988-1990 emergency workers) and hence it reflects increased, as compared control, incidence. As a control group we took the male population of Russia standardized by age.

As is seen from Table 4, the groups of increased radiation risk are emergency workers of 1986 (SIR=670%) and 1987 (SIR=590%). Of the 1986 emergency workers, the highest risk is observed in those who worked in the 30-km zone in April-July.

Let us now move to prediction of occurrence of thyroid cancer in emergency workers. Figure 2 shows observed incidence rate for emergency workers per 100 thousand people and expected incidence rate radiogenic plus spontaneous cancers. For calculation of spontaneous incidence we used statistical data for Russian Federation [7]. As was said above, a considerable discrepancy between the observed and expected values is observed 4-5 years since the disaster.

Figure 3 presents the dynamics of SIR in 8 observation years for the 1986-1987 emergency workers. As is seen from Figure 3, the value of SIR for the time period corresponding to the latent period of 4 years remains practically constant: 220-260%. If we assume that during this period no induction of radiogenic cancers occurs, then the difference of the observed SIR from 100% accounts for the screening effect (better medical examination). This plot also gives an estimate of expected input of radiogenic cancers to SIR under the assumption that the thyroid dose is only due to external exposure. It may be seen that the calculation model and risk coefficients [8, 9] account for only half of the real effect of the SIR increase after the latent period. If we exclude internal exposure of thyroid to iodine radionuclides, it can be inferred that either the model is not perfect or the external radiation dose is underestimated. Of course, it may be a combination of factors including the latest two.

TABLE 4. Main medico-dosimetric characteristics of EWs used in thyroid cancer incidence

Date of study	*1986*					*1986*	*1987*	*1986-1987*	*1988-1990*
	April - May	*June*	*July*	*August - Sept.*	*Oct. - Dec.*	*April - Dec.*	*Jan. - Dec.*	*April 86 - Dec. 87*	*Jan. - Dec.*
Population	19600	9800	11000	20200	17100	77700	58700	136400	31600
Mean age (years)	32	32	34	34	35	33	33	33	34
Mean dose (Gy)	0.16	0.19	0.16	0.17	0.17	0.17	0.10	0.14	0.04
Number of thyroid cancers	8	6	4	6	4	28	15	43	4
Expected number of thyroid cancers	1.32	0.74	0.82	1.60	1.31	5.80	3.64	9.44	1.71
SIR (latent) 95% CI		371 119, 865		150 16; 542		260 100; 540	260 80; 600	260 134; 452	180 80; 730
SIR (post-latent) 95% CI		844 449, 1440		508 218;1002		670 420;1030	590 280;1080	645 438;915	330 40;1180

3.3. ESTIMATION OF RADIATION RISKS

As was stated above, the main goal of this study is not just an epidemiological analysis in terms of SIR, but also determination of radiation risk coefficients for leukemia and thyroid incidence in emergency workers. By this the following coefficients are meant: excess relative risk per Gy (ERR/Gy), excess absolute risk per 10^4 PY Gy (EAR/10^4 PY Gy) and attributable risk percent (AR%) at 1 Gy.

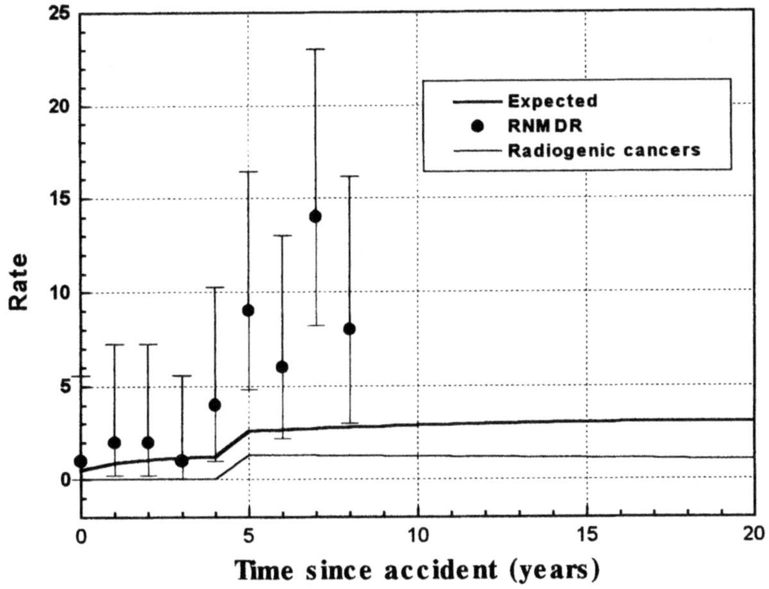

Figure 2. Thyroid cancer incidence rate among emergency workers as a function of time since the accident.

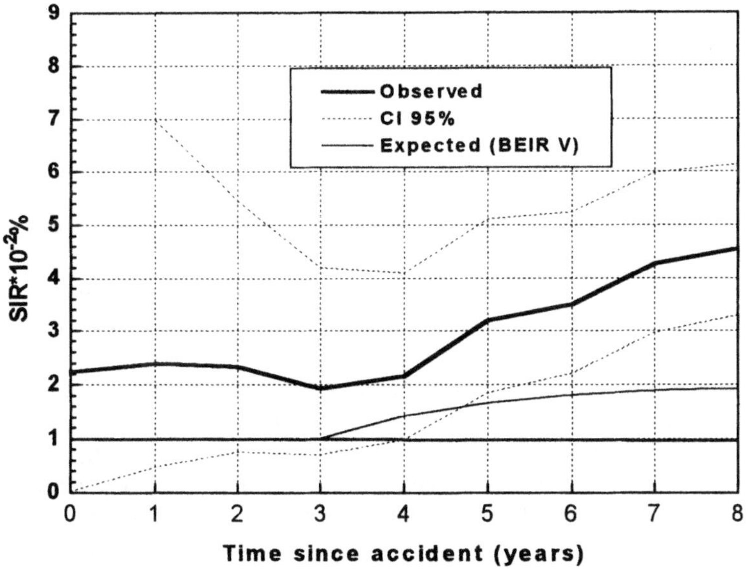

Figure 3. Dynamics of thyroid cancer incidence SIR among emergency workers (1986-1987 works).

In estimation of radiation risk coefficients it has to be assumed that the increase in leukemia and thyroid cancer occurrence is due to the dose factor only. At the same time, our calculations suggest that certain estimates of the dose relationship (using the method of epidemiological internal comparisons within cohorts) can be obtained only in the next 10-15 years. This requires about $2 \cdot 10^6$ PY of observations of the cohort (today this quantity is about 10^6 PY).

In a large-scale epidemiological study conducted some time ago to evaluate radiation risk of thyroid incidence in children exposed to external radiation these coefficients were estimated to be: ERR/Gy = 7.7; EAR/10^4 PY Gy = 4.4; AR = 88% [10].

At the same time, there is limited number of works in literature dealing with estimation of radiation risks in induction of thyroid cancer after exposure to iodine radionuclides [11].

As leukemia and thyroid cancer is a rare decease and the time of follow-up of the emergency workers cohort is relatively short, an external control group was used for estimation of radiation risks as an unexposed group.

To estimate excess absolute risk (EAR) normalized to unit dose the ratio was used:

$$EAR = \frac{O - E \cdot a}{PY \cdot D} \tag{1}$$

where
- O is observed number of cases;
- E is expected number of cases obtained with data of national medical statistics;
- α is coefficient allowing for the screening effect;
- PY is person-years of observation;
- D is external radiation dose [Gy].

Excess relative risk (ERR) per 1 Gy is calculated with the formula:

$$ERR = \frac{EAR \cdot PY}{E \cdot a}. \tag{2}$$

Attributable risk (AR) at 1 Gy dose was obtained from the ratio:

$$AR = \frac{ERR}{1 + ERR} \cdot 100\% \tag{3}$$

The confidence interval were calculated using the method of linearization of the function of random variables.

Estimates of radiation risk coefficients were made for the cohort of the 1986-1987 emergency workers. This was because the latent period in leukemia and thyroid cancer induction are over for the indicated cohort.

It is significant that in estimation of radiation risk coefficients in the present study we introduced parameter α accounting for the effect of in-depth screening of emergency workers. The thing is that in line with the regulations of the Ministry of Health Care of Russia the emergency workers are to undergo an in-depth medical examination on a yearly basis. This fact should be taken into account as a control comparison group we use male population of Russia standardized by age to the cohort of emergency workers.

For leukemia incidence the coefficient α=1. Indeed, as is seen from Table 3, during the latent period (1986-1989) SIR=113% and it does not differ with statistical significance from 100%.

For thyroid cancer incidence (Table 4) α=2.6. This is because during the latent period (1986-1990) SIR=260%, differs from 100% with statistical significance and reflects the effect of in-depth screening.

The indicated screening effect is well known from the literature. For example, the in-depth screening coefficient for the cohort of atomic bomb survivors (LSS) is: 2.4 for female and 3.5 for male population [12].

Tables 5, 6 present radiation risk estimates for the emergency workers cohort and their comparison with literature data by other cohorts.

As is seen from the tables, there is good agreement between the risk values obtained by us and coefficients published earlier in the literature. Further epidemiological study of the emergency workers cohort would permit deriving radiation risks from basic parameters: radiation dose, age at exposure and time since exposure.

The work presents data on incidence of solid malignant neoplasm among liquidators of the Chernobyl accident which have been gathered in RNMDR from 1986 to the beginning of 1996. Standardized incidence ratio (SIR) with 95% confidence intervals for all solid malignant neoplasms and cancers of the digestive system were 1.23 (1.15; 1.31) and 1.11 (1.01; 1.24) respectively. Estimation of radiation risks for the same disease classes has revealed a statistically significant increase in cancer incidence with increase in external radiation dose of liquidators.

TABLE 5. Radiation risk of leukemia incidence among Ews (1986-1993 observation period)

Data source	*EAR/10⁴ PY Gy (95% CI)*	*ERR/Gy*	*AR (at 1 Gy) %*
EWs	1.31 (0.23, 2.39)	4.30 (0.83, 7.75)	81
LSS cohort	2.38	7.8	88

TABLE 6. Radiation risk of thyroid cancer incidence among Ews (1986-1994 observation period)

Data source	*EAR/10⁴ PY Gy (95% CI)*	*ERR/Gy*	*AR (at 1 Gy) %*
EWs	1.15 (0.08, 2.22)	5.31 (0.04, 10.58)	84
BEIR V	1.25	5.8	85

4. Conclusion

We have presented unique information on radiation risk for the liquidators who worked in the Chernobyl 30-km zone and who now reside in Russia. The group reported on here can be considered to be a representative sample of the liquidators. Although the data we have for each individual liquidator are not very accurate in terms of received radiation dose, what is presented is considered to be the best estimate.

5. References

1. Ivanov, V.K., Tsyb, A.F. (1996) Chernobyl radiation risks: assessments of morbidity, mortality and disability rates according to the data of the National Radiation and Epidemiological Registry. In: *Nagasaki symposium "Radiation and Human Health"*, Elsevier, Amsterdam, pp. 31-48.
2. Ivanov, V.K., Tsyb, A.F., Rastopchin, Eu.M., Maksyutov, M.A., Gorsky, A.I., Biryukov, A.P., Chekin, S.Yu, Konogorov, A.,P. (1994) Planning of long-term radiation and epidemiological research on the basis of the Russian National Medical Dosimetric Registry. In: *Nagasaki symposium on Chernobyl update and future*, Elsevier, Amsterdam, pp. 203-216.
3. Ivanov, V.K., Tsyb, A.F., Maksyutov, M.A., Rastopchin, Eu.M., Gorsky, A.I., Konogorov, A.P., Chekin, S.Yu., Pitkevich, V.A., Mould, R.F. (1995) Cancer morbidity and mortality among Chernobyl accident emergency workers residing in the Russian Federation, *Current Oncology* **2**, 102-112.
4. Ivanov, V.K., Tsyb, A.F., Maksyutov, M.A., Pitkevich, V.A., Gorsky, A.I., Rastopchin, Eu.M., Korelo, A.M., Chekin, S.Yu., Konogorov, A.P., Nilova, E.V. (1996) Radiation epidemiological analysis of the data of the National Chernobyl Registry of Russia: prognostication and facts nine years after the accident, *Radiation Protection Dosimetry* **64**, 121-128.
5. Cardis, E., Anspaugh, L., Ivanov, V.K., Liktharíev, I., Mabuchi, K., Okeanov, A.E., Prisyazhniuk, A. (1996) Estimated long term health effects of the Chernobyl accident. In: *International Conference "One decade after Chernobyl: summing up the consequences of the accident"*, Background paper, session 3, Vienna.
6. Preston, D.L., Kusumi, S., Tomonaga, M., Izumi, S., Ron, E., Kuramoto, A., Kamada, N., Dohy, H., Matsui, T., Nonaka, H., Thompson, D.E., Soda, M., Mabuchi, K. (1994) Cancer incidence in atomic bomb survivors. Part III: leukemia, lymphoma and multiple myeloma, 1950-1987, *Radiat. Res.* **137**, 68-97.
7. Malignant tumours in Russian Federation in 1993. Collection of statistic materials (1995), Chissov, V.I., Starinsky, V.V., Remennik, L.V. (eds), Moscow, part I.
8. US NAS (National Academy of Sciences Committee on the Biological Effects of Ionizing Radiation) (1990) Health effects on populations of exposure to low levels of ionizing radiation, BEIR V Reports, US National Academy of sciences, Washington, DC.
9. ICRP (International Commission on Radiological Protection) (1990) Recommendations of the International Commission on Radiological Protection, ICRP Report 60, Pergamon Press, Oxford.
10. Ron, E., Lubin, J.H., Shore, R.E., Mabuchi, K., Modan, B., Pottern, L.M., Schneider, A.B., Tucker, M.A., Boice, J.D. (1995) Thyroid cancer after exposure to external radiation: a pooled analysis of seven studies, *Radiat. Res.* **141**, 259-277.
11. Shore, R.E. (1996) Human thyroid cancer induction by ionizing radiation: summary of studies based on external irradiation and radioactive iodines. In: *Proceedings of the first international conference*, Minsk, pp. 669-675.
12. Ivanov, V.K. (1996) Health status and follow-up of liquidators in Russia. In: *Proceedings of the first international conference*, Minsk, pp. 861-870.

A MODEL OF TWO REACTIONS AS DOSE-STOCHASTIC RADIOLOGICAL EFFECT RELATIONSHIP AND ESTIMATION OF RADIATION RISKS

M. M. KOMOCHKOV

Joint Institute for Nuclear Research, Division of Radiation and Radiobiological Research
141980 Dubna, Russia

Abstract

A defense system consists of two forms of organism reaction, which determine innate μ_n and adaptive μ_a radiosensitivities. The significance of μ_n is determined by host (inner) factors, and the significance of μ_a by external factors. The possibilities of adaptive reaction are determined by the coefficient of capabilities of the defense system. The formulas of the dose-effect relationship are the solutions of differential equations of assumed process in the defense system of organism. The model and formulas have been checked both for cells and humans. Based on the model and personal monitoring data the estimation of radiation risk is performed at the Joint Institute for Nuclear Research.

1. Introduction

The attempt to describe dose-effect relationships properly began many years ago [1–8]. However, possible consequences of radiation exposure of humans, particularly by low doses, are poorly understood [9]. The striking illustration of uncertainties concerning the problem of human radiation exposure consequences are two mutually exclusive predictions for humans exposed due to the Chernobyl accident [10, 11]. One of them predicts about 300,000 lethal outcomes from cancer as a result of the Chernobyl accident [10]; the second expects a prevention of 20,000 lethal outcomes from cancer in the ex-USSR republics [11]. Another example is connected with the 1990 Recommendations of the International Commission on Radiological Protection [12]. The ICRP has considered the dose limits insufficiently low and recommended to reduce them while a number of specialists considers this recommendation as unjustified and extravagant. Furthermore Keirim-Markus [13] considered the ICRP recommendation as speculative causeing danger for human health. These discrepancies are the consequence of the fact, that the predictions, opinions and recommendations expressed by the authors are based on different dose-effect relationships. In this connection a search of new more effective dose-effect relationships may be justified and actual.

C. Baumstark-Khan et al. (eds.),
Fundamentals for the Assessment of Risks from Environmental Radiation, 25-34.

This article presents such a new dose-stochastic radiobiological effect relationship and demonstrates its application to fit some of the most striking control results. Based on the two defense reactions (TDR) model the relationship gives the basis for the estimation of radiation risk at the Joint Institute for Nuclear Research.

2. Basis of the Model

The main considerations and conditions assumed as basis of the TDR model and formulas are following.

1) Radiobiological effects are the result of random events
2) Biological objects may be any organism capable of self-defense
3) A defense system consists of two types of organism reaction (response), which determine innate μ_n and adaptive μ_a radiosensitivities
4) The significance of μ_n is determined by host (inner) factors, and the significance of μ_a by external factors
5) The possibilities for an adaptive reaction are determined by the coefficient of capabilities of the defense system ν.
6) Ionizing radiation exposure decreases or increases the frequency F or the probability W of the effect

The considerations 3-6 need in argumentation, which partly may be adopted from published studies with cells [15, 16] and partly may be supported in successful fitting of experimental results and control. For higher organisms it is determined, that response of immune system on the action of any dangerous factors sum up from two components, innate and adaptive, divided by some time interval [17].

3. Equations for the TDA Model

We assume a function f, as a basis of the dose-effect relationship. It is defined thus, that the product $(1-W_c)$ f is the probability to avoid stochastic effect at dose exposure H; here W_c is the probability of the effect in unexposed controls (probability of spontaneous effect, background). Then the probability of the effect is

$$W = 1-(1-W_c)f \tag{1}$$

The excess of the effect above background is the difference of W and W_c:

$$\Delta W = W - W_c = (1-W_c)(1-f) \tag{2}$$

The relative risk RR is determined as a ratio:

$$RR = \frac{W}{W_c} = \frac{1-(1-W_c)f}{W_c} \tag{3}$$

The mean frequency of the expected effect in accordance to [8] and equation (1) is

$$F = \ln\frac{1}{1-W} = \ln\frac{1}{(1-W_c)f} \tag{4}$$

A function f is found by solving the next differential equations:

$$df_n = -\mu_n f_n dH \tag{5}$$

$$df_a = (\mu_n v f_n - \mu_a f_a)dH \tag{6}$$

where

- f_n is the fraction of the avoided effect (lesion) due to the innet defense reaction at radiosensitivity μ_n,
- f_a is the additional fraction of the avoided effect of individuals because of the adaptive defense reaction at radiosensitivity μ_a and
- v is the coefficient of capability of a defense system.

The equations can be solved for two particular cases that may help to describe control results adduced in the part 4.

In the first case, when significances of μ_n, μ_a, v were consider as independent on H, the solutions are [14]

$$f_n = \exp(-\mu_n H) \tag{7}$$

$$f_a = \frac{v\mu_n}{\mu_n - \mu_a}[\exp(-\mu_a H) - \exp(-\mu_n H)] \tag{8}$$

The total fraction of avoid effects (lesion) of individuals is considered as the sum of f_n and f_a :

$$f = \exp(-\mu_n H) + \frac{v\mu_n}{\mu_n - \mu_a}[\exp(-\mu_a H) - \exp(-\mu_n H)] \tag{9}$$

In the second case, when μ_n is considered as independent on H, and μ_a and v as dependent on H, the f_n is given as equation (7) and f_a is described by equation (10):

$$f_a = \exp(-\int_0^H \mu_a dH)\{\mu_n \int_0^H [v \exp(-\mu_n H)\exp(\int_0^H \mu_a dH)]dH\} \tag{10}$$

In all cases the significances μ_n, μ_a v have to satisfy condition (11)

$$f \leq \frac{1}{1 - W_c} \tag{11}$$

The analysis of equation (9) for the condition described in equation (11) shows the possibility of describing hormesis [11, 18, 19], as f may be greater than 1, and $W<W_c$ at $v>1$. At low doses when $\mu_n H<<1$, we may get equation (12) when the first two terms of expansion were restricted in the row of exponents:

$$f = 1 + \mu_n(v - 1)H \tag{12}$$

Substituting this value f into equation (2) gives

$$\Delta W = (1 - W_c)(1 - v)\mu_n H \tag{13}$$

and for derivation

$$W' = \Delta W' = (1 - W_c)(1 - v)\mu_n \tag{14}$$

Equation (13) shows, that the absolute excess of risk is straightly proportion to dose for low doses and the above-mentioned conditions. In this case for the derivation $\Delta W'$ the ICRP recommends the name "probability coefficient for stochastic effects". From equation (14) one can see, that the significances of $\Delta W'$ may to a wide range, that was noticed in the study [9]. Negative significances ΔW and $\Delta W'$ ($\nu>1$) correspond to hormesis.

4. Verification of the Model

At present time a verification of the validity of the considerations assumed as basis of the TDR model, may be only proven in forms of fitting different observed dose-effect relationships using the formulas presented in part 3. In the capacity of observed quantities in the epidemiological control W and RR are selected, furthermore f and F in the experiments with cells.

4.1. EPIDEMIOLOGICAL RESULTS.

The relative risk of cancer mortality for Japanese A-bomb survivors in Hiroshima and Nagasaki is presented in figure 1. The TDR model results are obtained by equation (3); the parameters at the consent condition of calculation and control results have been determined to be: μ_n=0.66 Sv^{-1}, μ_a=1.022 Sv^{-1}, ν =0.95 to solid cancer and μ_n=0.4 Sv^{-1}, μ_a=0.16 Sv^{-1}, ν=1.02072 to leukemia. The significances W_c are assumed equal to 0.2 for solid cancer [20] and 0.0018 for leukemia [18].

Figure 2 shows lung cancer mortality as function of radon level for two cohorts of humans: underground miners and dwellers of the USA counties. An effective dose H_E

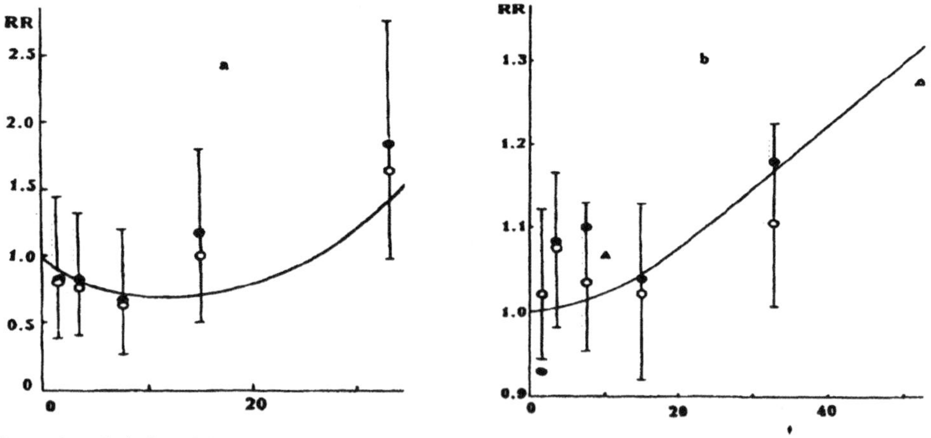

Figure 1. Relative risk RR of A-bomb survivors: a) mortality from leukemia; b) mortality from all cancer other than leukemia; • ρ □ – epidemiological studies results on all ages of survivors (• [18], ρ [20]) and those who were less than 40 years old at the time of bombing (□ – 95% confidence interval [18]), —— – TDR model result.

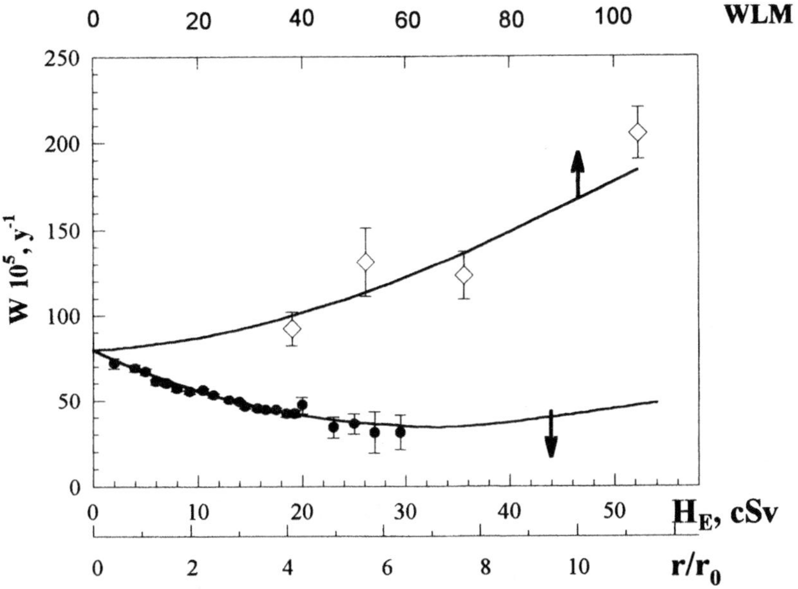

Figure 2. Lung cancer mortality W vs. radon level (r_0 – 37 Bq m^3; WLM – 0.5 cSv): • – in homes (Cohen B.L., 1995, from ref. [13]); ↓ – in mines (ICRP, 1993, from ref. [13]); —— fit to the data of TDR model.

for the dwellers is a cumulative dose over lifetime; an exposure of the miners at 1 WLM approximately creates the effective dose 0.5 cSv[13]. The TDR model results are calculated at the next conditions: W=W/70, where 70 is the lifetime significance, and W is calculated according to equation (1) at W_c=0.056 [8, 12, 20]; the lower curve is calculated at μ_n=1 Sv^{-1}, μ_a=0.5 Sv^{-1}, ν=1.22, and the upper curve is calculated at μ_n=1 Sv^{-1}, μ_a=0.644 Sv^{-1}, ν=0.98. The significances of the parameters reflect the fact that the capability of the miners defense (immune) system is lower than that of the dwellers. This is apparently connected with additional risk factors of miners. The epidemiological results of the lung cancer of Swedish dwellers without a hormesis are presented by [11].

4.2. EXPERIMENTAL RESULTS WITH MAMMALIAN CELLS.

Figure 3 presents the frequency of dicentric chromosomes in human lymphocytes as a function of the photon dose to which the cells were exposed [22–25]. The extremes of the experimental results are described better by the TDR model, as can see in figure 3; the TDR model parameters as calculated using equation (4) are: μ_a=100 Gy^{-1}, μ_a=0.1 Gy^{-1}, ν=1.00193.

Figure 4 shows the survival of human cells (line HT29). The data points are the experimental results [26] and the curves are fitting of the different models to the data. The values of TDR model parameter fitting (equation (7) and equation (10)) are: μ_n=0.625 Gy^{-1}, μ_a=0.606 Gy^{-1}; the values of ν vary from 0 to 1.1 close for the dose of 0.15 Gy^{-1} in

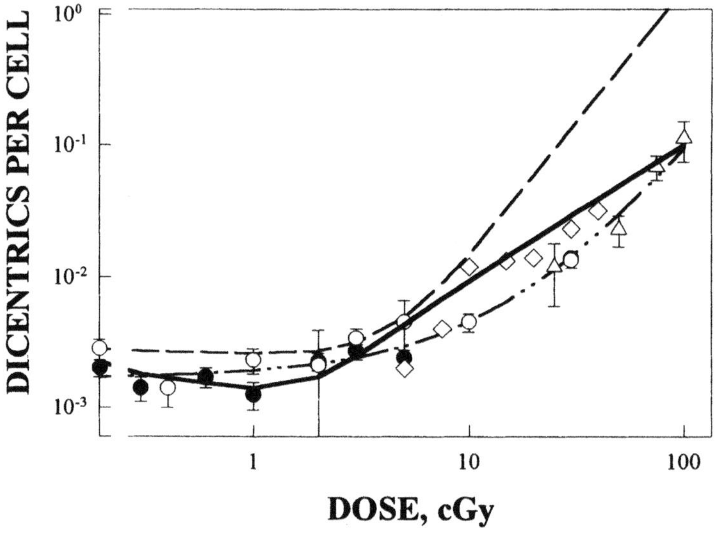

Figure 3. Dicentric yields as function of photon dose: ↓ • ρ □ – experimental results; — fit to the data of TDR model; — · , – – – fit to the data of linear-quadratic model.

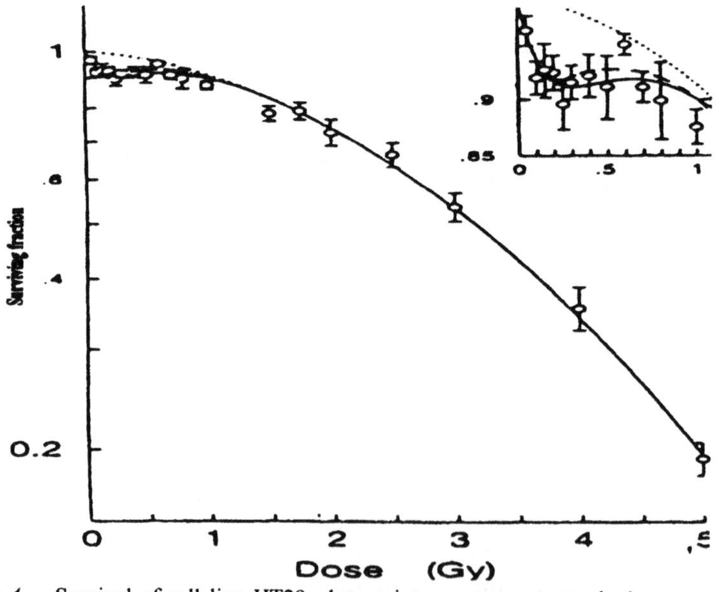

Figure 4. Survival of cell line HT29: data points are mean ± standard error ; ····· linear-quadratic model; —— induced repair model fit; - - - TDR model fit.

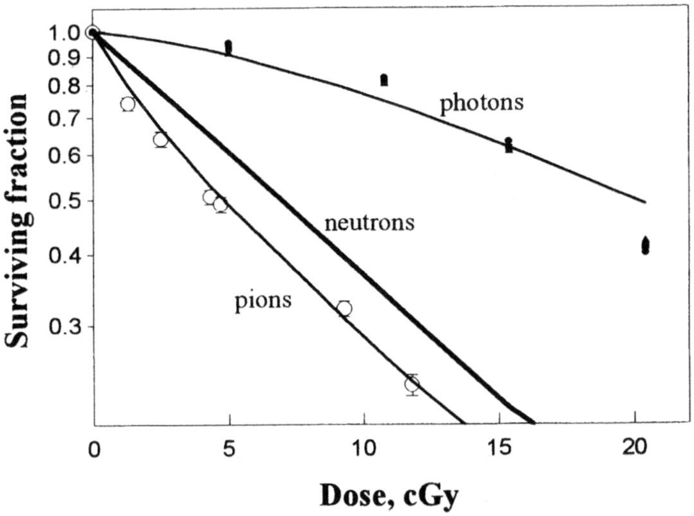

Figure 5. Survival of spermatogonia type B of mice after irradiation at peak dose in stopped negative pion, neutron and photon beams: data point – experiment, solid line – TDR model fit

accordance to the sigmoid shape of the curve. It is assumed that this reflects the fact that adaptive reactions require a minimal dose 0.15 Gy. Such phenomenon was already observed earlier [27]. At the doses <1 Gy the linear-quadratic model substantially underestimates the effect of X-rays.

The survival of spermatogonia type B of mice after exposure with different types of radiation [28] is shown in figure 5. The mice were irradiated using Pu-Be neutrons, 14-MeV neutrons and high-energy neutrons up to 600 MeV (data points are not indicated in figure 5). The parameter significances of the TDR model fitting (equation (9)) are: μ_n=10 Gy^{-1} for all types of radiation, μ_a=6.25 Gy^{-1}, ν=0.9 (photons), μ_a=50 Gy^{-1}, ν=-0.9 (pions) and ν=0 (neutrons). A negative value of ν indicates apparently that the cells defense systems are suppressed by the action of pion nuclear reaction products.

5. Approximation of the Results for Risk Estimation

The verification of the presented TDR model describing dose-stochastic relationships is done in order to realize the estimations of radiation risk R at determined significances μ_n, μ_a, ν for human cohorts exposed to radiations with known dose distributions. For the simplest case R is estimated as a product of derivation of the effect probability W′ (or ΔW′) at the dose H on its signification

$$R = W'H \tag{15}$$

For the cohort of N_0 individuals exposed to radiation with the dose distribution dN/dH the risk is

$$R = \int_0^{H_m} W' H \frac{dN}{dH} dH \qquad (16)$$

where H_m is maximum dose of the distribution. At low doses in accordance with equation (13) the risk is a product of the probability coefficient for the stochastic effect (W′) on H:

$$R = (1 - W_c)(1 - \nu)\mu_n H \qquad (17)$$

The application of the presented TDR model and its formulas is illustrated by some examples.

5.1. COMPARISON OF PROBABILITY COEFFICIENTS OF CANCER MORTALITY W′M

The most important outcomet of epidemiological surveillance of human exposure is the estimation of W'_m, which is based on extrapolation of risk from high to low dose studies. Such extrapolation on the base of the TDR model formulas and the parameters described in 4.1 gives the following.

5.1.1. Comparison Results of W′m on base of Cancer Mortality of Japanese A-Bomb Survivors

The values of W'_m for lifetime risk are:

Solid cancer (without leukemia)

- according to equation (14) at W_c=0.2 — $W'=2.6 \cdot 10^{-2}\ Sv^{-1}$
- according to ICRP [12] — $W'=4.5 \cdot 10^{-2}\ Sv^{-1}$

All cancer

- according to equation (14) — $W'=1.8 \cdot 10^{-2}\ Sv^{-1}$
- according to ICRP [12] — $W'=4.5 \cdot 10^{-2}\ Sv^{-1}$

5.1.2. Comparison Results of W′m on base of Lung CancerMortality of Underground Miners

The values of W_m' without correction for lifetime risk are:

- according to equation (14) — $W'=1.9 \cdot 10^{-2}\ Sv^{-1}$
- according to linear model — $W'=14 \cdot 10^{-2}\ Sv^{-1}$

5.2. ESTIMATION OF RADIATION RISK AT JINR

The estimation of risk significance is calculated according to equation (16). The dose distribution of the staff of the Joint Institute for Nuclear Research (JINR) is presented in [29] for the last ten years. The relationship W′ as function H is calculated based on equations (1) and (9) and part 5.1.1 for workers. The results give the following values for the radiation risk level R of 2500 workers of JINR using radiation sources:

$$\dot{R} = \int_0^{H_m} W' \dot{H} \frac{dN}{dH} dH = 0.1 \text{ cancer death/y.}$$

This value is more than 30% as compared with a product of collective dose per year of JINR workers (5 Sv/y) on W′ calculated with equation (13). The risk calculation based on the quantity of W′ recommended by ICRP [12] gives a value of 0.2 cancer death per year, which exceeds twice the risk value of the TDR model.

6. References

1. Zimmer R.G. (1961) *Studies on quantitative radiation biology*. Oliver and Boyd, Edinburgh and London.
2. Timofeev-Resovski N.V., Korogodin V.I. and Ivanov V.I. (1968) *Application of hit principle in radiobiology*. Atomizdat, Moscow (Russ.).
3. Hug O. and Kellerer A. (1966) *Stochastik der Strahlenwirkung. Sonderband zur Strahlentherapie*. Spinger-Verlag, New York.
4. Kapultchevich V.G. (1978) *Quantitative regularities of ray lesion of cell*. Atomizdat, Moscow (Russ.).
5. Filushkin I.V. and Petojan I.M. (1988) *Carcinogenic risk theory of ionizing radiation influence*. Energyatomizdat, Moscow (Russ.).
6. Krasavin E.A. (1989) *RBE problem and DNA reparation*. Energyatomizdat, Moscow (Russ.).
7. Krasavin E.A. and Kozubek S. (1991) *Mutagene action of radiation with different LET*. Energyatomizdat, Moscow (Russ.).
8. United Nations Scientific Committee on the Effects of Atomic Radiation. 1986 report to the General Assembly, with annexes. "*Genetic and somatic effects of ionizing radiation*." United Nation, New York.
9. Komochkov M.M. (1996) *About possible consequences of men irradiation by small dose of ionizing radiation*, JINR, P16-96-70, Dubna (Russ.).
10. Gofman J.W. (1994) *Chernobyl accident: radiation consequences for this and future generations*. CNR Books and Vysheishaya Shkola, Minsk (Russ.).
11. Luckey T.D. (1991) *Radiation hormesis*, CRC Press, Boca Raton, Florida, USA.
12. ICRP (1991) International Commission on Radiological Protection. 1990 Recommendation of the ICRP, Pub. 60. Annals of the ICRP, 21, No 1–3. Pergamon Press, Oxford.
13. Keirim-Markus I.B. (1995) New information about the consequences of radiation exposure in humans at low dose – a crisis of the conception of exposure regulation? *Atomic Energy*, **79**, 279–289.
14. Komochkov M.M. (1996) *Formulas of dose-stochastic radiobiological effect relationship (the first approach)*, JINR, P16-96-323, Dubna (Russ.).
15. Geraskin S.A. (1995) Critical analysis of up-to-date conceptions and approaches to estimation of biological effect of low-level radiation, *Radiation Biology. Radioecology*, **35**, 563–570 (Russ.).
16. Skov K.A. (1994) Molecular, cellular, and genetic basis of radiosensitivity at low doses: a case of inducible repair? *Radiation Research*, **138**, s1–s4.
17. Janeway C.A. and Travers P. (1994) *Immunobiology: The immune system in health and decease*. Garland publishing inc., New York and London.
18. Okada S. (1995) *Panel discussion on health effects of low-dose ionizing radiation: scientific findings and non-threshold hypothesis*. JAERI - Conf. 95-010, Tokyo, p. 4.
19. Kuzin A.M. (1995) *Ideas of radiation hormesis in atomic century*. Nauka, Moscow (Russ.).
20. Kellerer A.M. (1995) The new estimates of radiation risks. *Kerntechik*, **55**, N 4, 198–203.

21. Lagarde F., Pershagen G., Åkerblom G., Axelson O., Bäverstam V., Damber L., Enflo A., Svartengren M. and Swedjemark G.A. Residential radon and lung cancer in Sweden: risk analysis accounting for random error in the exposure assessment. *Health Physics*, **72**, 269–276.
22. Pohl-Ruling J., Fisher P., Haas O., Obe G., Natarajan A.T., van Buul P.P.W., Buckton K.E., Bianchi N.O., Larramendy M., Kucerova M., Poliková Z., Leonard A., Fabry L., Palitty F., Sharma T., Binder W., Mukherjee R.N. and Mukherjee V. (1983) Effect of low-dose acute X-irradiation on frequencies of chromosomal aberrations in human peripheral lymphocytes *in vitro*. *Mutation Research*, **110**, 71–82.
23. Lloyd D.C. Edwards A.A., Leonard A., Deknudt G.L., Vershaeve L., Natarajan A.T., Darraudi F., Obe G., Palitti F., Tanzarela C. and Tawn E.J. (1992) Chromosomal aberrations in human lymphocytes induced in vitro by very low doses of X-rays. *Int. J. Radiat. Biol.*, **61**, 335–343.
24. Luchnik N.V. and Sevan'kaev A.V. (1976) Radiation-induced chromosomal aberrations in human lymphocytes. 1. Dependence on the dose of gamma-rays and an anomaly at low doses. *Mutation Research*, **36**, 363–377.
25. Sevan'kaev A.V., Ankina M.A., Zavitaeva T.A., Zhloba A.A., Golub E.V., Epifanova N.V., Kozlov V.M., Mikhailova G.F., Moiseenko V.V., Pozdyshkina O.V., Potetnya O.I., Potetnya V.I., Pyatenko V.C. and Semenova T.G. (1995) Comparative studies of frequency of stable and unstable chromosome aberrations in human lymphocytes gamma-irradiated *in vitro*, *Radiation Biology. Radioecology*, **35**, 611–617.
26. Lambin P., Coco-Martin J., Legal J.D., Begg A.C., Parmentier C., Joiner M.C. and Malaise E.P. (1995), Intrinsic radiosensitivity and chromosome aberration analysis using fluorescence in situ hybridization in cells of two human tumor cell lines. *Radiation Research*, **138**, s40–s43.
27. Irushima T. (1989) Proceedings of the third Japan–US workshop on tritium radiobiology and health physics, Nov. 8–10, 1988, Kyoto, Japan, 189–197.
28. Baarly J., Bianchi M., Sallivan A.H. and Di Paola M. (1976) Proceedings of a symposium on "Biological and environmental effects of low-level radiation", Chicago, 3-7 November 1975, IAEA, Vienna, **1**, 195–204.
29. Komochkov M.M. and Mokrov U.V. (1994) *Personal dosimetry at JINR*. JINR P16-94-178, Dubna (Russ.).

EPIDEMIOLOGICAL STUDIES FOR THE ASSESSMENT OF RISKS FROM ENVIRONMENTAL RADIATION ON TUNDRA NENTSI POPULATION

L. P. OSIPOVA[1], O. L. POSUKH[1], K. P. KOUTZENOGII[2],
F. V. SUKHORUKOV[3], V. G. MATVEEVA[1], A. S. GRAFODATSKII[1],
N. A. KONOVALOVA[4], Y. G. SUKHOVEY[5], S. A. PETROV[5],
G. LEFRANC[6], M.-P. LEFRANC[6]

[1] *Institute of Cytology & Genetics, Novosibirsk, Russia*
[2] *Institute of Chemical Kinetics & Combustion, Novosibirsk, Russia*
[3] *Analytical Center of the Joint Institute of Geology, Geophysics & Mineralogy, Novosibirsk, Russia*
[4] *Regional Center of Eye Protection, Tyumen, Russia*
[5] *Tyumen Filial of the Institute of Clinical Immunology, Tyumen, Russia*
[6] *Institute of Molecular Genetics, Montpellier, France*

Abstract

We have studied a population of 1200 people of the Tundra Nentsi inhabiting Northwestern Siberia in the immediate proximity to the nuclear test ground Novaya Zemlya. As has been shown, cancer has increased over the last decade, and so too the number of secondary immune deficiency diseases. Chromosome analysis of peripheral blood (N=74) suggests that the overall percentage of chromosomal aberrations in the population is 4.59% (1.5% in the control), while that of "ring" and dicentric" type aberrations, which are the "markers" of radiation damage, is 0.53% (0.05% in the control). We have discovered a unique variant of an immunoglobulin gene structure combined with a small extranumerical chromosome. Total blood analysis has revealed such abnormalities as anemia, leucopenia and deformed erythrocytes in various combinations in 17% of the people examined. Analysis of soils has shown that the content of ^{137}Cs in the upper 5 cm layer is 59 Bq/kg and 65 Bq/kg at two locations which is higher than that following the global fall-outs at these latitudes. A correlation has been found between the content of ^{137}Cs in lichen and reindeer meat (61 Bq/kg and 62 Bq/kg, respectively). This suggests that the natives, who live largely on reindeer meat, may have been continuously exposed to radiation from inside.

1. Introduction

The effects of anthropogenic environmental contamination on humans, both contemporary and those yet to come, are one of the major ecological problems world-wide [1].

C. Baumstark-Khan et al. (eds.),
Fundamentals for the Assessment of Risks from Environmental Radiation, 35-42.

This complex problem cannot be solved unless test objects and regions are chosen correctly and skilled experts from different fields put their effort together.

The high latitudes of Western Siberia are experiencing powerful pressure from adverse technogenic factors that come about due to oil and gas exploration and not at least to the nuclear tests at the Severny testing ground (Novaya Zemlya) which is located in the immediate vicinity of the continental Siberia.

Deterioration of the ecological situation affects the numerically small aboriginal tribes of Siberia in such special way that their long-term existence appears rather questionable. In our opinion, these populations and the lands they inhabit can represent adequate objects for developing a model for a complex assessment of the effect of technogenic environmental contamination on the gene pool and the biological healthiness of man.

2. Test Object and Methods

2.1. TEST OBJECT

The Tundra Nentsi in the Pur Rayon of the Tyumen Region (Yamal-Nentsi Autonomous District) satisfy the conditions under which research can be conducted:

- immediate proximity to the producers of large-scale technogenic effects (oil-and-gas exploration industries, refineries, non-ferrous metallurgy, Novaya Zemlya testing ground);
- a rather small population size (about 2000) with ethnic homogeneity; not liable to migrate.
- a rather large extent to which the tribe has been studied on the population and genetic level, which will allow genetic and demographic monitoring to be performed and gradual genetic research to be conducted (in particular, family groups can be sampled for within-family analysis).
- rather a simple nutrition transfer: lichen-reindeer-human, water-fish-human.

A Brief Review

Linguistically, the Tundra Nentsi belong to the Northern Samoyed branch of the Uralic family; anthropologically, they are of the Uralic race displaying both Mongoloid and Caucasoid traits. Reindeer-breeders, fishermen and hunters, largely nomads, their *modus vivendi* allows them to be classed as "original". They inhabit a vast (about 100000 km^2) yet well-defined area. The population is ethnically homogenous with its "genetic core" well preserved, the inbound migration rate is low, and the outbound one alike. This is the so-called growing type population with a large percentage of underreproductive members and a large reproductive size [2] (Figure 1). Preliminary data on the polymorphism of some blood genetic systems have been obtained [3].

After the mortality rate was reported to sharply increase on account of cancer (Figure 2), which had hitherto been rare to the natives, and public health, in general, to be ruined, it has been decided to continue research.

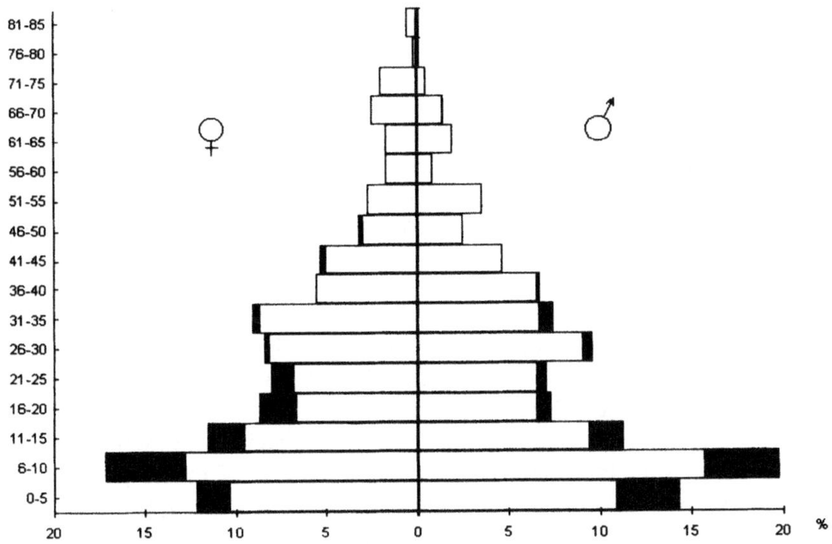

Figure 1. Distribution with respect to sex and age in the Samburg population. Abscissa - percentage inhabitants of the given age; ordinate - age in years. Mestizoes of the native ethnic groups and immigrants are where shaded

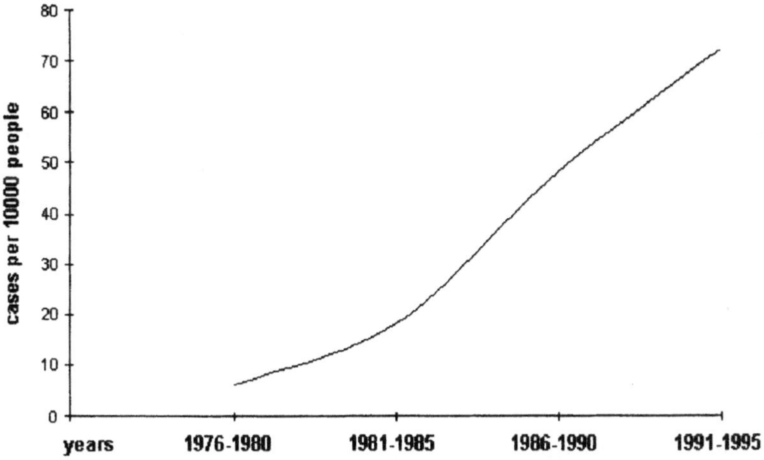

Figure 2. Mortality rate growth due to cancer among the Tundra Nentsi in Samburg (1976-1995). 50% - gastroenteric cancer; 14%- uterine cancer; 18% - lung cancer; 7% -leukemia; 10% - others (skin cancer, Williams tumor, ...pituitary cancer). Cumulative age-specific cancer mortality is an index approximately independent from changes in age distribution. For the calculation of this index, the cases per 10,000 people are calculated in each age group and then the mortality rate is summed up over all groups up to 75 years.

2.2. METHODS

2.2.1. Medical Studies

Medical studies included immunological and therapeutic examination for detecting immune deficiency diseases. Using the facilities of the local hospital in Samburg we took blood samples for further establishing total hemoglobin (Hb), leukocytes (L) and the erythrocyte sedimentation rate (ESR). Blood smears were assayed in the hospital and in Novosibirsk.

2.2.2. Cytogenetical Analysis

Blood samples of 10 ml each were taken in 1994-1996 by venipuncturing and were delivered to the laboratory in Novosibirsk within 3 days. Lymphocytes were cultivated after a standard macromethod [4]. On average, 100 metaphases were analyzed in each individual. The whole spectrum of structural chromatid and chromosome aberrations was taken into account, as these are regarded as the criteria of mutagenic effects [5]. Differential staining was performed on the chromosomes for karyotyping each individual.

2.2.3. Molecular-genetic Analysis of Immunoglobulin (IGH) Genes

Total DNA was extracted from peripheral blood leukocytes for further RFLP analysis using double BamHI-SacI restriction with the C-γ-probe [6]. Pulsed-field gel electrophoresis was performed on the "inserts" prepared after a standard technique [7] in the local hospital of Samburg followed by MluI restriction and hybridization of the C-γ-probe in Montpellier (France).

2.2.4. Nuclear-physical Methods of Research

Atmospheric aerosols were sampled on filters in 1995-1996. The samples were analyzed by roentgen-fluorescence using synchrotron radiation. Soil, lichen, meat were sampled in 1996-1997 from two sites, namely the surroundings of Samburg and the location of Brigade No7. Technogenic ^{137}Cs was measured using γ-spectrometry with a detection limit for ^{137}Cs of 2-3 Bq/kg applying on pure germanium semiconductor detectors. ^{90}Sr was measured by the radiochemical method with a detection limit of 1-2 Bq/kg.

3. Results and Discussion

3.1. RESULTS OF THE CYTOGENETIC STUDIES

A total of 113 blood samples has been collected, of which 10 were taken twice from the same people in order to ascertain the correctness of the diagnosis. Of the remaining 103 samples, 19 (18.4%) failed to display cell growth. The other 84 were divided into two groups: “natives” (N=74) and “immigrants” (N=10) (Table 1).

We compared our evidence with literature data, according to which the boundary of “normalcy” for the overall number of chromosomal abnormalities per metaphase is 0.5-

1.5% [5]. The upper limit of the control level for the marker aberrations ("dicentrics" and "rings") is 0.05%, or 1/2000 metaphases [8].

The overall chromosomal damage (4.59%) in the Samburg "natives" is 3 times as much as the upper limit, and the "rings" and "dicentrics" 10 times, which is comparable with the indices in the nuclear weapon testers or the Chernobyl emergency workers.

The sample of "natives" was divided into three subsamples depending on the percentage of observed chromosomal aberrations (x) (Figure 3). As can be seen, almost 50% of the population has a high level of chromosomal aberrations, and most of "dicentrics" and "rings" falls within this group.

Furthermore, we have discovered a unique variant of an immunoglobulin gene structure combined with a small extranumerical chromosome (to be published).

We analyzed the rates of chronic inflammatory and recurrent influenza-like diseases (the so-called SIDD - secondary immune deficiency diseases) in these people and found that the carriers of "dicentrics" occur 1.8 times as much among the SIDD-affected individuals as they do among those in the "risk group" with respect to chronic diseases. Besides, the share of SIDD cases is especially big in the people whose blood failed to set the lymphocytes growing. It is possible that lymphocyte growth failure reflects the so-called "immune depression" of the cells.

From within the sample being analyzed, we selected a group of children (N=15) that were born after 1980. The share of the total number of chromosomal aberrations in this group was 5.25%, and the share of rings and dicentrics 0.37% (Table 1). Although the sample size is not large, the data obtained may be indicative of both the possibility of essential genetic damage passed through generations and the possibility that extremely adverse ecological factors affecting chromosomal damage frequency may still exist.

TABLE 1. Chromosomal aberrations in the Samburg population (1% means 1 chromosomal abnormality in 100 metaphases)

	N	*Number of metaphases*	*Total number of aberrations* (%)	** Number of acentrics* (%)	** Number of rings + dicentrics* (%)
Total sample	84	11008	489 (4,44%)	390 (3,64%)	56 (0,52%)
Natives	74	9704	445 (4,59%)	352 (3,74%)	50 (0,53%)
Immigrants	10	1304	44 (3,37%)	38 (2,91%)	6 (0,46%)
Children born after 1980	15	2134	112 (5,25%)	104 (4,87%)	8 (0,37%)

* - one case with an abnormal karyotype is not included

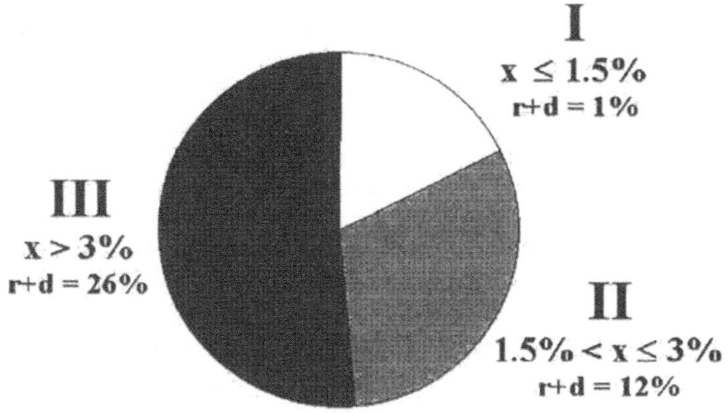

Figure 3. Distribution of chromosomal aberrations in the sample of the Tundra Nentsi population in Samburg (N=74). With x= - percentage chromosomal aberrations: I. normal (x≤ 1.5%) - 17.6%; II. medium (1.5% <x≤ 3%) - 31.1%; III. high (x >3%) - 51.3%. r - rings; d - dicentrics.

Our studies of the peripheral blood indices on 90 randomly selected people revealed that the "maximum" departure from the "regional" norm (low hemoglobin, low leukocytes, deformed erythrocytes) was observed in 17% of those examined, and only in 24% the indices are normal.

3.2. RESULTS OF ECOLOGICAL STUDIES

3.2.1. Analysis of the Content of Atmospheric Aerosols

It has been shown that the content of aerosols in the region under study is in good agreement with the few data that has been published on the content of aerosol particles in other areas of the Arctic basin [9]. Analysis of the multi element content of atmospheric aerosols suggests possible contamination of the area being studied following the transfer of admixtures of anthropogenic origin from hundreds and thousands kilometers away. Interestingly, the enrichment factor is very high for a considerable number of elements. This fact may indicate that anthropogenic impact on the Siberian North has now become notable. We have demonstrated that 78% of the elements of atmospheric aerosols in Samburg are due to human activity [9].

3.2.2. Results of Differential Analysis of Samples for the Presence of Radionuclides

As is known, a considerable part of the radioactive products from the Novaya Zemlya as well as from American and Chinese explosions was thrown into the atmosphere and then fell out thereby contaminating soil on a global background scale. The common feature is the latitudinal zonality with a peak at 40-50° N. Liubashevskiy et al. [10] report from the Extreme North contamination by long-span technogenic radionuclides exceeding the global background along the global contamination of all the components of the environment. Unfortunately, the observations are mostly focused on the seas and shores west of the Yamal peninsula, which itself has only been examined in 6 sites. In three of them, closest to Samburg, soils were "hot" (25, 35, 44.9 Bq/kg vs 17 Bq/kg in each of the other three).

Our data on soils, gathered in Table 2 are indicative of a higher content of ^{137}Cs in the upper 5 cm layer, 59 Bq/kg in Samburg and 65 Bq/kg in the location of Brigade No7 which is higher than that due to the global fall-outs of ^{137}Cs at these latitudes [10].

A correlation is observed between the content of ^{137}Cs in lichen and reindeer meat (156 Bq/kg and 131 Bq/kg, respectively, by Liubashevskiy et al. [11] and 61 Bq/kg and 62 Bq/kg, respectively, according to our data). This suggests that the meat-dieting aborigines are exposed to continuous internal radiation. Unfortunately, publications omit from their considerations the assessments of radioactive contamination by other radionuclides that have contributed the largest share of dose imposed by the explosions at the Severny testing ground.

TABLE 2. Content of Cs-137 in some samples

Sampled	*Depth*	*Sample weight*	*Radionuclide content*			
where	*cm*	*g*	*U (Ra) g/t*	*Th g/t*	*K g/t*	*^{137}Cs Bq/kg*
			Soil			
Samburg	0 - 5	45	7.2	4.3	1.11	59 ± 5
	5 - 10	87	6.4	5.9	1.78	n.t.
	10 - 15	97	5.1	6.2	1.76	n.t.
			Lichen			
		125	0.5	0.3	0.02	65 ± 3
			Reindeer meat (averaged for three samples)			
			n.t.	n.t.	0.02	62 ± 3
			Soil			
Brigade No7	0 - 5	25	20.2	4.2	n.t.	65 ± 7
	5 - 10	52	98	3.4	0.78	27 ± 3
			Lichen			
		125	0.3	0.5	0.08	56.3

3.3. RESULTS OF GENETIC AND MOLECULAR STUDIES

RFLP analysis of IGHG immunoglobulin genes in 128 DNA samples from the Tundra Nentsi has revealed an extraordinarily high frequency (41%) of the multi gene duplication of the GP-G2-G4 genes possibly involving neighboring genes too. To assess the phenomenon, pulsed-field gel electrophoresis was performed on 59 samples. As turned out, not only may genes be in duplicate but also in triplicate within a 350-kb IGH-super gene. The share of duplication was 35%, that of triplication 5%. This is the only population known to display such a high frequency of multi gene duplication.

In our opinion, these results could be helpful in searching for the molecular-genetic mechanisms that endow resistance upon some part of the Tundra Nentsi population to the adverse ecological factors, and primarily to radiation.

Acknowledgement. This work was supported in part by the Administration of the Pur Rayon of the Tyumen region, in part by an «integration» grant from the Siberian Division of the Russian Academy of Sciences and the INTAS grant 93-0035 Ext. We are in debt to S. Lambin, Head of the Pur Rayon Administration, for cordial help while the field portion of this work was in progress. We are also thankful to V. Filonenko for translating this manuscript from Russian into English.

4. References

1. Gofman, J.W. (1994) *Chernobyl accident: radiation consequences for this and future generations*, CNR Books and Vysheishaya Shkola, Publishing House, Minsk (in Russian).
2. Posukh, O.L., Osipova, L.P., Ivakin, E.A., Kryukov, Yu.A. (1996) Genetic-demographic analysis of the population of native inhabitants of the Samburg Tundra, *Russian J. Genetics* **32**, N6, 715-721.
3. Osipova, L.P., Posukh, O.L., Ivakin, E.A., Kryukov, Yu.A., and Karafet, T.M. (1996) The gene pool of native inhabitants of the Samburg Tundra, *Russian J. Genetics* **32**, N6, 722-728.
4. Methodical Recommendations on Cytological Techniques of Diagnosing Human Chromosomal Diseases (1976) *Meditsina Publisher House*, Moscow (in Russian).
5. Pohl-Ruling, J. (1990) Chromosome aberrations of blood lymphocytes induced by low-level doses of ionizing radiation, in G.Obe (ed.), *Advances in mutagenesis research 2.*, Springer-Verlag, Berlin Heidelberg, pp.155-186.
6. Dard, P., Sanchez-Mazas, A., Dugoujon, J.-M., De Lange, G., Langaney, A., Lefranc, M.-P., Lefranc, G. (1996) DNA analysis of the immunoglobulin IGHG loci in a Mandenka population from eastern Senegal: correlation with Gm haplotypes and hypothesis for the evolution of the Ig CH region, *Hum. Genet.* **98**, 36-47.
7. Bottaro, A., Cariota, U., DeMarchi, M., and Carbonara, A.O. (1991) Pulsed-field electrophoresis screening for immunoglobulin heavy-chain constant-region (IGHC) multigene deletions and duplications, *Am. J.Hum.Genet.* **48**, 745-756.
8. Vogel, F., Motulsky, A.G. (1990) *Human genetics. Problems and Approaches*, «Mir», Moscow.
9. Koutzenogii, K.P., Osipova, L.P., Posukh, O.L., Bufetov, N.S., Smirnova, A.I., Kryukov, Yu.A., Ivakin, E.A., Konovalova N.A. (1997) The potentialities of SRXFA technique in analyzing a technogenic impact on northern population, *Nuclear Instruments and Methods in Physics Research* A NIMA 3036 (in press).
10. Makhanko, K.P., Rabotnova, F.A. and Volokitina, A.A. (1990) Assessment of Soil Contamination by Cs-137 in the USSR in 1988, *Atomnaya Energiya* **68**, 262-264 (in Russian).
11. Liubashevskiy, N.M., Starichenko, V.I., Balonov, M.I. et al. (1993) Radioactive contamination of Yamal peninsula and radiation hazard assessment, *Ekologiya* **4**, 39-45 (in Russian)

EPIDEMIOLOGY.

A Bridge Between Radiobiology and Health Risk Assessment

J. R. GOLDSMITH, M. R. QUASTEL

Department of Epidemiology and Institute of Nuclear Medicine, Ben Gurion University of the Negev
Beer Sheva, 84 120 Israel

Abstract

Conventionally epidemiology has determined the numbers of cases of cancer which followed radiation exposures in 5 to 30 years. However, epidemiology can also tackle additional tasks reflecting the advances in radiobiology: such as chromosomal changes [1], persistent biochemical abnormalities, effects on growth, reproduction and development, and impact on survival. One weakness of traditional epidemiology is the combination of the years of latency and the relatively low probability that a given exposure will ultimately produce cancer. Into the black hole of latency, with its decades of apathy and anxiety, epidemiology of a new sort can bring light and hope for preventive intervention. We studied about 1,600 of the 120,000 immigrants from the Former Soviet Union (FSU) to Israel who were potentially affected by the Chernobyl disaster. [2] We observed positive glycophorin-A assays and plasma clastogenic factor (CF) increases in "liquidators" as well as elevated CF in children [3]. Emerit and her colleagues have shown that activity of the clastogenic factor in serum can be reduced by treatment with an oral anti-oxidant [4, 5] and Kordysh et al. [6] have shown that clastogenic activity is less in children eating a diet high in fresh fruit and vegetables. We need to determine whether those with positive tests differ from others, how to standardize them, and what may be their prognostic value. Clinical trials of oral antioxidant are needed. Residents in some areas of high natural background of radiation have increased prevalence of chromosomal abnormalities [7], even though excess cancer has not been established.

We will need to find out:

(a) How stable are these abnormalities and at what age do they occur?

(b) Do the populations who show such abnormalities differ in any important way from similarly exposed populations who do not?

(c) Are the persons with these changes more likely to have cancer or other illness?

The wisps of information available challenge us to define sufficiently large populations, to carry out controlled trials using valid tests to see to what extent these relationships offer leads for prevention of the possible long-term effects of the exposures. Other

C. Baumstark-Khan et al. (eds.),
Fundamentals for the Assessment of Risks from Environmental Radiation, 43-51.

problems are multiple factors and effects, and appropriate epidemiological models and methods. Similar questions arise for RF radiation exposure. The work ahead needs large scale cooperation and stable funding if the most is to be made of these challenges. Collaboration in radiation epidemiology is urgently needed for risk assessment.

1. Introduction

Thanks to the opportunity for overview which was based on the authors' co-chairmanship of an International meeting on Radiation and Health in November 1996, and the opportunity to edit a monograph based on the meeting, we offer some suggestions as to possible future contributions of collaborative epidemiologic studies.

Important questions regarding biological indicators of radiation exposure concern their correlation (if any) with late health effects. Despite formidable obstacles in radiation epidemiology, there are hopeful developments in the interaction with radio-biology. Szmigielski, who has followed Polish army personnel exposed to radiofrequency and microwave electromagnetic radiation is also active in radiobiological research [8, 9] Hagmar et al [10] have followed sufficient numbers of Scandinavian subjects with abnormal cytogenetic findings, and showed that those with stable changes have an excess of leukemia, whereas those with more transient chromatid changes do not show such excesses. Japanese investigators have had similar findings [11]. Stewart has shown that the history of acute radiation reaction symptoms is associated with a different pattern of excess mortality risk per Gray for Japanese A-bomb survivors, than those who had no such reactions [12].

An excellent model is the set of collaborative studies of liquidators (clean-up workers) exposed at Chernobyl, conducted by a group of Baltic country scientists under the leadership of Dr. Timo Hakulinen [13-17]. These workers have looked for but did not find excess leukemia in the workers. They have demonstrated relationships of exposure to chromosomal changes, but in contrast to Jensen et al [18], did not show significant effects using the somatic cell Glycophorin A (GPA) mutation assay. Hence exposure evaluation using GPA are still not clear cut. Lazutka et al [19], who studied Lithuanian workers exposed during clean-up operations at Chernobyl, have shown that a variety of variables may be associated with various types of chromosome aberrations.

In an analysis of variance [19], different sets of variables were found to be associated with total chromosomal aberration, chromatid breaks, chromatid exchanges, chromosome breaks and chromosome exchanges (Table 1).

Finkelstein is a leading investigator of miners in the study of interaction of silica exposure and radon daughters, and their interaction with smoking [20]. His work suggests that not all health effects attributed to radon in the past were due to radon but rather may have been due to silica. This is an example of the need for studies of interaction of radiation with other exposures. We are developing structural models for better treatment of the analysis of multiple independent variables and of confounding [21]. Goldsmith and Beeser have demonstrated methods for pooling studies [22].

In the examination of thyroid function of children exposed at Chernobyl, we have shown that girls from more exposed locations have significantly higher TSH than do girls from less exposed areas. Similar patterns affect TSH of boys but the differences do not reach significance. [23] In a recent longitudinal study of diet and cancer, Knekt et al. [24] have shown that increased consumption of dietary flavonoids significantly reduces the incidence of lung cancer, and non-significantly the incidence of other cancers. The effect is not due to effects of other dietary antioxidants such as Vitamin C, E or beta carotene. This finding is particular interest because Emerit et al have shown that in Armenian liquidators with elevated clastogenic factor, oral treatment with Ginko Biloba extract, a potent source of flavonoids [5].

Two studies, Shantyr et al [25] and Goldsmith et al [26] have demonstrated approaches to estimating exposures for liquidators by weighting the factors in their exposure history.

2. The Role of Epidemiology

The opportunities that the Chernobyl disaster presented for learning more about human reactions to radiation seem at times to be slipping through our fingers. Through lack of preparedness, the immaturity of cooperation, and the preoccupation with political or personal problems, we seem to be putting off to future generations the tasks we could have accomplished. Such an outcome shames us as scientists and as citizens. A vigorous program of collaboration in radiation epidemiology in the new mode can obviate this risk, and at the same time bring much benefit both to those who may be at unusual risk because of exposure and to scientific understanding of low-dose radiation.

TABLE 1. Results of analysis of variance for significant relationships between chromosomal changes and exposure and other variables in Chernobyl clean-up workers from Lithuania

Dependant Variable	*Independent Variable*	*Percent of "Explained Variance"*
All aberrations	Alcohol abuse	2.8
	Diagnostic X-rays	0.7
	Chernobyl exposure	4.9
Chromatid breaks	Alcohol abuse	2.5
Chromatid exchanges	Alcohol abuse	1.7
	Recent infection	0.9
	Smoking	11
	Age	1.3
	Chernobyl exposure	1.1
Chromosome breaks	Diagnostic X-rays	1.2
	Chernobyl exposure	5.5
Chromosome exchanges	Age	1.5
	Recent viral infection	0.9

Based on Lazutka et al.[15] Variables transformed, $Y = (4X + 1.5)^{0.5}$
All associations are significant at $p<0.05$

Table 2 gives some sources of possible radiation exposure and Table 3 types of exposures. Table 4 shows examples of some types of biological indicators and some criteria which may affect susceptibility to radiation effects. Given the background data, a sample of which has been cited, we can offer the following model:

$$H_t = H_0 + (a_1 s_1 + a_2 s_2 + \ldots + a_n s_n) + bE + cA + \qquad (1)$$
$$(d_1 r_1 + d_2 r_2 + \ldots + d_n r_n)e \cdot f_t(C) + g \cdot f_t(CR) + h \cdot f_t(L) + j \cdot f(U) + k \cdot f(PS) + M$$

where

H_t	is Health at any given time t in the future and H_o is health status at time 0
a_1, a_2, and a_n	are dummy variable coefficients for reflecting susceptibility variables
s_1, s_2, and s_n	reflecting susceptibility S,
E	is a matrix representing exposure type, intensity and time,
A	is age of the subject, (possibly represented as logarithmic, Gompertz or Weibull function)
d_1, d_2, d_n	are dummy variable coefficients reflecting initial reaction variables r_1, r_2, and r_n.
f_t	represents the change with time in the abnormalities of chromatids, C; chromosomes, CR and clastogenic or other factors, L.
f(U)	represents both habitual protective procedures, such as diet and specific protective procedures such as medication
f(PS)	represents a matrix of psychosocial factors and
M, b, c, e, g, h, j and k	are coefficients

The purpose of such a model is to offer a framework for multiple considerations in the conduct and interpretation of risk from epidemiological studies. However, specific studies can and should focus on components of the model and their relationship.

TABLE 2. Sources of Possible Radiation Exposure

	Kind of Exposure	*Source*
A	Background	Geological Cosmic (high altitude) Radon
B	Incidental	Proximity to Nuclear Installations Downwind to fall-out Clothing of workers
C	Diagnostic X-rays	
D	Therapeutic radiation	
E	Occupational	Military Clean-up workers
F	Accidents and disasters	
G	Incidental to communication (RF) or power transmission (ELF)	

TABLE 3. Types of Radiation Exposure with Possible Epidemiologic Relevance

	Type of Radiation
A	Ultra-violet
B	ELF (Extremely low frequency, e.g. power lines)
C	RF (Broadcast, Cellular phones, Radar for Air traffic controllers)
D	X-ray (Industrial)
E	Medical radiations (X-rays and radioisotopes)
F	Natural background (Cosmic, radon, naturally occurring radioisotopes in body)

We propose three general objectives for collaborative epidemiological studies. For each of these, there are preliminary findings, which can provide both guidelines and experience.

(1) What is the most realistic association of exposure to the various biological indicators, susceptibility factors, and the time course of the indicator strength ?

(2) Given the existence of an assessed biological indicator in an exposed population, what is its association with health effects over time ?

(3) Given the existence of an assessed biological indicator at a given time, to what extent does diet or specific therapy affect association with indicator level and health output over time ?

Table 5 lists some risks in need of assessment. While time does not allow and strategy contravenes the putting forth of specific protocols for the testing of these propositions, there are a few comments that may be useful.

3. Advantages of a Collaborative Approach to these Problems

(1) First, there are in the effort to evaluate these propositions, some advantages of multiple collaborative efforts. First, the special situation of one population is less likely to bias the result of many such population studies, assuming that some method for pooling is used.

(2) Secondly, inadequate power of studies is likely to occur if populations under study are too small. Collaborative effort allows a larger population, and there are various methods as shown in Goldsmith and Beeser [14] for prevention of the dominance of the biggest sized study.

(3) Thirdly, the involvement of multiple groups in a collaborative effort tends to allow a broad input to the task of interpretation of results.

(4) Fourthly, to involve multiple populations in such studies increases the likelihood that the results, if clear, will be accepted and applied.

4. Problems to be Faced in a Collaborative Approach

(1) First, there must be clear agreement on general objectives, and the role of various

TABLE 4. Types of Radiation Exposure with Possible Epidemiologic Relevance

Time of Occurrence	*Findings*	*Basis for Relevance*
Within weeks of exposure	Blood count changes	Reversibility Chromosomal changes
	Epilation	Associated chromosomal changes
	Other signs and symptoms	Dose-index
Years after exposure	Chromosomal or chromatid changes	Possible cancer or mutational index
	"Clastogenic Factor"	Possible cancer
	"FISH test"	Predictors
	Glycophorin A mutation	?Preventive Rx ?Clinical trials
Susceptibility	Genetic susceptibility	Special protection
	p53 Variants	
	HLA Groups	
	Other exposures Radiation Smoking AIDS Asbestos, etc. Alcohol abuse	?Confounding ?Prevention

personnel, especially including personnel whose task it is to bring about common methods and definitions.

(2) There must be uniform procedure, definitions, laboratory test criteria, data management, and quality control. These may be the basis of a procedure manual or protocol.
(3) Criteria for evaluation and interpretation must be agreed in advance and rigorously applied by all participants.
(4) There must be adequate funding of work of high quality, and this may vary among participants according to their basic situation.
(5) There will need to be agreement on language usage, confirmation of translation of questionnaires as being equivalent, and agreed-on procedures for resolution of uncertain or ambiguous interpretations.
(6) The collaborative agreement must allow for individual studies components to vary within limits and yet for results to be effectively pooled.

5. Some general considerations

(1) There always exists the temptation in research to follow up leads which may develop and this is important to encourage, in principle. Yet the primary goal must not be ignored, assessment of risk based on new developments in radiobiology.

TABLE 5. Some risks needing assessment

Cancer, site-specific
Overall shortened life span
Other chronic disease
Impaired well-being
Stigmatization and discrimination
Iatrogenic risks (Secondary to diagnosis or therapy)
Inappropriate constraints as to job, recreation and family
Unfavorable reproductive outcomes (Perceived)

(2) The three major questions each have a different experience, and may well have different strategies and epidemiological approaches. For example the third one may involve a retrospective evaluation of diet in a case-referent procedure, for which having a positive Biological Indicator would define as subject as a "case"; the evaluation of interventive medication may use a double-blind clinical trial model. The first two questions may best be approached by cohort study methods.

(3) Given that the larger number of participating groups the more complicated the planning and conduct of the study, there is every reason for pilot studies, and as the next phase, adding groups successively as they become ready. This has disadvantages of disconcordant operations, but can be compensated for.

(4) For the work to be most effective and for inherent reasons, the highest standards of scientific ethics must be used. It may be useful to designate one or more persons who have this overall responsibility.

(5) There exist a number of long-standing controversies as to the interpretation of existing scientific information, such as regarding the linear, no threshold assumptions for low level radiation effects. Caution must be exercised not to let the emotion of such disputes impair the collaborative work.

(6) Various stakeholders are likely to become involved who have non-scientific goals, such as status, profit, and political aspects. The work must be realistically planned to avoid the bias or prejudice which may be in the interest of various stakeholders. If conflicts are identified, they should be resolved to the extent possible in favor of the public good and public protection.

The most important single step toward realizing some of the work suggested would be the compilation of a comprehensive and critical review of radiation epidemiology, with particular attention to radiobiological findings.

6. References

1. Snigiryova G, Braselmann H, Salassidis K, Shevchenko V, and Bauchinger M. (1997) Retrospective Biodosimetry of Chernobyl Clean -Up Workers, using chromosome painting. *Int. J. Radiat. Biol.* **71**, 119-127.

2. Kordysh E, Goldsmith JR, Quastel MR, Poljak S, Merkin L, Cohen R, Gorodisher R (1995) Health effects in a casual sample of immigrants to Israel from areas contaminated due to the Chernobyl explosion. *Env. Health Persp.* **103**, 936-941.
3. Emerit I, Quastel MR, Goldsmith JR, Merkin L, Levy A, Cernjavski L, Alaoui-Youssefi A, Pogossian A, and Riklis E (1996) Clastogenic factors in the plasma of children exposed at Chernobyl. *Mutat. Res.* **373**, 47-54.
4. Emerit I, Oganesian N, Arutyunian R, Pogossian A, Sarkisian T, Cernjavski L, Levy A, Feingold J (1997) Oxidative stress-related clastogenic factors in plasma from Chernobyl liquidators: protective effects of antioxidant plant phenols, vitamins and oligoelements. *Mutat. Res.* **377**, 239-246.
5. Emerit I, Arutyunyan R, Oganesian N, Levy A, Cernjavsky L, Sarkisian T, Pogossian A, Asrian K (1995) Radiation-induced clastogenic factors: Anticlastogenic effect of Ginkgo biloba extract. *Free Radic. Biol. Med.* **18**, 985-991.
6. Kordysh EA, et al. (1997) Diet and clastogenic factor in children immigrating from Chernobyl contaminated zones. (Abst) Isr. Epidemiol. Soc.
7. Beir V (1990) Health Effects of Exposure to Low Levels of Ionizing Radiation. National Research Council, National Academy Press, pp 384-5.
8. Szmigielski S (1996) Cancer morbidity in subjects occupationally exposed to high frequency (radiofrequency and microwave) electromagnetic radiation. *Sci.Tot. Environ.* 1809-17.
9. Szmigielski S, Szudzinski A, Pietraszek A, Bielec M, and Wrembel JK (1982) Accelerated development of spontaneous and benzopyrene-induced skin cancer in mice exposed to 2450 MHz microwave radiation. *Bioelectromagnetics* **3**, 179-191.
10. Hagmar L, Brogger A, Hansteen IL, Heims, and Hogstedt B et al. (1994) Cancer Risks in Humans Predicted by Increased Levels of Chromosomal Aberrations in Lymphocytes. *Cancer Res.* **54**, 2919-2922.
11. Akiyama M, Umeki S, Kusunoki S, Nakamura N, Mori T, et al. (1995) Somatic Cell Mutations as a possible Predictor of Cancer. *Health Physics* **68**, 643-649.
12. Stewart A, Kneale GW. A-Bomb survivors: Tests of whether they form a single homogeneous population" Environ Health Perpect, Supplement (in print).
13. Bigbee WL, Jensen RH, Veidebaum T, Tekkel M, Rahu M, Stengrevics A, Kesminiene A, Kurtinaitis J, Auvinen A, Hakulinen T, Servomaa K, Rytomaa T, Obrams GI, and Boice Jr. JD (1996) Glycophorin A Biodosimetry in Chernobyl cleanup workers from the Baltic countries. *Brit. Med. J.* **312**, 1078-9.
14. Bigbee WL, Jensen RH, Veidebaum T, Tekkel M, Rahu M, Stengrevics AJ, Auvinen A, Hakulinen T, Servomaa K, RytomaaT, Obrams GI, and Boice Jr. JD (1997) Biodosimetry of Chernobyl clean-up workers from Estonia and Latvia using the Glycophorin A in vivo somatic sell mutation assay. *Radiation Res.* **147**, 215-224.
15. Granath F. Darroudi F, Auvinen A, Ehrenberg L, Hakulinen T, Natarajan AT, Rahu M, Rytooma T, Tekkel M, and Veidebaum T (1996) Retrospective Dose Estimates in Estonian Chernobyl clean-up workers by means of FISH. *Mutat. Res.* **369**, 7-12.
16. Inskip PD, Hartshorne MF, Tekkel M, Rahu M, Veidebaum T, Auvinen A, Crooks LA, Littlefield LG, Mc Fee AF, Salomaa S, Makinen S, Tucker JD, Sorenson KJ, Bigbee WL and Boice Jr JD. (1977) Thyroid Nodularity and Cancer among Chernobyl Cleanup Workers from Estonia. *Radiat. Res.* **147**, 225-235.
17. Rahu M, Tekkel M, Veidebaum T, Pukkala E, Hakulinen T, Auvinen A, Rytomaa T, Inskip PD, Boice Jr. JD. (1977) The Estonian Study of Chernobyl Cleanup Workers: II Incidence of Cancer and Mortality. *Radiat. Res.* **147**, 653-657.
18. Jensen RH, Langlois RG, Bigbee WL, Grant SG, Moore D 2nd, Pilinskaya M, Vorobtsova I, Pleshanov P. (1995) Elevated frequency of glycophorin A mutations in erythrocytes from Chernobyl accident victims. *Radiat Res.* **141**, 129-135.
19. Lazutska JR, Mierauskiene J, Slapsyte G, Dedonyte V, Kesminiene A, Rimdeika G (1997) Cytogenetic Study in Peripheral Blood Lymphocytes of Chernobyl Cleanup Workers. *Acta Medica Lituanica* **2**, 62-67.
20. Finkelstein MM (1995) Silicosis, Radon and Lung Cancer in Ontario Miners. *Health Physics* **69**, 396-399.

21. Goldsmith JR, Friger MD, and Abramson M (1996) Associations between Health and Air Pollution in Time-Series Analyses". *Arch. Environ Health* **51**, 359-367.
22. Goldsmith JR and Beeser S (1984) Strategies for Pooling Data in Occupational Epidemiologcial Studies" *Annals of the Singapore Academy of Medicine* **13**, Supp. 297-307.
23. Quastel MR, Goldsmith JR, Mirkin L, Poljak S, Barki Y, Levy J, and Gorodischer R (1997) Thyroid-stimulating Hormone Levels in Children from Chernobyl. *Environ. Health. Perspect.* **105S(Suppl 6),** 1497-8.
24. Knekt P, Jarvinen, R, Heliovaara M, Teppo L, Pukkala E, and Aromaa A (1997) Dietary Flavonoids and the Risk of Lung Cancer and Other Malignant Neoplasms. *Am. J. Epidemiol.* **146**, 223-230.
25. Shantyr II, Nikiforov AM, Romanovich IK, Schwartz VA, Makarova NV, Deryapa LN, and Saygina EB (1997) Estimation of Radiation Exposures of "Liquidators" of the Chernobyl Nuclear Power Station:Identification of Risk Groups. *Int. J. Occup.Environ Health* **3**, 45-50.
26. Goldsmith JR, Merkin L, Quastel MR, Poljak S, Kordysh EA, Wishkerman YV and Emerit I (1997) Evaluation of the radiation exposures of Chernobyl "liquidators": Exploratory studies of a sample of immigrants to Israel. *Int. J. Occup. Environ Health* **3**, 51-59.

GERMAN URANIUM MINERS.

Facts and Perspectives

H. RINK

Experimentelle Radiologie und Strahlenbiologie, Universität Bonn
Sigmund-Freud-Str. 25, 53105 Bonn, Germany

This paper considers the radon problem in general and focuses on the radon-induced lung cancers in underground uranium miners. It was the noble prize winner of 1908, Ernest Rutherford who said: „We must bear in mind that all of us are continuously inhaling the radium and thorium emanations and ionized air...". Radon, which derives from uranium (^{238}U) was first detected by Friedrich Dorn in 1890. It represents a radioactive alpha-particles emitting element (number ^{222}Rn) with a half-lifetime of 3.8 days, which further decays in a number of short living alpha-particle emitting elements having half-lifetimes in the range from 162 microseconds up to 27 minutes. Normally, when alpha-particles irradiate our bodies from outside there is no radiation danger or risk, because of the short range of the densely ionizing alpha-particles. The noble gas radon, however, and its different decay products become incorporated by inhalation together with aerosols and very fine distributed dusty particles. That is the reason, why it acts inside our bodies. It is the bronchial part of the lungs with their radiation-sensitive basal cells, where we have an inhomogeneous distribution of point-like microareas with very high local energy doses. This is the origin for the risk to get this very dangerous small cell lung cancer [7, 16].

Radon and its decay products belong to the naturally occurring radiation sources. This is the reason why for a long time they have been scarcely considered under the aspect of legal radiation protection. Since 1985, however, there are serious considerations and estimations saying that the radon burden will make up to 70% of the total natural radiation burden. Consequently some US researchers have tried to analyze the ratio of radon-induced lung cancers in the total lung cancer population and came to surprising high numbers (Table 1) [10].

A special chapter for the action of radon is given by the occupational diseases of underground uranium miners in different parts of the world (Czech Republic, Canada, United States, France and the former Yugoslavia) and also by other underground workers in comparable mining surroundings. In Germany we have a large district which is well known for its mining activities since the late middle ages (about 1500-1600). It is the area neighbored to Bohemia, in the eastern part of Saxonia and the south-eastern part of Thuringia. Here we also find a famous place called Joachimsthal. The silver coins once produced here, had been named „Joachimsthaler" and this is told to be the origin of our today's denotation „dollar". Madame Curie, who got the noble prize

C. Baumstark-Khan et al. (eds.),
Fundamentals for the Assessment of Risks from Environmental Radiation, 53-58.

TABLE 1. Death from Lung Carcinoma in USA (160,000 Persons/year) and Estimations for Radiation Induced Lung Carcinomas

Death	*Year*	*Report*
27,000	1980	BEIR III
5,000	1984	NCRP
13,000	1988	BEIR IV
21,600	1989	Oge et al.

(1903) for the purification of the element radium got the raw-material called „Pitchblende" from this area [17].

Miners working in this district, especially in Schneeberg, frequently suffered on an early deadly lung disease, which was named „Schneeberg miners disease" [12]. About 1540 famous medical doctors like Paracelsus and Agricola already knew this disease. It lasted, however, until 1879 when Haerting and Hesse found and proved that this disease was a lung cancer [2, 11]. As late as 1925 it was finally accepted as an occupational disease [16].Some ten years later (1945-55) it was also found that this cancer was a radon-induced lung cancer [13].For a long time, however, we had no further knowledge about the limit dose or risk dose. This situation changed in 1960-1980 when epidemiological follow-up data measured over a period of 20 years became available which derived from some ten thousands miners (Table 2) [3, 15, 18].

In 1945 the second world war was definitively finished by the A-bombs, which destroyed Hiroshima and Nagasaki. The era of atomic bombs made from uranium and 10 years later the nuclear reactor plants using uranium started. It was for this reason that the Soviet government, having occupied the middle and eastern part of Germany immediately started with uranium-mining in the well-known German districts reported just before. The SAG (Soviet AG) Wismut was the mining company in the years 1946-1955, afterwards the name changed to SDAG (Soviet-German AG)-Wismut. Until 1990 this company produced 220,000 tons of uranium. In other words the former German Democratic Republic was the third biggest uranium producer in the world. This business was done on an area of 40,000 square kilometers, in tunnels having a length of 1,500 km, and in more than 100 shafts. For the production of 1 ton of uranium, 1,100 tons of ore and 2,100 tons of other stones were needed. By this way 700 million tons of waste products were distributed into the environment of the shafts [9, 14].

Today a total of 303,000 miners and related workers are known by name (Table 3) and 160,000 are still alive. In the meantime they all were asked by letter to undergo medical investigations. 71,000 are interested in this program and accepted medical investigation. 65,000 have already been examined by about 150 medical doctors [1, 9].
Up to now radon-induced lung carcinoma was assumed in 7,163 cases. In 5,237 cases the diagnosis lung-carcinoma was accepted as an occupational disease. In 2,500 cases further examinations are necessary. For comparison, in the same population about 20,000 cases of silicosis have been established. Within 30,000 Colorado uranium miners about 3,000 lung carcinoma have been found.

TABLE 2. Epidemiological Data on the Lung Carcinoma Risk observed over a Time Period of 20 Years

Death	
3,000	Miners of Colorado
3,500	Miners of New Mexico
4,000	Miners of former CSSR
18,000	Miners of Canada

In the early period from 1946-1955, the so-called wild years, about 250,000 persons were employed by the SAG Wismut without any radiation protection (Table 3). Some of them worked only for a short time (5 years or less) some others, however, for very long times (30-40 years). They have been employed under different working conditions, in different professional positions and in different mining places.

Since the radon-induced lung carcinoma belongs to the stochastic events induced by radiation and because it is told to persist after a short a latency period over 20 to 30 years, all these peoples are under the risk to get lung carcinoma. In any case the mining companies and mining associations have to suppose that such a carcinoma may be caused by inhaling radon and its follow-up products during the time of occupation . The reviewers have then to decide, whether the probability to suffer from this disease originates to more than 50% from the occupying conditions. In this case lung carcinoma is accepted as an occupational disease (BK 2402). This is of great importance as far as the financial aspects of medical care are considered. In order to do these examinations, it is necessary for the reviewers to have data on the personal burden and hard data for risk coefficients coming from safe and accepted epidemiological studies.

During the first period (1946-55) when most of the people were employed, there were no personal radiation protection measurements at all and additionally we had the worst occupying conditions :

- Long working time with no or very bad hygienic conditions.
- Dry drilling and no supply of fresh air in the shafts or galleries.

Dry drilling increases the amount of dust in the air and thereby the inhalation of alpha-emitting particles. No supply with fresh air means, that the balance factor (F), which considers the balance between radon and air, is high in the order of F = 0.5 to 0.8, instead of 0.5 to 0.2 when sufficient fresh air is supplied.

Beginning with 1955 there were radon measurements at different places given in Bequerels/m^3. With 1964 the potential alpha-radiation energy (Joule/m^3) was measured (Markow method) monthly at defined local positions. Beginning with 1970/71 supply with fresh air started and wet drilling was introduced additionally. Starting with these years we also have the first personal dosimetry, simply achieved by multiplying the correspondent local dose (Lucas chamber) with the working time. Individual personal film dosimeters have been introduced as late as 1990.

TABLE 3. Number of Persons employed as miners in the SDAG Wismut

Time	*Number of Employees*
1946 - 1955	250,000
1955 - 1970	80,000
1971 - 1985	30,000

The radioactivity of radon is given as usually in Bq/m^3. The radiation energy coming from alpha-particles is best measured in J/m^3 , but mostly given in a traditional way as WLM (working level month). One WL represents the radon activity of 100 pCi/l in the balance with its radioactive decay products. 1 WL corresponds to 2.08 10^{-5} J/m^3. WLM results from WL times 160 hours (working hours/month). Roughly 1 WLM might be equivalent to 10 mSv. The mean additive relative risk to get radon-induced lung cancer is calculated as 0.02 per WLM, it may increase for young persons to 0.036 per WLM [8].

Because of the different and changing conditions the radiation burden differed enormously over the years, going from 30 -300 WLM/a in 1946-55 to 1-4 WLM/a in the period later as 1976 (Table 4). But there are other more specific differences which originate from the different geological conditions, from the uranium concentration of the ore (about 0.1-0.2%) in the different mines, from the depth of the shaft, from the type of professional work (miner, carpenter, dosimetrist) from the workers age and of

TABLE 4. Mean Radiation Level in WLM/Year in the SDAG Wismut

Years	*WLM / Year*	*Region*
1946 - 1955	30 - 300	
1956 -1960	10 - 100	
1960 - 1965	5 - 50	
1966 - 1970	3 - 25	
1971 - 1975	2 - 10	
1976 -	1 - 4	
There were big local differences for instance		in
1950	150 WLM/a 5 WLM/a	Zobes Johanngeorgenstadt
1938	100 - 200 WLM/a	Aue
1950	150 - 200 WLM/a	Aue
1964	15 - 25 WLM/a	Aue

TABLE 5. WLM numbers for the acknowledgement of lung cancer as an occupational disease

Time	*Number of Employees*
Former GDR in the beginning	450 WLM
later, 10 years in the underground	200 WLM
later, 3 years in the underground	200 WLM
later, individual checking with	150 WLM
Today	50 WLM

course from the duration of his employment. All these parameters were different from one place to the other, changed over the time and were different from person to person, dependent on the working time, the type of work (inhalation of air in m^3/h), the working conditions, the working places and the age of the worker.

Under these circumstances it was necessary to establish a job-exposure matrix in such a way that for each person knowing the time and the duration of work, its profession (292 different jobs have been registered and weighted for their proportional radiation), the age and the location of work an individual dose could be calculated. The last mentioned parameter, the location of work, with its local dose contains most of the difficulties. The dose rate depends not only from the geological conditions, the uranium concentration in the ore, the type of drilling (dry or wet) and the supply with fresh air, but also from the size of the radon-emitting area and the quantity of air in which this radon could be distributed. It is just for this reason, why sometimes the dose rate increased at one and the same place over the time when the size of the emitting area increased. All these conditions and parameters were seriously taken into consideration. Some of the necessary data have been available, others could be calculated from the yield of uranium and from the uranium concentration and others could be measured retrospectively under simulated conditions. Data for the alterations in the dose rate with increasing fresh air supply, were mostly taken from measurements in the nearby Bohemian area. Table 5 comprises the change in WLM-numbers in order to acknowledge an occupational disease.

The first part of the huge work has been completed. The job-exposure-matrix for the various time periods including the calculated WLM-data for the different localizations have been presented in 1997. With this data at hand it is now possible to calculate the individual radiation burden of the single miners not only for medical care and insurance reasons but also for scientific and epidemiological purposes.

References

1. Breuer, J. (1991) Die Last der Wismut - eine Herausforderung für die Berufsgenossenschaften. *Die BG* **9**: 508-518.
2. Haerting F.H. und Hesse W. (1879) Der Lungenkrebs, die Bergkrankheit in den Schneeberger Gruben. *Med. öffentl. Ges.wesen* **30**: 296-309

3. Hornung R.W. und Meinhardt T.J. (1987) Quantitative risk assessment of lung cancer in U.S. uranium miners. *Health Phys.* **52,** 417-430.
4. International Commission on Radiological Protection (ICRP Publication 24, 1976) Radiation protection in uranium and other mines.
5. International Commission on Radiological Protection (ICRP Publication 35, 1982) General principles of monitoring for radiation protection of workers.
6. International Commission on Radiological Protection (ICRP Publication 50, 1986) Lung cancer risk from environmental exposures to radon daughters.
7. Jacobi, W. (1986) Lungenkrebs nach Bestrahlung: Das Radon-Problem. *Naturwiss.* **73,** 661-668.
8. Jacobi, W. (1992) Verursachungs-Wahrscheinlichkeit von Lungenkrebs durch die berufliche Strahlenexposition von Uran-Bergarbeitern der Wismut AG. Institut f. Strahlenschutz der BG der Feinmechanik und Elektrotechnik.
9. Lehmann, F.(1997) Belastung durch ionisierende Strahlung im Uranerzbergbau der ehemaligen DDR. Berufsgenossenshaft Bergbau, Essen/Bochum.
10. Oge M.T. (1989) Environmental Protection Agency, Internal Memorandum.
11. Paracelsus (1925) Von der Bergsucht und anderen Bergkrankheiten. Bearbeitet von F.Koelsch: Schriften aus dem Gesamtgebiet der Gewerbehygiene, Neue Folge Heft **12**.
12. Rajewsky, B. (1940) Bericht über die Schneeberger Untersuchungen. *Z. Krebsforschung* **49,** 315-340.
13. Rajewsky, B., Schraub, A. und Kahlau, G. (1943) Experimentelle Geschwulsterzeugung durch Einatmung von Radiumemanation. *Naturwiss.* **31,** 170-171.
14. Richter, S.(1990) Grundlagen und Gesichtspunkte einer Aufarbeitung der Strahlenbelastung in der SAG SDAG Wismut. Tagung :"Der Uranbergbau in der DDR und seine Folgen", Ronneburg.
15. Samet, J.M., Pathak, D.R., Morgan, M.V., Key, C.R. and Valdivia, A.A. (1991) Lung cancer mortality and exposure to radon progeny in a cohort of New Mexico underground uranium miners. *Health Phys.* **61,** 745-752.
16. Schüttmann, W. (1987) Die Anerkennung der Schneeberger Lungenkrankheit als Berufskrankheit in der Ersten Berufskrankheitenverordnung 1925. *Z. ges.Hyg.* **33,** 662-665.
17. Schüttmann, W. (1988) Die Rolle des sächsich-böhmischen Erzgebirges bei der Entdeckung des Radiums. *Naturwiss. Rundsch.* **41,** 435-441.
18. Sevc, J., Tomasek L. and Kunz, E. (1993) A survey of the Czechoslovak follow-up of lung cancer mortality in uranium miners. *Health Phys.* **64,** 335-369.

CYTOGENETIC RISKS IN GOLD-ORE MINERS IN SLOVAKIA

M. BENO

Institute of Preventive and Clinical Medicine
Limbova 14, SK-833 01 Bratislava, Slovak Republic

1. Introduction

Radon causes lung cancer in the miner population occupationally exposed underground to mine air of uranium mines [1-3] or ore mines [4, 5]. There were reports about higher incidence of lung cancer also from ore mines of Slovakia and there was a tendency to ascribe the higher incidence of this cancer to the radon exposure underground [6-8].

The chromosomal aberrations (CA) are considered to be indicators of the risk of cancerogenic effects [9, 10]. A positive dependence of numbers of CA from lifetime underground exposure in uranium ore miners has been reported [11]. Similarly, a dependence of numbers of CA from lifetime exposure to indoor radon at very low levels has been observed in inhabitants of dwellings [12].

The above information has lead us to perform a pilot study in which a trial for assessment of the clastogenesis at the professional underground exposure was made by counting the CA in lymphocytes of miners of the gold ore mine in Hodruša-Hámre located in central Slovakia and by relating these counts to the lifetime underground exposure to radon and to lifetime smoking.

2. Materials and Methods

Twenty-one miners (average age 33.3 ± 7.4 years) volunteering after an informed consent served as donors of venous blood. Similarly, twenty clerks of non-significantly different average age (38.1 ± 8.3 years) donated blood for control samples.

Cultures from one-hour buffy coat were started using RPMI 1640 (USOL Praha) medium supplemented with phytohaemagglutinin (MUREX, 0.025 ml/ml) and antibiotica (Penicillin BIOTIKA, 100 i.U./ml, Streptomycin GALENIKA, 100 microg/ml) at 37 °C. The cells were harvested after 48 hours of cultivation. The chromosomal aberrations (chromatid and chromosomal breaks, fragments, minutes, terminal deletions, gaps, dicentrics, and chromatid exchanges) were counted in 200 well stretched mitoses.

The underground exposure time of the miners was from two to 14 years. The average lifetime underground number of shifts was 1457 ± 884 with a range from 676 to 3730 shifts. The lifetime potential alpha energy intake was calculated from mine air measurements made monthly by calibrated devices from the begin of the employment in

C. Baumstark-Khan et al. (eds.),
Fundamentals for the Assessment of Risks from Environmental Radiation, 59-64.

the mine. The average lifetime alpha energy intake was 0.012 Joule ±. 0.014 Joule with a range from 0.0032 to 0.049 Joule. The frequency distributions of the miners according to the potential alpha energy intake and according to the underground shifts are on Fig.1.and Fig.2, respectively. The exposure to smoking has been estimated according to the questionnaire filled-up individually before the blood withdrawal.

For biometrical evaluation the Fisher's exact test and regression analyses were used.

3. Results and Discussion

The numbers of CA found in individual donors showed a broad interindividual variation. The frequency distribution of the numbers of CA deviated less from log-normal than from normal distribution so that for regressions the natural logarithms of numbers of CA have been used.

A statistically significant higher count in miners ($p=0.011$) than in controls at comparing the sums of all grouped (at least from $n \geq 4000$ mitoses) cells containing chromosomal aberrations were found by Fisher's exact test (Table 1). Further testing showed that this difference could be ascribed to the miner smokers subgroup. In comparing the counts of cells containing CA in smoking controls with those in smoking miners a significantly higher count ($p=0.009$) in smoking miners was found. Comparing the counts in subgroups of non-smokers did not show a statistically significant difference. Comparing the counts of cells with CA excluding those with sole gap-type aberrations

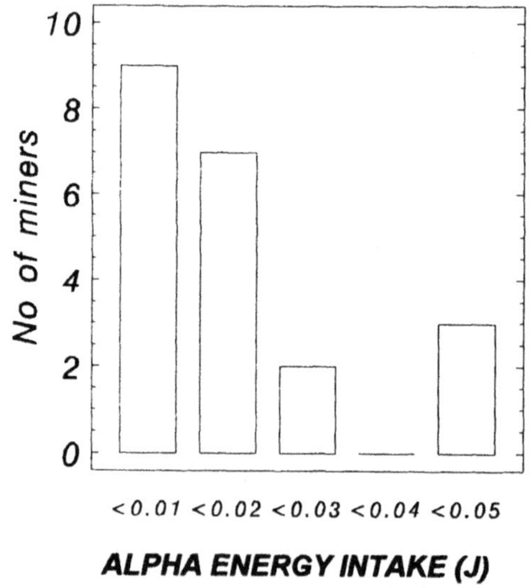

Figure 1. The frequency distribution of the miners according to the lifetime potential alpha energy intake

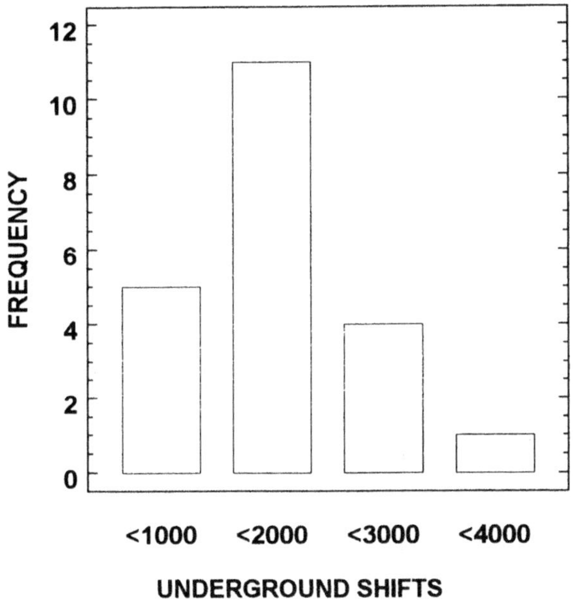

Figure 2. The frequency distribution of the miners according to the lifetime underground shifts

showed that the difference was caused by a higher number of gaps in the group exposed simultaneously to underground environment and smoking.

Regressions of the logarithms of numbers of CA upon either the numbers of lifetime underground shifts or upon the individual lifetime potential alpha energy intake (Fig. 3) calculated from radon air concentration measurements showed a small slope but the deviation of these slopes was not significantly different from zero slope ($p>0.05$). Also, the slope of the regression of the numbers of CA upon the numbers of lifetime cigarettes smoked up to the time of blood withdrawal did not significantly differ from zero.

No significant difference in counts of CA containing cells in comparing non-smoker and smoker samples either within control or within miner samples was found. Taking this into account, the significant difference in counts of CA containing cells between pooled control and miner samples showed a higher clastogenic risk in miners that might be ascribed to the exposure resulting from underground mining in the gold ore mine in Hodruša-Hámre. However, the higher counts in miners could not be ascribed exclusively to the irradiation by radon because more clastogens, as diesel exhaust fumes, mineral dust containing clastogenic metal compounds, mycotoxins etc. are present in mine environment. First, the significant differences in counts of CA containing cells in comparing control non-smoker and smoker samples with those of smoker miners, and second, the non-significant difference between non-smoker control and miner

TABLE I. Total counts, and, in parentheses, counts per 100 mitoses of chromosomal aberrations (CA) in control and miner lymphocytes grouped according to smoking status. In the lower part of the table a comparison of the number of cells containing CA by the Fisher's exact test was made.

	Total		*Non-smokers*		*Smokers*	
	Control	*Miners*	*Control*	*Miners*	*Control*	*Miners*
mitoses counted	4000	4200	2400	1400	1600	2800
gaps	156 (3.9)	221 (5.3)	103 (4.3)	66 (4.7)	53 (3.3)	153 (5.5)
breaks	37 (0.9)	56 (1.3)	21 (0.8)	17 (1.2)	16 (1.0)	39 (1.4)
fragments	60 (1.5)	48 (1.14)	33 (1.4)	14 (1.0)	27 (1.7)	32 (1.1)
exchanges	0	3 (0.07)	0	0	0	3 (0.1)
sum of C.A.	253 (6.3)	328 (7.8)	157 (6.5)	97 (6.9)	96 (6.0)	227 (8.1)
sum of C.A., no gaps	97 (2.4)	107 (2.5)	54 (2.25)	31 (2.2)	43 (2.7)	74 (2.6)
sum of cells with C.A.	233 *@	297 *	144 #	90	89 &	207 #&@
cells with C.A. - gaps $	92	103	50	31	42	72
multiple C.A. >>	19 (0.4)	29 (0.7)	11 (0.005)	7 (0.5)	8 (0.5)	21 (0.8)

$ - sum of cells with C.A., cells containing gaps only are excluded;
>> - sum of cells containing more than one CA;
* # & - difference significant ($p<0.05$) by the Fisher's exact test. At testing the No of cells with CA excluding the cells containing only gaps no significant differences between subgroups were found.

samples showed that the clastogenic risk of the underground exposure in smokers is higher. Smoking therefore might be considered as potentiating the clastogenic effect of other noxae encountered in the mine environment, including radon. In order to assess the clastogenic risk of underground radon more precisely we will in future correlate the endpoints of clastogenesis to the estimates of radiation effective doses from personal dosimeter readings [13]. Nevertheless, this study indicates that the clastogenic risk from underground professional exposure to radon in ore miners may be effectively confounded by that from smoking, and, it would be not surprising if the risk from smoking would exceed that from underground exposure to air containing low radon concentration. To elucidate this problem higher counts of probands in non-smoker and smoker in both control and miner groups should be examined.

4. Conclusions

1) Significantly higher counts of chromosomal aberrations in lymphocytes of 20 miners than in an age matched control group of twenty clerks were found and showed an increased clastogenic risk at the underground exposure;

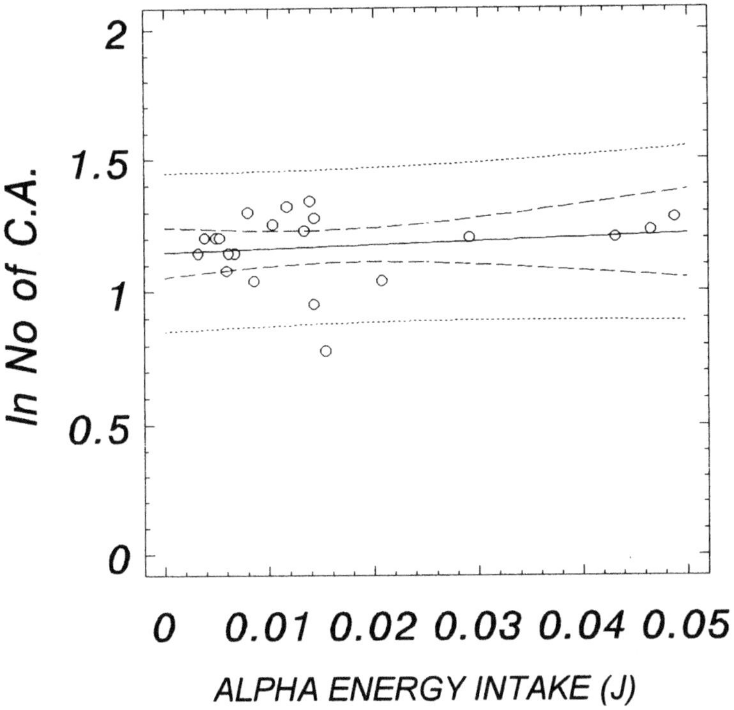

Figure 3. Regression of ln of No of chromosomal aberrations on lifetime potential alpha energy intake of miners

2) The higher counts of chromosomal aberrations were caused mainly by the increase of counts of chromatid and chromosomal gaps in the miner samples and, they could be ascribed to the underground exposure of miners, and to smoking;
3) A dependence of chromosomal aberration counts from the underground exposure to radon could not be assessed. The results showed that confounding of such dependence by smoking might have taken place. To elucidate this problem higher counts of probands in non-smoker and smoker in both control and miner groups should be examined and the personal dosimetry of radon should be used.

5. References

1. Sevc, J., Kunz, E., Placek, V. (1988) Cancer in man after exposure to Rn daughters. Health Phys. **54**, 27-46.
2. Sevc, J., Tomasek, L., Kunz, E., Placek, V., Chmelevsky, A., Barclay D., and Kellerer, A.M. (1993) A survey of the Czechoslovak follow-up of lung cancer mortality in uranium miners, *Health Phys.* **64**, 355-369.

3. Tomasek, L., Swerdlow, A.J., Darby, S.C., Placek, V., and Kunz, E. (1994) Mortality in uranium miners in West Bohemia: A long term cohort study. *Occup. Environ. Med.*, **51**, 308-315.
4. StClair Renard, K. G. (1974) Respiratory cancer mortality in an iron ore mine in northern Sweden, *Ambio*, **2**, 67-69.
5. Pekarek, V., Lazar, C., and Rychlik, J. (1982) Radon and its fission products in the air of ore mines of North-Bohemian region and the occurrence of occupational cancer of lungs in exposed miners, *Pracov. Lek.* **34**, 270-272.
6. Buchancova, J., Zigova, A., Galikova, E., Plank, L., Hutka, Z., and Fridrich, J. (1988) First cases of lung carcinoma related to radioactive substances in ore mines of middle Slovakia, *Folia Medica Martiniana*, **15**,253-259.
7. Buchancova, J., Zigova, A., Plank, L., Vrlik, M., and Galikova, E. (1988) Occupational exposure to a carcinogen and tumours of the respiratory system, *Studia pneumol. et phtiseol. Cechoslov.*, **48**, 512-525.
8. Icso, J., and Szollosova, M. (1984) The incidence of lung cancer in the miners of iron-ore mines, *Pracov. Lek.*, **36**, 294-298.
9. UNSCEAR (1988) *Radiation carcinogenesis in man.* In : *Sources, effects and risks of ionizing radiation.* United Nations Scientific Committee on the Effects of Atomic Radiation Report to the general Assembly, United Nations, New York.
10. Miller, Y.E., and Franklin, W.A. (1997) Molecular events in lung carcinogenesis, *Hematol. Oncol. Clin. North America*, **11**, 215-234.
11. Brandom, W.F., Bloom, A.D., Archer, P.G., Archer, V.E., Bistline, R.W., and Saccomanno, G. (1978) Somatic cell genetics of uranium miners and plutonium workers. A biological dose-response indicator, in: *Proc. Symp. on Late Biological Effects of Ionizing Radiation*, Vol.1., IAEA Vienna, pp.507-517.
12. Bauchinger, M., Schmid, E., Braselmann, H., and Kulka, U. (1994) Chromosome aberrations in peripheral lymphocytes from occupants of houses with elevated indoor radon concentration. *Mutation Research - Fundamental and Molecular Mechanisms of Mutagenesis*, **310**, 135-142.
13. Nikodemova, D., Vicanova, M. and Havlik, F: (1993) Indoor radon investigation in Slovakia, in Winter, M., Wicke, A. (eds.), *Umweltradioaktivität, Radiooekologie, Strahlenwirkungen*, Bd.I.; TÜV Rheinland GmbH, Köln, pp.131-135.

RADON LEVELS IN NEW AND OLD BUILDINGS

F. S. EREES, G. YENER

Ege University, Institute of Nuclear Sciences
35100 Bornova-Izmir, Turkey

Abstract

^{222}Radon is a colorless, odorless, tasteless, radioactive gas that occurs naturally in soil, rocks, underground water, and air. It is produced by the natural breakdown of ^{226}Radium in soil and rocks. In outdoor air, radon is usually presents at such low levels that there is very little risk. However, when radon enters a building , it and its decay products can accumulate to high concentrations. Radon, itself, naturally breaks down and forms radioactive decay products. As one breathes, the radon decay products can become trapped in lungs. As these decay products break down further, they release small bursts of energy which can damage lung tissue and lead to lung cancer. In this study, the indoor radon concentrations of some buildings of different ages in Bornova town of Izmir were determined. Lucas Cells and nuclear track detectors (CR-39) were used for the measurements of ^{222}Rn levels in the rooms of 15, 25 and 150 years old buildings. The annual effective dose equivalent from ^{222}Rn was calculated in these buildings.

1. Introduction

Man has always been exposed to ionizing radiation from various natural sources. The exposure to natural radiation sources varies substantially from place to place and even locally [1].

^{222}Rn is one of the atmospheric sources that causes radiological health risk. ^{222}Rn is a naturally occurring, odorless and tasteless inert gas. The radon and its daughters are present everywhere in the atmosphere, especially those at sufficiently elevated levels are known to cause lung cancer. It has been estimated that between 5,000 and 20,000 lung cancer fatalities occur every year in the U.S. because of high levels of radon existing in indoor air [2].

The bar chart in Figure 1 shows the annual effective dose equivalents from the four components of natural radiation for each of the seventeen countries[3].

C. Baumstark-Khan et al. (eds.),
Fundamentals for the Assessment of Risks from Environmental Radiation, 65-68.

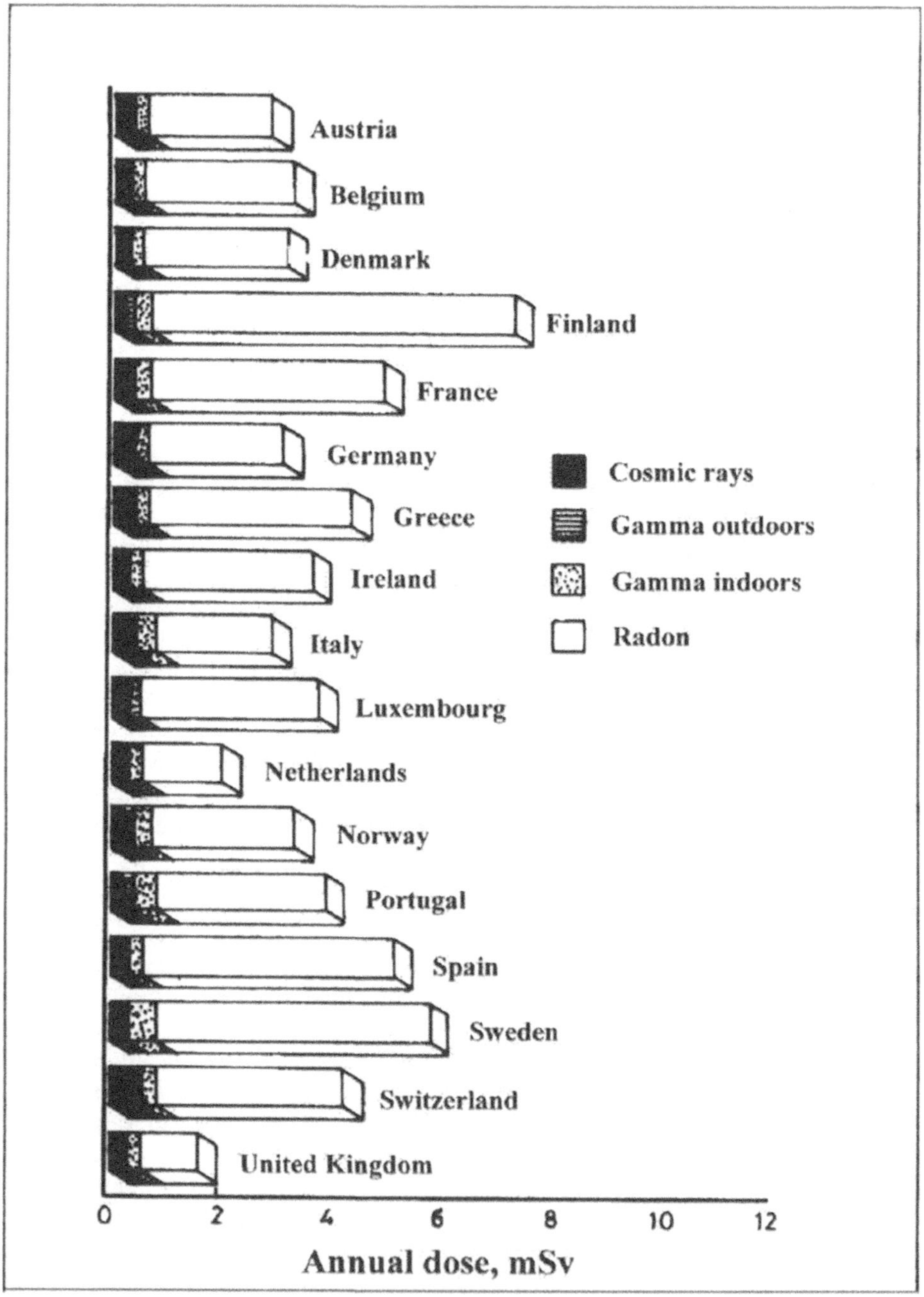

Figure 1. Average doses from natural radiation sources

2. Experimental

^{222}Rn levels in the rooms of 15, 25, and 150 years old buildings in Bornova Town of Izmir were measured with Lucas cells and with one month exposures of nuclear track detectors(CR-39).

In the Lucas cell methods, a filter paper is placed in one of the entering valves of the cell for stopping the radon daughters in the air and the vacuum pump is connected to the other valve. The sampling was done using Gelman Membrane filters with effective diameter of 2.5 cm at a flow rate of 10 L/min for a collection period of 10 min. After the sampling, the filter paper and the pump were taken away from the cell. The measurement of each cell is done after 3 hours waiting period for attaining the equilibrium between radon and its daughters.

Calibrating of the cells was done by means of a calibration plot obtained by storing a known activity of ^{226}Ra solution in a tank and monitoring the ^{222}Rn concentration at different periods of time. The calibration factor for Lucas cells were found to be 70.75 Bq m-3 /cpm. For the method of nuclear track etching, track detectors(CR-39) were etched in 20% NaOH at 70 °C for one hour, then they were placed in the plastic vessels for one month exposure. After the track detectors were collected, they were etched in 30% NaOH at 70 °C for 17 hours and the tracks were counted with an optical microscope.

3. Results and Discussion

Indoor radon concentrations are determined in the basements of 15 and 150 years old buildings. The mean values of the activity concentrations in ventilated and unventilated rooms of the basements of 150 years old building were found to be 330 Bq/m^3 and 1000 Bq/ m^3, respectively by track etching method and 416 Bq/ m^3 and 1164 Bq/ m^3 by Lucas cell measurements. This high radon concentration observed in one of the rooms is obviously due to the fact that there is no window and no ventilation in that room whereas the other room has two windows and is ventilated occasionally. Average radon concentration in basement of 15 years old building were found to be 90 Bq/ m^3 and 120 Bq/ m^3 with track detectors and Lucas cells, respectively. Average radon concentrations in the used room on first floors were found to be 53 Bq/m^3, 69 Bq/m^3 and 86 Bq/m^3 with cells in 15, 25 and 150 years old buildings, respectively.

150 years old building are made of stone, mud, clay and lime. 15 and 25 years old buildings are made of brick wall and cement. The fly ash ratio of cement was not known in 25 years old building. Because the fly ash of cement (35%) is determined in 1975 from Institute of Turkish Standards [4]. 50 years old building and old buildings more than 50 years made of muddy, stone, clay and lime. The fly ash ratio of the new cement in 1994 is 30% [5]. The building materials of this region are used in the buildings of this region.

From the calculated annual effective dose equivalent for given amounts of the radon concentrations it is known that 1 Bq/m^3 results in 0.05 mSv/year [3]. Annual effective dose equivalent from ^{222}Rn was calculated 4.5 mSv/year and 6 mSv/year for average radon concentrations of 90 Bq/m^3 and 120 Bq/m^3 respectively in basement of 15 years old building. Annual effective dose equivalent from Rn-222 was calculated 2.69 mSv/year, 3.45 mSv/year, and 4.3 mSv/year for average radon concentrations 53 Bq/m^3, 69 Bq/m^3, 86 Bq/m^3 room on first floors of 15, 25, 150 years old buildings, respectively (Fig. 2).

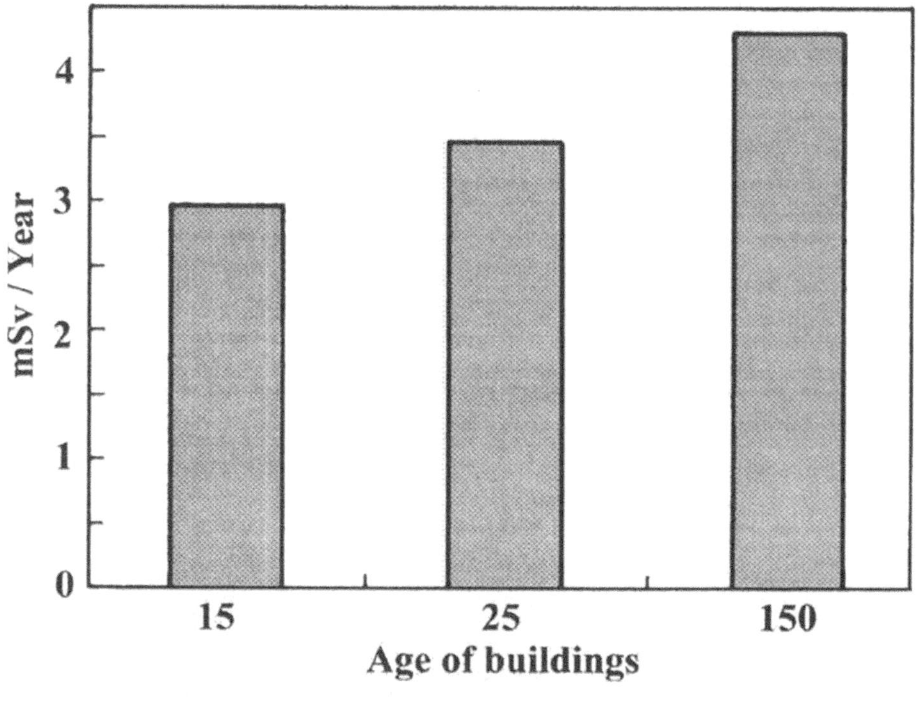

Figure 2. Annual effective dose equivalent from ^{222}Rn in buildings

4. References

1. UNSCEAR (1982). Ionizing Radiation: Sources and Biological Effects. UNSCEAR Report, United Nations, New York.
2. USEPA (1986). A Citizens' Guide to Radon. Office of Radiation Programs (ANR-460) Washington, DC.
3. Green, B.M.R, Hughes, J.S., and Lomas, P.R Radiation Atlas, Natural Sources of Ionising Radiation in Europe, Commission of the European Communities, pp.30.
4. Türk Standartlar Enstitüsü (1975). Uçucu Küller. TS-639/Nisan, 3.
5. Türk Standartlar Enstitüsü (1994). Çimento-Uçucu Küllü Çimento. TS-640/Nisan.

MEASURING EQUIPMENT FOR PHOTOBIOLOGICALLY EFFECTIVE RADIATION

T. Q. KHANH

Gigahertz-Optik
Fischerstr. 4, 82178 Puchheim, Germany

1. Summary

With the strong depletion of the stratospherical ozone layer and the consequently over-proportional increase of UV-B radiation and with new application fields of optical radiation in medicine, biology, chemistry and industry, it is necessary to measure photobiologically effective radiation (actinic radiation). This paper describes the most frequently used photobiological functions, the actinic radiant quantities, the two main measuring methods, the measuring arrangements and equipments, the calibration procedures and the state-of-the art knowledge on the measuring uncertainties. An outlook on future works will be also shown.

2. Photobiological Effects and General Aspects

Optical radiation causes different photobiological effects according to their wavelength range. These effects depend on the number and energy of the photons absorbed by the receivers and the kind of actinic effects (e.g. photolysis, photodimerisation etc.). Therefore the actinic effects depend on:

- irradiance E_e and exposure time t
- spectral power distribution $S_{e,\lambda}(\lambda)$ of the radiation sources
- action spectrum $s(\lambda)_{biol,rel}$ of the receiver

For practical measurements and assessments the following photobiologically effective (actinic) radiation quantities are important:

- actinic radiance (in $W/m^2 \cdot sr$) L_{biol} (where $L_{e\lambda}(\lambda)$ is the spectral radiance of the radiation source)

$$L_{biol} = \int_0^\infty L_{e\lambda}(\lambda) \cdot s(\lambda)_{biol,\,rel} \cdot d\lambda \qquad (1)$$

- actinic irradiance (in W/m^2) E_{biol} (where $E_{e\lambda}(\lambda)$ is the spectral irradiance of the radiation source)

C. Baumstark-Khan et al. (eds.),
Fundamentals for the Assessment of Risks from Environmental Radiation, 69-74.

$$E_{biol} = \int_0^\infty E_{e\lambda}(\lambda) \cdot s(\lambda)_{biol,\,rel} \cdot d\lambda \qquad (2)$$

- actinic radiant exposure (so called dose, in $J/m^{2)}$ H_{biol}

$$H_{biol} = \int_0^{t_1} E_{biol} \cdot dt \qquad (3)$$

Actinic effects can generally be divided into:

- effects on the human skin (erythema, DNA, non-melanoma-skin cancer, vitamin-D-formation, immediate pigmentation, PUVA, hyperbilirubin dissociation etc.)
- effects on the human eye (photoconjunctivitis, photokeratitis, blue-light-hazard) and the general health damage on work places (ACGIH-actinic spectrum, s.[4])
- effects on plants (e.g. photosynthesis), on materials (e.g. color change of materials, ect.)

Some well-known actinic spectra are defined in [1-4] and are shown in Figure 1.

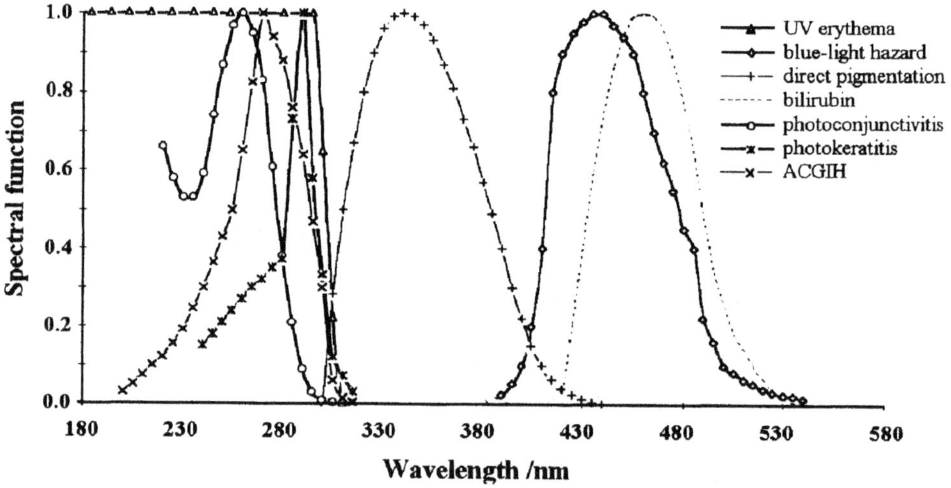

Figure 1: Actinic spectral functions of some biological end-points important for human health

3. Measuring methods, arrangements and equipment

According to the equations (1) and (2) two measuring methods can be used for measuring actinic radiation:

- ***Spectral methods***: measurements of the spectral quantities to be needed (e.g. spectral irradiance $E_{e\lambda}(\lambda)$ or spectral radiance $L_{e\lambda}(\lambda)$), evaluation with the action spectrum $s(\lambda)_{biol,rel}$ to be observed and integrating this product.
- ***Integral methods:*** measurements of actinic radiant quantities can be done with „radiometers for the measurement of actinic radiant quantities", using a radiometer head with a relative spectral responsivity $s(\lambda)_{rel}$, which is spectrally matched to the action spectrum of the considered effect.

3.1. EQUIPMENTS FOR SPECTRAL MEASUREMENTS

Generally, a spectroradiometer for measuring actinic irradiance consists of:

- input optics
- monochromator
- detector unit and signal processing unit
- calibration unit.

Spectroradiometers for atmospheric research have to fulfill the highest requirements according to stability, reproducibility, dynamic range and measuring certainty. In order to measure the global spectral irradiance on a horizontal surface, the input optics consist of a diffuser evaluating the incident radiation according to the cosine of the incidence angle and a fiber bundle transporting the radiation to the input of the monochromator. Because the spectral distribution of the global radiation in UV-B region between 290 nm and 320 nm increases over a dynamic range of 6 decades and the action spectrum of induced erythema for humans, mostly used in biological studies, decreases in the same wavelength range over three orders of magnitude, only a bandwidth of 1 nm, a resolution of maximum 0.1 nm, a wavelength setting uncertainty of 0.1 nm and a straylight rejection of better than 10^{-8} are allowed.

Some spectroradiometers of a UV monitoring network in Germany are delivered and calibrated by Gigahertz-Optik (Puchheim, Germany) and have the following specifications:

- double monochromator, focal length 600 mm, grating 2400 lines/mm
- wavelength-reproducibility: 0.02 nm, wavelength setting uncertainty: better than ±0.1 nm
- detection limit: less than 1 $\mu W / (m^2 \cdot nm)$
- cosine error: ± 5 % at incidence angle of 60°
- calibrated against working standard lamps for spectral irradiance of the German Calibration Services (DKD) accredited laboratory of Gigahertz-Optik

The measuring uncertainty is frequently checked and improved by national and international measuring campaigns [5].

3.2. INTEGRAL METHOD AND RECENT DEVELOPMENTS

3.2.1. On the Uncertainties of Existing Measuring Devices

The measuring uncertainties of up-to-day existing devices were examined in Germany by the German Federal Institute of Physics and Metrology (Braunschweig) and by some studies made at the Technical University of Berlin. The publication [6] reported on the quality of some existing UV meters which show uncertainties up to +79% (UV-A meters) and +98% (UV-B meters). The publication [7] reported on the uncertainties of UV-A meters in the range between -52% and -18% and of UV-B meters in the range between -57% and 250%. That means, that the existing devices are not suitable for radiation measurement. For serious scientific documentations and industrial quality controls it is necessary to use measuring instruments with much better quality, developments of which we report here.

3.2.2. Recent Developments - Contruction

Generally, an instrument integrally measuring the actinic quantities consists of:

- radiometer head
- display unit

The basic construction [8] of a radiometer head is illustrated in Figure 2. The following development steps are made:

a) Theoretically and practically the cosine-correction for the respective wavelength range has been optimized. Figure 3 shows the quality for the cosine-correction of an erythema radiometer with a maximum uncertainty of 5 % at 65° angle of incident.
b) Using an optimization program based on the multicriteria-method the spectral transmittance of optical components (optics, filters, diffusers) and the spectral responsivities of semiconductor-sensors have been optimized in such a way that the integral difference between the optimized spectral responsivities and the action spectra to be considered is a minimum (Figures 4-9). In each diagram the target and realized curves are presented.
c) Development of two display devices allowing actinic measurements with radiometers of different types (e.g. UV-A, UV-B, erythema etc.) and also the possibility of measurement of actinic radiant exposure H_{biol}.

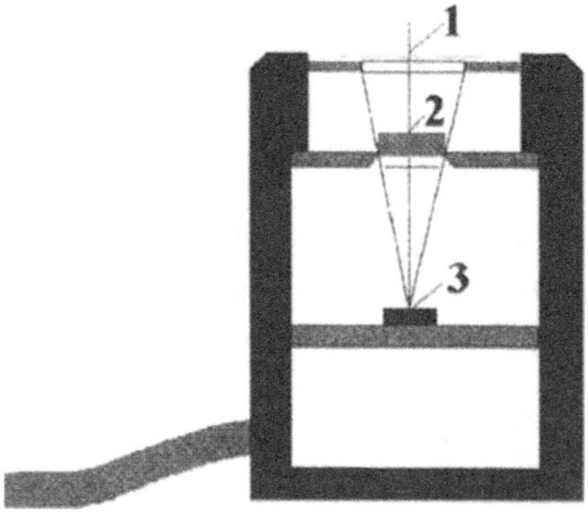

Figure 2: Radiometer construction: (1): cosine diffusor, (2): filter combination, (3): detector

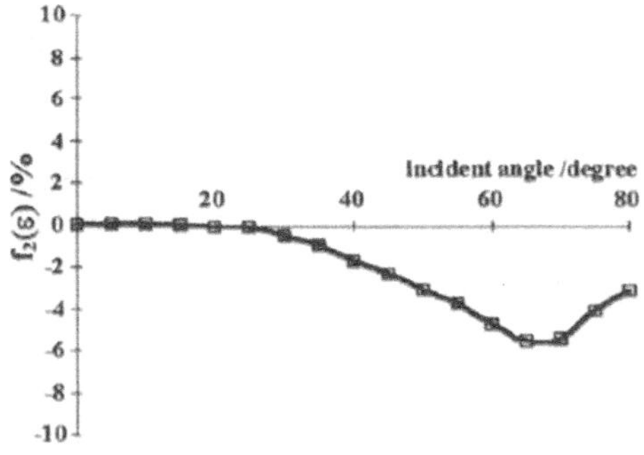

Figure 3: Cosine correction for erythema

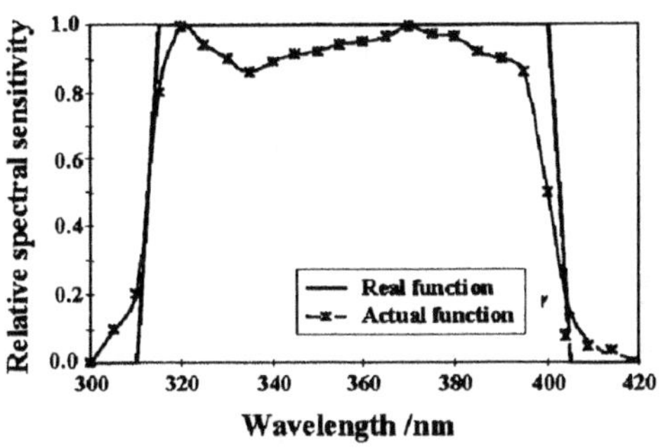

Figure 4: Spectral responsivity of the UV-A detector 3701

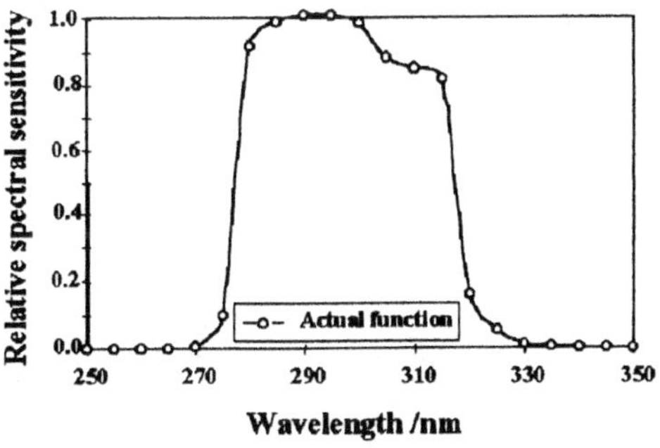

Figure 5: Spectral responsivity of the UV-B detector 3702

These devices are calibrated using standard lamps of national laboratories (PTB/Germany, NPL/Great Britain). On a sunny day with clear sky the global irradiance values of a UV-A meter were compared to those values measured with an absolutely calibrated spectroradiometer. The difference between the measured values was within 5 %. Of course, these deviations include uncertainties of calibration and spatial evaluation of both instruments.

The uncertainties of UV-B and erythema radiometer heads was checked using a well calibrated spectroradiometer as reference device. The UV irradiances displayed by integral instruments and this spectroradiometer were compared when some UV lamps for therapeutical purposes were measured. With UV-B and erythema detectors relative deviations of about ± 15% were calculated and present the state-of-the art quality of

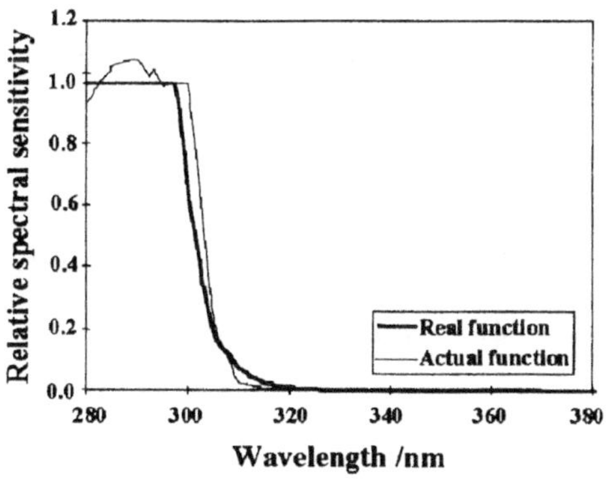

Figure 6: Spectral responsivity of the detector 3704 for the erythema function

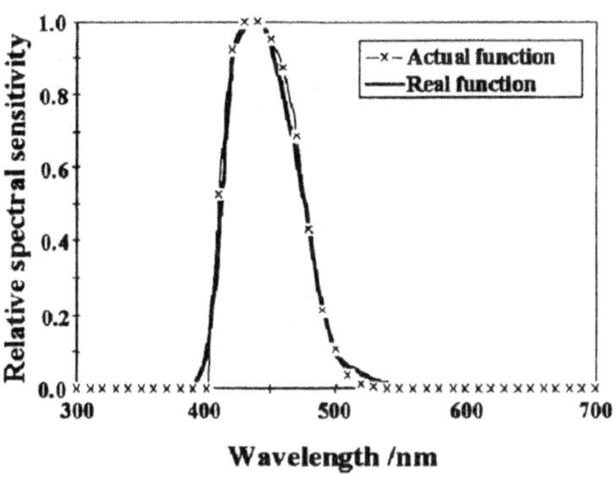

Figure 7: Spectral responsivity of the detector 3709 for the blue-light-hazard function

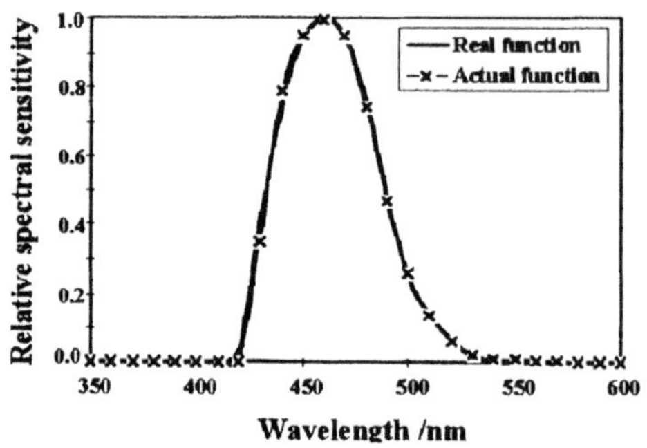

Figure 8. Spectral responsivity of the detector 3705 for the immediate pigmentation function

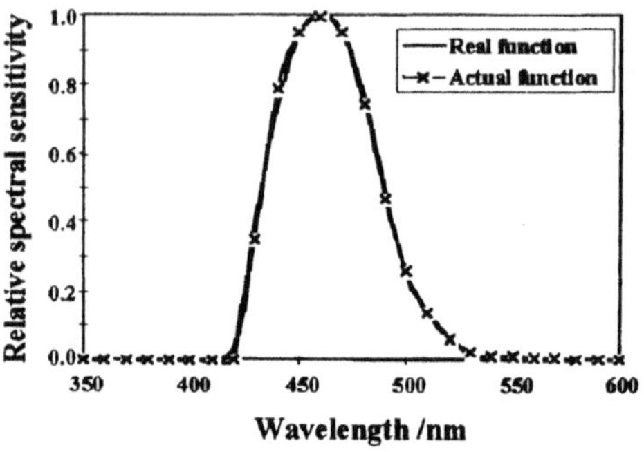

Figure 9. Spectral responsivity of the detector 3706 for the bilirubin dissociation function

optical measuring techniques. Of course, further studies are needed in order to improve and to stabilize this quality.

The development of the detectors for measuring the health risk on outdoor and indoor work places according to the ACGIH actinic spectrum and detectors for assessment of vitamin-D-formation were finished some weeks ago. The testing results of these detectors will be reported in the near future.

4. References

1. DIN 5031-10 (1996), Strahlungsphysik im optischen Bereich und Lichttechnik: Photobiologisch wirksame Strahlung, Größen, Kurzzeichen und Wirkungsspektrum.
2. IEC 825-1(1993) Safety of laser products Part 1: Equipment, classification, requirements and user's guide.
3. Amendement 2 to IEC 825-1 (1997): Safety of laser products – Maximum Permissible Exposure to incoherent optical radiation.
4. IRPA/ICNIRC (1985) Guidelines on limits of exposure to ultraviolet radiation of wavelengths between 180 nm and 400 nm. *Health Physics* **49**, 331-340.
5. H.Nordmeyer (1995) Meßgerätevergleich von Spektralradiometern zur Vermessung der bodennahen solaren UV-Globalstrahlung, Bundesamt für Strahlenschutz, Jahresbericht.
6. H. Kaase, J. Metzdorf, H. Berger (1988) Standards for UV Instrument and risk of existing devices. *Photochem. Photobiol.* **4**, 439-443.
7. P. Zwick (1990) Messung und Bewertung aktinischer UV-Strahlungsanteile, Dissertation, Institut für Lichttechnik, TU Berlin.
8. T.Q. Khanh et al. (1997) Messung von Ultralviolett-Strahlung, *Die BG* **6**, 296-302.

INDIVIDUAL EXTENT OF SOLAR UV EXPOSURE IN EVERY DAY LIFE

P. KNUSCHKE, A. KRINS, M. MEURER

Department of Dermatology, Dresden University of Technology
Fetscherstr. 74, D-01307 Dresden, Germany

Abstract

In order to evaluate the possible risk of increasing solar UV exposure in normal persons and in patients with photosensitive skin disorders it is necessary to collect base-line data on individual UV exposures in every day life. The personal daily UV exposure is mainly influenced by environmental factors (e.g. season, weather, degree of latitude) and by behaviour (e.g. profession, social status, leisure time, sports activities). In addition, the constant change of position and direction to the sun requires personal UV monitoring. Using polysulphone film (PSF), the individually biological-effective UV exposure was measured in the course of the year 1996/1997. By a weighting procedure the dosimeter readings were transformed to erythema-effective data. 180 volunteers participated in this study: the age ranged from kindergarten children (5-6 a) to retirement home people (72-85 a) with students (18-25 a) and factory workers (20-60 a) in between. The factory worker group was further divided into individuals with normal or with increased outdoor activities during work or weekends, respectively. In September, December 1996 and February/March, May/June 1997 the photobiological personal UV dosimetry was performed in Dresden/Germany (51° N) in four periods: either throughout two working weeks (10 days) or at the three weekends before, during or after. The individually accumulated UV exposure was compared with the total erythema-effective solar UV exposure by PSF dosimetry and a Robertson-Berger-Meter. Two groups of indoor workers (Dresden, Köln/51° N) and a group of students (Dresden) were controlled by PSF and Biofilm dosimeters simultaneously (part of a common project with the DLR, Institute of Aerospace Medicine Köln/Germany for advanced development of personal UV dosimetry).

1. Introduction

In Europe the spectral irradiance of the solar UV radiation is well recorded by an established measurement network. The actual extent of the environmental UV exposure on ground is determined by various factors such as solar elevation (season; time of the day), geographical factors (latitude; altitude) and meteorological situation (weather; ozone layer; air pollution).

C. Baumstark-Khan et al. (eds.),
Fundamentals for the Assessment of Risks from Environmental Radiation, 75-80.

In contrast to the risks from environmental ionizing radiation, the extent of solar and artificial UV exposure is influenced by the individual behaviour. Besides environmental effects, the level of the individual accumulated fraction of the total solar UV exposure depends on personal behaviour e.g. profession, social status, age, spare time activities. Quantitative data on the range of the individual UV exposure, however, are missing. It is well known that the incidence of UV induced skin cancer has increased over the last decades and the ozone layer thickness has decreased since the beginning 1980's. The quantitative evaluation of the risks from environmental UV radiation requires epidemiological investigations on individual biological-effective UV exposure in daily life. Furthermore quantitative data on UV exposure levels in different sections of population will support public education programs about the risks of UV radiation.

In 1996, we have started the study „Personal UV dosimetry by polysulphone films" (supported by the BMB+F / German Federal Ministry of Education, Science, Research and Technology). The aim of this two-years study was to collect first epidemiological data on individual biological-effective UV exposure in every day life. This study aim required to establish a routine personal UV monitoring. The personal UV monitoring required personal UV dosimetry (polysulphone films [1-4]) connected with a PC-database to handle and to cross link all exposure-related personal variables. The DLR/ILRM Köln/Germany (German Aerospace Research Establishment/Institute of Aerospace Medicine) developed a new type of personal UV dosimeter based on the Biofilm-technique [5, 6]. In co-operation with the DLR/ILRM Köln Biofilm(BF)-dosimeters will be also used for personal UV monitoring.

With reference to results from previous investigations [7, 8] epidemiological investigations of individual biological-effective UV exposure in every day life should be conducted

- in several groups of the population representative for their different behaviour towards solar exposure
- separately for working day activities, weekend activities and holidays, respectively
- in different seasons
- during several years (to eliminate meteorological influences)

2. Method of Personal UV Monitoring

To realise a complex personal UV monitoring for epidemiological investigations of the individual UV exposure level in every day life a PC-data base software was developed. With the consent of the volunteers that were monitored, we collected personal data relevant to solar exposure. These included profession, way to work, hobbies, spare time activities, facts to characterise the UV sensitivity of the skin (skin type), use of sunscreens. In addition, for each measurement campaign we store the protocol on time intervals spent outside and seasonal differences in the behaviour. All data including the personal exposure data can be cross linked in separate tables.

The personal UV dosimeters are based on polysulphone films (PSF). The dosimeter preparation, dosimeter read-out and erythema-effective weighting of the dosimeter reading [8] is computer-aided. The results can be imported into the PC-data base of the personal UV monitoring program.

TABLE 1. Periods of the measurement campaigns (coupled to fixed solar elevations as points of reference)

Solar elevations γ_s (at noon in Dresden / 51° N)	*Measurement period*
41.7° (Sept. 16)	Sept. 14 - 29, 1996 Sept. 6 - 21, 1997
15° (Dec. 21)	Dec. 7 -22, 1996 Dec. 6 -21, 1997
30° (Feb. 26)	Feb. 22 - Mar. 9, 1997 Feb. 21 -Mar. 8, 1998
60° (May 26)	May 24 - Jun. 8, 1997 May 23 - Jun. 7, 1998

The period of two years of the study shall decrease meteorological influences. We control the individual UV exposure level in Dresden/Germany (51° N) every season. The measurement campaigns are coupled to fixed noon solar elevations as points of reference for measurements in other years (or for measurements at other geographical latitudes) (Tab. 1). The design of a single measurement campaign is presented in Fig. 1. To measure the accumulated fraction of the erythema-effective solar UV exposure each volunteer is given one dosimeter for workingdays and a second one for the weekend days. Each volunteer has to wear the personal UV dosimeters on the chest. He or she has to use the dosimeter indoor and outdoor, from the morning to the evening. Simultaneously the total solar UV exposure will be measured by PSF-dosimeters, Biofilm-(BF)-dosimeters and a Robertson-Berger-Meter (erythema-effective broadband UV meter with data logger).

In this first screening we included 13 groups of person different in regards to their behaviour concerning solar exposure (Tab. 2). In the case of the indoor worker - the most typical UV behaviour - we selected two groups: one group of persons with distinct outside activities in spare time and on weekends and a second group with normal or less outside activities. In the current study 162 volunteer are included. Volunteers of two groups (indoor worker, students) wear PSF- and BF-dosimeters simultaneously.

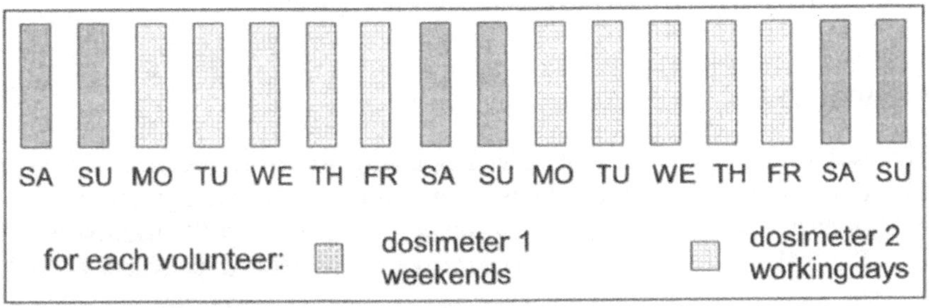

Fig. 1. Design of a single measurement campaign

TABLE 2. Investigated groups of persons different in regard to their behaviour towards solar radiation

Group	*Feature*	*Age*	*Participants*
Children	Kindergarten	5 - 6 y	n = 15
Pupils	Primary school	6 - 7 y	n = 8
Students	School class	15 -16 y	n = 24
(grammar school)	Rowers	13 - 16 y	n = 13
Students	University	18 - 27 y	n = 15
Indoor workers	Outdoor not active in spare time	26 - 59 y	n = 11
	Outdoor active in spare time	28 - 59 y	n = 11
Outdoor workers	Road construction worker	23 - 48 y	n = 8
	Roofer	18 - 38 y	n = 10
	Gardener	23 - 45 y	n = 10
Not in employment	Housewife	26 - 54 y	n = 10
	Beginning retirement age	62 - 70 y	n = 10
	Old people's home	72 - 85 y	n = 13

Furthermore we estimate the exposure peaks in holidays in selected regions in Europe (Mediterranean, high mountains, North Sea/Baltic) and the body distributions (21 dosimeter positions per volunteer, 3 days) in relation to the standard measurement position at the chest in different seasons (different solar elevations).

3. Results

For the initial period of the study from September 1996 to June 1997 a first set of individual UV exposure data is available. To demonstrate the possibilities of the established personal UV monitoring preliminary data of five selected groups of persons are presented in figure 2a - 2d. The erythema-effective UV exposure rates are given in minimal erythema doses per day (MED/d, 1 MED = 250 J/m^2 effective). The four graphs present the median of the individually accumulated UV exposure of the selected groups on workingdays and on weekends in the course of a year. In figure 2d - in addition - the mean daily exposure level of German holiday makers in the Mediterranean region is given in relation to the daily exposure level in Germany.

4. Summary

We have established a personal UV monitoring system to realise a screening as a first step to capture epidemiological basic data on individual biological-effective UV exposure in every day life. The personal monitoring includes a PC-data-base to handle and cross link the individual UV exposure data, UV related personal variables and the measurements of the total biological-effective solar UV exposure.

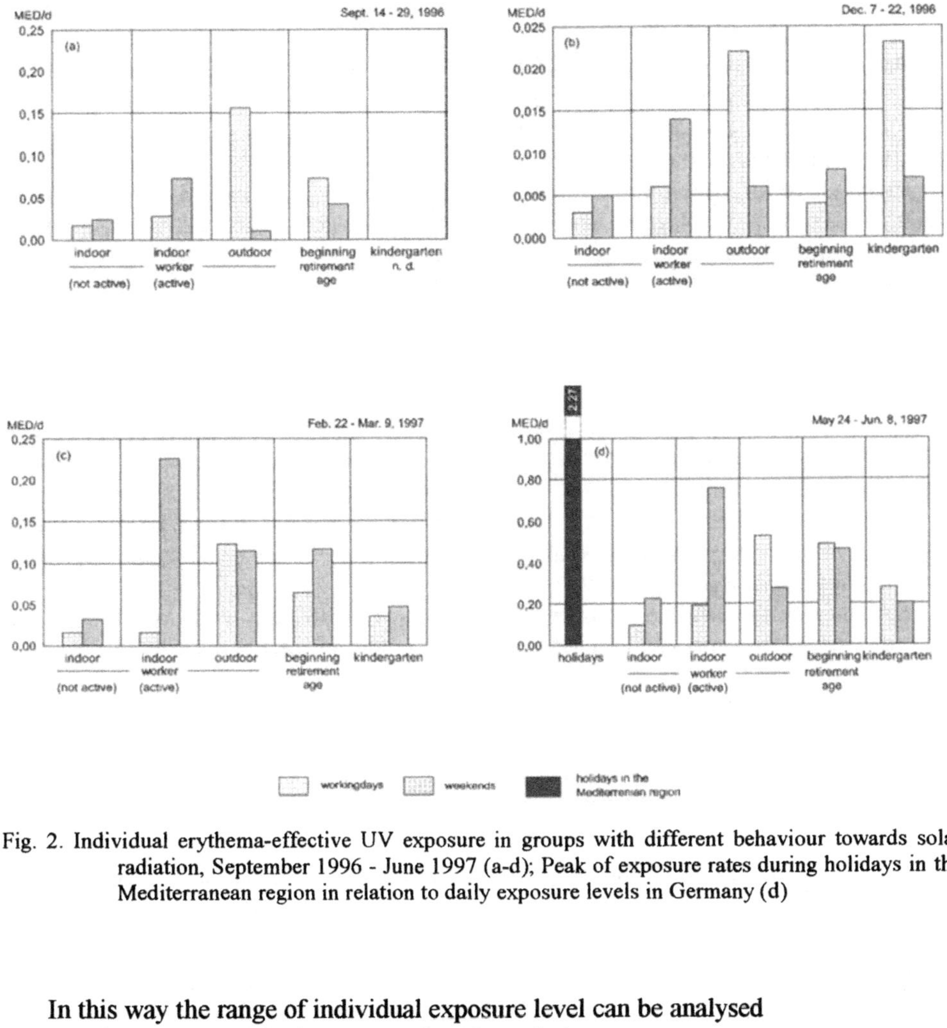

Fig. 2. Individual erythema-effective UV exposure in groups with different behaviour towards solar radiation, September 1996 - June 1997 (a-d); Peak of exposure rates during holidays in the Mediterranean region in relation to daily exposure levels in Germany (d)

In this way the range of individual exposure level can be analysed

- for one person in the course of study period
- between person of one behaviour group (the composition of the behaviour groups is variable by the data base program)
- between groups of person different in respect to their behaviour towards solar exposure.

The exposure level and the individual behaviour can be cross linked. It will be possible to compare e.g. the necessity of individual care against solar exposure dependent on skin type and the actual behaviour of person to solar radiation - especially person of skin type I and II.

In conclusion of the preliminary data given it is obvious that - in contrast to working time activities - mostly the weekend and spare time activities lead to higher UV exposure levels. This is not valid for outdoor workers. In the course of the year the relations between workingday and weekend exposure within the controlled groups seems to be roughly the same.

An interesting result presents the quantification of the exposure peak in holiday time. The mean exposure rate of holidays at the Mediterranean region is about 20x (10x) higher than the mean exposure rate of an indoor worker in Germany on working days (weekend days) in this period of time.

Prerequisite for substantial results of such epidemiological investigations on individual UV exposure levels are a good compliance of the volunteers and careful wearing of the personal UV dosimeters. Continuous information of the volunteers on their personal results stimulated the compliance.

5. Conclusion

The established personal UV monitoring has proven to be a suitable method to build up an epidemiological data base of individual biological-effective UV exposure and related behaviour facts.

It presents a valuable method to support environmental programs on UV exposure risks, to evaluate the possible risk of normal persons, on patients with photosensitive skin disorder and at working places.

Furthermore the detailed measurement results in the different behaviour groups and in holiday time will support public education campaigns on solar care.

Acknowledgement. The work was supported by the German Federal Ministry of Education, Research, Development and Technology.

6. References

1. Davis, A., Deane, G.H.W., Diffey, B.L. (1976) Possible dosimeter for ultraviolet radiation, *Nature* **261**, 169-170.
2. Davis, A., Kockott, D., Schroepl, F. (1986) Measuring UV-dosage with the help of polysulphone-films, First European Congress of Photobiology, Sep. 7-12 1986, Grenoble/France
3. Diffey, B.L. (1989) Ultraviolet radiation dosimetry with polysulphone film, in Diffey, B.L. (ed.), Radiation measurement in photobiology, Academic Press Limited, New York, pp. 136-159.
4. CIE Technical Report (1992) Personal dosimetry of UV radiation, Pub. N°. CIE 98, 1st Edition
5. Quintern, L.E., Horneck, G., Eschweiler, U., Bücker, H. (1991) A biofilm used as ultraviolet-dosimeter, *Photochemistry and Photobiology* **55**, 389-395
6. Rettberg, P., Horneck, G. (1997) Messung der UV-Strahlenbelastung mit dem DLR-Biofilm, *Derm* **3**, 1-4
7. Herligy, E., Gies, P.H., Roy, C.R., Jones, M. (1994) Personal dosimetry of solar UV radiation for different outdoor activities, *Photochem. Photobiol.* **60**, 288-294.
8. Knuschke, P. and Barth, J. (1996) Biologically weighted personal UV dosimetry, J. *Photochem. Photobiol.: B Biol.* **36**, 77-83

RISK ASSESSSMENT.

Recommendations for Research Needed

R.E. SHORE [1], L.P. OSIPOVA [2] AND
W. ATWELL [3], F.S. EREES [4], R.J.M. FRY [5], V.K. IVANOV [6],
T.Q. KHANH [7], P. KNUSCHKE [8], M.M. KOMOCHKOV [9],
M.R. QUASTEL [10], L. TOMASEK [11], R.M. TYRRELL [12],
L.O. BJÖRN [13], F.R. DE GRUIJL [14], R. GREINERT [15], G. OBE [16],
P. RETTBERG [17]

[1] *Department of Environmental Medicine, New York University Medical School, New York, USA.*
[2] *Institute of Cytology & Genetics, Novosibirsk, Russia*
[3] *Boeing North American, Space Systems Division, Houston, USA*
[4] *Ege University, Institute of Nuclear Science, Bornova-Izmir, Turkey*
[5] *Life Science Division Oak Ridge National Laboratory, Oak Ridge, USA*
[6] *Medical Radiological Research Centre of RAMS, Obninsk,Russia*
[7] *Gigahertz-Optik, Puchheim, Germany*
[8] *Dept. of Dermatology, University of Technology, Dresden, Germany*
[9] *Joint Institute for Nuclear Research, Division of Radiation and Radiobiological Research, Dubna, Russia*
[10] *Department of Epidemiology and Institute of Nuclear Medicine, Ben Gurion University of the Negev, Beer Sheva, Israel*
[11] *National Radiation Protection Institute, Prague, Czech Republic*
[12] *Dept. of Pharmacy and Pharmacology University of Bath, Bath, UK*
[13] *Plant Physiology, Lund University, Lund, Sweden*
[14] *Dermatology; University Hospital Utrecht, Utrecht, The Netherlands*
[15] *University Institute of Medical Physics and Biophysics, Göttingen, Germany*
[16] *Dept. of Genetics, University of Essen, Essen, Germany*
[17] *DLR, Inst. of Aerospace Medicine, Radiobiology Unit, Köln, Germany*

1. Introduction

The *Task Group on Needs for Future Research in Risk Assessment* considered both ionizing radiation and ultraviolet radiation. The discussion and recommendations regarding ionizing radiation are limited mainly to epidemiological and other human studies that currently form the primary basis for human risk assessment. Needs regarding radiobiology and mechanisms of radiation-induced effects are referred to only in general terms; they are developed in more depth by other task groups in this Workshop. The

C. Baumstark-Khan et al. (eds.),
Fundamentals for the Assessment of Risks from Environmental Radiation, 81-90.

recommendations regarding UV radiation focus on gaps in knowledge pertaining to skin cancer, UV action spectra and UV measurement problems.

In order to provide a coherent organization by subject matter, various unsolved problems or gaps in knowledge are stated along with the recommended research to address them. In addition, since epidemiological studies are sometimes proposed or undertaken by researchers who are not highly trained in epidemiology, this chapter also provides a brief didactic summary of methodological criteria that are important to consider in planning or evaluating radiation epidemiology studies.

2. Ionizing Radiation

2.1. MAGNITUDE OF RADIATION EFFECTS

2.1.1. Risk at Low Doses.

The public-health and radiation-protection communities have an obvious interest in the magnitude of risk at low radiation doses, since most current radiation exposures are at a low level and are fractionated. There are important scientific questions relating to uncertainties about the shape of the dose-response curves for various cancer and genetic endpoints at low doses. Are the dose-response curves linear non-threshold, curvilinear (i.e., quadratic with upward curvature), or threshold in nature? And if there is a dose-threshold, at what dose does it occur? Does radiation hormesis play a role at low doses with regard to important health endpoints?

It seems unlikely that epidemiological data alone will provide adequate answers to these questions because of the intrinsic limitations in the sensitivity and the control of confounding factors in such studies (which will be further discussed below). A thorough understanding of the mechanisms of radiation-induced cancer is needed to provide satisfactory answers regarding risk at low doses, but an adequate delineation of mechanisms is probably some years away. Hence, the best possible answers from low-dose epidemiological data are useful in defining the bounds on low-dose risks. Specifically, the analysis of large, combined, valid sets of epidemiological data provides a much more precise answer to the low-dose question than individual studies can do, as the recent collaborative analyses by Cardis et al. [1] regarding nuclear-facility workers and by Lubin et al. [2-3]. Analyses of combined low-dose studies are to be encouraged as new data become available.

The potential quality and contribution of any proposed low-dose study should be carefully reviewed before deciding to embark upon it. Issues of the quality of dosimetric data, the magnitude of dose, the ability to follow the population adequately, the sample size and statistical power, and the availability of information on potential confounding factors should all be evaluated to determine whether the study is likely to be informative enough to merit the allocation of the considerable resources that will be needed to conduct it.

2.1.2. Dose-Protraction and Dose-Fractionation Effects.

Since most current radiation exposures are of a protracted and fractionated nature, it is essential to develop and carefully synthesize as much epidemiological data as possible on the associated risks. Protracted exposures may differ in their effect from acute exposures by virtue of producing chronic up-regulation of certain genes, allowing time for interim DNA repair, etc. The biological bases for the Dose and Dose-Rate Effectiveness Factor (DDREF) for low-LET radiation and the inverse dose-protraction effect for alpha radiation need to be better understood, including in what circumstances these factors do and do not apply. The contrasting epidemiological findings regarding the effects of fractionated exposures to the lung and breast strongly suggest there are different biological features of the radiogenic response in these two organs that need to be elucidated. Biological studies oriented toward determining

1) the primary gene targets in particular organs or tissues, and
2) the association of protracted exposures with effects upon gene expression, DNA repair, signal transduction, etc. are needed to provide a mechanistic understanding of radiation risks in general and protraction effects in particular.

2.1.3. Radon Research.

The recent landmark publication of combined analyses of the major cohort studies of radon-exposed miners [4-5] has provided a much firmer basis for assessing lung cancer risk from radon. However, even though the low-dose portion of the miner data was specifically analyzed [3], extrapolation is still required down to the exposure levels and exposure rates characteristic of most residential radon exposures. In addition, there are concerns about their applicability of the miner data to residential exposures (e.g., co-exposures in mines to arsenic and diesel exhaust, differences in particle size and unattached fraction, etc). More research on these issues is recommended.

Regarding residential radon exposures, combined analyses of the major studies should be expedited. Opportunities should be sought for cohort studies of lung cancer and residential radon exposure that can avoid the potential biases in case-control studies (e.g., control selection biases, retrospective surrogate reports of smoking histories) and that periodically obtain concurrent radon measurements and information on smoking. Data are needed on the degree of variability in personal radon exposure levels because of variations among individuals in time spent at home and elsewhere to provide a more realistic picture of the magnitude of measurement error.

New ecological studies of average radon levels and lung cancer rates are not encouraged. The great potential for uncorrectable confounding by smoking history in such studies has been thoroughly described [6] and means that such studies are not credible.

2.1.4. Importance of Dosimetry

Dosimetry is often a major weakness in human radiation studies. Without adequate dosimetry an epidemiological radiation study is of little scientific value, because radiation research has advanced beyond merely finding effects to quantitative risk estimation. The best radiation studies have measurements or reconstructed estimates of individual doses. In lieu of individual doses, estimates of average doses for various subgroups in the study can be used.

2.1.5. Uncertainty in Dosimetry Estimates.

The standard epidemiological measure of uncertainty, i.e., a confidence interval on the estimated risk, takes into account only the basic statistical uncertainty in the health endpoint under study. Ideally, estimates of measurement error in the dosimetry should also be included as part of the assessment of total uncertainty. Methods have been developed to incorporate dosimetric measurement error, although they are not yet routinely available in statistical packages. As one might expect, incorporating estimates of errors in the dosimetry will widen the confidence interval, and it may also increase the slope of the dose-response curve. It is recommended that uncertainties in dosimetry be taken into account when possible.

2.1.6. Statistical Models of Cancer Induction by Radiation.
Current mechanistic-quantitative models of cancer induction (e.g., Moolgavkar model) to apply to epidemiological data are very general and use only limited biological insights. The development of richer mechanistic models is to be encouraged. This will require collaboration between biologists and statisticians, but may have to await further understanding of the mechanisms to be fruitful.

2.2. GAPS IN THE RADIATION EPIDEMIOLOGY KNOWLEDGE BASE

2.2.1. Intelligence Deficits from in utero Irradiation.
The fairly recent analyses of Japanese atomic bomb data on mental retardation and IQ (Intelligence Quotient) measurements in children exposed *in utero* [7] raised the question of whether intelligence deficits might be a non-threshold phenomenon. However, a more recent publication has suggested that a threshold is likely, at least for severe mental retardation [8]. Additional analyses of the Japanese IQ data for the *in utero* exposed and the development of valid data from other *in utero* irradiation studies regarding IQ would be of value.

2.2.2. Adult Cancer Risks from in utero Irradiation.
A number of studies have been conducted of childhood cancer in relation to *in utero* diagnostic radiation, but only the Japanese atomic bomb survivor study has information on adult cancer following *in utero* irradiation. Because of the limited number of Japanese children exposed *in utero*, the data are sparse and additional data are needed from other cohorts irradiated *in utero*.

2.2.3. Adult Cancer Risks from Childhood Irradiation.
The relative risks from childhood irradiation tend to be larger than those following irradiation in adulthood. However, the irradiated childhood populations are only now reaching the ages when the major expression of cancers will appear, and there is a question as to whether the relative risks will be maintained at a constant level at older ages or will decline. This question will have a substantial impact on radiation risk estimates, so continued support for follow-up of the major childhood-irradiation cohorts is recommended.

2.2.4. Psychosocial Effects of Radiation Accidents.
Experience with population groups who lived near the Chernobyl reactor accident and the clean-up workers has provided evidence for post-traumatic stress disorder (PTSD) [9]. Symptoms of PTSD, and their expression through various physical complaints, were found to persist for a long time and must therefore be taken into account when

carrying out epidemiological studies on the long-term effects of radiation accidents on human populations.

2.2.5. Coordinated Studies of Chernobyl Populations to Examine Health Effects.
It has been suggested that a coordinated series of studies be conducted on a carefully matched sample of low-, medium- and high-dose persons from the Chernobyl region regarding a variety of biomedical parameters, such as the function of the immune, neuroendocrine, reproductive and cardiovascular systems. The merit of the specifics of such a proposal should be evaluated by those who are expert in the relevant biomedical specialties and who are familiar with the existing research on radiation effects in these systems.

2.3. BIOLOGICAL AND EPIDEMIOLOGICAL INTERFACE

2.3.1. Biodosimetry.
There are several open issues pertaining to biodosimetric markers.
Can new or improved methods of biodosimetry be developed that have greater sensitivity, in order to estimate low doses reliably? This will require markers with a low background signal but with high inducibility by ionizing radiation.

Development of automated methods for scoring chromosome aberrations or other biomarkers is needed, so that scoring is less costly and can be applied to larger populations. Such methods need to be carefully evaluated for their sensitivity and specificity, as compared to manual methods.

Further research is needed on the usefulness of microsatellite markers.

2.3.2. Banks of Biological Samples.
With the rapid advances in molecular and cell biology, biological samples from populations with significant radiation exposures will become increasingly valuable. There is a need to bank samples expeditiously, while highly exposed populations are still accessible, in order to have samples available for testing new biomarkers or addressing new mechanistic questions in the future. Long-term storage of lymphocytes under optimal conditions will provide opportunities for studies pertaining to biodosimetry, mechanisms of radiation damage, variations in cellular radiation effects as a function of genetic polymorphisms, etc. Storage of tumor tissue from irradiated populations is also of value in order to characterize various molecular aspects of radiation-induced tumors as compared to tumors of unknown etiology. Because of the expense of routine acquisition, preparation and long-term storage, biological sample storage should be limited to populations or subgroups that will be most informative, i.e., those with relatively high doses or other special characteristics of interest.

2.3.3. Heritable Effects of Ionizing Radiation.
Based on recent comprehensive re-analyses of genetic effects in the F_1 generation born to Japanese atomic bomb survivors, the magnitude of genetic risk for Mendelian diseases/conditions in humans is in question. Likewise, improved modeling of the contribution of radiation to genetic risk for polygenic diseases is causing a reassessment of the magnitude of polygenic risks. Valid information on the magnitude of radiation-induced

genetic risks in other human populations besides the atomic bomb survivors would be a very valuable contribution toward these reassessments.

2.3.4. Synergism between Radiation and Germline Cancer Mutations.

There has long been a concern as to whether subsets of a population may be highly sensitive to cancer induction by radiation. Advances in genetics are making it possible to identify subsets with germline mutations in tumor suppressor genes that confer high susceptibility for cancer. In a few cases (e.g., hereditary retinoblastoma, Li-Fraumeni syndrome, nevoid basal cell syndrome) preliminary evidence indicates these subsets are also highly susceptible to radiation-induced cancer. A number of the more common germline mutations conferring cancer risk (e.g., *BRCA1*, *BRCA2*, *APC*) need to be investigated. Experimental studies in transgenic mice with null mutations at loci corresponding to these human genes can also provide important confirmatory information regarding oncogenic radiosensitivity.

2.3.5. Evaluation of Antioxidants in Ameliorating the Late Effects of Radiation.

Biomarkers of oxidative stress can be inhibited by oral administration of antioxidants such as Ginko biloba [10]. One study reported that administration of an oral antioxidant (flavonoids) reduced lung cancer incidence [11]. These data suggest that treating people exposed to ionizing radiation with dietary antioxidants might be of value. However, the paucity of knowledge in this area suggests that further investigations of the biomedical effects of antioxidants with respect to radiation exposure be undertaken before embarking on a trial of the prophylactic effects.

2.4. CONSIDERATIONS AND CRITERIA FOR EPIDEMIOLOGICAL STUDY DESIGN AND METHODS

2.4.1. Statistical Power and Sample Size.

Statistical power is essential for epidemiological (or laboratory) studies because, if the statistical power is low:

1) a null result has little meaning (i.e., it can rule out only implausibly high risks);
2) any statistically significant result is likely to be a "false positive" result (i.e., the probability is high that the result is due to chance); and
3) the risk estimate associated with a statistically significant result is likely to be a large overestimate of risk.

In short, a study with low statistical power is likely to be either uninformative or misinformative.

It is generally believed that a dose-response curve for radiation-induced cancer will be either linear or curvilinear (i.e., quadratic with upward curvature) with respect to dose. Even for a linear association, and more so for a curvilinear one, the sample size needed to achieve adequate statistical power to detect the expected relative risk escalates rapidly as the average dose decreases. For example, if the linear radiation risk coefficient for some type of cancer induction is such that a sample size of 500 would be adequate to detect the expected risk in a population that received 1 Gy, a sample of over 29,000 would be required if the population instead had received only 100 mGy, or a sample of about 2.7 million would be needed if the dose was only 10 mGy. Having a range of doses and performing a dose-response analysis diminishes the sample size requirement somewhat, but the requirement is still strongly driven by the mean dose level. It is therefore important that investigators realistically consider their sample-size needs, given the doses involved, before embarking on a study. If the statistical power based on the dose distribution and size of the available sample is poor, the study is of doubtful scientific merit.

2.4.2. Design of Epidemiological Studies so as to Minimize Biases.
It is important for investigators, especially when planning low-dose studies, to give careful attention to study-design considerations, so as to minimize selection bias and information bias. Following is a brief outline of considerations.
Cohort Study. In a cohort study care should be taken to assure:

1) that subject selection procedures are equivalent and are unbiased across the dose range (including the control group);
2) if the study involves active subject participation (e.g., disease screening procedures), that sufficient resources are allocated to assure that participation rates will be maintained at a uniformly high level across the dose range;
3) that a high rate of disease ascertainment will be achieved that is equivalent across the dose range; the completeness of case-ascertainment should be verified;
4) that appropriate data on potential confounding factors is obtained.

Case-control Study. In a case-control study, care must be taken that several pitfalls are avoided:

1) Cases and controls should be drawn from a well-defined population base with unbiased selection procedures, rather than selected on an *ad hoc* basis.
2) Procedures should be implemented to assure a high rate of participation. This will take a significant allocation of resources.
3) Objective measures of exposure(s) should be obtained where possible, or respondents' reports should be verified for a subset of respondents. Great caution should be observed in study interpretation if the agreement between the reported and documented exposures is not high.

Ecological Study. In an "ecological" (aggregate) study a correlation is obtained between average amounts of exposure in various geographic areas and the disease rates in those areas -- as opposed to obtaining information on individuals' exposure and disease status. If such a study is undertaken, important caveats should be recognized. Because of the large intrinsic potential for undetectable or uncorrectable confounding in ecological studies[12], they are primarily a tool for hypothesis generation rather than hypothesis testing. Results should therefore be interpreted very cautiously.

3. Ultraviolet Radiation

3.1. SKIN CANCER EFFECTS OF UVR

3.1.1. Effects of UV-A Exposure.
The case-control studies of UV-A sunlamp exposure and risk of malignant melanoma that are currently available provide a weak basis for assessing risk. It would be valuable to conduct a large cohort study of UV-A sunlamp exposure in relation to melanoma risk

if a suitable opportunity for such a study can be found. A second issue pertains to the mode of action of UV-A. Experimental work is needed to determine whether it acts as a tumor promoter, as some evidence suggests, or as an initiator.

3.1.2. Mechanisms that define UVR-skin cancer susceptibility.

More needs to be learned about the mechanisms that define the oft-observed variations in skin cancer susceptibility.

More understanding of intrinsic protective mechanisms is needed; these include DNA repair, antioxidation and immune mechanisms.

New information about germline or somatic mutations, e.g., in the *patched* gene for basal cell carcinoma or the *CDKN2A/B* and *β-catenine* genes for malignant melanoma, may provide new leads regarding mechanisms.

3.1.3. Synergism between UV-B, UV-A, Ionizing Radiation and Chemical Exposures with Respect to Skin Cancer Risks.

There are limited experimental and epidemiological data indicating a synergistic relationship between UVR and ionizing radiation in inducing non-melanotic skin cancer (NMSC). Are there other synergisms: between UV-B and UV-A, or between UV exposure and certain chemical exposures (e.g., those with photosensitizing properties) with respect to NMSC? Do synergisms occur with regard to malignant melanoma induction? Information on these joint effects will advance scientific knowledge and is important for radioprotection purposes.

3.1.4. Information on the UVR Action Spectrum.

There is a need for a detailed action spectrum for melanoma based on a mammalian model; the best currently available are the Monodelphis domesticus model and the SCUP-mouse model. Information is also needed on action spectra for relevant biological endpoints in various organisms. Basic studies are required at several levels to understand the form of the action spectrum, wavelength dependence for damage, modulation of gene excision, and pathways for cellular defense. Similarly, to improve human protection, more information is needed at the fundamental level regarding:

Action spectra for expression of genes crucial to the process of carcinogenesis, and the nature and role of protective mechanisms.

Identification of the role and mechanisms of endogenous sensitizers involved in UV-B, and particularly UV-A, damage, e.g., porphyrins, melanin.

3.1.5. Measurement of UVR Exposures.

The photobiological effects in the UV region show a very strong wavelength dependence which is described by the action spectra. A number of basic biological investigations of UV photoeffects are currently in progress to determine the molecular and cellular pathways between incident UVR and risk effects. Often the irradiance and fluence measurement methods and documentation are not sufficient in such studies. For instance, the use of measurement values in unweighted physical units does not take into account the spectral dependencies of incident UVR, its biological efficiency and the spectral response of the broadband radiometer used. A related problem is that measurements of UV-A/UV-B exposures in dermatologic practice are often crude, and cumulative exposures are poorly documented. We recommend careful measurement procedures by devices calibrated to the erythema action spectra (CIE/1987), photodiagnosis using an identical spectrum to the UV therapy equipment, and a long-term documentation of exposures in order to be able to monitor carefully patients who are at high risk for skin cancer or other effects by virtue of high cumulative exposure.

If baseline investigations for UVR-risk assessment purposes are conducted with polychromatic, rather than monochromatic UVR, biological dosimeters or spectral measurements that permit weighting by the action spectrum are needed. It is necessary to inter-calibrate and standardize the various types of biodosimeters with regard to the action spectrum for DNA-related UV effects (i.e., mutation induction, single strand breaks, etc.) so as to be able to compare and integrate the results of studies that use various biodosimetric systems.

3.1.6. Personal UV Monitoring.

Because individual behavior patterns with respect to environmental and artificial UV sources affect the distribution of mean biologically-effective UV exposure levels, we recommend epidemiological studies based on personal UV monitoring.

Polysulphone film dosimeters have been used in early epidemiological investigations. New biofilm dosimeters are of considerable interest because they better match the biologically-effective UV exposure to human skin, especially for UV-A. Biodosimeters will be useful both for epidemiological applications and for assessing incremental risk from solar exposure in patients being treated with phototherapy. We recommend their further development.

For some research, it is important to be able to obtain UV measurements with a fine time resolution. For this purpose electronic data loggers that permit biologically-effective weighting are just being developed, but they are currently too costly for large-scale use. A cheaper, more miniaturized version would be valuable for research purposes.

3.1.7. Impact of Changes in the Ozone Layer.

It should be recognized that the rapid increases in non-melanoma, and especially melanoma, skin cancer over the last several decades have occurred primarily because of changes in people's sun-exposure habits. However, ozone depletion is a long-term concern that must be studied for its future impact.

Currently in the northern hemisphere, ozone layer depletion takes place only in late winter and early spring, when UV exposures are low. The degree to which recent and future emissions may lengthen the season of ozone depletion is unknown. A prolonged duration of ozone depletion will increase total UV exposure, but, more importantly, the spectrum of UV irradiance will change during the extended period. Small spectral changes can have potentially large biological effects because they occur in the UV spectral region where the action spectrum increases exponentially.

The standardized analysis of the data sets on spectral solar UV irradiance, ozone layer thickness, air pollution and meteorological data needs to be coordinated internationally to enable better prediction of future ozone-depletion effects. Further measurements are needed to predict the impact of changes in the ozone layer upon humans, animals, plants, microorganisms and ecosystems.

4. References

1. Cardis E., Gilbert E.S., Carpenter L., Howe G., Kato I., Armstrong B.K., Beral V., Cowper G., Douglas A., Fix J., (1995) Effects of low doses and low dose rates of external ionizing radiation: cancer mortality among nuclear industry workers in three countries. *Radiat. Res.* **142**, 117-32
2. Lubin J.H., Boice J.D. Jr., (1997), Lung cancer risk from residential radon: meta-analysis of eight epidemiologic studies. *J. Natl. Cancer. Inst.* **89**, 49-57
3. Lubin J.H., Tomasek L., Edling C., Hornung R.W., Howe G., Kunz E., Kusiak R.A., Morrison H.I., Radford E.P., Samet J.M., Tirmarche M., Woodward A., Yao S.X., (1997), Estimating lung cancer mortality from residential radon using data for low exposures of miners. *Radiat. Res.* **147**, 126-34
4. Lubin J.H., Boice J.D. Jr., Edling C., Hornung R.W., Howe G.R., Kunz E., Kusiak R.A., Morrison H.I., Radford E.P., Samet J.M., (1995), Lung cancer in radon-exposed miners and estimation of risk from indoor exposure. *J. Natl. Cancer Inst.* **87**, 817-27
5. Lubin J.H., Boice J.D. Jr., Edling C., Hornung R.W., Howe G., Kunz E., Kusiak R.A., Morrison H.I., Radford E.P., Samet J.M., (1995) Radon-exposed underground miners and inverse dose-rate (protraction enhancement) effects. *Health Phys.* **69**, 494-500
6. Greenland S., Robins J., (1994), Invited commentary: ecologic studies--biases, misconceptions, and counterexamples. *Am. J. Epidemiol.* **139**, 747-60
7. US NAS (National Academy of Sciences Committee on the Biological Effects of Ionizing Radiation) (1990) Health effects on populations of exposure to low levels of ionizing radiation, BEIR V Reports, US National Academy of sciences, Washington, DC.)
8. Otake M., Schull W.J., Lee S., (1996), Threshold for radiation-related severe mental retardation in prenatally exposed A-bomb survivors: a re-analysis. *Int. J. Radiat. Biol.* **70**, 755-63
9. Cwikel J., Abdelgani A., Goldsmith J.R., Quastel M., Yevelson I.I., (1997) Two-year follow up study of stress-related disorders among immigrants to Israel from the Chernobyl area. *Environ. Health Perspect.* **105**, Suppl 6, 1545-50
10. Emerit I., Oganesian N., Arutyunian R., Pogossian A., Sarkisian T., Cernjavski L., Levy A., Feingold J., (1997), Oxidative stress-related clastogenic factors in plasma from Chernobyl liquidators: protective effects of antioxidant plant phenols, vitamins and oligoelements. *Mutat. Res.* **377**, 239-46
11. Knekt P., Jarvinen R., Seppanen R., Hellovaara M., Teppo L., Pukkala E., Aromaa A., (1997), Dietary flavonoids and the risk of lung cancer and other malignant neoplasms. *Am. J. Epidemiol.* **146**, 223-30
12. Greenland S., Morgenstern H., (1989), Ecological bias, confounding, and effect modification. *Int. J. Epidemiol.* **18**, 269-74

RADIATION CHEMISTRY OF DNA.

Damage Induced by Ionizing and UV Radiations

J. CADET, M. BARDET, T. DELATOUR, T. DOUKI,
D. GASPARUTTO, D. MOLKO, J.-P. POUGET, J.-L. RAVANAT,
N. SIGNORINI AND S. SAUVAIGO.

Département de Recherche Fondamentale sur la Matière Condensée,
SCIB/Laboratoire "Lesions des Acides Nucléiques", CEA/Grenoble
F-38054 Grenoble Cedex 9, France

1. Introduction

Relevant structural and mechanistic information has been gained during the last decade on the main photo- and radiation-induced damage to isolated DNA [for recent comprehensive reviews, see 1-7]. This was mostly achieved using nucleosides and short oligonucleotides as DNA model compounds. Extrapolation to cellular DNA is still a matter of debate, particularly for ionizing radiation. In fact, there is still a paucity of relevant and accurate data on the formation of radiation-induced base damage to DNA in cells and tissues. In addition, there is a lack of both qualitative and quantitative information on the spectrum of UV-B-induced DNA lesions. This may be accounted for by the still challenging analytical problem associated with the measurement of photo- and radiation-induced damage to DNA [8]. In this respect, it should be mentioned that most of the determination of radiation-induced base damage using the gas-chromatography-mass spectrometry (GC-MS) assay developed initially by Dizdaroglu *et al.* [9-11], has to be reassessed. This also applies to the bulk of DNA repair studies of radiation-induced base damage where the specificity of recognition of several DNA-glycosylases was evaluated using the above GC-MS method. Emphasis was placed in this survey on mechanistic aspects of the photo- and radiation-induced decomposition of the base moieties of DNA and model compounds that recently became available. These mostly include the reaction pathways associated with exposure of the guanine (Gua) base of DNA and related model systems to both the direct and indirect effects of ionizing radiation. Relevant information is also provided on both the type I and type II photosensitized reactions of the guanine moiety of DNA which are involved in the effects of the UV-A component of solar light [7]. As a second major topic to be addressed, the main available methods aimed at measuring individual damage within DNA and isolated cells are critically reviewed.

C. Baumstark-Khan et al. (eds.),
Fundamentals for the Assessment of Risks from Environmental Radiation, 91-102.

2. Effects of Solar Radiation

The effects of solar radiation on cellular DNA may be rationalized in terms of direct excitation process provided by UV-B radiation ($290 < \lambda < 320$ nm) and indirect photo-oxidation reactions induced by UV-A ($320 < \lambda < 400$ nm) and visible lights.

2.1. DIRECT EFFECTS OF UV-B RADIATION

The interaction of UV-B radiation with DNA leads to the excitation of both purine and pyrimidine bases. The resulting singlet excited bases or more likely, the related triplet excited intermediates, may react in different competitive ways, depending partly on the DNA sequence context, giving rise either to dimeric or monomeric lesions.

2.1.1. Dimeric Photoproducts

Adjacent pyrimidine bases are the main DNA targets of the direct effect of UV radiation. As consequences of UV-B excitation of thymine and cytosine, cyclobutadipyrimidines and, to a lesser extent, pyrimidine (6-4) pyrimidone photoproducts are generated in a strong sequence dependence manner [12, 13]. In addition, UV-B irradiation promotes the isomerization of the (6-4) photoproducts into the related Dewar valence isomer. It should be noted that quantitative information is still lacking on the formation of the four individual base lesions for each of the three classes of photodamage in both isolated and cellular DNA. Efforts should also be made to characterize the putative (6-4) photoproduct at dicytosine sequences [14] and its related Dewar valence isomer. This also concerns the (6-4) photoproducts of 5-methylcytosine whose formation in double stranded DNA was ruled out on the basis of indirect evidence. However, it has to be mentioned that several of the latter photoproducts have been characterized in far-UV irradiated solutions of appropriate dinucleoside monophosphates [15]. A minor class of dimeric photoproduct is represented by adducts involving adenine and thymine. The main TpA photoproduct was recently identified as arising from the [2 + 2] cycloaddition of the 5,6-pyrimidine bonds of both the adjacent thymine and adenine moieties, giving rise after expansion of the resulting cyclobutane ring into an eight-membered-ring valence isomer [16].

2.1.2. Purine and Pyrimidine Single Photolesions

6-Hydroxy-5,6-dihydrocytosine [17], the so-called "cytosine photohydrate" on one hand and oxidative guanine lesions [7] on the other hand are generated as minor UV-C and UV-B DNA photoproducts. The yields of their formation are about two orders of magnitude lower than those of cyclobutadipyrimidines. 8-Oxo-7,8-dihydroguanine (Fig. 1, **9**) (8-oxoGua) and an oxazolone derivative (Fig. 1, **7**) have been characterized as guanine oxidation products in isolated DNA [18] and 2'-deoxyguanosine [7] respectively upon UV-C and UV-B irradiation. The main mechanism of induction of the latter photoproducts is likely to involve the guanine radical cation (Fig. 1, **2**) as a common precursor. Ionization of the purine ring through photoexcitation of the guanine base is not a likely process. A better mechanistic alternative for the latter one-electron oxidation reaction is a charge transfer process between an excited guanine base and a second guanine residue in the ground state [19]. It should be added that oxidation of the guanine moieties may also occur through a singlet oxygen ($^1O_2$) oxidation (*vide infra*) mechanism (Fig. 2). However, the relative contribution of charge transfer (one-electron mechanism) and energy transfer ($^1O_2$ reaction) [20, 21] associated to UV-B radiation remains to be established particularly within DNA. It should be added that direct ionization of the guanine moieties of isolated DNA and model compounds with subsequent formation of 8-oxoGua (Fig. 1, **9**) and oxazolone (Fig. 1, **7**) was found to occur upon both far-UV nano- and picosecond laser irradiation [22, 23].

2.2. PHOTO-OXIDATIVE DAMAGE INDUCED BY UV-A RADIATION

The effects of UV-A radiation on cellular DNA are mostly accounted for by indirect oxidative processes mediated by endogenous photosensitizers. However, there is still a possibility to induce, however in very low yields, the formation of cyclobutadipyrimidines through direct excitation of pyrimidine bases. A photosensitizer reacts usually in a triplet excited state according to two main processes including type I (electron transfer or hydrogen abstraction involving the DNA substrate) and type II (energy transfer with oxygen) mechanisms [1]. The type I mechanism is almost specific of guanine which exhibits the lowest ionization potential among DNA components (guanine < adenine < cytosine ≈ thymine < 2-deoxyribose moiety). Interestingly, $^1O_2$, the reactive oxygen species generated in the type II mechanism reacts exclusively with guanine [1]. Altogether, these observations explain why guanine is considered to be the critical cellular DNA target of UV-A and visible radiation.

2.2.1. Type I Photosensitization Reactions of the Guanine Bases of DNA and Model Compounds

The guanine radical cation (Fig. 1, **2**) was found to exhibit two main competitive reactions, at least, in double stranded DNA. Deprotonation of **2** is likely to give rise to the highly oxidizing oxyl radical (Fig. 1, **3**) which may also exist as several tautomers. Addition of oxygen to the likely C5 centered related tautomer (Fig. 1, **4**) takes place at a low rate ($k < 10^6$ M^{-1} sec^{-1}), explaining that competitive reduction (chemical repair) of the latter intermediate by ferrous ions for example, may occur. Formation of the oxazolone (Fig. 1, **7**) as the final oxidation product of guanine (Fig. 1, **1**) is rationalized in terms of a complex sequences of reactions involving the peroxyl radical (Fig. 1, **5**), (decarboxylation, addition of a water molecule, loss of formylamine) giving rise to the imidazolone precursor (Fig. 1, **6** [24-26]). The latter unstable intermediate **6** was found to quantitatively hydrolyze into **7** in aqueous solutions (half life ≈ 8h at 20°C). A major competitive reaction of **2** (≈ 40%), at least, in double stranded DNA, is hydration [27]. This leads to the formation of the 8-hydroxy-7,8-dihydroguan-7-yl radical (Fig. 1, **8**) which is subsequently converted in aerated aqueous solutions to about 90% of 8-oxoGua (Fig. 1, **9**, oxidation process) and 10% of the formamidopyrimidine or FapyGua (Fig. 1, **10**, reduction process).

2.2.2. Type II Photosensitization Reactions of the Guanine Moiety (Singlet Oxygen)

Reaction of $^1O_2$ with the guanine moiety ($k \approx 5.3 \times 10^6$ M^{-1} sec^{-1} for 5'-dGMP) is likely to proceed through a Diels-Alder [4 + 2] cycloaddition of the dienophile $^1O_2$ involving the C4 and C8 carbons of the purine ring (Fig. 2). The resulting 4,8-purine endoperoxide (Fig. 2, **12** [28]) which was recently characterized by NMR analysis at

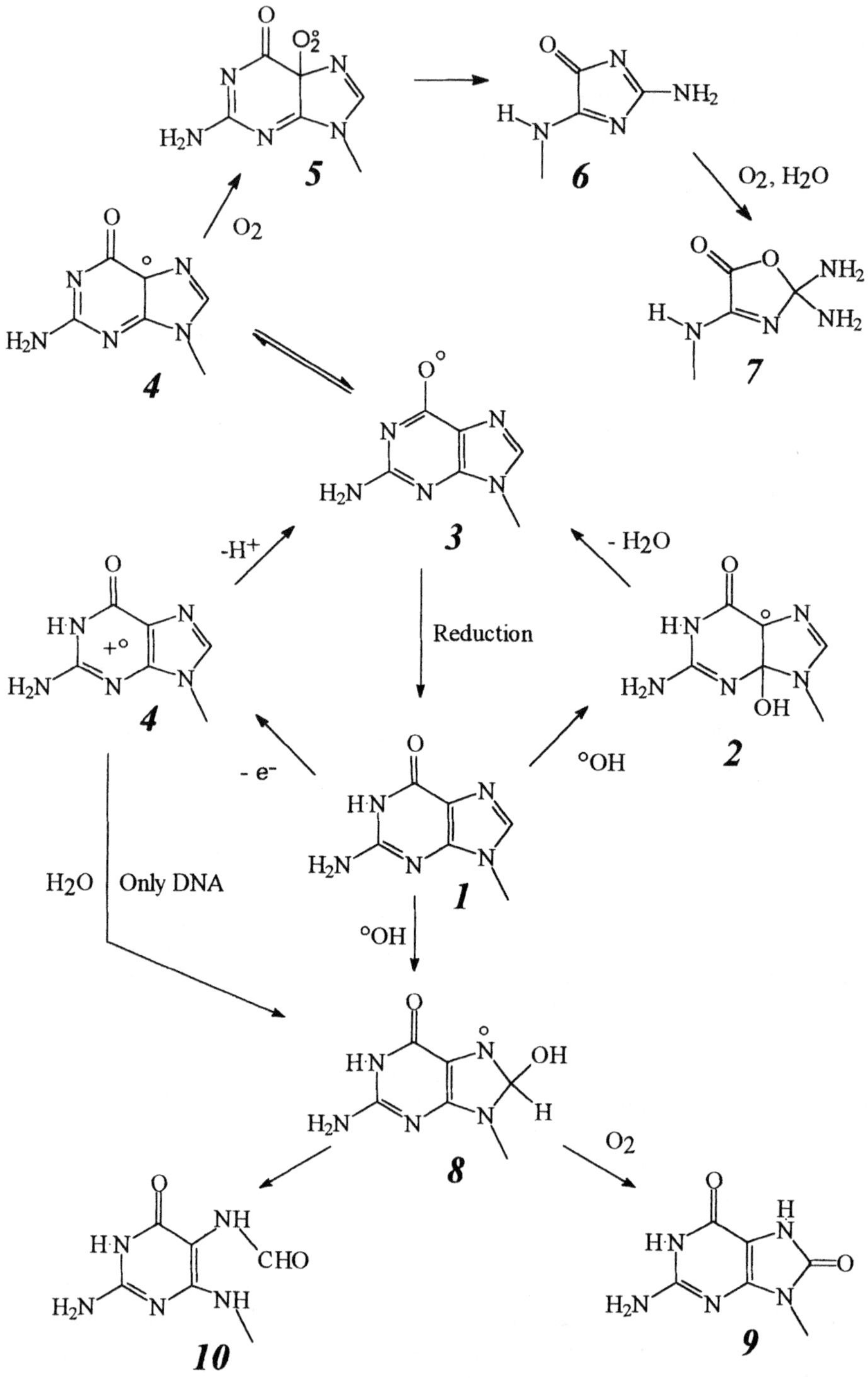

Figure 1. Main pathways of the ·OH and one-electron oxidation of the guanine moiety of DNA.

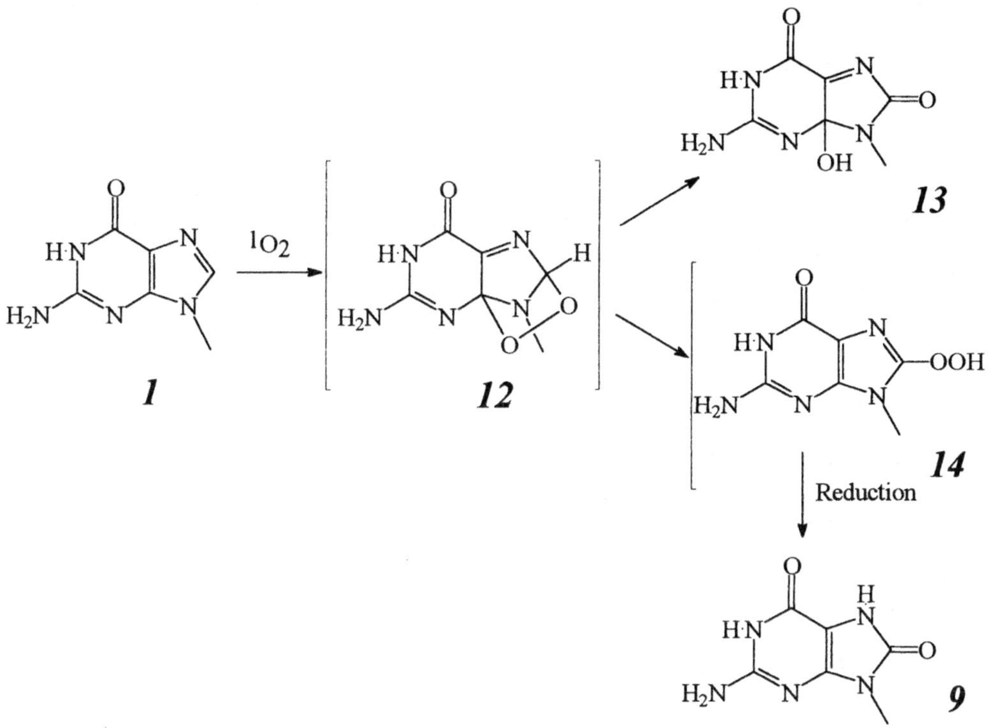

Figure 2. Singlet oxygen oxidation reactions of the guanine moiety of DNA and model compounds.

low temperature of a photosensitized solution of a *tertio*butyldimethylsilyl derivative of 8-methylguanosine is rather unstable. The decomposition of **12** derived from 2'- deoxyguanosine and short single stranded oligonucleotides was found to lead to the predominant formation of the 4*R* and 4*S* diastereomers of keto-alcohol (Fig. 2, **13** , ≈ 85% yield) with a low amount of 8-oxoGua (Fig. 2, **9** ≈ 15% yield) [29-32]. The situation is completely different within double stranded DNA since the formation of **9** is enhanced at the expense of **13** whose presence was even not detected [33, 34]. Under the latter conditions, the endoperoxide (Fig. 2, **12**) is likely to isomerizc into 8-hydroperoxyguanine (Fig. 2, **14**) [35] prior to undergo decomposition into **9**.

2.2.3. *UV-A- and Visible Light-mediated Photooxidation Reactions of Cellular DNA and RNA.*

The UV-A-induced formation of 8-oxoGua (Fig. 2, 9) has been measured in cellular DNA [36-40] and RNA [38] using the accurate high performance liquid chromatographic-electrochemical (HPLC/EC) assay. Further information on UV-A- and visible light-mediated photooxidation base damage in the DNA of various mammalian cell lines was obtained in an indirect way by applying the highly sensitive alkaline elution assay in combination with specific DNA repair enzymes [41, 42]. One of the most striking result deals with the predominance of formamidopyrimidine-DNA glycosylase (Fpg) sensitive sites over cleavages provided by other enzymes which reveal the presence of cyclobutadipyrimidines, oxidized pyrimidine bases and abasic sites [43, 44]. These data clearly show that 8-oxoGua which is the usual substrate of Fpg protein [45, 46] is the main photooxidative lesion induced in cellular DNA by UV-A and visible light. It is likely that type II photosensitization mechanism ($^1O_2$) is mostly involved in the formation of **9**. However, further work is required to determine more precisely that could be the contribution of ·OH radical, generated by Fenton type reactions from initially produced superoxide radicals, in the overall photooxidation of cellular DNA. Another major point of concern is the still unknown nature of the endogenous photosensitizers involved in the UV-A and visible light-induced oxidation reactions.

3. Radiation-induced Base Damage to DNA

An almost complete description of the radical reactions mediated by both ·OH radicals (indirect effects) and one-electron oxidation (direct effects) is now possible for guanine compounds in aerated aqueous solution. This was inferred from both the earliest investigations on the transient radicals involved in these reactions [for a review, see 7] and the recent characterization of the related final decomposition products [for a review, see 6].

3.1. INDIRECT EFFECTS OF IONIZING RADIATION ON GUANINE COMPONENTS.

The indirect effects of ionizing radiation on DNA in aerated aqueous solutions are almost exclusively mediated by ·OH radicals as the result of radiolysis of H_2O molecules. Interestingly the two main addition reactions of ·OH to the guanine moiety (Fig. 1, **1**) were found to give rise to the two predominant oxidation products [6], namely the oxazolone (Fig. 1, 7) and 8-oxoGua (Fig. 1, 9) already characterized as the degradation products of the radical cation (direct effect). This may be rationalized in terms of initial ·OH addition at C4 and C8 respectively. Dehydration of the C4 adduct (Fig. 2, **11**) is likely to generate the oxyl radical (Fig. 1, **3**) which subsequently gives rise to **7** through the intermediary of the imidazolone (Fig. 1, **6**). On the other hand, further reaction of the reducing radical 8 leads to the predominant formation of 8-oxoGua (Fig. 1, **9**) together with a small amount of FapyGua (Fig. 1, **10**).

3.2. DIRECT EFFECTS OF IONIZING RADIATION ON GUANINE COMPOUNDS

Interaction of ionizing radiation with the guanine base (Fig. 1, **1**) leads to the formation of **3**. In addition, the guanine moiety (Fig. 1, **1**) within native double stranded DNA is considered as the sink of hole migration when other purine and pyrimidine bases are initially ionized, at least, over a distance of a few base pairs [48]. On the other hand, the pyrimidine bases are the final targets of electron addition reactions. The two main final degradation products of the guanine moiety (Fig. 1, **1**) of DNA in aerated aqueous solutions are the oxazolone (Fig. 1, **7**) and 8-oxoGua (Fig. 1, **8**) through competitive deprotonation and hydration of precursor **2**, as previously described for the type I mechanism (Fig. 1).

3.3. TANDEM LESIONS

The most biologically harmful radiation-induced DNA lesions are likely those which are generated by multiple ionization and excitation events, the so-called multiple damage local sites (MDLS) [49]. Double-strand breaks which, however, constitute an heterogeneous class of radiation damage represent well-established examples of these complex types of DNA lesions. Obviously two or several base modifications and DNA strand breaks are expected to be generated in a close spatial environment along the radiation track. Efforts are currently made to gain insight into the chemical nature of the latter lesions. In addition, work is in progress in our laboratory to insert complex base damage at defined sites in either the same or the opposite DNA strand by varying the distances between the lesions. This should facilitate the evaluation of the biological impact (mutagenesis, DNA repair) of such complex radiation DNA modifications.

Several tandem lesions which arises from one radical event has already been characterized. Cyclopurine nucleosides, whose formation involves initial hydrogen abstraction at the C5' of the sugar moiety, exhibit a modification in both the sugar and the base moieties [59]. Examples of tandem base damage are illustrated by the isolation and characterization of the formylamine/8-oxoGua on one hand [51] and the 5-(hydroxymethyl)uracil/oxidized guanine double lesions on the other hand [52]. In both cases, initial radical oxidation of the pyrimidine moiety is the likely initiating event of the clustered lesions.

4. Measurement of Photo- and Radiation-induced DNA Base Lesions

Several methods including the rather unspecific but highly sensitive comet assay [53] were developed during the last decade for monitoring the formation of radiation-induced and photosensitized base damage within cellular DNA.

4.1. SPECIFIC METHODS FOR SINGLING OUT BASE DAMAGE

Various methods which involve either HPLC or GC as the separative analytical methods together with a sensitive technique of detection have been designed to monitor the formation of radiation-induced base damage within cellular DNA. It has to be reminded that the assays should be able to detect, at least, one modification per 10^5 normal bases in a sample size of less than 30 μg. The most popular and accurate method which has received several hundred applications is represented by the HPLC/EC assay. The method which was initially aimed at monitoring the formation of 8-oxoGua (Fig. 2, **9**) is able to measure other electroactive radiation-induced base damage including 8-oxo-7,8-dihydroadenine, 5-hydroxycytosine, 5-hydroxuracil and Fapy derivatives [for comprehensive reviews, see 8, 54].

4.1.1. Artificats Associated with GC-MS, [^{32}P]-Postlabeling and Immunological Assays. The GC-MS assay, operating in the electron impact ionization mode, is a versatile method for measuring a wide spectrum of radiation-induced base lesions. The sensitivity reaches the requested level of detection defined above when the measurement is achieved in the selective ion monitoring mode. However, it was recently shown that application of the earliest version of the GC-MS assay [9, 10] led to a 10- to 40-fold overestimation of 8-oxoGua (Fig. 2, **9**) and other oxidized purine and pyrimidine bases. This was found to be due to a partial oxidation (within the 10^{-4}-10^{-5} range efficiency) of the overwhelming normal bases during the silylation step aimed at making the targeted lesion volatile for the GC separation [55, 56].

A [^{32}P]-postlabeling assay [57] and an immunological method [58], both designed for monitoring the formation of 8-oxoGua (Fig. 2, **9**) in cellular DNA have recently become available. Unfortunately, major drawbacks which lead to the significant artifactual formation of 8-oxoGua make both methods not applicable. Autoxidation and self-radiolysis processes of 8-oxo-7,8-dihydro-5'-dGMP are at the origin of the about 40-fold increase in the level of 8-oxoGua (Fig. 2, **9**) observed in the former assay [54]. This could be prevented by pre-purifying 8-oxo-7,8-dihydro-3'-dGMP prior to the polynucleotide kinase phosphorylation step following the protocol proposed for the HPLC/[^{32}P]-postlabeling measurement of adenine N^{1} oxide [59]. The lack of specificity of the antibody directed against 8-oxoGua is likely to explain the high level of the latter lesion (1 oxoGua residue per 10 guanine bases) in the 0.1 mM H_2O_2 treated samples of cellular DNA.

4.1.2. Improved GC-MS Assay

Interestingly the artifactual formation of oxidized base lesions (*vide supra*) can be prevented in an improved version of the GC-MS assay by pre-purifying the targeted damage to be measured prior to the derivatization step [55, 56, 60]. This could be efficiently achieved by either HPLC or immunoaffinity chromatography prior to GC-MS analysis.

As a more general objective, efforts are currently made to minimize the formation of the artifactual oxidation of the overwhelming normal purine and pyrimidine components during extraction and work-up of DNA on one hand [54] and to optimize conditions of hydrolysis on the other hand [61].

4.2. DERIVED COMET ASSAYS.

4.2.1. DNA-Glycosylases Associated with the Single Cell Electrophoresis (Comet) Assay

The use of DNA-repair glycosylases including the *E. coli* Fpg and Endonuclease III proteins to reveal 8-oxoGua (Fig. 2, **9**) and a few oxidized pyrimidine bases as additional DNA strand breaks, constitutes an interesting approach to make the comet assay more specific [62]. Interesting application of this derived comet assay is illustrated by the measurement of Fpg-sensitive sites in isolated mammalian cells after exposure to doses as low at 2 Gy of gamma rays.

4.2.2. Comet Assay Combined with a Fluorescence Immunodetection

A second possibility to confer specificity to the comet assay is to use specific antibody with an appropriate fluorescent dye to monitor the formation of defined base damage in isolated cells [63]. This recently received an interesting application with the measurement of cyclobutadipyrimidines using the highly specific monoclonal antibody raised by Roza *et al.* [64]. Extension of the assay to both photo-and radiation DNA damage is expected as soon as specific antibodies are available.

5. Conclusion

Basic information is still required on both qualitative and quantitative aspects of the photo- and radiation chemistry of nucleic acids. However, major efforts should be devoted to the search and measurement on the main lesions induced in cellular DNA upon exposure to solar and ionizing radiations. For this purpose, accurate and sensitive methods should be still developed. In this respect, further progress is expected by associating performant analytical methods including HPLC and capillary electrophoresis with tandem mass spectrometric techniques (atmospheric pressure ionization, electrospray).

6. References

1. Cadet and Vigny, P. (1990). The photochemistry of nucleic acids. In: Bioorganic Photochemistry, H. Morrison, ed. (New York: Wiley and Sons) pp. 1-272.
2. Cadet, J., Anselmino, C., Douki, T., and Voituriez, L. (1992). Photochemistry of nucleic acids in cells. *J. Photochem. Photobiol. B* **15,** 277-292.
3. Cadet, J. (1994). DNA damage caused by oxidation, deamination, ultraviolet radiation and photoexcited psoralen. In: DNA adducts: Identification and biological significance, K. Hemminki, A. Dipple, D.G.E. Shuker, F.F. Kadlubar, D. Segerbäck and H. Bartsch., eds. (Lyon: IARC Scientific Publications) No. 125, pp. 245-276.
4. Taylor, J.-S. (1994). Unraveling the molecular pathway from sunlight to skin cancer. *Acc. Chem. Res.* **27**, 76-82.
5. Taylor, J.-S. (1995). DNA, sunlight and skin cancer. *Pure Appl. Chem.* **67,** 183-190.
6. Cadet, J., Berger, M., Douki, T., and Ravanat, J.-L. (1997) Oxidative damage to DNA: Formation, measurement and biological significance. *Reviews Physiol. Biochem. Pharmacol.,* **131**, 1-87.
7. Cadet, J., Berger, M., Douki, T., Morin, B., Raoul, S., Ravanat, J.-L., and Spinelli, S. (1997) Effects of UV and visible radiation on DNA - Final base damage. *Biol. Chem.,* **378,** 1275-1286.
8. Cadet J., and Weinfeld, M. (1993). Detecting DNA damage. *Anal. Chem.,* **65,** 675A-682A.
9. Dizdaroglu, M. (1991) Chemical determination of free radical-induced damage to DNA. *Free Radical Biol. Med.*, **10**, 225-242.
10. Halliwell, B., and Dizdaroglu, M. (1992). The measurement of oxidative damage to DNA by HPLC and GC/MS techniques. *Free Radic. Res. Commun.*, **16**, 15-28.
11. Dizdaroglu, M. (1993). Quantitative determination of oxidative base damage in DNA by stable isotope-dilution mass spectrometry. *FEBS Lett.*, **315,** 1-6.
12. Douki, T., Zalizniak, T., and Cadet, J. (1997). Far-UV induced dimeric photoproducts in short oligonucleotides: sequence effects. *Photochem. Photobiol.*, **66**, 171-179.
13. Sage E. (1993). Distribution and repair of photolesions in DNA: genetic consequences and the role of sequence context. *Photochem. Photobiol.*, **57**, 163-174.
14. Douki, T., Voituriez, L. and Cadet, J. (1995). Measurement of pyrimidine(6-4)pyrimidone photoproducts in DNA by a mild hydrolysis-HPLC-fluorescence assay. *Chem. Res. Toxicol.* **8**, 244-253.
15. Douki, T., and Cadet, J. (1994). Formation of cyclobutane dimers and (6-4) photoproducts upon far-UV photolysis of 5-methylcytosine containing dinucleoside monophosphates. *Biochemistry* **33**, 11942-11950.
16. Zhao, X., Kao, J.L.-F., and Taylor, J.-S. (1995). Preparation and characterization of a deoxyoligonucleotide 19-mer containing a site-specific thymidylyl-(3',5')-deoxyadenosine photoproduct. *Biochemistry* **34**, 1386-1392.

17. Zhang, X., Rosenstein, B.S., Wang, Y., Lebwohl, M., Mitchell, D.M., and Wei, H. (1997). Induction of 8-oxo-7,8-dihydro-2'-deoxyguanosine by ultraviolet radiation in calf thymus DNA and HeLa cells. *Photochem. Photobiol.* **65**, 119-124.
18. Wei, H., Cai, Q., and Rahn, R.O. (1996). Inhibition of UV light- and Fenton reaction-induced oxidative DNA damage by the soybean isoflavone genistein. *Carcinogenesis* **12**, 73-77.
19. Wood, P.D., and Redmond, R.W. (1996). Triplet state interactions between nucleic acid bases in solution at room temperature: intermolecular energy and electron transfer. *J. Am. Chem. Soc.* **118,** 4256-4263.
20. Mohammad, T., and Morrison, H. (1996). Evidence for the photosensitized formation of singlet oxygen by UV-B irradiation of 2'-deoxyguanosine 5'-monophosphate. *J. Am. Chem. Soc.* **118,** 1221-1222.
21. Bishop, S.M., Malone, M., Phillips, D., Parker, A.W., and Symons, M.C.R. (1994). Singlet oxygen sensitisation by excited state DNA. *J. Chem. Soc. Chem. Commun.* 871-872.
22. Spassky, A., and Angelov, D. (1997). Influence of the local helical conformation on the guanine modifications generated from one-electron oxidation. *Biochemistry*, **36,** 6571-6576
23. Angelov, D., Spassky, A., Berger M., and Cadet J. (1997). High-intensity UV laser photolysis of DNA and purine 2'-deoxyribonucleosides: formation of 8-oxopurine damage and oligonucleotide strand cleavage as revealed by HPLC and gel electrophoresis studies. *J. Am. Chem. Soc.*, **119,** 11373-11380
24. Cadet, J., Berger, M., Buchko, G.W., Joshi, P.C., Raoul, S., and Ravanat, J.-L. (1994). 2,2-Diamino-4-[(3,5-di-O-acetyl-2-deoxy-β-D-*erythro*-pentofuranosyl) amino]-5-(2*H*)-oxazolone: a novel and predominant radical oxidation product of 3',5'-di-O-acetyl-2'-deoxyguanosine. *J. Am. Chem. Soc.* **116,** 7403-7404.
25. Buchko, G.W., Cadet, J., Morin, B., and Weinfeld, M. (1995). Photooxidation of d(TpG) by riboflavin and methylene blue. Isolation and characterization of thymidylyl-(3',5')-2-amino-5-[(2-deoxy-β-D-*erythro*-pentofuranosyl)amino]-4*H*-imidazol-4-one, and its primary decomposition product, thymidylyl-(3',5')-2-diamino-4-[(2-deoxy-β-D-*erythro*-pentofuranosyl)amino]-(2*H*)-oxazolone. *Nucleic Acids Res.*, **19**, 3954-3961.
26. Raoul, S. Berger, M., Buchko, G.W., Joshi, P.C., Morin, B., Weinfeld, M., and Cadet, J. (1996). ^{1}H, ^{13}C and ^{15}N NMR analysis and chemical features of the two main radical oxidation products of 2'-deoxyguanosine: Oxazolone and imidazolone nucleosides. *J. Chem. Soc. Perkin Trans.* **2**, 371-381.
27. Kasai., H., Yamaizumi, Z., Berger, M., and Cadet, J. (1992). Photosensitized formation of 7,8-dihydro-8-oxo-2'-deoxyguanosine (8-hydroxy-2'-deoxyguanosine) in DNA by riboflavin. A non singlet oxygen mediated reaction. *J. Am. Chem. Soc.* **114,** 9692-9694.
28. Sheu, C., and Foote, C.S. (1993). Endoperoxide formation in a guanosine derivative. *J. Am. Chem. Soc.* **115,** 10446-10447.
29. Ravanat, J.-L., Berger, M., Benard, F., Langlois, R., Ouellet, R., van Lier, J.E., and Cadet, J. (1992). Phthalocyanine and naphthalocyanine photosensitized oxidation of 2'-deoxyguanosine: distinct type I and type II products. *Photochem. Photobiol.*, **55**, 809-814.
30. Buchko, G.W., Cadet, J., Berger, M., Ravanat, J.L. (1992) Photooxidation of d(TpG) by phthalocyanines and riboflavin. Isolation and characterization of dinucleoside monophosphates containing the 4R* and the 4S* diastereoisomers of 4,8-dihydro-4-hydroxy-8-oxo-2'-deoxyguanosine. *Nucleic Acids Res.*, **20**, 4847-4851
31. Ravanat, J.-L. and Cadet, J. (1995). Reaction of singlet oxygen with 2'-deoxyguanosine and DNA. Isolation and characterization of the main oxidation products. *Chem. Res. Toxicol.* **8**, 379-388.
32. Raoul S. and Cadet, J. (1996). Photosensitized reaction of 8-oxo-7,8-dihydro-2'-deoxyguanosine: Identification of 1-(2-deoxy-β-D-*erythro*-pentofuranosyl)-cyanuric acid as the major singlet oxygen oxidation product. *J. Am. Chem. Soc.* **118,** 1892-1898.
33. Devasagayam, T.P.A., Steenken, S., Obendorf, M.S.W., Schulz, W.A., and Sies, H. (1991). Formation of 8-hydroxy(deoxy)guanosine and generation of strand breaks at guanine residues in DNA by singlet oxygen. *Biochemistry* **30**, 6283-6289

34. Cadet, J., Ravanat, J.-L., Buchko G,W., Yeo, H.C., and Ames, B.N. (1994). Singlet oxygen DNA damage: Chromatographic and mass spectrometric analysis of damage products. *Methods Enzym.*, **234,** 79-88.
35. Sheu, C., and Foote, C.S. (1995) Reactivity toward singlet oxygen of a 7,8-dihydro-8-oxoguanosine ("8-hydroxyguanosine") formed by photooxidation of a guanosine derivative. *J. Am. Chem. Soc.*, **117,** 6439-6442.
36. Fischer-Nielsen, A., Loft, S., and Jensen, K.G. (1993). Effect of ascorbate and 5-aminosalicylic acid on light-induced deoxyguanosine formation in V79 Chinese hamster cells. *Carcinogenesis,* **14,** 2431-2433.
37. Rosen, J.E., Prahalad, A.K., and Williams, G.M. (1996). 8-Oxodeoxyguanosine formation in the DNA of cultured cells after exposure to H_2O_2 alone or with UV-B or UV-A irradiation. *Photochem. Photobiol.* **64,** 117-122.
38. Wamer, W.G., and Wei, R.R. (1997). *In vitro* photooxidation of nucleic acids by ultraviolet A radiation. *Photochem. Photobiol.* **65**, 560-563.
39. Zhang, X., Rosenstein, B.S., Wang, Y., Lebwohl, M., Mitchell, D.M., and Wei, H. (1997). Induction of 8-oxo-7,8-dihydro-2'-deoxyguanosine by ultraviolet radiation in calf thymus DNA and HeLa cells. *Photochem. Photobiol.*, **65**, 119-124.
40. Kvam E., and Tyrrell, R.M. (1997) Artificial background and induced levels of oxidative base damage in DNA from human cells. *Carcinogenesis,* **18,** 2281-2283.
41. Epe, B (1995). DNA damage profiles induced by oxidizing agents. *Reviews Physiol. Biochem. Pharmacol.* **127**, 223-233.
42. Epe, B., and Hegler, J. (1994). Oxidative DNA damage: endonuclease fingerprint. *Methods Enzymol.* **234,** 122-131.
43. Kielbassa, C., Roza, L., and Epe, B. (1997). Wavelength dependence of oxidative DNA damage induced by UV and visible light. *Carcinogenesis* **18,** 811-816.
44. Pflaum, M., Boiteux, S. and Epe, B. (1994). Visible light generates oxidative DNA base modifications in high excess of strand breaks in mammalian cells. *Carcinogenesis* **15,** 297-300.
45. Boiteux S. (1993). Properties and biological functions of the NTH and FPG proteins of *Escherichia coli*: two glycosylases that repair oxidative damage in DNA. *J Photochem Photobiol B: Biol*: **19**, 87-96.
46. Tchou, J., Bodepudi, V., Shibutani, S., Antoshechkin, I., Miller, J., Grollman, A.P., and Johnson, F. (1994). Substrate specificity of Fpg protein. *J Biol. Chem.*, **269,** 15318-15324.
47. Steenken, S. (1989). Purine bases, nucleosides, and nucleotides: aqueous solution redox chemistry and transformation reactions of their radical cations and e^- and OH adducts. *Chem. Rev.* **89,** 503-520.
48. Hall , D.B., Holmlin, R.E., and Barton, J.K. (1996). Oxidative DNA damage through long-range electron transfer. *Nature,* **382**, 731-735.
49. Ward, J.-F. (1994). The complexity of DNA damage: relevance to biological consequences. *Int. J. Radiat. Biol.*, **66,** 427-432.
50. Cadet, J., and Berger, M. (1985). Radiation-induced decomposition of purine bases within DNA and related model compounds. *Int. J. Radiat. Biol.*, **74,** 127-143.
51. Box, H.C., Budzinski, E.D., Dawidzik, J.B., Gobey, J.S., and Freund, H.G. (1997). Free radical-induced tandem base damage in DNA oligomers. *Free Radical Biol. Biol.*, **23,** 1021-1030.
52. Delatour, T., Douki, T., Gasparutto, D., and Cadet, J. (1997). Isolation and charactrization of a clustered lesion obtained by photosensitization of TpdG. *Photochem. Photobiol.*, **65s,** 22-23.
53. Collins, A.R., Dusinska, M., Gedik, C.M., and Stetina, R. (1996). Oxidative damage to DNA: do we have a reliable biomarker? *Environ. Health Perspect.* **104,** 465-469.
54. Cadet, J., Douki , T., and Ravanat, J.-L. (1997). Artifacts associated with the measurement of oxidized DNA bases. Environ. *Health Perspect.*, **105**, 1034-1039.
55. Ravanat, J.-L, Turesky, R.J., Gremaud, E., Trudel, L.J., and Stadler, R.H. (1995). Determination of 8-oxoguanine in DNA by gas chromatography-mass spectrometry and HPLC-electrochemical detection. Overestimation of the background level of the oxidized base by the gas chromatography mass spectrometry assay. *Chem. Res. Toxicol.* **8**, 1039-1045.

56. Douki, T., Delatour, T., Bianchini, F., and Cadet, J. (1996a). Observation and prevention of an artefactual formation of oxidized DNA bases and nucleosides in the GC/EIMS method. *Carcinogenesis*. **17**, 347-353.
57. Devanaboyina, U.-s., and Gupta, R.C. (1996). Sensitive detection of 8-hydroxy-2'-deoxyguanosine in DNA by ^{32}P-postlabeling assay and the basal levels in rat tissues. *Carcinogenesis,* **17**, 917-924.
58. Musarrat, J., and Wani, A.A. (1994) Quantitative immunolo-analysis of promutagenic 8-hydroxy-2'-deoxyguanosine in oxidized DNA. *Carcinogenesis,* **15**, 2037-2043.
59. Mouret, J.-F., Odin, F., Polverelli, M., and Cadet, J. (1990). ^{32}P-Postlabeling measurement of adenine N-1-oxide in cellular DNA exposed to hydrogen peroxide. *Chem. Res. Toxicol.*, **3**, 102-110
60. Douki, T., Delatour, T., Paganon, F., and Cadet, J. (1996b). Measurement of oxidative damage at pyrimidine bases in γ-irradiated DNA. *Chem. Res. Toxicol.* **9**, 1145-1151.
61. Douki, T., Martini, R., Ravanat, J.-L., Turesky, R.J., and Cadet, J. (1997). Measurement of 2,6-diamino-4-hydroxy-5-formamidopyrimidine and 8-oxo-7,8-dihydroguanine in gamma irradiated DNA. *Carcinogenesis* **18**, 2385-2391.
62. Dusinska, M., and Collins, A. (1996). Detection of oxidised purines and UV-induced photoproducts in DNA of single cells by inclusion of lesion-specific enzymes in the comet assay. *ALTA, 24*, 405-411.
63. Sauvaigo, S., Signorini, N., Emonet, N., Richard, M.-J., and Cadet, J. (1997). Immunofluorescent detection of cyclobutane pyrimidine dimers by a modification of the comet assay. *Mutat. Res.*, **379s**, 134.
64. Roza, L., van der Wulp, K.J.M., MacFarlane, S.J., Lohman, P.H.M., and Baan, R.A. (1988). Detection of cyclobutane thymine dimers in DNA of human cells with monoclonal antibodies raised against a thymine dimer-containing tetranucleotide. *Photochem. Photobiol.*, **48**, 627-633.

DNA REPAIR.

Basic Functions for Cellular Survival and Recovery

C. BAUMSTARK-KHAN

Deutsches Zentrum für Luft- und Raumfahrt, Institut für Luft- und Raumfahrtmedizin, Abteilung Strahlenbiologie
D-51170 Köln, Germany

Abstract

DNA lesions are capable to change the ways how cells normally function; resulting consequences are cell death, mutation, misdifferentiation and unlimited proliferation or metastatic potential. Hence, nature has evolved mechanisms to minimize damage: DNA repair deals with the mechanisms invoved in avoidance (error free) and/or fixation of genetic mutations (error-prone). The potentially pathogenic consequence of genotoxic exposure includes somatically acquired alterations sometimes provoked by inherited deficiencies. According to various DNA damages, which were induced up to a certain extent by radiation of distinct quality, multiple repair mechanisms have been evolved, but all mechanisms involve the processes of lesion recognition followed by withdrawal or modification of the damaged site.

1. Introduction

Non-damaged DNA is the prerequisite for cellular life as DNA is:

- the repository of hereditary information
- the blueprint for operation of individual cells.

Nearly all DNA damage, we have seen so far, is harmful for cells. Therefore, it is essential to reduce this damage to a tolerable level. The importance of repair can be seen from the facts that

- DNA is the only biomolecule that is specifically *repaired*. All others are *replaced*.
- More than 100 genes participate in various aspects of DNA repair, even in organisms with very small genomes

Although all cells possess a large number of different types of repair system, each relatively specific for a certain type of DNA damage, those repair systems can be grouped into 3 broad categories [1]:

- Direct reversal of damage
- Excision of damaged region, followed by replacement

C. Baumstark-Khan et al. (eds.),
Fundamentals for the Assessment of Risks from Environmental Radiation, 103-114.

- Damage tolerance: attempts to minimize the effects of damage that cannot be repaired.

2. Direct Reversal of Damage

Although direct reversal of damage would be the simplest way to correct DNA damage, the reverse reaction is only possible in few cases. The best studied repair reaction of this type is photoreversal of cyclobutane pyrimidine dimers by the enzyme photolyase [1]. Photolyase contains two chromophores responsible for absorbing light energy (Fig. 1). In all photolyases, one of the chromophores is $FADH^-$, and the other is either methenyltetrahydrofolate (MTHF) or 8-hydroxy-5-deazaflavin (8-HDF). MTHF and 8-HDF act as primary light gatherers, transferring their energy to $FADH^-$. The energy from $FADH^-$ is then used to split the dimer. Photolyases are found in bacteria, fungi, plants and animals, but not in all species. *Saccharomyces cerevisiae* has a photolyase, but *Schizosaccaromyces pombe* does not. Photolyase activity has been detected in goldfish, but not in mammals.

Another example of direct damage reversal is repair of O^6-methyl guanine by transfer of the alkyl group from the DNA to a cysteine in a protein, an O^6-methylguanine-DNA methyltransferase. This enzyme appears to be present in all living organisms.

A final example of direct damage reversal is the sealing of a subset of nicks in DNA by DNA ligase. Of course, DNA ligase can only seal nicks having 5'-phosphates and 3'-hydroxyls. Nicks with other configurations, or nicks accompanied by additional backbone or base damage, require more complicated repair mechanisms [1].

3. Mismatch Repair

3.1. CORRECTION OF REPLICATION ERRORS

The mechanism of mismatch repair [2] has been studied most thoroughly in *E. coli,* where the sequence of events in the repair process has been re-constituted the from purified proteins.

The mismatch arising from a replication mistake leading to a distorted double helix is recognized by MutS protein which binds to such mismatches (true mismatches plus insertions/deletions of up to 4 nucleotides). Additional binding of MutL stabilizes the complex. The MutS-MutL complex activates MutH, which locates a nearby methyl group and nicks the newly synthesized strand opposite the methyl group. The fact that the newly synthesized strand strand is not immediately methylated at GATC sequences in bacterial DNA, while the old strand carries methyl groups in proximity to the replication fork allows *E. coli* cells to distinguish the presumably correct (old strand) from the presumably incorrect (newly-synthesized) strand. This use of methylation to distinguish the parental strand is probably peculiar to *E. coli*. Data from yeast and mammalian *in vitro* mismatch repair experiments suggest that single-strand nicks, which are present in nascent DNA strands, between Okazaki fragments, provide a signal for strand-specificity in these organisms.

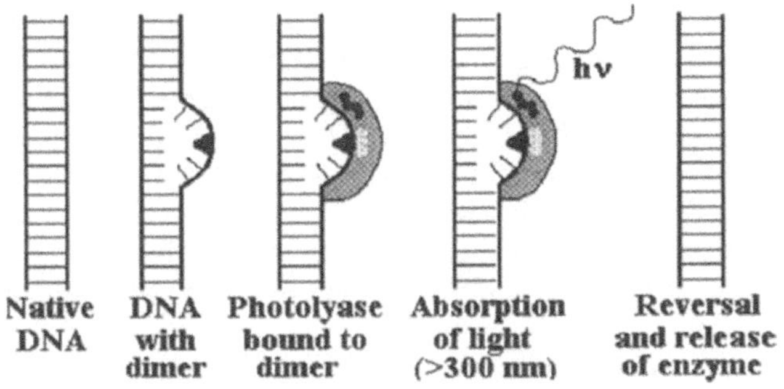

Figure 1: Reversal of UV induced DNA damage by bacterial photolyase

Excision is accomplished by cooperation between the UvrD (Helicase II) protein, which unwinds from the nick in the direction of the mismatch, and a single-strand specific exonuclease of appropriate polarity (one of several in *E. coli*), followed by resynthesis (Polymerase III) and ligation (DNA ligase).

Although lacking homologs of MutH and uvrD, eukaryotic organisms possess numerous homologs of MutS and MutL (Table 1). The eukaryotic genes appear to be homologs of the corresponding *E. coli* genes not only in terms of amino acid sequence but also and in terms of functional similarities. While the bacterial proteins are active in

TABLE 1: Genes encoding enzymes of mismatch repair [1, 2]

E. coli	*S. cerevisiae*	*Function in S. cerevisiae (human homologs have similar functions)*	*Human*
MutS	MSH1	Found in mitochondria	?
	MSH2	loop mismatch repair (with MSH3) single mismatch repair (with MSH6)	MSH2
	MSH3	loop mismatch repair (with MSH2)	?
	MSH4	Meiosis	?
	MSH5	Meiosis	?
	MSH6	single mismatch repair (with MSH2)	GTBP
MutL	MLH1	Mismatch repair (with PMS1)	MLH1
	PMS1	Mismatch repair (with MLH1)	PMS2
	?	?	PMS1
MutH	?	?	?
uvrD	?	?	?

the monomeric form, the eukaryotic proteins form homo- or hetero-dimers. The dimers of the MutS homologs, such as the dimer of MSH2 and MSH6/GTBP, are obviously responsible for initial recognition of mismatches, and dimers of the MutL homologs of the MutS homologs, such as the dimer of MSH2 and MSH6/GTBP, are obviously responsible for initial recognition of mismatches, and dimers of the MutL homologs (MLH1 and PMS1/PMS2) interact with the resultant complex. In eukaryotes some mismatch repair proteins also assist in detecting mismatches that may occur during recombination, for example, *S. cerevisiae* MSH4 and MSH5 probably play roles during meiotic recombination.

The hereditary non-polyposis colon cancer (HNPCC) form is characterized by early age of onset and autosomal dominant inheritance with high penetrance. It is always associated with defects in DNA repair evidenced by "microsatellite instability" (variation in the number of repeat units of short tandemly repeated sequences) and it is frequently associated with defects in the genes encoding MSH2 and MLH1. Since the development of cancer is a result of mutations in multiple genes, it is not surprising that defects in a DNA repair pathway should increase cancer frequency. However, it is surprising that such defects should selectively increase one kind of cancer (colon cancer) and that mutations in some of the mismatch repair genes (MSH2 and MLH1) should appear to be more important for development of cancer than mutations in other mismatch repair genes (GTBP, PMS1 and PMS2).

4. Base Excision Repair (BER)

The pathway most commonly employed to remove incorrect bases (like uracil) or alkylated bases (like 3-methyladenine) or other damaged bases is called "base excision repair" (BER). Actually, there are numerous variations of BER, each specific for a different type of incorrect base. Nevertheless, all of the variants have features in common, which can be considered to consist of 3 steps [3, 4]:

1. Removal of the incorrect base by an appropriate DNA N-glycosylase creates an AP site.
2. Nicking of the damaged DNA strand by AP endonuclease upstream of the AP site creates a 3'-OH terminus adjacent to the AP site.
3. Extension of the 3'-OH terminus by a DNA polymerase is accompanied by excision of the AP site.

DNA N-glycoslyases hydrolyze the N-glycosylic bond between the damaged or incorrect base and the deoxyribose, thus creating a AP site identical to that created by spontaneous DNA depurination or depyrimidination (Fig. 2A). A large number of DNA N-glycosylases has been identified. Most of them are small (<30 kd), monomeric enzymes with no cofactor or divalent cation requirement. They are highly specific for a certain type of altered base, only some have more relaxed specificity. The DNA N-glycosylases characterized to date include enzymes that recognize uracil, hydroxymethyl uracil, 5-methyl cytosine, hypoxanthine, thymine in G-T mispairs, adenine in G-A mispairs, 3-methyl adenine, formamidopyrimidine (result of purine oxidation), and thymine glycol. Most enzymes have been characterized in *E. coli*, but similar activities have been detected in extracts of eukaryotic cells. Cloned uracil DNA glycosylases from various organisms are very similar; human and *E. coli* genes show about

50% homologies. The crystal structure of the human enzymes suggest that, the uracil must flip out of the DNA double helix into a deep pocket on the enzyme to be recognized.

Excision of the AP site is catalyzed by an AP endonuclease, which hydrolyses the phosphodiester bond on the 5' side of the AP site. The best characterized AP endonuclease is the *E. coli* enzyme exonuclease III. Beneeth the 3' to 5' exonuclease and phosphatase activities of this enzyme the AP endonuclease activity of exonuclease III accounts for about 90% of the total AP activity in *E. coli*. The remaining 10% of the activity is accounted for by a different enzyme, endonuclease IV. In *S. cerevisiae*, the major AP endonuclease is related to *E. coli* endo IV, but in mammalian cells, exonuclease III related AP endonucleases seem to be abundant.

After strand cleavage by AP endonuclease, the nicked strand contains a 3'-OH terminus that can easily be extended by a polymerase, and a 2'-deoxyribose-5'-phosphate terminus that must be removed to permit ligation. Current evidence suggests that this non-nucleotide terminus is removed by one of two alternative pathways.

In human cells the major pathway appears to be mediated by DNA polymerase β and an activity which has been only crudely characterized called DNA deoxyribophosphodiesterase (dRPase). DNA polymerase β generates a repair "patch" (stretch of newly synthesized DNA) only a single nucleotide in length by extending the 3'-OH terminus by a single nucleotide. The displaced deoxyribose phosphate moiety is then removed by dRPase. Which of the 3 mammalian ligases is responsible for the final ligation step is not yet clear (Fig. 2B). The alternative repair pathway which generates longer repair patches appears to depend on DNA polymerase δ or ε, and it requires PCNA (a cofactor for both polymerases δ and ε). The mechanism (Fig. 2C) involves synthesis of a stretch of DNA several nucleotides long, which results in displacement of the strand terminated by the deoxyribosephosphate group. The displaced strand is then removed by a structure-specific nuclease (DNase IV or FEN-1 in mammals, Rad27 in *S. cerevisiae*, and Rad2 in *S. pombe*), to create a ligatable substrate.

5. Nucleotide Excision Repair (NER)

Although base excision repair is clearly important, it appears to be insufficient, as there must be a DNA glycosylase for a given type of damage to be dealt with. There is a variety of DNA-reactive chemicals in the environment as well as a huge variety of alterations that can be produced by radiation and by oxidative and free radical attack on DNA. So many types of damage can be generated that coping with by evolutionary development of damage-specific DNA glycosylases would be difficult if not impossible. A more flexible damage repair mechanism has evolved in living organisms, which recognizes damaged regions based on their abnormal structure, then excises and replaces them. In all organisms, this nucleotide excision repair (NER) involves the steps [5-7]:

1. Damage recognition and binding of a multi-protein complex at the damaged site
2. Double incision of the damaged strand several nucleotides away from the damaged site, on both the 5' and 3' sides
3. Removal of the damage-containing oligonucleotide from between the two nicks
4. Filling in of the resulting gap by a DNA polymerase and ligation

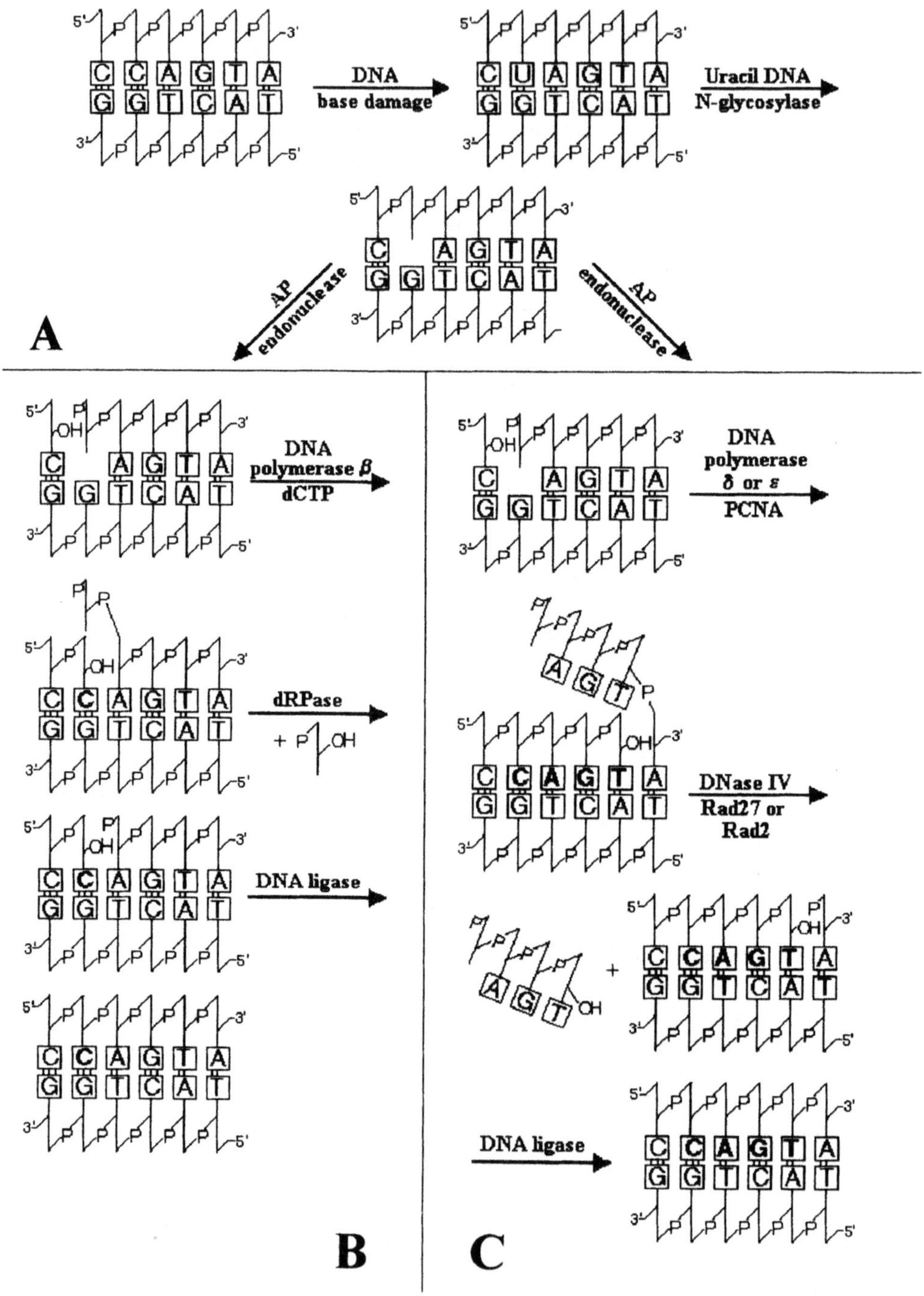

Figure 2: Two modifications of the Base Excision Repair (BER) Pathway

5.1. NER IN *E. COLI*

In *E. coli*, 3 proteins, the products of the *uvrA*, *uvrB*, and *uvrC* genes, are responsible for damage recognition and DNA nicking [5, 6]. The current model is that, in a ATP dependent reaction, a dimer of the UvrA protein forms a complex with a single UvrB molecule. The UvrAB complex may (i) bind to undamaged DNA and then slide along the DNA a short distance to its preferred damaged DNA binding site or (ii) the complex may recognize the damaged site by its distortion of the double helix bind directly to it. Further ATP hydrolysis may permit the release of UvrA and stable binding of UvrB at the damage, concomitant with increased bending of the DNA (Fig. 3A).

Next, the UvrC protein binds next to the UvrB protein therby activating it to nick the DNA approximately 4 nucleotides 3' to the damaged site. This, in turn, activates UvrC to nick the DNA approximately 7 nucleotides 5' to the damage. It is possible that activation of UvrC is a consequence of a conformational change in the DNA after nicking by UvrB (Fig. 3B). These steps all require ATP binding but not ATP hydrolysis.

A fourth protein, the the *uvrD* gene product, known as DNA helicase II, displaces the damage-containing oligonucleotide. At the same time the UvrC protein is released and afterwards the UvrB protein is displaced, when DNA polymerase I (or II) fills in the gap. Finally, the remaining nick is sealed by DNA ligase (Fig. 3C).

5.2. NER IN EUKARYOTIC CELLS

Although less is known about the mechanism of many of the steps of NER in eukaryotic cells [7, 8], the available evidence suggests that the process resembles that in *E. coli*, but there are many differences in detail. Information about eukaryotic NER has emerged from genetic and biochemical studies [7]. As a result of these studies, proteins important for the first steps of NER have been identified in yeast and human cells (Table 2).

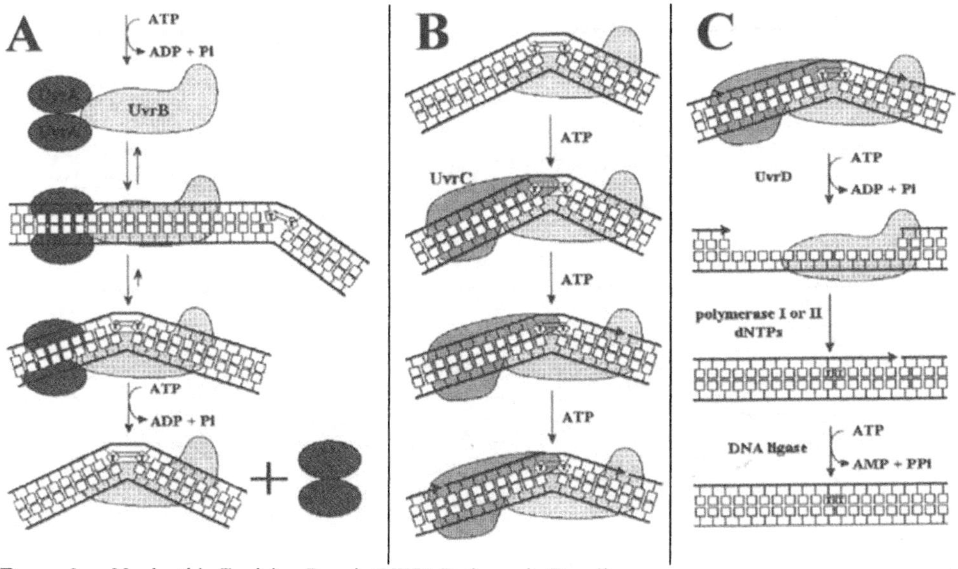

Figure 3: Nucleotide Excision Repair (NER) Pathway in E. coli

Available evidence suggests that in the initial stages of NER in eukaryotic cells the XPA protein plays a major role, together with RPA, the trimeric single-stranded-DNA-binding protein complex essential for replication. XPA has a DNA-binding domain, binding relatively well to damaged DNA and a second domain binding to RPA. The RPA/XPA complex binds with enhanced affinity to damaged DNA (Fig. 4A).

Another part of XPA binds to the basal transcription factor TFIIH [8-10], which contains 9 subunits. TFIIH and XPA interact with and recruit XPC+HHR23B, which help to stabilize the DNA and the protein-protein interactions in the complex (Fig. 4B). Two of the subunits of TFIIH are the XPB and XPD proteins, helicases with opposite polarities. It is suspected that these helicases unwind the DNA outwards from the site of XPA binding, permitting subsequent binding of XPF/ERCC1 and XPG, which are structure-specific nucleases. Binding of the structure-specific nucleases is enhanced by the fact that another domain of XPA interacts with ERCC1, and RPA interacts with XPG.

The XPF/ERCC1 heterodimer nicks DNA with the polarity illustrated below near junctions of single-stranded and double-stranded DNA. XPG recognizes similar junctions, but cuts with opposite polarity. The result is the introduction of nicks into the damaged strand, approximately 22-24 nucleotides 5' to the damage and approximately 5 nucleotides 3' to the damage (Fig. 4C). The mechanism by which the incised oligonucleotide is displaced is not clear. Perhaps the XPB/XPD helicases assist in this function. In any case, after the oligonucleotide is removed, the resulting gap is filled in by DNA polymerase ε or δ, assisted by PCNA, and the final nick is sealed by DNA ligase I.

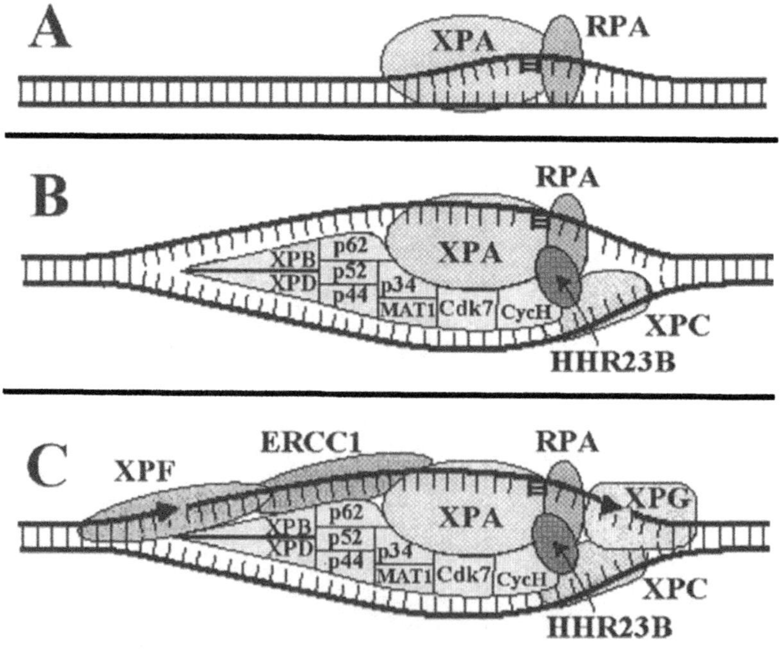

Figure 4: Nucleotide Excision Repair (NER) Pathway in Human Cells

TABLE 2: Proteins Required for Eukaryotic Nucleotide Excision Repair [1, 8-10]

S. pombe protein	*S. cerevisiae protein*	*Human protein*	*Probable function*
?	RAD14	XPA	Binding to damaged DNA
Ssb1,2,3	RFA1,2,3	RPAp70, p32, p14	Damage recognition (with Rad14/XPA)
ERCC3[sp]	SSL2 (RAD25)	XPB	3' to 5' helicase; TFIIH component; required for repair
Rad15/rhp3	RAD3	XPD	5' to 3' helicase; TFIIH component; required for repair
?	TFB1	p62	TFIIH component; required for repair
?	TFB2	p52	TFIIH component
?	SSL1	p44	TFIIH component; required for repair
?	?	p34	TFIIH component; required for repair
?	TFB3	MAT1	CDK assembly factor
?	KIN28	Cdk7	CDK; C-terminal domain kinase
?	CCL1	CycH	Cyclin
?	RAD4	XPC	Binds ssDNA; helps stabilize DNA in pre- and post-incision complexes
?	RAD23	HHR23B	Contains ubiquitin domain; helps stabilize XPA/TFIIH complex
Rad16	RAD1	XPF	Part of endonuclease (5' incision)
Swi10	RAD10	ERCC1	Part of endonuclease (5' incision)
Rad13	RAD2	XPG	Endonuclease (3' incision)

5.3. NER AND HUMAN GENETIC DISEASES

The genes encoding many of the human NER proteins (Table 2) were first identified in genetic complementation studies of the human DNA repair disease, *Xeroderma pigmentosum* (XP), which suggested that mutations in any of 7 genes (*XPA*-*XPG*) could give rise to the disease. In addition to XP, two other human genetic diseases appear to involve nucleotide excision repair defects [11]. These two additional diseases are Cockayne's syndrome (CS) and trichothiodystrophy (TTD).

Patients suffering from XP have severe light sensitivity, pigmentation irregularities, early onset of skin cancer at high incidence, and elevated frequency of other forms of cancer. TTD patients display light sensitivity only in some cases. CS gives rise to light sensitivity in some cases and early death. The clinical symptoms of these diseases are largely non-overlapping. In particular, CS and TTD patients do not have elevated rates of skin cancer, and their skins appear relatively normal (in contrast to XP patients).

TABLE 3: Overlap between Complementation Groups in Xeroderma pigmentosum, Cockayne's Syndrome andTrichothiodystrophy

Xeroderma pigmentosum (XP)	*Cockayne's Syndrome (CS)*	*Trichothiodystrophy (TTD)*	*Factor of*
		TTDA	
	CSA		
	CSB		
XPB	XPB/CS	XPB/TTD	TFIIH
XPD	XPD/CS	XPD/TTD	TFIIH
XPG	XPG/CS		

Although patients suffering from defects in NER display different clinical symptoms, they could be grouped within different complementation groups which show partial overlapping (Table 3). Two of the 3 complementation groups of TTD and 3 of the 5 groups of CS overlap with 3 XP groups. Biochemical studies have proved, that these two TTD complementation groups overlap with CS and correspond to the XPB and XPD proteins. These complementation groups correspond to proteins that are components of TFIIH, an essential transcription factor. The identify of the third (TTDA) awaits its cloning. The relationship to TFIIH is extended, though not completely, by the complementation groups of CS. Neither CSA nor CSB is known to be a component of TFIIH. However, CSA interacts with the p44 subunit of TFIIH *in vitro*, CSB interacts with XPG *in vitro*, and CSA and CSB interact with each other.

Defects in CSA and/or CSB lead to loss of preferential NER of the transcribed strand. In normal cells (prokaryotic and eukaryotic), transcribed regions are repaired more rapidly after damage than are non-transcribed regions, and, within the transcribed region, the transcribed strand (template strand) is repaired more rapidly than the non-transcribed strand [1, 12]. The latter preference is abolished by defects in CSA and CSB. A reasonable hypothesis is that CSA and CSB may cooperate to recruit proteins needed for NER (such as TFIIH and XPG) to sites where the transcription apparatus is blocked by DNA damage. The interactions between XPG and CSB, CSB and CSA, and CSA and p44 (Fig. 4.) may possibly explain how certain mutations in XPG lead to additional defects in transcription. The different phenotypes of XP, CS and TTD may be explained by the interaction of NER and transcription. Both processes require the action of TFIIH. The fact that transcribed strands are preferentially repaired makes it not unlikely that proteins involved in these interactions (TFIIH components, proteins involved in recruiting the repair machinery to transcribed regions) might display two types of mutations:

- mutations that preferentially affect NER
- mutations that preferentially affect transcription

Mutations of the first type would give rise to the XP phenotype, while mutations of the second type would produce CS or TTD (depending on the kind of the transcriptional defect). The failure of CS and TTD to display elevated cancer frequencies might be explained by a possible requirement for normal transcription for cell progression toward cancer.

6. Repair of DNA Strand Breaks

DNA strand breaks are associated with various endogenous processes (transcription, recombination, replication) and with the process of active cell death, which aims to eliminate cells. In addition, DNA strand breaks can be induced by external radiation, exposure to chemicals, and during repair of other types of DNA damage. DNA strand breaks are potentially mutagenic and must be recognized and repaired. Single-strand breaks (SSB) are efficiently repaired within minutes by simple rejoining processes, double-strand break (DSB), however, constitute potentially lethal lesions. DNA DSB induction and their repair are considered to play a major role as determinants of cellular sensitivity to ionizing radiation. Two molecular end-points were identified which often correlated with radiosensitivity: (i) the apparent number of DSB induced per Gy per DNA unit and (ii) the half-time of the DSB rejoining kinetics. DSBs are repaired by mechanisms that involve recombination proteins. For repairing a DSB a homolog of the broken molecule is needed, which serves as the template in a recombination reaction. DNA polymerase ε, Ku, and proteins such as DNA-PKcs play key roles in damage recognition pathways [13-16].

7. Outlook: Control of Cell Cycle Progression and Radiosensitivity

The cellular response against radiation in addition to repair of DNA lesions include other safe-guards [1, 10], such as cell cycle regulation and mechanisms responsible for scavenging of free radicals, which are produced by radiation as well as by many endogenous metabolic pathways. For example, ataxia telangiectasia, an autosomal recessive human disorder with increased sensitivity to ionizing radiation, defective DNA repair is probably not responsible for the increase in cell killing after ionizing radiation, but rather the defective cell cycle regulation in response to DNA damage [17].

Biochemical studies have provided important insights into the mechanisms of DNA repair and how DNA damage signals are transduced to transcription and the cell cycle machinery. Cell cycle checkpoints represent integral components of DNA repair that coordinate the cell cycle and several biochemical pathways that respond to damage and restore DNA structure. By delaying progression through the cell cycle, checkpoints provide more time for repair before the critical phases of DNA replication and of mitosis. There are two checkpoints which control the regulation of the cell cycle progression after induced DNA damage, in G_1 or G_2 phase. The G_1/S and S checkpoint ensures that damaged DNA is not replicated, thus avoiding genetic changes. The G_2/M checkpoint prevents cell division to occur until the DNA damage has been repaired.

Mutations affecting the cell cycle control may influence genomic stability and lead to cancer development (Fig. 5). Loss or attenuation of checkpoint function may increase spontaneous and induced gene mutations and chromosomal aberrations by reducing the efficiency of DNA repair. Defects in checkpoint control have been seen at early stages of cell transformation and in certain hereditary cancer syndromes.

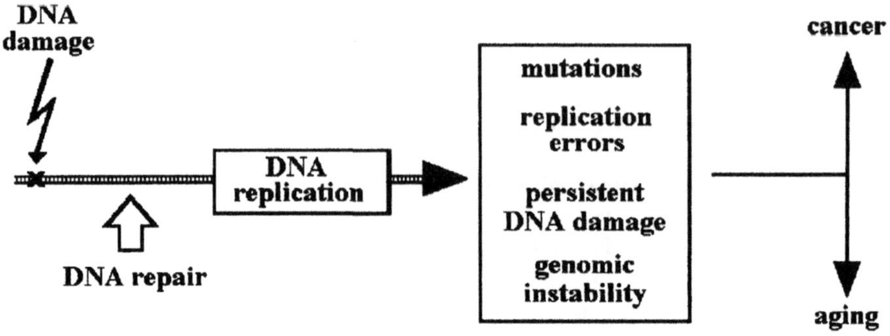

Figure 5: Consequences arising by replication of damaged DNA

8. References

1. Friedberg E.C., Walker G.C., Siede, W. (1995) *DNA Repair and Mutagenesis,* ASM Press, Washington, USA.
2. Modrich P., Lahue R. (1996) Mismatch repair in replication fidelity, genetic recombination, and cancer biology, *Ann. Rev. Biochem.* **65**,. 101-133.
3. Sancar A. (1996) DNA excision repair, *Ann. Rev. Biochem.* **65**, 43-81.
4. Sobol R.W., Horton J.K., Kuhn R., Gu H., Singhal R.K., Prasad R., Rajewsky K., Wilson S.H. (1996) Requirement of mammalian DNA polymerase-beta in base-excision repair. *Nature* **379**,183-186.
5. Gordienko I., Rupp W.D. (1997) UvrAB activity at a damaged DNA site: is unpaired DNA present? *EMBO J* **16**, 880-888.
6. Gordienko I., Rupp W.D. (1997) The limited strand-separating activity of the UvrAB protein complex and its role in the recognition of DNA damage. *EMBO J* **16**, 889-895.
7. Aboussekhra A., Biggerstaff M., Shivji M.K.K., Vilpo J.A., Moncollin V., Podust V.N., Protic M., Huebscher U., Egly J.M., Wood R.D. (1995) Mammalian DNA nucleotide excision repair reconstituted with purified protein components. *Cell* **80**, 859-868.
8. Svejstrup J.Q., Wang Z., Feaver W.J., Wu X., Bushnell D.A., Donahue T.F., Friedberg E.C., Kornberg R.D. (1995) Different forms of TFIIH for transcription and DNA repair: holo-TFIIH and a nucleotide excision repairosome. *Cell* **80**, 21-28.
9. Svejstrup J.Q., Vichi P., Egly J.-M. (1996) The multiple roles of transcription/repair factor TFIIH. *Trends Biochem. Sci.* **21**, 346-355.
10. Wood R.D. (1996) DNA repair in eukaryotes, *Ann. Rev. Biochem.* **65**, 135-167.
11. Chu G., Mayne L. (1996) Xeroderma pigmentosum, Cockayne syndrome and trichothiodystrophy: do the genes explain the diseases? *Trends Genet.* **12**, 187-192.
12. Friedberg E.C. (1996) Relationships between DNA repair and transcription, *Ann. Rev. Biochem.* **65**, 15-42.
13. Weaver D.T. (1995) What to do at an end: DNA double-strand break repair. *Trends in Genetics* **11**, 388-392.
14. Jackson S.P., Jeggo P.A. (1995) DNA double-strand break repair and V(D)J recombination: involvement of DNA-PK. *Trends in Biochemical Science* **20**, 412-415.
15. Feldmann H., Winacker E.L. (1993) A putative homologue of the human autoantigen Ku from Saccharomyces cerevisiae. *J. Biol. Chem.* **268**, 12895-12900.
16. Gottlieb T.M., Jackson S.P. (1993) The DNA-dependent protein kinase: requirement for DNA ends and association with Ku antigen. *Cell* **72**, 131-142.
17. Meyn M.S. (1995) Ataxia telangiectasia and cellular responses to DNA damage. *Cancer Res.* **55**, 5991-6001.

DNA DAMAGE AND MUTATIONS INDUCED BY SOLAR UV RADIATION.

Solar UV mutagenesis

E. SAGE

CNRS UMR 218, LRC-1 du CEA, Institut Curie, Section de Recherche
26 rue d'Ulm, 75248 Paris Cedex 05, France

Abstract

Exposure of individuals to sunlight creates DNA damage, and subsequently mutations, in actively dividing cells of the basal layer of the epidermis. Mutations in oncogenes and tumor suppressor genes are crucial early events in the development of non-melanoma skin tumors. The precise identification of DNA lesions and DNA sequence alterations induced by environmental solar UV radiation in cultured cells represents a powerful approach for elucidating the mechanism of sunlight mutagenesis, as well as the genotoxic basis of multistage photocarcinogenesis in humans. Such investigations are analysed here, along with data obtained in human and experimentally UV-induced murine skin cancer.

1. Introduction

Epidemiological, experimental and molecular evidence demonstrates that DNA damage and subsequent mutations induced by the ultraviolet (UV) component of sunlight are critical factors in the induction of skin cancer [1, 2]. Most of our knowledge on the nature, repair and mutagenic consequences of DNA photolesions comes from the numerous investigations using 254 nm UV-C. However, UV-C and short UV-B radiation, the most energetic UV wavelengths emitted by the sun, do not reach the surface of the earth, whereas UV-B (λ>295 nm) and UV-A (320-400 nm), the environmentally relevant UV radiation, constitute about 0.4 and 5.5% of the terrestrial solar energy respectively. Indeed, sunlight is a complex mixture of mutagenic wavelengths. DNA damage caused by direct absorption of UV-B photons, i.e. bipyrimidine photoproducts, is a crucial event in the initiation of skin tumors [1]. The UV-A portion of solar radiation, poorly absorbed by DNA, excites other cellular chromophores, generating reactive oxygen species (ROS), which in turn have pleiotropic effects that may be involved in the carcinogenic process. It is of particular importance to determine the relative contribution of the different regions of the terrestrial solar spectrum to sunlight induced mutagenesis and carcinogenesis. Furthermore, the knowledge of the relative proportion of different classes of DNA lesions induced by the various UV wavelengths is essential in understanding early events involved in such processes.

C. Baumstark-Khan et al. (eds.),
Fundamentals for the Assessment of Risks from Environmental Radiation, 115-126.

2. Mutational Specificities of UV-C, UV-B, UV-A and Solar Radiations

A large data base of UV-C-induced mutations has been produced over the past twenty years, using diverse mutational target genes in various species, from *Escherichia coli* to cultured human cells and transgenic mice, including yeast and rodent cells, under different repair status [3-4 for review]. Even though the damage distribution may differ slightly depending on the sequence of the selected gene and on the irradiation conditions, the overall collection of mutation spectra contributed to characterize the main

TABLE 1: Classes of mutations induced by UV-C radiation

	lacI	*supF*	*Aprt*	*hprt*		*hprt*		*lacZ*	
	E. coli	*human*	*hamster*	*hamster*		*human*		*E. coli*	
		NER+	*NER+*	*NER+*	*NER-*	*NER+*	*XP NER-*	*254 nm vac*	*160 nm vac*
BASE SUBSTITUTION	77.5%	97%	98%	91%	100%	94%	93%	75%	92%
TRANSITION									
GC→AT	48%	45%	56%	33%	38.5%	39.5%	40%	21 (40%)	31 (52%)
AT→GC	7%	4%	7%	9.5%	7%	14.5%	0	7 (13%)	7 (12%)
TRANSVERSION									
GC→TA	9%	7.5%	0%	10%	0	5%	9%	1 (2%)	3 (5%)
GC→CG	2%	3%	8%	0%	0%	5%	0	3 (6%)	1 (1.5%)
AT→TA	5%	5%	5%	8.5%	9%	10%	18%	4 (8%)	6 (10%)
AT→CG	3%	2%	5%	8.5%	4.5%	0	5%	0	0
DOUBLE									
TANDEM	3.5%	16.5%	10%	18.5%	41%	20%	21%	3 (6%)	7 (11.5%)
Non-TANDEM		24%	7%	3%	0	0	0		
CC→TT	3.5%	8%	5%	12.5%	23.5%	13.3%	14%	2 (4%)	5 (6.5%)
OTHERS									
FRAMESHIFT	13.5%		2 %	9%	0	6%	7%	12 (23%)	0%
DELETION	8%	3%	-	-	-	-	-	1 (2%)	5 (8%)
INSERTION	-	-	-	-	-	-	-		
DUPLICATION	1%	-	-	-	-	-	-		
Reference	[6]	[7]	[8]	[9]	[9]	[10]	[10]	[11]	[11]

tandem double mutations comprise CC to TT events.

features of UV-C mutational specificity (Table 1). Nearly all mutations occur at bipyrimidine sites, with a predominance of GC to AT transitions and 5-10% of tandem double CC to TT events, and it is consistent with a role for bipyrimidine photoproducts as premutagenic lesions. This unique specificity of UV-C mutagenesis has been defined as a «UV mutagenic signature» [5]. More generally, molecular signatures of mutagenic agents allow to perform molecular epidemiology of human cancer risk. Indeed, the UV hallmark described above has been observed in human skin carcinoma [5; see below].

As expected from the known UV-B photoproducts, UV-B radiation induces a large majority of GC to AT transitions among base substitutions (Table 2). The class of GC to

TABLE 2: Classes of mutations induced by UV-B radiation

	broad UV-B		*313 nm UV-B*	*broad UV-B*		*broad UV-B*
	aprt *rodent*	*lacZ* *human*	*SupF* *monkey*	*Sup4-0* *yeast*	*supF* *E. coli*	*LacZ* *Transgenic mice*
TRANSITION						
G:C→A:T	49 (66%)	23 (50%)	17%	166 (80%)	28%	21 (78%)
A:T→G:C	0	2 (4%)	1%	13 (6%)	0	0
TRANSVERSION						
G:C→T:A	1 (1.5%)	3 (7%)	5%	4 (2%)	4%	2 (7%)
G:C→C:G	1 (1.5%)	2 (4%)	3%	3 (1.5%)	4%	0
A:T→T:A	2 (3%)	3 (7%)	1%	2 (1%)	5%	0
A:T→C:G	7 (9%)	1 (2%)	0.5%	1 (0.5%)	0	0
DOUBLE						
TANDEM	5 (7%)	0	8 (3.5%)	16 (8%)	5 (5%)	1 (4%)
Non-TANDEM	6 (8%)	0	57 (25%)	0		0
CC→TT	4 (5%)			13 (6%)		1 (4%)
OTHERS	3 (4%)	12 (26%)	102 (44%)	3 (1%)	41 (54%)	3 (11%)
INSERTION	0					
DELETION	0					
FRAMESHIFT	2 (3%)			2 (1%)		3 (11%)
DUPLICATION	1 (1%)					
TOTAL MUTANTS	74	46	230	208	76	27
Reference	[14]	[15]	[16]	[12]	[13]	[17]

AT transitions represents 62% to 87% of total single base substitutions, whereas it com prises 48% to 69% of UV-C-induced single base changes, depending on the target gene and the species. However, statistical analysis performed on individual experiments [12, 13] indicates that the small differences observed in the UV-C and UV-B mutation spectra are not significant, at least when point mutations are considered. Even though UV-B may produce ROS, GC to TA transversions, the hallmark of 8-oxo-7,8-dihydro-2'-deoxyguanosine (8-oxodGuo), are infrequent.

The mutagenicity and carcinogenicity of UV-A is rather weak in comparison with UV-B, when expressed per unit of energy fluence [18, 19]. However, a large part of human exposure to UV comes from UV-A and the contribution of UV-A and even visible wavelengths in the mutagenic effect of sunlight has to be considered [4, 19]. The molecular nature of UV-A-induced mutations has been poorly investigated until now. Broad UV-A mutagenesis has been studied at the adenine phosphoribosyltransferase (aprt) locus in chinese hamster ovary (CHO) cultured cells and in the lacZ gene carried on a shuttle vector in embryonic kidney human cells. The GC to AT transitions at bipyrimidine sites are still recovered, but at a drastically reduced frequency (Table 3). A new class of events, AT to CG transversions, predominates the UV-A spectrum at both the

TABLE 3: Classes of mutations induced by UV-A radiation

		broad UV-A		*365 nm UV-A*
		aprt	*LacZ*	*p53*
		rodent cells	*human cells*	*rodent skin tumours*
TRANSITION	G:C→A:T	12 (22%)	12 (33%)	7 (100%)
	A:T→G:C	5 (9%)	5 (14%)	
TRANSVERSION	G:C→T:A	2 (4%)	1 (3%)	
	G:C→C:G	5 (9%)	0	
	A:T→T:A	1 (2%)	4 (11%)	
	A:T→C:G	20 (37%)	0	
DOUBLE	TANDEM	6 (11%)		
	Non-TANDEM	2 (4%)		
	CC→TT	3 (5%)		
	OTHERS	1 (2%)	14 (39%)	
TOTAL MUTANTS		54	36	7
Reference		[14]	[15]	[18]

aprt and the hprt (hypoxanthine phosphoribosyltransferase) locus in rodent cells, but not in human cells (Drobetsky, personal communication). Nevertheless, in both human and rodent repair-proficient cultured cells, the general trends in UV-A mutagenesis are the presence of 20-30 % of GC to AT transitions and a general increase in mutations at AT base pairs [20]. Interestingly, GC to TA transversions which could be due to expected 8-oxodGuo, are poorly represented. Since UV-A radiation induces DNA strand breaks, it is remarkable that only point mutations, originating from base modifications, are observed at the aprt locus. In the mutated p53 gene in skin tumors of mice irradiated with 365 nm UV-A, the GC to AT mutation hotspots are located at bipyrimidine sites implicated in PyCpG and may originate from enhanced deamination of cytosine which is involved in the formation of a cyclobutane pyrimidine dimer [18].

Table 4 shows that the mutation spectra produced by simulated (SSL) and natural sunlight are not superposable with those induced by UV-B, and largely differ from those exhibited after UV-A exposure. Mainly base substitutions are recovered. Except for a single-stranded DNA bacteriophage in E. coli where transversions largely predominate, about 40% of events after sunlight exposure result in GC to AT transitions at bipyrimidine sites, up to 61% in human skin tumors. The decrease of this class of mu tations, in comparison to those produced by UV-B (60-80%), is accompanied by a significant increase in tandem double CC to TT mutations in rodent cells and human skin tumors. In rodent cells, the AT to CG occurrence is thought to result from the contribu-

TABLE 4: Classes of mutations induced by Simulated (SSL) and natural Sunlight

		SSL	*natural sunlight*		
		aprt	*p53*	*Sup4-0*	*lacZ*
		rodent	*human skin tumor*	*yeast*	*E. coli*
TRANSITION	G:C→A:T	13 (36%)	75 (61%)	65(36.5%)	7 (9%)
	A:T→G:C	1 (3%)	5 (4%)	14 (8%)	9 (11%)
TRANSVERSION	G:C→T:A	0	13 (11%)	49(27.5%)	17 (21%)
	G:C→C:G	0	8 (7%)	35 (19%)	41 (51%)
	A:T→T:A	2 (6%)	6 (5%)	7 (4%)	5 (6%)
	A:T→C:G	9 (25%)	2 (2%)	4 (2%)	1 (1%)
DOUBLE	TANDEM	11 (31%)	13 (11%)	4 (2%)	0
	Non-TANDEM	0	0	0	0
	CC→TT	9 (25%)	13 (11%)	4 (2%)	0
TOTAL MUTANTS		36	122	178	80
Reference		[21]	[22]	[23]	[24]

tion of the UV-A portion of SSL [14, 21]. This UV-A trace is not observed after natural sunlight in yeast or at the mutated p53 gene in human skin tumors. In contrast, the frequency of GC to TA transversions, which likely originate from 8-oxodGuo, is readily increased in these two last examples. In addition, the comparison of mutations induced in the p53 gene from squamous cell carcinoma in mice exposed to UV-B and in human non-melanoma skin tumors, reveals differences similar to those observed in cultured eucaryotic cells (Table 5).

So far, the UV mutagenic signature which has been defined from mutation spectra and damage distribution [5, 26], is recovered after UV-B, SSL, and after sunlight in human skin tumors. Interestingly, it is also partly obtained after UV-A (occurrence of GC to AT transitions at bipyrimidine sites). Such an hallmark is sufficiently precise and different from those of other chemical or physical mutagenic agents to allow the detection of the trace of sunlight in the genetic material of individuals [27]. However, it is insufficient to determine the role of the different UV wavelengths in solar mutagenesis and the implication of each type of photoproduct in such a process. Indeed, some differences appear among mutation spectra obtained for the various regions of solar UV and sunlight. Furthermore, data obtained with several systems are compared here and bias could be introduced into the mutation spectra, due to the sequence of the selected gene and specific repair mechanism in each species. In order to eliminate this caveat, mutational specificities of UV-C, UV-B, UV-A and SSL have been compared at the aprt locus of nucleotide excision repair (NER)-proficient and -deficient CHO cells [14, 21, 28].

TABLE 5: Mutations on *p53* gene in murine SCC and in human non-melanoma skin tumors

		UV-B	*natural sunlight*	
		p53 *murine skin*	*p53* *human skin non-XP*	*p53* *human skin XP*
TRANSITION	G:C→A:T	140 (81%)	57 (68%)	39 (87%)
	A:T→G:C	9 (5%)	3 (4%)	1 (2%)
TRANSVERSION	G:C→T:A	9 (5%)	12 (14%)	2 (4%)
	G:C→C:G	2 (1%)	9 (11%)	0
	A:T→T:A	8 (5%)	3 (4%)	1 (2%)
	A:T→C:G	4 (2%)	0	2 (4%)
Reference		[25]	[25]	[25]

base substitutions (single + tanden) are taken as 2 events
CC to TT mutations represent 50% of all tandem double events after UV-B, and 90% or 100% after sunlight in human non-XP and XP tumors respectively

Altogether a collection of about 400 independent aprt- mutants is presented in Table 6. In repair-proficient cells, the UV-C- and UV-B-mutation spectra are similar. The major class of events after UV-B and UV-C irradiation, the GC to AT transitions at bipyrimidine sites, represents 1/3 of the mutations induced by SSL and 1/5 of thoses produced by UV-A. After SLL, the lower occurrence of GC to AT transitions is associated with a statistically significant increase in the frequency of tandem double CC to TT mutations, which appears to be the real hallmark of sunlight. In addition, AT to CG transversions, infrequently recovered after UV-C or UV-B, predominate in the UV-A spectrum and have been considered as the UV-A hallmark [14]. These events constitute a significant proportion of the SSL-induced mutant collection and likely reflect the mutagenic contribution of the UV-A component of SSL [14, 21].

In repair-deficient cells, the frequency distributions of UV-A- and UV-C- induced mutational classes are surprisingly comparable, thus demonstrating the mutagenic

TABLE 6: Classes of mutations induced by UV-C, UV-B, UV-A and SSL at the aprt locus in NER+ and NER- CHO cells [28]

	NER+				*NER-*			
	UV-C	*UV-B*	*UV-A*	*SSL*	*UV-C*	*UV-B*	*UV-A*	*SSL*
TRANSITION								
GC→AT	22(56%)	49(66%)	12(22%)	13(36%)	29(57%)	26(51%)	32(65%)	23(48%)
AT→GC	4 (7%)	0	5 (9%)	1 (3%)	1 (2%)	0	3 (6%)	0
TRANSVERSION								
GC→TA	0	1 (1%)	2 (4%)	0	2 (3%)	0	2 (4%)	0
GC→CG	5 (8%)	1 (1%)	5 (9%)	0	3 (6%)	0	0	0
AT→TA	3 (5%)	2 (3%)	1 (2%)	2 (6%)	1 (2%)	0	1 (2%)	0
AT→CG	3 (5%)	7 (9%)	20(37%)	9 (25%)	2 (4%)	2 (4%)	0	1 (2%)
DOUBLE								
TANDEM	6 (10%)	5 (7%)	6 (11%)	11(31%)	10(20%)	19(37%)	7 (14%)	23(48%)
Non-TANDEM	4 (7%)	6 (8%)	2 (4%)	0	3 (6%)	1 (2%)	1 (2%)	1 (2%)
CC→TT	3 (5%)	4 (5%)	3 (5%)	9 (25%)	7 (14%)	18(35%)	2 (4%)	20(42%)
OTHERS								
FRAMESHIFT	1 (2%)	2 (3%)	1 (2%)	0	0	0	0	0
DELETION	0	0	0	0	0	3 (6%)	3 (6%)	0
INSERTION	0	0	0	0	0	0	0	0
DUPLICATION	0	1 (1%)	0	0	0	0	0	0
TOTAL	59	74	54	36	51	51	49	48

weight of bipyrimidine photoproducts in UV-A-induced mutagenesis. In addition, the UV-B and SSL induced mutation spectra are virtually identical and are marked by a significant increase in tandem CC to TT mutations. A similar increase in CC to TT events is also observed for the mutated p53 gene from non-melanoma skin tumors in repair-deficient Xeroderma pigmentosum (XP) patients relative to normal individuals [29], as well as in uvrB Salmonella typhimurium exposed to sunlight [19].

Studies on the sequence specificity of mutations induced by UV and solar radiations presented above give some clues to solar mutagenesis.

1. So far, the results indicate an inadequacy of using irradiation with 254 nm UV-C, or even UV-B, as a physiologically relevant model for sunlight carcinogenesis.
2. The mutational specificity of UV-A radiation is characterized by a significant occurrence of UV type GC to AT transitions likely due to bipyrimidine photoproducts, by a relative increase in events at AT base pair in comparison to UV-B or UV-C, and a low frequency of GC to TA transversions possibly due to 8-oxodGuo.
3. After simulated or natural sunlight, GC to AT transitions are well produced, together with CC to TT mutations and/or GC to TA transversions, whose frequencies are increased in comparison with UV-B or UV-C and UV-A.
4. The marked increase in tandem CC to TT mutations observed in the absence of repair reflects a different distribution of premutagenic lesions and/or translesion synthesis in repair -proficient and -deficient cells.

3. Main DNA Damage Produced by Solar UV Radiation

Knowledge of the relative distribution and repair of the different classes of photolesions produced by UV-B, UV-A and simulated or natural sunlight is essential in understanding the mechanism of solar mutagenesis. Cyclobutane pyrimidine dimers (CPD) are produced by UV-C, UV-B, sunlight, and also by UV-A, though DNA weakly absorbs radiation in this region [29, 30]. Our unpublished data (Perdiz & Sage) reveal that UV-B radiation is 10^{-2}, SSL 10^{-4} and UV-A 10^{-5} less efficient than UV-C, in producing CPD in mammalian cells. Using DNA sequencing methodology, an enhancement in the relative induction of C-containing CPD has been observed after irradiation of isolated DNA with UV-B and SSL (Table 7).

TABLE 7: Distribution of CPD in DNA fragments

	TT	*TC*	*CT*	*CC*	*Reference*
UV-C	68	16	13	3	[21, 32, 33]
UV-B	52	21	19	7	[21, 32, 33]
UV-B/UV-C	**0.77**	**1.3**	**1.5**	**2.3**	
UV-C	56	20	14	10	
SSL	36	26	17	21	
SSL/UV-C	**0.67**	**1.3**	**1.3**	**2.1**	

Pyrimidine (6-4) pyrimidone (6-4 photoproducts) are readily produced after UV-C and UV-B and SSL in a dose-dependent manner (our unpublished data). A small induction of 6-4 photoproducts is evident following 1-h sunlight irradiation of cultured cells, which, however, remains constant with further exposure [34]. The Dewar photoproduct, a valence photoisomeric form of (6-4) photoproducts, is well formed after UV-B and SSL [35-36], but barely detected after UV-C and UV-A (Perdiz & Sage, unpublished data). Furthermore, the ratio of Dewar to (6-4) photoproducts is largely enhanced after SSL (our data) and sunlight [34]. Indeed, the last study indicates that most of (6-4) produced in cultured cells after sun exposure are promptly converted into Dewar photoproducts. The exact yields of formation of (6-4) and Dewar photoproducts still need to be clarified. The overall data clearly demonstrates that the relative proportion of the bipyrimidine photolesions varies greatly between UV-C, UV-B, UV-A and sunlight. In addition, the level of DNA lesions decline with increasing epidermal depth in human skin, depending on the wavelength of irradiation [37].

The rate of elimination of photolesions is an important factor in the mechanism of solar mutagenesis. The removal of CPD is slow, incomplete, and occurs mainly through transcription-coupled repair (TCR), whereas (6-4) photoproducts are quickly and efficiently removed and are thus a better substrate for global genome repair (GGR) than for TCR [38]. Dewar photoproducts do not seem to be as efficiently repaired as (6-4) in human cells [36].

We (Douki, Cadet, Perdiz & Sage, unpublished data) and others [39, 40] have measured the formation of 8-oxodGuo in DNA of cells irradiated with solar UV radiation (Table 8). The discrepancies in the ratios of CPD to oxidized guanine presented in the table may be partly due to the various light sources used in the 3 reports and also to the different methods employed to detect 8-oxodGuo. However, a large relative increase in the induction of 8-oxodGuo is observed after UV-A, in comparison with UV-B or UV-C. In addition, the action spectra of formation of CPD and Fpg-sensitive sites (Fpg protein recognizes oxidized purines, including 8-oxodGuo) show tremendous variations in the ratio between these two lesions in the UV-A region, depending on the wavelength [40].

TABLE 8: Ratio of CPD to oxidized G in mammalian cells

	UV-C	*UV-B*	*UV-A*	*SSL*	*Reference*
CPD/8-oxodGuo	300	45	14	170	own data*
CPD/8-oxodGuo	2000	2000	0.23		[39]
CPD/Fpg-SS		100	0.7		[40]

* CHO cells, (UV-A: OSRAM 400W lamp plus Schott WG345 and KG1 filters)

[39] in human HeLa cells, (UV-A: Cosmolux A1-11 plus Schott WG335 filters

[40] CHO cells, (UV-A: $\lambda > 360$ nm xenon arc lamp plus Schott WG360filters

4. Conclusion : Trends in Solar Mutagenesis and Opened Questions

Based on data recently obtained at physiologically relevant doses of UV-A, the role of UV-A radiation in solar UV mutagenesis and photocarcinogenesis should not be neglected [14, 20]. It is probably an important risk factor for individuals who are regularly exposed to sunbeds [40]. Indeed, using *S. typhimurium hisG46* strains, it has been demonstated that light from a tanning salon bed (largely UV-A/B wavelengths) is considerably more mutagenic than sunlight, based on unit per energy fluence [19]. The data presented above suggest that CPD produced by UV-A are likely critical premutagenic lesions, whereas 8-oxodGuo play a minor role. It becomes obvious that the UV-B portion of solar spectrum only partly explains the distribution of mutations observed in the *p53* gene from human non-melanoma skin tumors. In this respect, the CC to TT mutational events seem to be a sunlight hallmark. It is worth questioning if the higher frequency of tandem CC to TT mutations observed after exposure to sunlight than after UV-B correlates with the relative increase in the formation of Dewar photoproduct over (6-4) (after sunlight in comparison to UV-B). In this regard, the role of Dewar photoproduct in solar mutagenesis is still unknown. In addition, is the origin of the relatively high frequency of GC to TA mutation due to 8-oxodGuo which would be produced by visible light [41] or spontanously?

The relative contribution of the 3 bipyrimidine photoproducts in solar mutagenesis, as well as the implication of oxidative damage, is still a matter of debate. The mechanism of translesion synthesis, generating mutations, is also an open field. Since sunlight is a mixture of various wavelengths, the interaction between them may be of importance in the carcinogenic process, not only at the level of damage or repair, but also at the level of signal transduction and gene induction.

Acknowledgement. I thank my collaborators, Drs J. Cadet, E. Drobetsky, T. Douki, Z. Kuluncsics and D. Perdiz for letting me present unpublished data, and Dr. J. Smith for critical reading of the manuscript. The recent work is supported by Ligue Nationale Française contre le Cancer, CNRS and ARC.

5. References

1. Brash, D.E. (1997) Sunlight and the onset of skin cancer, *Trends In Genetics* **13**, 410-141.
2. Kraemer, K.H. (1997) Sunlight and skin cancer : another link revealed, *Proc. Natl. Acad. Sci. USA* **94**, 11-14.
3. Sage, E. (1993) Distribution and repair of photolesions in DNA: genetic consequences and the role of sequence context, *Photochem. Photobiol.* **57**, 163-174.
4. Sage, E. (1996) UV-induced mutations and transformation, in B. Volkmer, H. Heller (eds), *Environmental UV radiation, risk of skin cancer and primary prevention*, Gustav Fischer, Stuttgart, pp. 107-116.
5. Brash, D., Rudolph, J., Simon, J., Lin, A., McKenna, G., Baden, H., Halperin, A. and Ponten, J. (1991) A role for sunlight in skin cancer: UV-induced p53 mutations in squamous cell carcinoma, *Proc. Natl. Acad. Sci. USA* **88**, 10124-10128
6. Schaaper, R. M., Dunn, R.L. and Glickman, B.W. (1987) Mechanisms of ultraviolet-induced mutation: mutational spectra in the *Escherichia coli lacI* gene for a wild type and an excision-repair deficient strain, *J. Mol. Biol.* **198**, 187-202.

7. Bredberg, A., Kraemer, K. and Seidman, M.M. (1986) Restricted ultraviolet mutational spectrum in a shuttle vector propagated in xeroderma pigmentosum cells, *Proc. Natl. Acad. Sci. USA* **83,** 8273-8277.
8. Drobetsky, E.A., Grosovsky, A.J., and Glickman, B.W. (1987) The specificity of UV-induced mutations at an endogenous locus in mammalian cells, *Proc. Natl. Acad. Sci. USA* **84**, 9103-9107.
9. Vrieling, H., van Rooijen, M.L., Groen, N.A., Zdzienicka, M.Z., Simons, J.W., Lohman, P. and van Zeeland, A.A. (1989) DNA strand specificity for UV-induced mutations in mammalian cells, *Mol. Cell. Biol.* **9**, 1277-1283.
10. Mc Gregor, W.G., Chen, R.H., Lukash, L., Maher V. and Mc Cormick, J.J. (1991) Cell-cycle dependent strand bias for UV-induced mutations in the transcribed strand of excision repair-proficient human fibroblasts but not in repair-deficient cells, *Mol. Cell. Biol.* **11**, 1927-1934.
11. Wehner, J., and Horneck, G. (1995) Effects of vacuum UV and UV-C radiation on dry *Escherichia coli* plasmid pUC19. II. Mutational specificity at the *lacZ* gene, *J. Photochem. Photobiol. B: Biol.* **30**, 171-177.
12. Amstrong, J.D. and Kunz, B. (1992) Photoreactivation implicates cyclobutane dimers as the major promutagenic UV-B lesions in yeast, *Mutat. Res.* **268**, 83-94.
13. Okaichi, K., Nagashima, K., Nozu, K. and Ohnishi, T. (1992) Mutagenic specificity in DNA base sequence by irradiation of health lamp light (UV-B) in *Escherichia coli, Mutat. Res.* **282**, 183-190.
14. Drobetsky E.A., Turcotte J., and Chateauneuf A. (1995) A role for ultraviolet A in solar mutagenesis. *Proc. Natl. Acad. Sci. USA* **92**, 2350-2354.
15. Robert, C., Muel, B., Benoit, A., Dubertret, L., Sarasin, A.and Stary, A. (1996) Cell survival and shuttle vector mutagenesis induced by UV-A and UV-B radiation on a human cell line, *J. Investig. Dermatol.* **106**, 721-728.
16. Keyse, S.M., Amaudruz, F., and Tyrrell, R.M. (1988) Determination of the spectrum of mutations induced by defined-wavelength solar UV-B (313-nm) radiation in mammalian cells by use of a shuttle vector, *Mol. Cell. Biol.* **8**, 5425-5431.
17. Frijhoff, A., Rebel, H., Mientjes, E., Kelders, M., Steenwinkel, M.J., Baan, R., van Zeeland, A., and Roza, L. (1997) UV-B-induced mutagenesis in hairless λ*lacZ*-transgenic mice, *Environm. Mol. Mut.* **29**, 136-142.
18. van Kranen, H., de Laat, A.M., van de Ven, J., Wester, P., de Vries, A., Berg, R., van Kreijl, C., and de Gruijl, F. (1997) Low incidence of *p53* mutations in UV-A (365-nm) induced skin tumors in hairless mice, *Cancer Res.* **57**, 1238-1240.
19. DeMarini, D.M., Shelton, M.L., and Stankowski, L.F. (1995) Mutation spectra in Salmonella of sunlight, white fluorescent light, and light from tanning salon beds: induction of tandem mutations and role of DNA repair, *Mutat. Res.* **327**, 131-149.
20. Stary, A., Robert, C., and Sarasin, A. (1997) Deleterious effects of ultraviolet A radiation in human cells, *Mutat. Res.* **383**, 1-8.
21. Drobetsky, E.A., Moustacchi, E., Glickman, B.W., and Sage, E. (1994) The mutational specificity of simulated sunlight at the *aprt* locus in rodent cells, *Carcinogenesis* **15**, 1577-1583.
22. Hollstein, M., Shomer, B., Greenblatt, M., Soussi, T., Hovig, E., Montesano, R., and Harris, C.C. (1996) Somatic point mutations in the *p53* gene of human tumors and cell lines: updated compilation, *Nucleic Acids Res.* **24**, 141-146.
23. Amstrong, J.D. and Kunz, B. (1992) Excision repair influences the site and strand specificity of sunlight mutagenesis in yeast, *Mutat. Res.* **274**, 123-133.
24. Negishi, K. and Hao, W. (1992) Spectrum of mutations in single-stranded DNA phage M13mp2 exposed to sunlight: predominance of G-to-C transversion, *Carcinogenesis* **13**, 1615-1618.
25. Dumaz, N., van Kranen, H., de Vries, A., Berg, R.J.W., Wester, P.W., van Kreijl, C.F., Sarasin, A., Daya-Grosjean, L & de Gruijl, F. (1997), The role of UV-B light in skin carcinogenesis through the analysis of *p53* mutations in squamous cell carcinomas of hairless mice. *Carcinogenesis* **18**, 897-904.
26. Brash, D.E. and Haseltine W.A. (1982) UV-induced mutation hotspots occur at DNA damage hotspots, *Nature* **298**, 189-192.

27. Dumaz, N., Stary, A., Soussi, T., Daya-Grosjean, L., and Sarasin, A. (1994) Can we predict solar ultraviolet radiation as the causal event in human tumours by analysing the mutation spectra of the *p53* gene ? *Mutat. Res.* **307**, 375-386.
28. Sage, E., Lamolet, B., Brulay, E., Moustacchi, E., Chateauneuf, A., and Drobetsky, E.A. (1996) Mutagenic specificity of solar UV light in nucleotide excision repair-deficient rodent cells, *Proc. Natl. Acad. Sci. USA* **93**, 176-180.
29. Dumaz, N., Drougard, C., Sarasin, A., and Daya-Grosjean, L. (1993) Specific UV-induced mutation spectrum in the *p53* gene of skin tumors from DNA repair deficient xeroderma pigmentosum patients, *Proc. Natl. Acad. Sci. USA* **90**, 10529-10533.
30. Tyrrell, R.M. (1973) Induction of pyrimidine dimers in bacterial DNA by 365 nm radiation, *Photochem. Photobiol.* **17**, 69-73.
31. Ley, R.D. and Fourtanier, A. (1997) Sunscreen protection against ultraviolet radiation-induced pyrimidine dimers in mouse epidermal DNA, *Photchem. Photobiol.* **65**, 1007-1011.
32. Mitchell, D.L., Jen, J., and Cleaver, J.E. (1992) Sequence specificity of cyclobutane pyrimidine dimers in DNA treated with solar (ultraviolet B) radiation, *Nucleic Acids Res.* **20**, 225-229.
33. Sage E, Drobetsky EA, Brulay E, Moustacchi E (1995) Molecular analysis of mutations induced by simulated solar light at an endogenous locus in mammalian cells. Detection of lesions and genetic consequences. Les Editions I.N.S.E.R.M., eds Dubertret L, Santus R, Morlière P, 145-151.
34. Clingen, P.H., Arlett, C.F., Roza, L., Mori, T., Nikaido, O., and Green, M. (1995) Induction of Cyclobutane pyrimidine dimers, pyrimidine(6-4)pyrimidone photoproducts, and Dewar valence isomers by natural sunlight in normal human mononuclear cells, *Cancer Res.* **55**, 2245-2248.
35. Clingen, P.H., Arlett, C., Cole, J., Waugh, A., Lowe, J., Harcourt, S., Hermanova, N., Roza, L., Mori, T., Nikaido, O., and Green, M. (1995) Correlation of UV-C and UV-B specific photoproducts in T-lymphocytes and fibroblasts from normal human donors, *Photochem. Photobiol.* **61**, 163-170.
36. Rosenstein, B.S. and Mitchell, D.L. (1991) The repair of DNA damages induced in normal human skin fibroblasts exposed to simulated sunlight, *Radiation Res.* **126**, 338-342.
37. Chadwick, C.A., Potten, C.S., Nikaido, O., Matsunaga, T., Proby, C., and Young, A.R. (1995) The detection of cyclobutane thymine dimers, (6-4) photolesions and the Dewar photoisomers in section of UV-irradiated human skin using specific antibodies, and the demonstration of depth penetration effects, *J. Photochem. Photobiol. B: Biol.* **28**, 163-170.
38. Mullenders, L., van Hoffen, A., Vreeswijk, M., Ruven, H.J., Vrieling, H., and van Zeeland, A. (1996) UV-induced repair and mutagenesis in mammalian cells, in B. Volkmer, H. Heller (eds), *Environmental UV radiation, risk of skin cancer and primary prevention*, Gustav Fischer, Stuttgart, pp. 95-106.
39. Zhang, X., Rosenstein, B.S., Wang, Y., Lebwohl, M., Mitchell, D.M., and Wei, H. (1997) Induction of 8-oxo-7,8-dihydro-2'-deoxyguanosine by ultraviolet radiation in calf thymus DNA and HeLa cells, *Photochem. Photobiol.* **65**, 119-124.
40. Kielbassa, C., Roza, L., and Epe, B. (1997) Wavelength dependence of oxidative DNA damage induced by UV and visible light, *Carcinogenesis* **18**, 811-816.
41. Autier, P. (1996) Exposure to UV-radiation in sunbeds, in B. Volkmer, H. Heller (eds), *Environmental UV radiation, risk of skin cancer and primary prevention*, Gustav Fischer, Stuttgart, pp. 311-315.
42. Pflaum, M., Boiteux, S., and Epe, B. (1994) Visible light generates oxidative DNA base modifications in high excess of strand breaks in mammalian cells, *Carcinogenesis* **15**, 297-300.

THE APPLICATION OF COMET AND MICRONUCLEUS ASSAY FOR THE ASSESSMENT OF INDIVIDUAL RADIATION RISK

W.-U. MÜLLER, T. BAUCH, C. STREFFER

Institut für Medizinische Strahlenbiologie, Universitätsklinikum Essen D-45122 Essen, Germany

Abstract

One of the central goals of biological dosimetry must be the assessment of individual radiosensitivity. Physical dosimetry is only able to give the information on the energy deposited in the human body. This is not sufficient in radiation protection, because the response of the body to radiation dose is much more important than dose itself. This response strongly depends on the individual radiosensitivity, which is not determinable by physical means. Several biological indicator systems have been introduced meanwhile from which answers are hoped to be obtained. One of the major factors affecting individual radiosensitivity is the repair capacity of a person. The comet assay offers the possibility to monitor repair at the level of single cells. Thus, one obtains not only information on the average repair capacity but also on the range. Individual cells are irradiated in vitro and checked for DNA damage either immediately or after various times at 37 °C. The speed of damage removal and the amount of residual damage after a fixed time gives the desired information on the repair capacity. Pronounced differences among people have been found up to now, whereas samples taken from one person at different times show remarkably similar results. Since the introduction of the CB technique, micronuclei in lymphocytes are successfully used as a biological dosemeter after radiation exposure. Their analysis is much faster and easier than that one of dicentrics. The comparatively poor sensitivity in the low dose range can be overcome by isolating sensitive subpopulations of lymphocytes. In addition or alternatively, centromere identification within the micronuclei improves sensitivity of the assay. Very promising is the fact that automatic scoring of micronuclei is feasible and can be done routinely meanwhile.

1. Introduction

Slowly, but with increasing impetus the fact is accepted that physical dosimetry has a very serious limitation, when it comes to the assessment of the individual radiation risk: there is no way to take into consideration the differences in individual radiation sensitivities by physical means. It is not the physical dose alone that determines the extent of health problems, but a major contribution comes from the individual radiation sensitiv-

C. Baumstark-Khan et al. (eds.),
Fundamentals for the Assessment of Risks from Environmental Radiation, 127-134.

ity. Thus, much more important than radiation dose itself is the response of the individual to this dose.

This aspect is most obvious in the case of some hereditary diseases like ataxia telangiectasia [1, 2]. This recessive disease affecting the vascular system and coordination of movements is accompanied by a markedly increased radiation risk. Unfortunately, also the heterozygotes, which do not show the symptoms of the disease and therefore cannot be identified by uncoordinated movements and telangiectasias, have a significantly higher radiation risk than the unaffected population [3-6]. And it has been speculated that some unexpected, serious side effects observed during radiation therapy may be explained by the fact that an ataxia heterozygote had been exposed.

Besides radiotherapy [7], there are at least two other fields in which differences in individual radiation sensitivity play a crucial role: radiation accidents and radiation exposures at working places. In particular, the latter one may turn out to be a real problem, because right now the dose limits are designed to comply with an "average" radiation sensitivity; thus, for individuals with a high radiation sensitivity even observance of the dose limits may result in unacceptably high risks.

Therefore, assays are urgently required that allow the assessment of individual radiation sensitivity. Two of these assays, that may be useful in the future, will be introduced here: the micronucleus and the comet assay.

2. The Micronucleus Assay

An overview of micronucleus assays has been given by Müller and Streffer [8]. The most pertinent features for the topic discussed here will be presented in the following.

Micronuclei are small, round-shaped particles that show up in the cytoplasm after nuclear division, and they stain in almost the same way as the nucleus itself. They consist of complete chromosomes or of chromosome fragments. In the first case, either defective kinetochores or spindle fibres are responsible for the occurrence of micronuclei, in the latter case the missing kinetochore is the major reason; sometimes disrupture of multicentric aberrations may be involved. This "classical" micronucleus should not be mixed up with apoptotic bodies which are generated during interphase by nuclear fragmentation. Morphologically both structures are frequently indistinguishable.

The fact, that micronuclei can be generated during nuclear division (karyokinesis) only, requires the identification of those cells that actually divided, because it is only these cells that got the opportunity to express a micronucleus. Those cells that did not divide must be excluded from micronucleus scoring, because one cannot tell whether they included a micronucleus or not. The problem of identifying those cells that divided was solved by Fenech and Morley in 1985 [9, 10] by the application of cytochalasin B, a fungal toxin, that permits karyokinesis, but not cytokinesis. This effect thus results in "binucleated cells" (cells with 2 nuclei). Only these cells are used for micronucleus scoring.

Virtually thousands of papers have been published meanwhile, which demonstrate the usefulness of micronuclei for the assessment of risks due to radiation or due to chemicals. Therefore, it is obvious to think of micronuclei, when it comes to the assessment of individual radiation sensitivity. The background idea is to expose lymphocytes in vitro to increasing radiation doses and to determine individual radiosensitivity from the micronucleus frequency.

Figure 1 shows one of the prerequisites that are necessary to qualify an assay as a candidate for the assessment of individual radiation sensitivity: there must be donor dependent differences in the dose response relations. Obviously, this requirement is met by the micronucleus assay, because markedly different dose response curves are shown in Fig. 1. However, quite a number of caveats must be mentioned that may limit the usefulness of the micronucleus assay in this context:

- Some of the variability is definitely due to methodological problems (culture conditions, quality of phytohemagglutinin and cytochalasin B etc.).
- Very prominent is the influence of the person who scores micronuclei. We routinely test the scoring efficiency of all of our co-workers involved in micronucleus counting once per year using standard slides, because this efficiency is not only dependent on the individual, but there is also a time-dependent "learning" and "forgetting" effect. This means, an experienced person, who is scoring micronuclei frequently, shows a certain trend to count more micronuclei with time, whereas a newcomer and someone who is counting micronuclei only from time to time tends to count comparatively few micronuclei. An experienced individual counts up to about 50% more micronuclei than an unexperienced one.
- There is some hope that this serious problem can be overcome or at least be diminished by using automatic micronucleus scoring [11-13]. The machine will not change the criteria that it has been told. The problem that remains, however, is that the quality of the slides may change, so that some of the micronuclei drop out of the window that the machine was informed to contain the micronuclei.

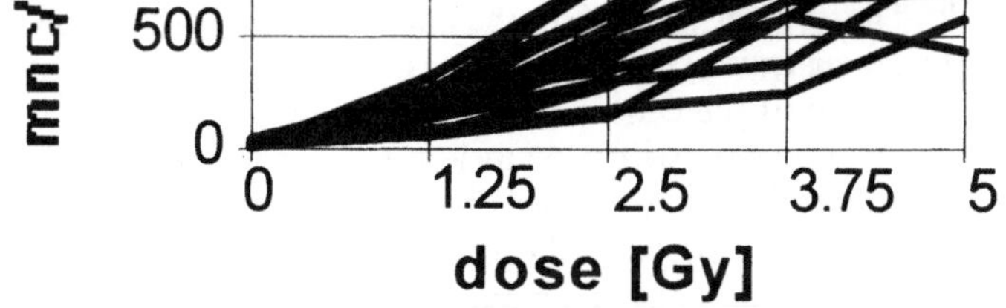

Figure 1. Radiation-induced micronucleus formation in 19 healthy donors (mnc = micronuclei, BNC = binucleated cells)

- Even phytohemagglutinin stimulated lymphocytes consist of different subpopulations with different radiosensitivities [14]. If the contribution of the lymphocyte subsets is different from donor to donor or for the same donor at different times, this will markedly affect micronucleus frequency. The solution to this problem may come from a restriction of scoring micronuclei in lymphocytes of the same type. This procedure, however, is very tedious.
- The same donor screened for in vitro radiation induced micronuclei at different times shows different dose response curves (data not shown). This may mean that radiation sensitivity is time-dependent (which is not so unreasonable), but it could also mean that the intra-individually methodological variability is very high.
- The frequency of micronuclei that is observed after radiation exposure is the result of the cooperation of several factors (initial DNA damage, DNA repair, cell proliferation). Thus, one has to expect a comparatively high variability, because the variabilities of all the affecting factors will play a role. This is different for other assay systems which determine just one endpoint. On the other hand, individual radiosensitivity of a person is also the sum of many factors affecting each other, so that micronuclei may be superior with regard to this aspect, when compared with assays that measure only one endpoint.

In conclusion, a lot more studies will be necessary before one can decide, whether micronuclei in lymphocytes exposed in vitro are a reliable indicator for individual radiosensitivity.

3. The Comet Assay

In 1984, Östling and Johanson [15] introduced a method that went almost unnoticed for some time, but boomed since the early nineties: the detection of DNA damage in entire cell nuclei by electrophoresis of the DNA of these nuclei in agarose on slides. In dependence on the amount of DNA damage induced (or left unrepaired), "comets" are formed in the gel. This technique opens the unique opportunity to obtain quantitative information on the amount of DNA damage and DNA repair in individual cells.

Various modifications of the original method have been published meanwhile (e.g. [16-18]). We use a method that is comparatively close to that one of Östling and Johanson and is described in detail in [19]. Briefly, single cells to be checked for DNA damage are embedded in melted agarose on a slide. After gelling, most non-DNA compounds are lysed by SDS (sodium dodecyl sulfate), the slides are exposed to a weak electric field in a specific buffer system and after washing, the DNA is stained with propidium iodide. The amount of DNA in the tail and in the head region of the comet is determined by image analysis. The more DNA is contained in the tail, the higher is the amount of DNA damage. If the cells had the opportunity to repair some of the damage at 37° C before lysis, the tails will contain less DNA compared with cells that were fixed immediately after radiation exposure, i.e. without the opportunity to repair.

The comet assay makes possible the determination of initial DNA damage, the speed of repair, and the amount of unrepaired damage (residual damage) after a specific time of repair. In all cases, the assay does not only give the information on mean values, but it allows the representation of distributions based on the results on the single cell level. In the following, our data on the repair capacity of lymphocytes of different donors will be discussed.

Similarly to Fig. 1 for micronuclei, Fig. 2 shows individual repair kinetics of lympho-cytes of 24 healthy donors. Again, considerable differences are observed for individual persons. In the case of the comet assay, however, the repair kinetics are remarkably similar when measured for one person at different times (Fig. 3).

That a slow repair in radiation exposed lymphocytes connected with a high residual damage after three hours is indicative of a higher radiosensitivity of the donor can be concluded from results obtained for radiotherapy patients, who showed severe side effects after some fractions. Most, but not all of these patients, had lymphocytes whose repair kinetics were either in the range of those healthy people with the slowest repair or they performed even worse.

Compared with the micronucleus assay, the comet assay has some advantages for the determination of individual radiosensitivity:

- It is markedly faster.
- No cell proliferation is necessary (only vitality of the cells is required).
- Intra-individual variability is less.
- A clearly defined endpoint is measured.
- Most of the people with a demonstrable increased radiosensitivity (e.g. patients with severe side effects during radiation therapy) were conspicuous in the comet assay, but not in the micronucleus assay.

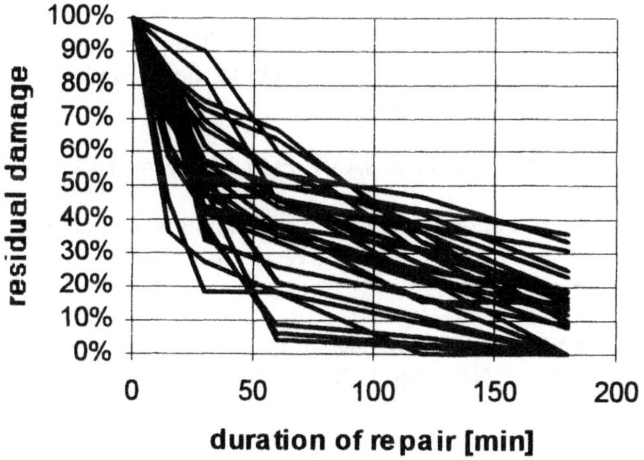

Figure 2. Repair kinetics in lymphocytes of 24 healthy donors

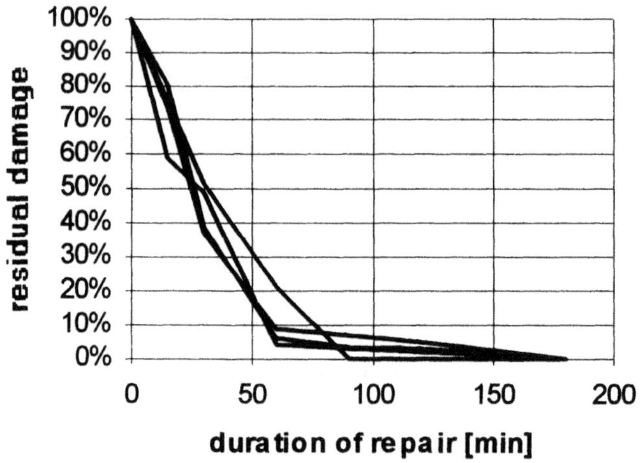

Figure 3. Repair kinetics in lymphocytes of 1 healthy donor at different times (About 1 month difference each between the determinations)

4. Questions That Must be Answered in the Future for the Assessment of Individual Radiation Sensitivity

We do know that there are marked differences in radiation sensitivity among individuals. Right now, we do not know the final answers to the following questions (although in some cases there are first indications):

4.1. ARE THE SENSITIVITIES THE SAME FOR ALL RADIATION RISKS?

This means that we do not know, whether an individual whose cells are radiosensitive, when cell death is measured, is also sensitive as to, for example, transformation, i.e. with regard to tumour risk.

4.2. WHICH CELL TYPE IS THE OPTIMAL ONE FOR THE ASSESSMENT OF INDIVIDUAL RADIATION SENSITIVITY?

The few studies that have been carried out up to now seem to indicate that an individual with radiosensitive lymphocytes must not have necessarily also sensitive fibroblasts [20-23]. Very careful studies are required in the future, taking into account the various parameters that determine radiosensitivity (oxgen supply, cell cycle distribution, subpopulations etc.).

4.3. WHICH ENDPOINT IS MOST APPROPRIATE AND, THUS, WHICH ASSAY SYSTEM SHOULD BE USED?

Some assays determine just one specific endpoint (e.g. double strand breaks), others measure the result of damage to various endpoints (e.g. micronuclei). We do not know which endpoint gives the optimum information with regard to radiosensitivity, and therefore it is not possible to predict right now which assay system is most suitable.

4.4. HOW LARGE IS THE RANGE OF THE INDIVIDUAL RADIOSENSITIVITIES?

As "radiosensitivity" is a very ill-defined term, it is not surprising, that it is difficult to quantify the range of the sensitivities. "Ill-defined" primarily refers to the fact that people think in different terms, when they talk about radiosensitivity; some refer to DNA- or chromosomal damage, some to cell death, to cell transformation or to still other endpoints. The problem is aggravated by differences obtained in in vitro- and in vivo-studies.

5. Summary

We will need assays that determine the individual radiation sensitivity. This information is important for three groups of people:

1. tumour patients that receive a radiation therapy;
2. victims of a radiation accident;
3. people being employed at a radiation exposed working place.

Most probably, several complementary assays will be required. A lot of studies have to be carried out, before one can decide, which assays are appropriate. From the two assays introduced here, the comet assay seems to be slightly more promising than the micronucleus assay as a source of information of individual radiosensitivity.

6. References

1. Taylor, A.M.R., Harnden, D.G. and, Arlett, C.F. (1975) Ataxia telangiectasia: a human mutation with abnormal radiation sensitivity, *Nature*, **258**, 427.
2. Boder, E. (1985) Ataxia-telangiectasia: An overview, in R.A. Gatti and M. Swift (eds.), Ataxia-Telangiectasia: Genetics, Neuropathology, and Immunology of a Degenerative Disease of childhood; Kroc Foundation Series 19, A.R.Liss, New York, pp. 1-63.
3. Arlett, C.F. and Priestley, A. (1985) An assessment of the radiosensitivity of ataxia-telangiectasia heterozygotes, in R.A. Gatti and M. Swift (eds.), Ataxia-telangiectasia: Genetics, Neuropathology, and Immunology of a Degenerative Disease of Childhood; Kroc Foundation Series 19, Alan R. Liss, New York, pp. 101-109.
4. Rosin, M.P. and Ochs, H.D. (1986) In vivo chromosomal instability in ataxia-telangiectasia homozygotes and heterozygotes, *Hum. Genet.* **74**, 335-340.
5. Rosin, M.P., Ochs, H.D., Gatti, R.A. and, Boder, E. (1989) Heterogeneity of chromosomal breakage levels in epithelial tissue of ataxia-telangiectasia homozygotes and heterozygotes, *Hum. Genet.* **83**, 133-138.
6. West, C.M.L., Elyan, S.A.G., Berry, P., Cowan, R. and, Scott, D. (1995) A comparison of the radiosensitivity of lymphocytes from normal donors, cancer patients, individuals with ataxia-telangiectasia (A-T) and A-T heterozygotes, *Int. J. Radiat. Biol.* **68**, 197-203.

7. Streffer, C. (1997) Genetische Prädisposition und Strahlenempfindlichkeit bei normalen Geweben, *Strahlentherapie*, **173**, 462-468.
8. Müller, W.-U. and Streffer, C. (1994) Micronucleus assays, *Adv. Mutag. Res.* **5**, 1-133.
9. Fenech, M. and Morley, A. (1985) Solutions to the kinetic problem in the micronucleus assay, *Cytobios*, **43**, 233-246.
10. Fenech, M. and Morley, A.A. (1985) Measurement of micronuclei in lymphocytes, *Mutat. Res.* **147**, 29-36.
11. Böcker, W., Streffer, C., Müller, W.-U. and, Yu, C. (1996) Automated scoring of micronuclei in binucleated human lymphocytes, *Int. J. Radiat. Biol.* **70**, 529-537.
12. Böcker, W., Müller, W.-U. and, Streffer, C. (1995) Image processing algorithms for the automated micronucleus assay in binucleated human lymphocytes, *Cytometry*, **19**, 283-294.
13. Schreiber, G.A., Beisker, W., Bauchinger, M. and, Nüsse, M. (1992) Multiparametric flow cytometric analysis of radiation-induced micronuclei in mammalian cell cultures, *Cytometry*, **13**, 90-102.
14. Wuttke, K., Streffer, C. and, Müller, W.-U. (1993) Radiation induced micronuclei in subpopulations of human lymphocytes, *Mutat. Res.* **286**, 181-188.
15. Ostling, O. and Johanson, K.J. (1984) Microelectrophoretic study of radiation-induced DNA damages in individual mammalian cells, *Biochem. Biophys. Res. Commun.* **123**, 291-298.
16. Müller, W.-U., Bauch, T., Streffer, C., Niedereichholz, F. and, Böcker, W. (1994) Comet assay studies of radiation-induced DNA damage and repair in various tumour cell lines, *Int. J. Radiat. Biol.* **65**, 315-319.
17. Olive, P.L., Banáth, J. and, Durand, R.E. (1990) Heterogeneity in radiation-induced DNA damage and repair in tumor and normal cells measured using the "comet" assay, *Radiat. Res.* **122**, 86-94.
18. Singh, N.P., McCoy, M.T., Tice, R.R. and, Schneider, E.L. (1988) A simple technique for quantitation of low levels of DNA damage in individual cells, *Exp. Cell Res.* **175**, 184-191.
19. Pöller, F., Bauch, T., Sauerwein, W., Böcker, W., Wittig, A. and, Streffer, C. (1996) Comet assay study of DNA damage and repair of tumour cells following boron neutron capture irradiation with fast d(14)+Be neutrons, *Int. J. Radiat. Biol.* **70**, 593-602.
20. Cole, J., Arlett, C.F., Green, M.H.L., Harcourt, S.A., Priestley, A., Henderson, L., Cole, H., James, S.E. and, Richmond, F. (1988) Comparative human cellular radiosensitivity: II. The survival following gamma-irradiation of unstimulated (G0) T-lymphocytes, T-lymphocyte lines, lymphoblastoid cell lines and fibroblasts and normal donors, from ataxia telangiectasia patients and from ataxia telangiectasia heterozygotes, *Int. J. Radiat. Biol.* **54**, 929-943.
21. Kushiro, J., Nakamura, N., Kyoizumi, S., Nishiki, M., Dohi, K. and, Akiyama, M. (1990) Absence of correlations between radiosensitivities of human T-lymphocytes in G0 and skin fibroblasts in log phase, *Radiat. Res.* **122**, 326-332.
22. Green, M.H.L., Arlett, C.F., Cole, J., Harcourt, S.A., Priestley, A., Waugh, A.P.W., Stephens, G., Beare, D.M., Brown, N.A.P. and, Shun-Shin, G.A. (1991) Comparative human cellular radiosensitivity: III. Gamma-radiation survival of cultured skin fibroblasts and resting T-lymphocytes from the peripheral blood of the same individual, *Int. J. Radiat. Biol.* **59**, 749-765.
23. Geara, F.B., Peters, L.J., Ang, K.K., Wike, J.L., Sivon, S.S., Guttenberger, R., Callender, D.L., Malaise, E.P. and, Brock, W.A. (1992) Intrinsic radiosensitivity of normal human fibroblasts and lymphocytes after high- and low-dose-rate irradiation, *Canc. Res.* **52**, 6348-6352.

ENVIRONMENTAL RADIATION DOES NOT ENHANCE THE TOLERANCE OF YEAST CELLS TO ACUTE RADIATION EXPOSURE

D. FRANKENBERG [1], K. HELMSTAEDTER [2], J. BOEHM [2], W. PESSARA [2], M. FRANKENBERG-SCHWAGER [1]

[1] *Universitaet Goettingen, Klin.Strahlenbiologie und Klin.Strahlenphysik von-Siebold-Str. 3, D-37075 Goettingen, Germany.*
[2] *Physikalisch-Technische Bundesanstalt Bundesallee 100, D-38116 Braunschweig, Germany*

Abstract

The aim of the investigations was to clarify whether cells adapted to the normal background radiation show enhanced resistance to the damaging effects of ionizing radiation. The underground laboratory for dosimetry and spectrometry (UDO) of the Physikalisch-Technische Bundesanstalt (PTB) at a depth of 925 m allowed to grow cells at a dose rate of 2.2 nSv/h compared with about 110 nSv/h of the normal background radiation. The yeast mutant rad54-3 which is temperature conditional for the repair of DNA double-strand breaks (dsb) was cultivated at 2.2 and 110 nSv/h, respectively, and irradiated in UDO using electrons ($^{90}Sr/^{90}Y$). This mutant rad54-3 allows to determine separately induction and repair of dsb, solely by cultivating irradiated cells at 36 °C (no dsb repair) or at 23 °C (with dsb repair). The results show that the normal background radiation does not adapt yeast cells for reduced initial dsb and elevated repair of dsb.

1. Introduction

Radiation-induced resistance (adaptive response) is observed in cells or organisms pretreated with a low inducing dose of radiation [1, 2, 3]. In mammalian cells the inducing dose is usually a few cGy [4] and the time interval between the inducing dose and the challenging dose is about several hours [5]. Radioresistance is not induced when de novo protein synthesis during this interval is inhibited [6, 7], suggesting that the induction of enzymatic activity plays an important role. Radiation-induced resistance is observed under appropriate conditions for various endpoints such as the induction of chromosome aberrations [1, 2], mutations [8], cell survival [9, 10], neoplastic transformation [11] and micronucleus formation [11]. Radiation-inducible repair of DNA double-strand breaks was reported for a mouse cell line [12] but no such effect was reported for the repair of DNA single-strand breaks [13].

C. Baumstark-Khan et al. (eds.),
Fundamentals for the Assessment of Risks from Environmental Radiation, 135-140.

In the unicellular eukaryotic yeast Saccharomyces cerevisiae inducible radioresistance at the level of cell survival and mutation was shown to result from radiation-induced DNA damage mediated by OH radicals [14]. Based on these findings we were interested to answer the question: does environmental radiation enhance the tolerance of yeast cells to acute radiation exposure? Or, in other words, does a significantly reduced level of environmental radiation enhance the sensitivity of yeast cells against high doses of radiation? A similar question was asked by an italian group studying the influence of a reduced background of environmental radiation on the sensitivity of diploid yeast cells against the radiomimetic compound methyl-methane sulfonate (MMS) [15]. These experiments were performed in an underground laboratory in the Gran Sasso mountain where the background radiation was reduced 7-fold (dose rate: 26 nSv/h). Cells cultivated under this condition exhibited the same capacity to repair MMS-induced damage as cells grown at the normal environmental radiation background. However, they also showed an enhanced frequency of recombination events which was interpreted as a lower protection against the damaging effect of MMS compared to cells grown at normal radiation background. On the other hand it is known that DNA dsb play an important role in recombination processes in yeast. Therefore, these findings prompted us to investigate the effect of the natural environmental radiation background with specific emphasis on DNA dsb induction and repair.

For this purpose the underground laboratory for dosimetry and spectrometry (UDO) of the PTB Braunschweig was used which is located at a depth of 925 m in the research mine "Asse" run by the GSF-Forschungszentrum für Umwelt- und Gesundheitsforschung. A photon equivalent dose rate of 2.2 nSv/h was measured in the incubation chamber where the cells were cultivated. Thus, the environmental radiation background in the UDO is reduced by a factor of 50 [16]. Induction and repair of DNA dsb was measured in the mutant rad54-3 of the yeast Saccharomyces cerevisiae. This mutant is temperature conditional with respect to dsb repair [17].

2. Materials and Methods

2.1. YEAST CELLS

Diploid cells of the temperature conditional yeast mutant rad54-3 are radiosensitive and deficient in dsb repair at the restrictive temperature of 36 °C [17]. They show an exponential survival curve where the dose D_0 (i.e. one lethal event yielding a surviving fraction of 37 %) corresponds to about 1 dsb per cell [18]. Therefore, at the restrictive temperature the induction of dsb can be indirectly monitored simply by measuring the survival response. At the permissive temperature of 23 °C cells are able to repair dsb and this results in a much better survival after radiation exposure [18]. Since the difference between survival curves obtained at 36 °C and 23 °C is caused by dsb repair, the rad54-3 survival assay provides a simple method to assess induction and repair of dsb by assaying the colony forming ability of cells.

Starting from a single yeast colony, cells were cultivated on YPD growth agar plates (1% yeast extract, 2% peptone, 2% glucose, 1.5% agar) at 30°C for more than

120 generations both in our laboratory in Goettingen at normal radiation background and in the UDO at a 50-fold reduced environmental radiation background.

2.2. RADIATION EXPOSURE AND DOSIMETRY

Yeast cells were irradiated with electrons from a Sr-90/Y-90 source of 0.7 cm^2 (Amersham-Buchler) with an activity of about 2.5 GBq. The source is located within a container which serves for transport when closed and which can be used as a radiation source in different radiation facilities when opened [19]. Cells were kept at ice temperature during irradiation in order to inhibit repair processes. The distance between radiation source and cells in monolayer was 4 cm. The tissue equivalent dose rate at the location of cells was 17.85 Gy/h ± 3% (flat ionization chamber calibrated in the PTB) and 17.20 Gy/h ± 5% (thin thermoluminescence detectors: TLD 700). The densitometric evaluation of exposed films showed a homogeneity better than ± 0.5% over the irradiation field (Ø 30 mm) at an exposure distance of 4 cm. The experimental set-up is illustrated in Figure 1.

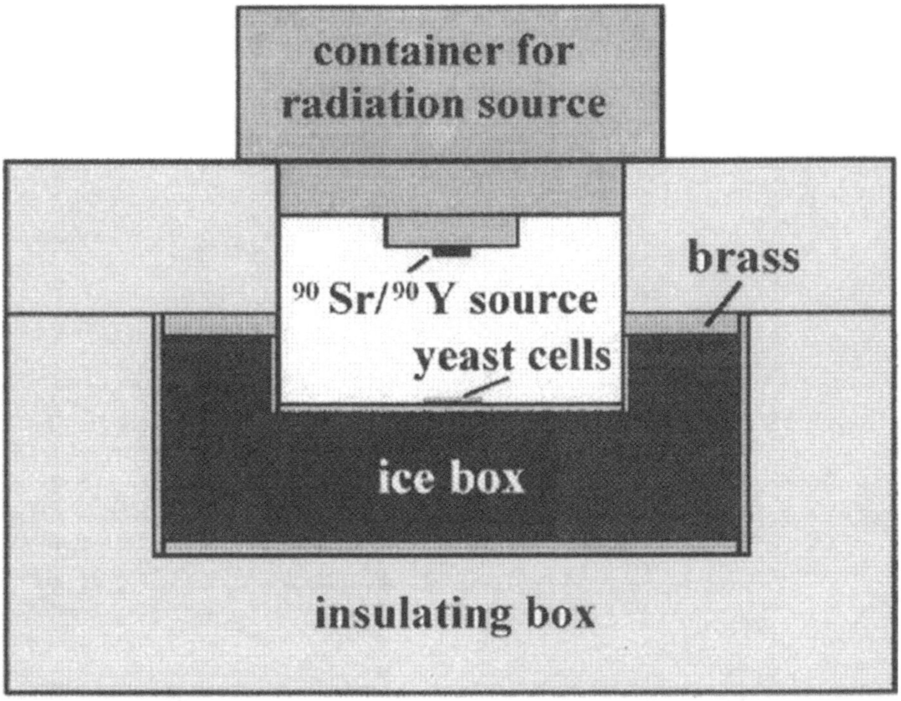

Figure 1. Schematic drawing of the irradiation facility in the underground laboratory

2.3. SURVIVAL STUDIES

Cells of one yeast colony were homogeneously resuspended in 67 mM phosphate buffer, pH 7.0. A volume containung $5 \cdot 10^4$ viable cells was spread on the surface of a halved membrane filter (Ø 11 mm, Millipore, pore size 0.8 µm) which was transferred onto an icecold moist glass fiber filter. The glass fiber filter accommodated 6 halved membrane filters with cells grown in Goettingen and 6 halved filters with cells grown in the underground laboratory. After loading the glass fibre filter was transferred into the irradiation facility where cells were irradiated simultaneously. After irradiation each filter was transferred into icecold phosphate buffer and cells were resuspended by vigorous vortexing for one minute. Cells were appropriately diluted to yield about 100 macrocolonies per growth agar plate after 7 days (incubation temperature 36 °C) or 4 days (incubation temperature 23 °C).

3. RESULTS AND DISCUSSION

Ten experiments were performed with cells grown for at least 120 generations either at normal radiation background in Goettingen or at a 50-fold reduced radiation background in the underground laboratory. Cells from Goettingen were transported to the underground laboratory where they were irradiated with various doses of electrons from the Sr-90/Y-90 source together with cells grown in the underground laboratory. The resultant survival curves are presented in Figure 2 for cells incubated at the permissive

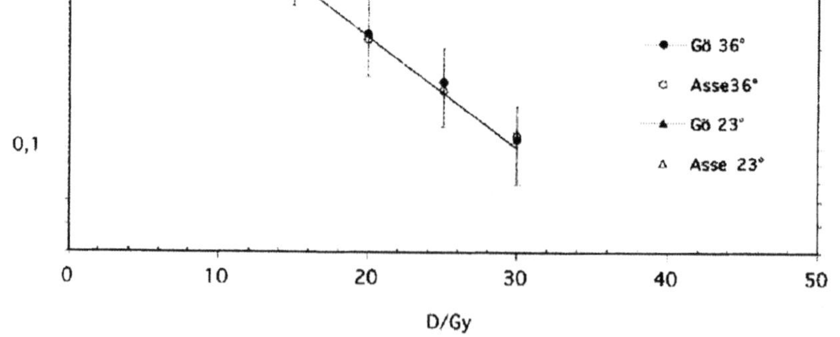

Figure 2. Survival curves after exposure to a Sr-90/Y-90 source of rad54-3 cells grown for at least 120 generations at normal radiation background in Goettingen (Gö) or at a 50-fold reduced radiation background in the underground laboratory (Asse). Survival was assessed after incubation of plated cells at 36 °C (dsb repair restricted) or at 23 °C (dsb repair permitted).

(23 °C) and at the restrictive (36 °C) temperature for dsb repair. It is evident that cells pregrown in Goettingen or in the underground laboratory exhibit virtually identical survival curves at the corresponding temperatures. The different levels of environmental radiation background in which cells were grown have, therefore, neither an effect on the induction of dsb (survival curves at 36 °C) nor on the repair of these lesions (survival curves at 23 °C). At the dose rate of 110 nSv/h measured for the normal environmental radiation background we have calculated an induced number of $7 \cdot 10^{-9}$ double-strand breaks [20, 21]. $7 \cdot 10^{-7}$ single-strand breaks and about $5 \cdot 10^{-6}$ damaged bases per cell per hour [22]. Thus, the yield of DNA damage per cell induced by environmental radiation seems to be far too low to cause any stimulatory effect on cellular radiation resistance, especially when the relatively higher background of spontaneous endogeneously produced oxidative DNA damage per cell is taken into account [see 22].

Our results are only in partial agreement with data reported by Satta et al. [15] who studied in the yeast Saccharomyces cerevisiae the effect of environmental radiation background on induction and repair of lesions induced by MMS. They found an increased level of MMS-induced total and reciprocal recombinations for cells adapted to a 7-fold reduced background radiation. However, the survival curves observed for increasing concentrations of MMS were identical for cells pregrown at normal and reduced environmental background radiation. From these results the authors concluded that normal environmental radiation has a protective effect on the induction of MMS-lesions, while repair of these lesions is unaffected. While the latter statement is supported by our results, we could, however, not observe a protective effect for the induction of dsb induced by acute radiation.

In summary, it is concluded from our results that the normal environmental background radiation does not enhance the tolerance of the yeast Saccharomyces cerevisiae to acute radiation exposures.

4. References

1. Olivieri, G., Bodycote, J., Wolff, S. (1984) Adaptive response of human lymphocytes to low concentrations of radioactive thymidine. *Science* **223**, 594-597.
2. Shadley, J.D., Wolff, S. (1987) Very low doses of x-rays can cause human lymphocytes to become less susceptible to ionizing radiation. *Mutagenesis* **2**, 95-96.
3. Cai, L., Liu, S.-Z. (1990) Induction of cytogenetic adaptive response of somatic and germ cells in vivo and in vitro by low doses of x-irradiation. *Int. J. Radiation Biology* **58**, 187-194.
4. Shadley, J.D., Wiencke, J.-K. (1989) Induction of the adaptive response by x-rays is dependent on radiation intensity. *Int. J. Radiation Biology* **56**, 107-118.
5. Shadley, J.D., Afzal, V., Wolff, S. (1987) Characterization of the adaptive response to ionizing radiation induced by low doses of x-rays to human lymphocytes. *Radiation Research* **111**, 511-517.
6. Wolff, S., Wiencke, J.-K., Afzal, V., Joungblom, J., Corts, F. (1989) The adaptive response of human lymphocytes to very low doses of ionizing radiation: A case of induced chromosomal repair with the induction of specific proteins. In K.F. Baverstock and J.W. Stather (eds.) *Low Dose-Radiation: Biological Bases of Risk Assessments*. Taylor and Francis, London, pp. 446-454.
7. Wiencke, J.K., Afzal, V., Olivieri, G., Wolff, S. (1986) Evidence that the [^{3}H] thymidine-induced adaptive response of human lymphocytes to subsequent doses of x-rays involves the induction of a chromosomal repair mechanism. *Mutagenesis* **1**, 375-380.

8. Kelsey, K.T., Memisoglu, A., Frenkel, D., Liber, H.L. (1991) Human lymphocytes exposed to low doses of x-rays are less susceptible to radiation-induced mutagenesis. *Mutation Research* **263**, 197-201.
9. Shadley, J.D., Dai, G. (1992) Cytogenetic and survival responses in G_1 phase human lymphocytes. *Mutation Research* **265**, 273-281.
10. Azzam, E.I., deToledo, S.M., Raaphorst, G.P., Mitchel, R.E.J. (1992) Radiation-induced radioresistance in normal human skin fibroblast cell line. In T. Sugahara, L.A. Sagan, T. Aoyama (eds.) *Low Dose Irradiation and Biological Defense Mechanisms*, Elsevier Science Publishers, Amsterdam, pp. 291-294.
11. Azzam, E.I., Raaphorst, G.P., Mitchel, R.E.J. (1994) Radiation-induced adaptive response for protection against micronucleus formation and neoplastic transformation in C3H 10T1/2 mouse embryo cells, *Radiation Research Suppl.* **138**, S28-S31.
12. Zhou, P.-K., Sun, W.-Z., Liu, X.-Y, Zhang, Y.-P.,Wei, K. (1992) Adaptive response of mutagenesis and DNA double-strand break repair in mouse cells induced by low doses of γ-rays. In T. Sugahara, L.A. Sagan, T. Aoyama (eds.) *Low Dose Irradiation and Biological Defense Mechanisms*, Elsevier Science Publishers, Amsterdam, pp. 271-274.
13. Koch, C.J., Giandomenico, A.R. (1994) Do low doses of radiation induce a response modulating induction or repair of DNA single-strand breaks? *Radiation Research Suppl.* **138**, S89 - S92.
14. Boreham, D.R. and Mitchel, R.E.J. (1991) DNA lesions that signal the induction of radioresistance and DNA repair in yeast, *Radiation Research* **128**, 19-28.
15. Satta, L., Augusti-Tocco, G., Cundari, E., Esposito, A., Fiore, M., Paggi, P., Poggesi, F., Ricordy, R., Scarsella, G. (1992) Biological response of the yeast Sacch. cerev. to chemical radiomimetic agents: Influence of low background environment. In: see lit. 10. pp. 383-386.
16. Peßara, W., Lauterbach, U. (1993) Ein extrem strahlungsarmes Untergrundlaboratorium für Dosimetrie und Spektrometrie (UDO). Tagung des Fachverbandes für Strahlenschutz in Binz auf Rügen, FS-93-67-T, 455-460.
17. Budd, M., Mortimer, R.K. (1982) Repair of double-strand breaks in a temperature conditional radiation-sensitive mutant of Sacch. cer. *Mutation Research* **33**, 157-164.
18. Frankenberg, D., Frankenberg-Schwager, M., Harbich, R. (1984) Split-dose recovery is due to the repair of DNA double-strand breaks. *Int. J. Radiation Biology* 46, 541-553.
19. Böhm, J., Hillion, Ph., Simoen, J.-P. (1976) Intercomparison of the PTB and LMRI standards in beta dosimetry. PTB-Dos-1.
20. Frankenberg-Schwager, M., Frankenberg, D., Harbich, R. (1987) Potentially lethal damage repair is due to the difference of DNA double-strand break repair under immediate and delayed plating conditions. *Radiation Research* **111**, 192-200.
21. Frankenberg-Schwager, M., Frankenberg, D., Bloecher, D., Adamczyk, C. (1979) The influence of oxygen on the survival and yield of DNA double-strand breaks in irradiated yeast cells. *Int. J. Radiation Biology* **36**, 261-270.
22. Ward, J.F. (1994) DNA damage as the cause of ionizing radiation-induced gene activation. *Radiation Research Supplement* **138**, S85-S88.

BIOLOGICAL UV DOSIMETRY OF ENVIRONMENTAL RADIATION BASED ON DNA DAMAGE

A. BÉRCES, S. GÁSPÁR, G. RONTÓ

Institute of Biophysics, Semmelweis University of Medicine
Budapest, H-1444, P.O.Box 263, Hungary

Abstract

Biological UV dosimetry gets an increasing importance in the biological risk assessment due to environmental radiation. Achievements obtained with different biological dosimeters, based mainly on DNA damage, make possible in the future the development of a measuring network. This paper summarizes the most important criteria for biological UV dosimeters elaborated by BIODOS group in 1996. For the phage T7 dosimeter these criteria are discussed. In addition, application of the phage T7 dosimeter in surface and underwater measurements is demonstrated.

1. Introduction

The leading environmental factors of physical nature are ionizing radiations and solar (environmental) radiation, the latter one is called frequently as non-ionizing radiation. This distinction has been made on the basis of energy content of the photons/particles present in the radiation. Considering the history of the efforts to quantify these environmental factors, one can conclude that by the increasing knowledge on biological damage due to these physical agents the development of quantification was enhanced. However, the greatest development have been promoted by shocking accidents, like the nuclear weapon explosions in Japan or nuclear plant disaster, recently in Chernobyl [see 1]. In the mid-eighties a similarly shocking as well as stimulating information was for the scientific community dealing with UV radiation the evidence for the depletion of ozone layer in the stratosphere of the Southern Hemisphere [2, 3]. Although the rate of depletion has been found smaller, a significant decrease was detected [4-6] in the Northern Hemisphere as well.

2. Methods and Theoretical Considerations

Development of the biological/health damage due to radiations (both to ionizing and non-ionizing ones) can generally be characterized by a chain of reactions presented schematically in Fig. 1. The first step is the primary interaction between radiation and

C. Baumstark-Khan et al. (eds.),
Fundamentals for the Assessment of Risks from Environmental Radiation, 141-148.

the target molecule/molecules of the biological system, which through multiple events leads to manifestation of the biological damage.

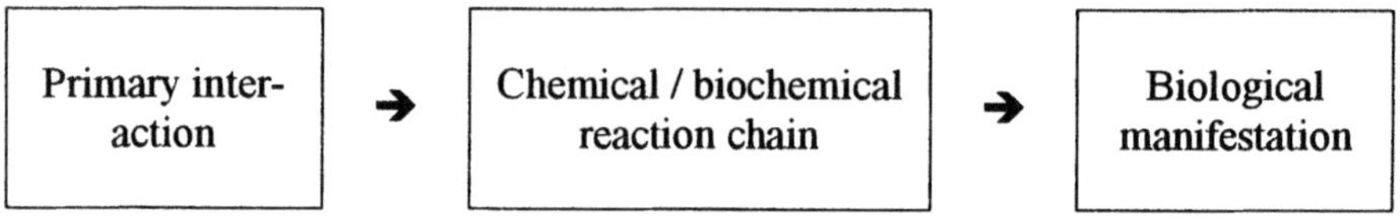

Figure 1. Scheme of the development of biological damage due to radiation

Biological UV dosimetry looks at the biological manifestation, the last step in the Fig. 1., and measuring a given biological effect, partly the biological hazard, partly the radiation will be assessed. Among the simple biological systems proposed for UV dosimetry, bacteria, bacterial viruses, bacterial spores [8-13] have of special interest because their DNA components are mainly the target molecules in the living cells. Thus these dosimeters can be considered as models of cellular chromosome damage.
BIODOS group makes efforts to elaborate and to use a few promising DNA-based dosimeters. As biological dosimeter, *bacteriophage T7 dosimeter* has been developed in our laboratory. The biological effect to be detected/measured is the phage killing/inactivating effect of radiation. The phage activity is evaluated on E. coli B/r host bacteria. Our earlier results [13, 14] made it possible to quantify the biological effect. It has been demonstrated that in accordance to the CIE recommendations [7]

$$\left|\ln(n/n_0)\right| = \sum_{\lambda,t} E(\lambda) S^{T7}(\lambda)\Delta\lambda\,\Delta t \qquad (1)$$

where $E(\lambda)$ is the spectral distribution of the irradiance/radiant flux and $S^{T7}(\lambda)$ means the spectral sensitivity of phage T7 inactivation, n/n_0 is the survival ratio of bacteriophage. The spectral sensitivity has been determined with monochromatic light and as an action spectrum is presented in Fig. 2, together with other prominent action spectrum MED and the sensitivity spectrum of a Robertson-Beger meter [15]. All sensitivities are expressed in absolute values in cm^2/mJ units. From equation (1) is obvious, that through determining the survival ratio (n/n_0) of bacteriophage T7 the BED can be determined directly. Phage survival ratio is measured using traditional or automatized microbiological methods [12, 16]. Thus, the accuracy of the BED does not depend on the values in the right side of equation (1) (i.e. $E(\lambda)$ and $S(\lambda)$ values).

Based on earlier results obtained with UV-C (λ = 254 nm) radiation [17], the absolute value of the phage survival ratio gives the average number (expected value) of the lethal damages in one phage particle (in the phage nucleoprotein) assuming a Poisson distribution of these damages. This aspect defines the phage T7 dose unit as well: one lethal damage/particle.

3. Results and Discussion

3.1.CRITERIA OF BIOLOGICAL DOSIMETERS

The most important criteria have been elaborated by the BIODOS group using DNA-based dosimeters [9, 12, 13, 18, 19]. In terms of these requirements phage T7 dosimeter will be discussed.

3.1.1. Biological Relevance

The molecular basis of the UV effect on living (cellular) systems is very frequently formation of DNA photoproducts. This photochemical event is an essential reaction in the biological UV damage (see Fig. 1.). Its correlation to cancer induction is widely accepted. Phage T7 particle contains a single DNA molecule of small size (40000 base pairs) with a protein core in its center which is covered with a protein envelope [20]. Thus the DNA-protein structure can be considered as a model of chromosomes.

Phage inactivation due to environmental radiation is also connected with photoproduct formation in the phage-nucleoprotein. Thus, without going into details of the different photoproducts formed by different wavelength components of the solar radiation, one can conclude that the phage T7 dosimeter does fulfill the criterion to have relevance to the biological/health damage due to environmental radiation.

3.1.2. Statistical Evaluation of the Damage, Zero Point Definition

Phage T7 dosimeter contains a high number (500-50000/ml) of phage particles in solution and it is exposed to the solar radiation in quartz cuvettes [21]. Inactivation of the particles follows one-hit kinetics, it is a real statistical event and the evaluation of the number of active particles is also a statistical one.

For phage T7 dosimeter the number of the active particles at the beginning of the exposure (n_0) is the zero point of the measurement. Using a phage solution with an active phage number of 500-50000/ml, depending on the experimental conditions, this number can be measured with an error not higher than 10 %.

3.1.3. Quantification of the BED and Linear Response

The quantitative description of the phage T7 inactivation dose is given by equation (1). Using this dose definition the phage T7 dose scale becomes linear corresponding to the fact that the phage inactivation follows an exponential function of the fluence/radiant exposure.

3.1.4. Validity of the Reciprocity Law and Wavelength Interaction

For phage T7 dosimeter the reciprocity law has been checked both for artificial and environmental radiation. The law was found to be valid for a 100-fold intensity/irradiance range at least. The validity of the reciprocity law allows the rational planning of the measuring tasks and economical use of the samples in field monitoring for different experimental conditions.

The additivity of the inactivating effect due to different wavelengths was not tested, however, earlier at 254 nm direct reversion of the phage inactivation was demonstrated [17]. The interpretation of this effect is the dimer-monomer equilibrium which is shifted to the direction of monomers under the irradiation conditions (at shorter wavelengths) used. The physical background of the direct photoreversion is the spectral change of the nucleotides damaged by UV radiation. For pyrimidine dimers identified as leading photoproducts a significant absorption peak has been detected at wavelengths lower than 220 nm [22].

Phage T7 dosimeter is a buffer solution of phage particles without any metabolic activity, thus enzymatic reactivation of the photoproducts in this system by longer wavelength components of radiation can not be expected. Thus from one hand the antagonistic effect, on the other hand synergistic effect of the different wavelengths can be excluded because of the lack of metabolism of the phage particles during irradiation.

3.1.5. Stability of the UV Sensitivity of the Dosimeter

The UV sensitivity of phage inactivation, more precisely its wavelength dependence is demonstrated in Fig. 2. For phage T7 the experimental errors are also indicated. However, any future improvement in the accuracy of the action spectrum will not influence the values of the doses measured by phage T7 dosimeter, because in the evaluation process the sensitivity spectrum is not used.

A spontaneous change in the phage inactivation sensitivity can be also excluded because it is genetically determined. Thus storing these data for long-term monitoring, they will be applicable at any time for evaluation of the tendencies of biological/health risk.

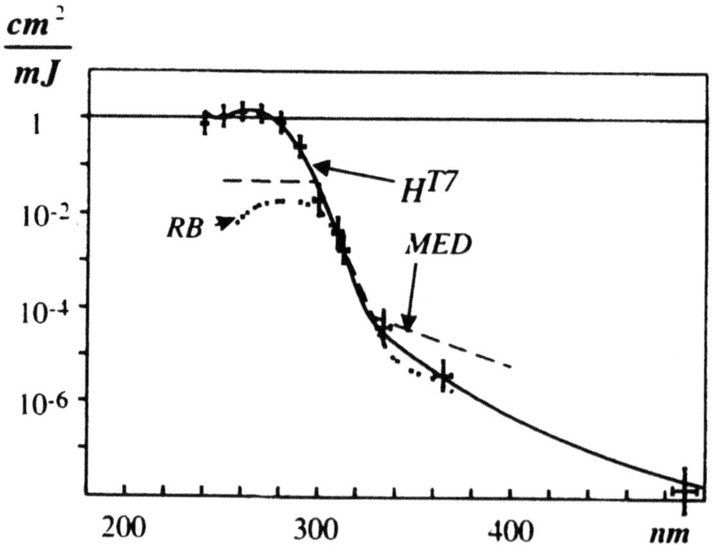

Figure 2. Action spectra of phage T7 inactivation (—), Minimal Erythema Dose (MED; - - -) according to CIE recommendations and sensitivity spectrum of Robertson-Berger meter (RB; ···) on vertical axis absolute sensitivity values are presented in cm^2/mJ. The bars indicate the experimental errors of phage T7 inactivation.

3.2. APPLICATION OF THE DOSIMETER IN FIELD MEASUREMENTS

In the following the different types of measurements will be presented with the aim to demonstrate the many-sided applicability of phage T7 dosimeter. It has been applied in both surface and underwater measurements. Fig. 3. shows a *diurnal profile* obtained by phage T7 dosimeter from sunrise till to sunset on 26. August, 1994, in Hungary [23]. The lower curve represents the profile from direct, while the upper one that from global radiation. In addition the RB-meter data are also included corrected by a factor of *1.7* for the sake of comparison.

Phage T7 dosimeter has been used also for determining the *penetration* of biologically effective UV radiation *into the water*. An example for this measurements is given in Fig. 4., where the cumulated daily doses are presented in three different nice sunny days above the water surface and in two different depths (at 5 and 55 cm under the water) [24]. Penetration depth of 150 cm has been found in this water type.

Phage T7 biological dosimeter was applied since 1992 to monitor the *long-term trend* of the change in BED from environmental radiation [25]. Measurements have been performed in Budapest, downtown, at noon of nice sunny days. Fig. 5. summarizes the results in the late spring-summer months of the years 1992-1995. Data comprising four years are too few for evaluation of a real long-term tendency, however, a doubling of the measured dose rates has been found between the years 1992 and 1993. This can be interpreted by the decrease of industrial emission in the Middle-Eastern European region.

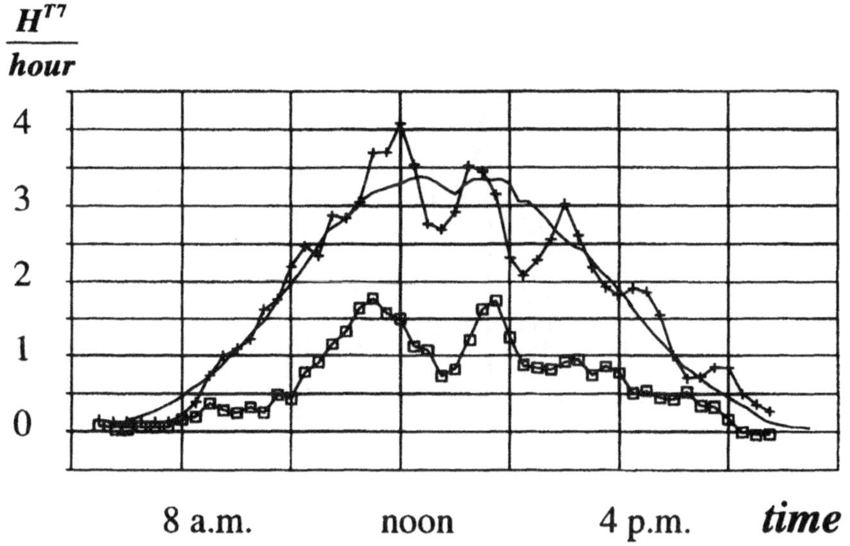

Figure 3. Diurnal profile measured with phage T7 dosimeter on August, 26, 1994 in Hungary from direct (□—□) and global (+ — + — +) radiation. Full line represents the same profile measured by RB meter and corrected by a factor of 1.7

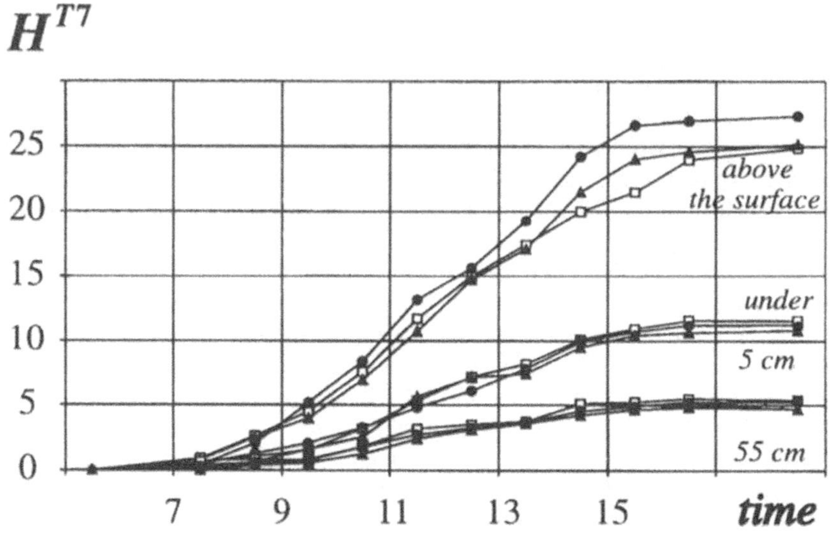

Figure 4. Cumulated daily doses at the lake Gyékényes (Hungary) measured on three nice sunny days (□ 6th, • 8th, Δ 10th August, 1994) above the water surface and in a depths of 5 cm and 55 cm.

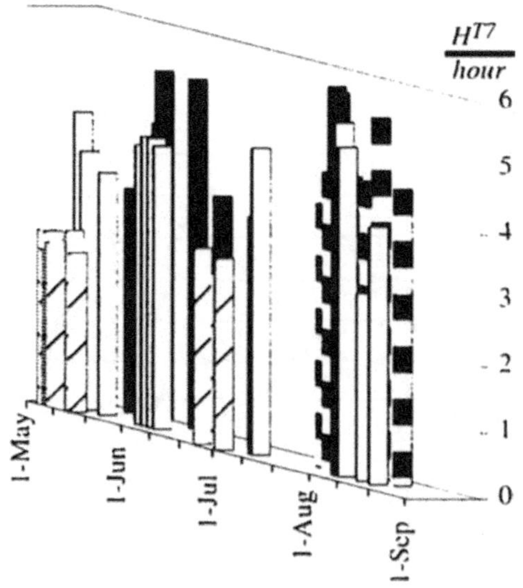

Figure 5. Monitoring data from 1992 (▨), 1993 (□), 1994 (⬒) and 1995 (■)

Summarizing the results presented here and obtained from other different measuring tasks, according to our opinion, biological UV dosimetry provides a sensitive, authentic method for measuring the biologically effective UV dose and using the long-term monitoring data also for the biological risk assessment.

Acknowledgment. This work has been supported by the Europen Commission, Contracts No.: EV5V-CT91-0034, ENV4-CT95-0044, ENV4-CT95-0174.

References

1. S. Nagataki, ed. (1994) Nagasaki Symposium on Chernobyl: Update and Future, Elsevier, Amsterdam
2. Farman, C.J., Gaedimar, B.G., Shamklin, J.D. (1985) Large losses of total ozone in Antarctica reveal seasonal Cl O_x/NO_x interaction, *Nature*, **315**, 207-210.
3. McKenzie, R. (1997) A Southern hemisphere perspective on global UV radiation, Special Lecture, held in Stresa, Italy, at the 7th Congress of ESP
4. Stolarski, R., Bojkov, R., Bishop, L., Zerefos, Ch.,. Staeheline, J., Zawodni, J. (1992) Measured trends in stratospheric ozone, *Science*, **256**, 342-349
5. Németh, P., Tóth, Z., Nagy, Z.(1996) Effect of weather conditions on UV-B radiation reaching the earth's surface, *J. Photochem. Photobiol. B.Biol.* **32**, 177-181
6. Bojkov, R., Bishop, L.B. Hill,.W.J., Reinsel, G.C., Tiao, G.C. (1990) A statistical trend analysis of revised Dobson total ozone data over the Northern Hemisphere. *J. of Geophys. Res.* **95**, 9785-9807.
7. CIE Research Note (1987) A reference action spectrum for ultraviolet induced erythema in human skin, *CIE J.*, **6**, 17-22
8. Tyrrell, R.M. (1978) Solar dosimetry with repair deficient bacterial spores. Action spectra, photoproduct measurements and comparison with other biological systems, *Photochem. Photobiol.* **27**, 571-570
9. Munakata, N. (1981) Killing and mutagenic action of sunlight upon Bacillus subtilis spores: a dosimetric system. *Mutat. Res.* **82**, 263-268
10. Karentz, D., Lutze, L.H. (1990) Evaluation of biologically harmful ultraviolet radiation in Antarctica with biological dosimeter designed for aquatic environments, *Limnol. Oceanogr.* **35**, 549-561
11. Furusawa,Y., Susuki, K., Sasaki, M.(1990) Biological and physical dosimeters for monitoring solar UV-B light, *J. Radiat. Res.* **31**, 189-206
12. Lindberg, C., Horneck, G. (1991) Action spectra for survival and spore photoproduct formation of Bacillus subtilis irradiated with short wavelength (200—300 nm) UV at atmospheric pressure and in vacuo, *J. Photochem. Photobiol. B.Biol* **11**, 69-80
13. Rontó, Gy., Gáspár, S., Bérces, A. (1992) Phage T7 in biological UV dose measurement, *J. Photochem. Photobiol. B. Biol.* **12,** 285-294
14. Rontó, Gy., Tóth, K., Gáspár, S., Csik, G. (1992) Phage nucleoprotein—psoralen interaction: quantitative characterization of dark and photoreactions, *J. Photochem. Photobiol. B.Biol.* **12,** 9-27
15. Rontó, Gy., Gróf, P., Gáspár, S. (1995) Biological UV dosimetry — a comprehensive problem, *J. Photochem. Photobiol. B.Biol.***31,** 51-56
16. Módos, K., Gáspár, S., Gróf, P. (1996) Measuring system and evaluation procedure for environmental damaging agencies, *Central Eur. J. Occupational and Environmental Med.***2,** 259-270
17. Rontó, Gy., Sarkadi, K., Tarján, I. (1967) Zur Analyse der UV-Dosiswirkungskurven der T7-Phagen, *Strahlenther.* **134**, 151-157
18. Horneck, G. (1995) Quantification of the biological effectiveness of environmental UV radiation, *J. Photochem. Photobiol. B.Biol.* **31,** 43-49
19. Munakata, N. (1995) Continual increase in biologically effective dose of solar UV radiation determined by spore dosimetry from 1980 to 1993 in Tokyo, *J. Photochem. Photobiol. B.Biol.* **31**, 63-68

20. Rontó, Gy., Agamalyan, M.M., Drabkin, G.M., Feigin, L.A., Lvov, Yu.M.(1983) Structure of bacteriophage T7. Small angle X-ray and neutron scattering, *Biophys. Journal*, **43**, 309-314
21. Rontó, Gy., Gáspár, S., Gróf, P., Bérces, A., Gugolya, Z. (1994) Ultraviolet dosimetry in outdoor measurements based on bacteriophage T7 as a biosensor (1994) *Photochem. Photobiol.* **59**, 209-214
22. Fekete, A., Földvári, I. (1978) Application of VUV absorption spectroscopy for the detection of UV-induced pyrimidine dimers in DNA, *Studia biophys.* **73**, 47-53
23. Bérces, A., Gáspár, S., Fekete, A., Kuluncsics, Z. (1995) Biological effectiveness of environmental radiation in surface measurements by phage T7, *J. Photochem. Photobiol. B.Biol.* **31**, 87-90
24. Gáspár, S., Bérces, A., Rontó, Gy., Gróf, P. (1996) Biological effectiveness of environmental radiation in aquatic systems, measurements by T7-phage sensor, *J. Photochem. Photobiol. B.Biol.* **32**, 183-187
25. Rontó, Gy., Gróf, P., Gáspár, S., Bérces, A. (1996) Biologic dosimeters in photomedicine. In: M.F. Holick, E.G. Jung (eds) Biologic Effects of Light 1995, W.de Gruyter, Berlin, New York

REPAIR OF UV-B-INDUCED DAMAGE IN HUMAN KERATINOCYTES

R. GREINERT [1, 2], O. BOGUHN [2], B. VOLKMER [1], D. HARDER [2], E.W. BREITBART [1]

[1] *Dermatologisches Zentrum Buxtehude, DZB Am Krankenhaus 1, 21614 Buxtehude, Germany*
[2] *Institut für Medizinische Physik und Biophysik, Universität Göttingen Goßlerstr. 10f, 37073 Göttingen, Germany*

Abstract

The incidence of skin cancers has been increased during the last decades with the steepest gradient compared to other cancers under investigation. Cumulative UV-exposure has been shown to be one of the main risk factors in developing (non-melanocytic) cutaneous malignancies, but the elucidation of molecular mechanisms leading to skin cancer is far of being complete. However, a number of investigations refer to the importance of enzymatic repair of UV induced DNA lesions as a key position in the initiation of skin cancer. This is dramatically documented for patients suffering from Xeroderma pigmentosum, a syndrome characterized by a deficiency in nucleotide excision repair (NER), who show a 5000 fold higher risk for developing skin cancer. We therefore investigated the repair of induced DNA lesions in human keratinocytes (HaCaT), by measuring unscheduled DNA repair-synthesis (BrdU-method) and using monoclonal antibodies against cyclobutane pyrimidine dimers (CPD) and pyrimidine-pyrimidone (6-4) photoproducts ((6-4)PP). Fluorescence microscopy and flow cytometry was used to examine the repair kinetics of both UV induced lesions. We report about remarkable differences in repair kinetics for the two photolesions under investigation. The meaning of measurements of repair kinetics as a significant and individual risk estimator for developing skin cancers is discussed.

1. Introduction

During the last decade skin cancer incidence is worldwide increasing with the steepest gradient compared to other cancers under investigation in the Caucasian population. The United Nations Environment Programme (UNEP) has estimated that over 2 million nonmelanoma skin cancers and 200,000 malignant melanomas occur globally each year [1]. In the case of nonmelanoma skin cancers (basal cell carcinoma, BCC, and squamous cell carcinoma, SCC) this substantial rise in skin cancer has been mainly attributed to earlier changes in lifestyle of large sections of the population (e. g. spending more

C. Baumstark-Khan et al. (eds.),
Fundamentals for the Assessment of Risks from Environmental Radiation, 149-154.

leisure time and holidays in sunny environments), resulting in an increased exposure to both UV-B and UV-A.

Beyond this, all other factors beeing unchanged, the depletion in stratospheric ozone and a corresponding increase in groundlevel (highly genotoxic) UV-B-radiation in sunlight has to be considered as a further seriously risk factor for higher incidences of skin cancer in the future [2]. In the event of a 10 % decrease in stratospheric ozone, with current trends and behaviour, UNEP expects an addiational 300,000 non-melanoma and 4,500 melanoma skin cancers globally each year [1]. However, these numbers have further to be corroborated by highly standardized simultaneous measurements of ozone concentration and UV-B irradiation at ground level [3].

Nevertheless, it is now widely accepted that UV-B plays an important role in the induction of multistage carcinogenesis of (at least) non-melanoma skin cancers (NMSC) due to direct DNA damages, which are formed after absorption of radiation in the wavelength region of 280-320 nm. In this way, UV-B induces characteristic DNA lesions like the cyclobutane pyrimidine dimere (CPD) and the pyrimidine-pyrimidone (6-4) photoproduct ((6-4)-PP), which are thought to be the preferential sides for UV-B-induced mutagenesis [4-5]. The role of UV-A in induction of non-melanomic skin cancers, as well as the role of both UV-B and UV-A in induction of malignant melanomas of the skin is less well understood. Yet, it becomes more and more clear, that the mutagenic potential of UV-A (320-400 nm) has been underestimated in the past [5]. Despite these and other important findings, the elucidation of molecular mechanisms leading to skin cancers is far of being complete. However, a number of investigations referr to the importance of enzymatic repair of UV-induced DNA lesions as a key position in the initiation of skin cancer, (e.g.) [6-9]. This is dramatically documented for patients suffering from Xeroderma pigmentosum, a syndrom characterized by a deficiency in nucleotide excision repair (NER), who show an up to 2000 fold higher risk in developing UV-induced skin cancer [9].

Therefore it appears necessary to establish methods for quantitative measurements of induction and removal of UV-induced photolesions (CPD, (6-4)-PP) in normal human cells and cells derived from skin tumors, in order to proof, wether a reduced repair capacity might be correlated to the malignant phenotype. One can than start to plan investigations which shine more light on the question, wether the proof of DNA repair deficiency can be used as a risk estimator for developing skin cancers in the normal population.

For this purpose we started to investigate the induction and the repair of UV-B-induced photolesions in human cells by using monoclonal antibodies against CPD's as well as (6-4)-PP's. Fluorescence microscopy and digital image analysis was used to examine the induction and the repair kinetics of both UV-induced lesions.

2. Material and Methods

2.1. CELL LINE

All experiments were carried out with spontaneously immortalized human keratinocytes, HaCaT (kindly provided by Prof. Boukamp, DKFZ Heidelberg). Cells were grown in DMEM, 10 % FCS at 37 C°, 5% CO_2.

2.2. IRRADIATION

Cells, exponentially growing on glass slides, were irradiated with UV-B (Phillips TL 20, 280-315 nm, λ_{max}=306 nm) in a distance of 30 cm at a dose rate of 9.8 $J/m^2/s$, measured with an IL 1700 dosimeter, further calibrated by reference measurements using a (Bentham M 300) spectro-radiometer. Culture medium was removed before irradiation. For measurements of repair kinetics of UV-B-induced DNA damage, slide-cultures were preincubated (30 min) and irradiated on ice, to prevent any enzymatic repair during the time of iradiation. After irradiation, cells were fixed either immediately or after incubation for repair intervals of 0-6 h.

2.3. STAINING

After fixation (methanol, -20 °C, 5 min; Aceton -20 °C,5 min), and denaturation of the DNA (0.1 N NaOH in 70% EtOH, 3 min) the cells were dehydrated (70%, 90%, 100% EtOH, 1 min) and air dried. Cells were then incubated with proteinase K (10 µg/ml) for 10 min at 37 °C, washed with PBS (5 times 5 min) and incubated with the first antibody (a-(6-4) PP 1:1000 or a-CPD 1:2000) at 4 °C overnight. After washing the slides with PBS (5 times 5 min) cells were incubated with an FITC-labeled anti-mouse-antibody (1:100), washed again with PBS (5 times 5 min) and mounted with antifade.

2.4. FLUORESCENCE MICROSCOPY

Fluorescence intensities of stained nuclei were determined with a LEICA DM fluorescence microscope equipped with a digital (air cooled) color CCD camera (Kappa CF 20 DXC/air). Digital image analysis was performed by using the software package OPTIMAS. Cell nuclei were marked and the mean fluorescence intensity/nucleus was calculated. At least 100 nuclei were analyzed for each data point.

3. Results

Figure1 shows the survival of HaCaT cells after UV-B-irradiation in a dose range from 0-200 J/m^2. A D_{37} of about 35 J/m^2 can be determined. Figure 2 shows the effect of UV-B irradiation (0 vs. 800 J/m^2) on the induction of CPD's as visualized by immunostaining using monoclonal antibodies against this specific DNA lesion as detected with fluorescence microscopy. Whereas nuclei of (unirradiated) control cells remained unstained, those of irradiated cells (e.g. 800 J/m^2) show a nucleus-specific fluorescence due to antibody-binding to CPD's. As can be seen from Figure 3, the fluorescence intensity of antibody-detected UV-B-induced DNA-lesions incresases in a dose proportional way both for CPD's as well as for (6-4)-PP's. It turns out that CPD's are induced more effectively than (6-4)-PP's by UV-B iradiation, which is in agreement with results published earlier [11].

After irradiation of human spontaneously immortalized keratinozytes (HaCaT) with a UV-B-dose of 800 J/m^2, CPD's are removed due to enzymatic repair. As can be seen from Figure 4 this repair process can be characterized by a time constant τ=40 min.

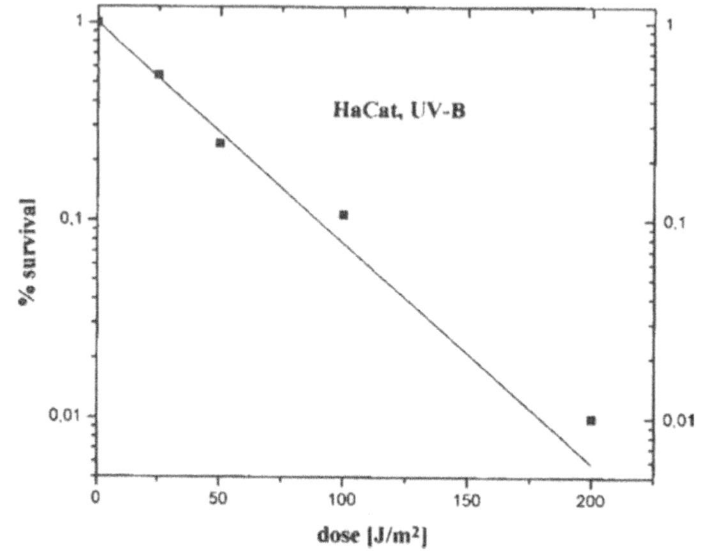

Figure 1. Survial of human keratinocytes, HaCaT, after UV-B-irradiation

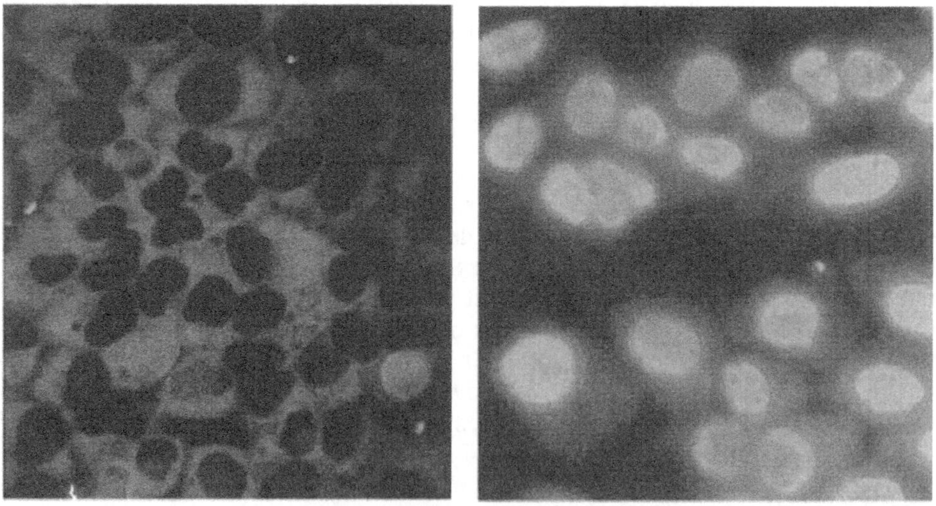

Figure 2. Induction of CPD's in human keratinocytes, HaCaT, by UV-B (800 J/m^2), as visualized by fluorescently labelled monoclonal antibodies. Left: control; right: irradiated cells

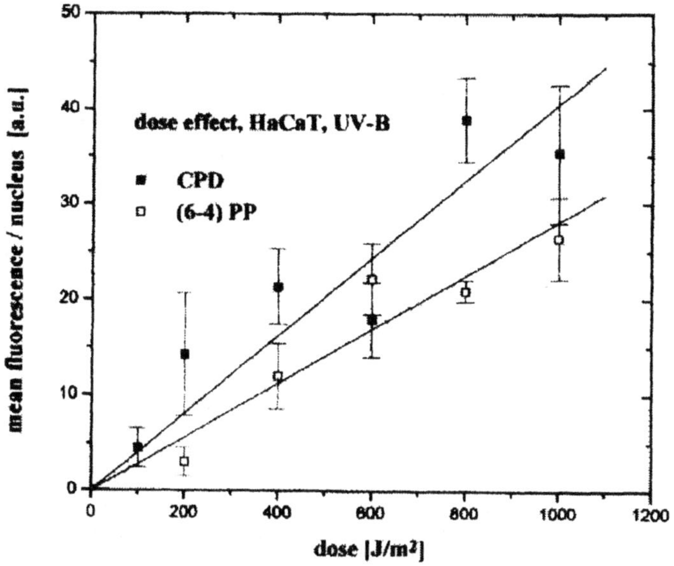

Figure 3. Dose dependent induction of CPD's and (6-4)-PP's after UV-B-irradiation of human keratinocytes. CPD's and (6-4)-PP's were quantified by the

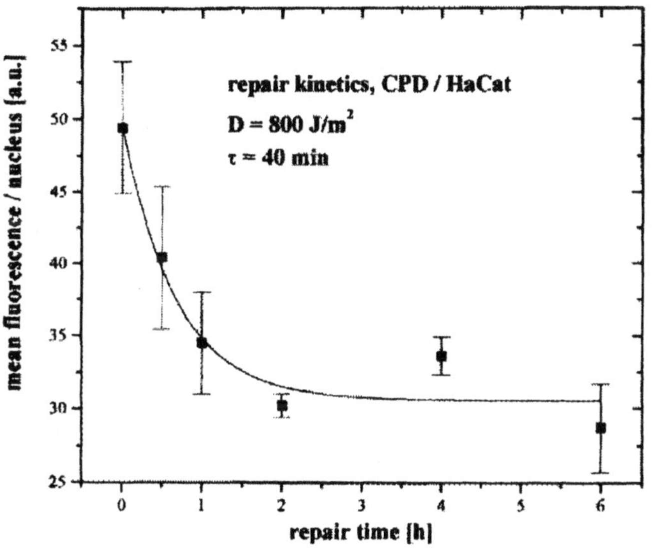

Figure 4. Kinetics of the repair of UV-B-induced CPD's in human keratinocytes. At each time point CPD's were quantified by the fluorescence intensity of cell nuclei, which were stained by specific monclonal antibodies and analyzed in a fluorescence microscope. At least 100 nuclei were analyzed per data point.

4. Conclusions

Our results show that immunostaining of UV-B-induced DNA-leasons like CPD's and (6-4)-PP's with fluorescently labelled monoclonal antibodies can be used to quantify DNA damage in human keratinocytes in experiments using fluorescence microscopy and digital image analysis. The method is further suitable to measure the kinetics of repair of UV-B-induced photolesions. We have just started to perform further experiments in primary cell cultures (keratinocytes, melanocytes, melanoma cell lines) as well as cell suspensions from biopsies of human skin to characterize DNA repair after UV-B-irradiation in these cell lines. This experiments will help to answer the question, wether the measure of repair characteristics in UV-irradiated cell may serve as a risk estimator for developing skin cancers.

5. References

1. Kjellström, T., Repacholi, M.H. (1996) The Intersun project of WHO. In: *Environmental UV-Radiation, Risk of Skin Cancer and Primary Prevention, Veröffentlichungen der Strahlenschutzkommission,* **Band 34** (B. Volkmer, H. Heller eds.), Verlag Gustav Fischer, Stuttgart, 333-346.
2. Slaper, H., Velders, G.J.M., Daniels, J.S., De Gruijl, F.R. and Van der Leun, J,C. (1996) Estimates of ozone depletion and skin cancer incidence to examin the achievements of the Vienna convention, *Nature* **384**, 256-258.
3. Seckmeyer, G., Mayer, B., Bernhard, G. Albold, A. Erb, R., Jäger, H. Stockwell, W.R. (1996) Measurement and calculation of UV-radiation (models and prognoses). In: *Environmental UV-Radiation, Risk of Skin Cancer and Primary Prevention, Veröffentlichungen der Strahlenschutzkommission,* **Band 34** (B. Volkmer, H. Heller eds.), Verlag Gustav Fischer, Stuttgart, 35-58.
4. Brash, D.E., Rudolf, J.A., Simon, J.A., Lin, A., McKenna, G.J., Baden, H.P., Halparin, A.J. and Ponten, J. (1991) A role for sunlight in skin cancer: UV-induced p53 mutations in squamous cell carcinoma, *Proc. Natl. Acad. Sci. USA* **88**, 10124-10128.
5. Sage, E. (1996) UV-induced mutation and transformation. In: *Environmental UV-Radiation, Risk of Skin Cancer and Primary Prevention, Veröffentlichungen der Strahlenschutzkommission,* **Band 34** (B. Volkmer, H. Heller eds.), Verlag Gustav Fischer, Stuttgart, 107-116.
6. Ananthaswamy, H.N. and Pierceall, W.E. (1990) Molecular mechanisms of ultraviolet radiation carcinogenesis. *Photochem. Photobiol.* **52**, 1119-1136.
7. Sage, E. (1993) Distribution and repair of photolesions in DNA: genetic consequences and the role of sequence context. *Photochem. Photobiol.* **57**, 163-174.
8. Mullenders, L.H.F. and Smith, C.A. (1994) DNA repair in specific sequences and genomic regions. In: *Methods to assess DNA damage and repair. Interspecies comparison* (R.G. Tardiff, P.H.M. Lohman, G.N. Wogan, eds.), John Wiley and Sons, Chicester UK, 141-156
9. Mullenders, L.H.F., Van hoffen, A., Vreeswijk, M.P.G., Ruven, H-J-, Vriesling, H, Van Zeeland, A.A. (1996) UV-induced repair and mutagenesis in mammalian cells. In: *Environmental UV-Radiation, Risk of Skin Cancer and Primary Prevention, Veröffentlichungen der Strahlenschutzkommission,* **Band 34** (B. Volkmer, H. Heller eds.), Verlag Gustav Fischer, Stuttgart, 95-106.
10. Hoeijmakers, J.H.J. (1994) Human nucleotid excision repair syndromes: Molecular clues to unexpected intricacies. *Eur. J. Cancer* **30A**, 1912-1921
11. Mitchell, D.L. (1988) The biology of the 6-4 photoproduct. *Photochem. Photobiol.* **49**, 805-819

REPAIR-INDUCED DNA STRAND BREAKS IN UV-IRRADIATED MAMMALIAN CELLS

Y. NIKANDROVA [1], C. BAUMSTARK-KHAN [2], G. HORNECK [2]

[1] *A.A. Bogomolez Institute of Physiology, Academy of Science Kyiv, Ukraine*
[2] *DLR, Institute of Aerospace Medicine, Radiobiology Section D-51170 Köln, Germany*

Abstract

Using the FADU technique (**F**luorescent **A**nalysis of **D**NA **U**nwinding) the occurence of repair-induced DNA strand breaks was measured in repair deficient and repair proficient chinese hamster ovary cells after irradiation with solar light and artificially produced UV-B/A light. Comparisons were made with cells irradiated with UV-C or X-rays. After a recovery period of 1 hour, the occurence of DNA single-strand breaks in UV-C irradiated cells was explained by removal of photoproducts by excision repair. In presence of the polymerase inhibitor aphidicolin (1 µg/ml) DNA repair synthesis was inhibited, leading to an enlarged yield of DNA strand breaks in repair competent cells (CHO AA8) but not in excision repair deficient cells (CHO UV5). After solar irradiation or exposure to UV-B/A or UV-A the number of repair-induced DNA single-strand breaks is reduced.

1. Introduction

It is well accepted that solar exposure is a major cause in the development of skin cancer. Of special concern is the spectral composition of solar light on terrestric level and the biological effectiveness of wavelengths in the UV-B (290-320 nm) and UV-A (320-400 nm) regions reaching the Earth's surface, in contrast to UV-C (100-280 nm) which is filtered out by the dissociation of ozone in the stratosphere.

Detection and measurement of DNA damage in mammalian cells belongs without any doubt to the most important subjects in radiobiology, especially when induction and repair of damage has to be correlated with biological endpoints, such as cell inactivation, cellular recovery and mutagenesis [1].The biological effect of exposure with UV light up to 320 nm is based on the induction of cyclobutane pyrimidine dimers (Py<>Py) and of (6-4)-photoproduct, at higher wavelengths active species of oxygen may account for increased amounts of strand breaks [2]. With the FADU technique it is possible to detect DNA strand breaks after X- irradiation in a convenient dose range [3]. Therefore, the repair of exposure-induced primary single-strand breaks and the occurrence of secondary single-strand breaks caused by incision enzymes during excision repair was investigated.

C. Baumstark-Khan et al. (eds.),
Fundamentals for the Assessment of Risks from Environmental Radiation, 155-160.

2. Materials and Methods

Chinese hamster ovary cells (cell lines: AA8 and UV5) were cultured under standard conditions (DMEM medium, 5% foetal calf serum, pH 7.0-6.8, 5% CO_2, 37°C. 3 days after inocculation medium was renewed and cells were incubated for an additional period of 3 days till confluency.

DNA damage was induced by exposure of confluent cell layers with UV-C (254 nm, germicidal lamp, NN8/15, Heraeus), and UV-B/A (Sol 2, Dr. Höhnle, petri dish lid as cut-off filter) as well as solar radiation. UV or solar dosimetry was performed using a calibrated UVX-radiometer (UVP Inc.) and a Robertson-Berger-radiometer (Solar Light Inc.), respectively. For calibration purpose DNA strand breaks were induced by X-irradiation (X-ray unit, 150 kV, 10 mA, dose rate of 2 Gy/min). Spectral distributions of solar and UV-B/A light are shown in Figure 1 and Figure 2, respectively.

For exposure with UV, medium was sucked off from the petri dishes and the cells were rinsed twice with phosphate buffered saline (PBS). Irradiation was performed on ice in order to avoid any DNA repair. Immediately after UV exposure, cells were incubated under standard conditions (DMEM-medium, 5% FCS, 37°C, 5%CO_2, 95% air).

Overall DNA strand breaks were measured using the fluorescent analysis of DNA unwinding (FADU) [3]. The cells were either analyzed immediately after irradiation or after a repair period of 1 hour in absence or presence of the polymerase inhibitor aphidicolin (1 µg/ml medium) [4].

All experimental data were standardised to $F_D/F_{D=0}$, where F_D and $F_{D=0}$ were the DNA fractions of treated and untreated cells, respectively. The experimental data (fraction of intact DNA: *FiDNA*) were fitted by a weighted least square regression to

$$FiDNA = n \cdot e^{-D/D_0} \quad (1)$$

where D is the dose (J/m^2), D_0 is -1/slope and n is the extrapolation number of the corresponding regression curve.

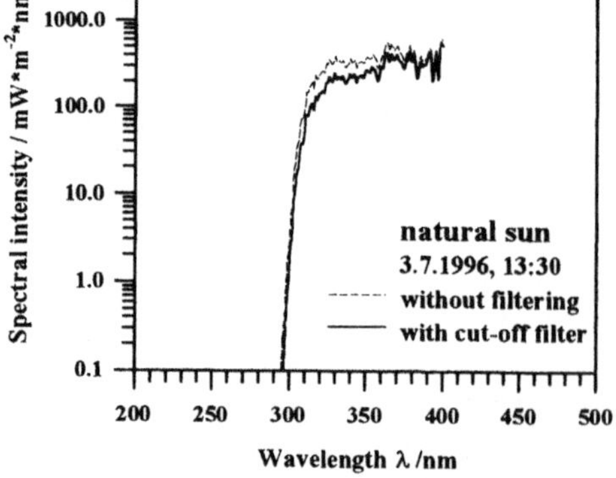

Figure. 1. Spectral distribution of solar light without filtering and filtered through a petri dish lid used as a cut-off filter, measured at typical exposure conditions.

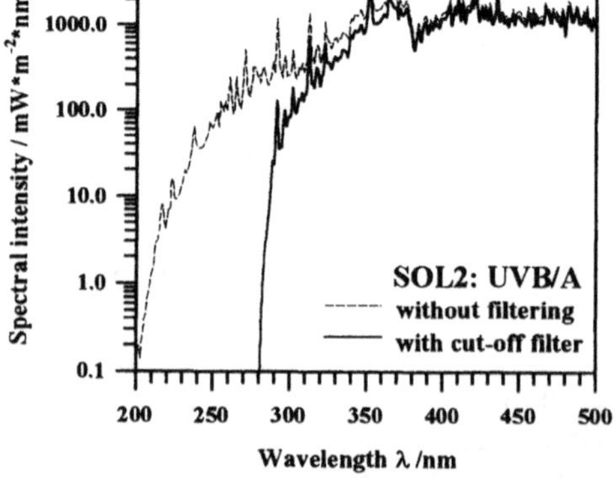

Figure. 2. Spectral distribution of the solar simulator SOL2 without filtering and filtered through a petri dish lid used as a cut-off filter.

3. Results

Induction of overall DNA strand breaks in repair proficient (AA8) and deficient (UV5) CHO cells and their rejoining was measured by fluorescent analysis of DNA unwinding (FADU). DNA induction and repair is described in terms of purely exponential dose effect curves (Figure 3 and 4) with D_0 values given in Table 1.

For X-irradiated cells, a high number of DNA strand breaks is present immediately after exposure. After a 1 hour recovery period most of the breaks have been repaired for both cell lines. In presence of aphidicolin, repair of X-ray induced DNA strand breaks is not significantly impaired, reflecting, that aphidicolin primarily inhibits polymerase-dependent DNA repair synthesis, while X-ray induced DNA strand breaks are essentially repaired by rejoining processes involving the action of DNA ligases.

Only a small amount of strand breaks is induced directly by UV-C-exposure, in both cell lines. After 1 hour of incubation the number of DNA strand breaks was significantly higher for repair proficient AA8 cells. In presence of aphidicolin the amount of secondary (repair-induced) DNA strand breaks is twice as high as without aphidicolin. For repair deficient UV5 cells repair-induced DNA strand breaks do not pretend to appear after repair incubation neither without nor with aphidicolin.

For the exposure conditions with UV-B/A given in Fig. 2 a dose rate of 8.474 $J/(m^2 \cdot s)$ was calculated. After exposure strand breaks were detected in limited amounts only for AA8 cells. 1 hour after exposure the amounts of strand breaks for repair proficient cells raised slightly. For repair deficient UV5 cells neither UV-B/A-induced primary strand breaks nor repair-induced secondary strand breaks were detectable.

For the exposure conditions with solar light given in Figure 1 a dose rate of 2.312 $J/(m^2 \cdot s)$ was calculated. For AA8 cells (Figure 4) solar light-induced DNA strand breaks were detected immediately after exposure. After a repair incubation of 1 hour the amounts of strand breaks for repair proficient cells was not significantly changed.

4. Discussion

Epidemiologic studies suggest that exposure to UV radiation increases the risk for the development of human skin cancers. Sensitivity of mammalian cells to exposure with ultraviolet and/or solar light is the result of initially induced DNA damage and of the cells ability to recover from this damage. Thus, the capacity of the individual to deal with UV-induced lesions is a crucial component in carcinogenesis.

Since the amount of initial damage is a function of UV-dose, it can be minimized by sun protection devices. The repair processes are dependent on cellular parameters, e.g. the genetic out-fit of the individual cell. This has been demonstrated by the existence of genetic diseases correlated with repair defects such as XP and CS. A number of different human DNA repair related genes have been cloned and functionally related to distinct steps in excision repair. XP-D cells are defective in the ERCC2 gene product, which is responsible in the incision step of nucleotide excision repair (NER). The hamster homologue is defect for UV5 cells. In this study the responses of repair proficient and incision-defect cell lines to UV wavelengths of the terrestrical UV solar spectrum were determined using the fluorescent analysis of DNA unwinding. Results were considered in comparison with the well-known effects of X-rays and UV-C.

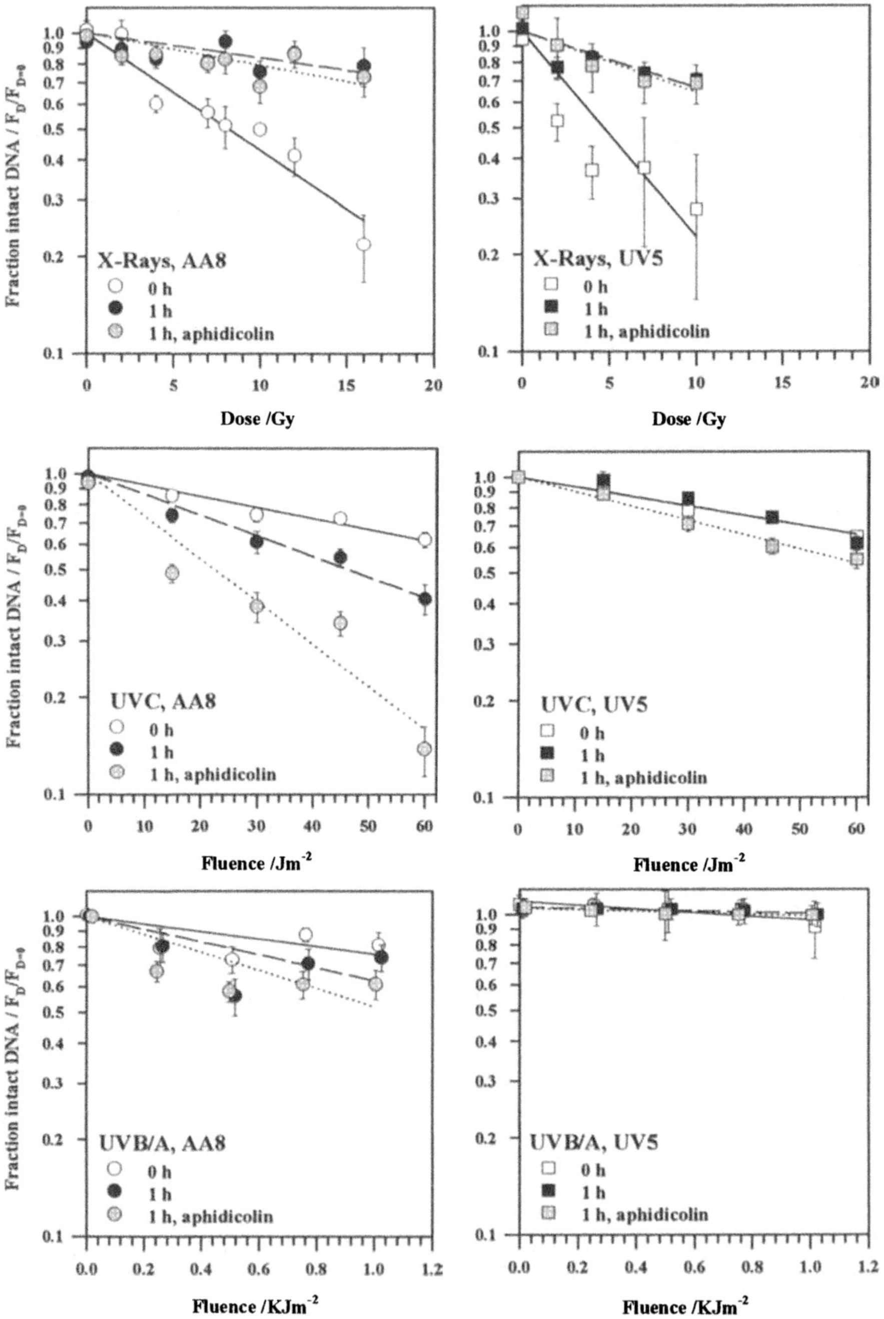

Figure. 3. Fraction of intact DNA immediately after exposure with X-rays, UV-C (254 nm) or UV-B/A and after a recovery period of 1 hour in absence or presence of 1 μg/ml aphidicolin.

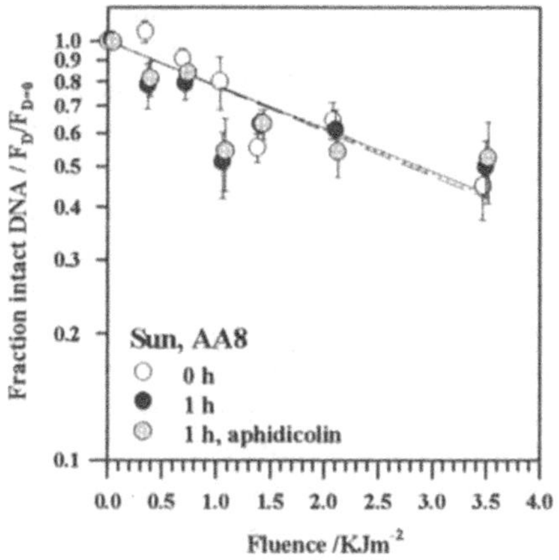

Figure. 4. Fraction of intact DNA immediately after solar-exposure and after a recovery period of 1 hour in absence or presence of 1 μg/ml aphidicolin.

TABLE. 1. DNA strand breaks of in vitro cultivated Chinese hamster ovary cells with different repair capacity for induced strand breaks (0 hour repair) and for 1 hour of repair (in absence or presence of aphidicolin).

Exposure	*Cell Line*	*Repair Condition*	D_0	*n*
X-rays (Gy)	AA8	None	11.45± 1.83	1.00±0.10
		1 h	88.87±76.97	0.90±0.03
		1 h + aphidicolin	53.78±22.85	0.92±0.03
	UV5	None	8.06± 3.64	0.72±0.15
		1 h	33.55±10.48	0.92±0.03
		1 h + aphidicolin	40.05±27.84	0.87±0.09
UV-C (J/m^2)	AA8	None	131.7±18.7	0.94±0.03
		1 h	69.8±10.6	0.91±0.05
		1 h + aphidicolin	32.5± 4.2	0.86±0.09
	UV5	None	117.0±17.7	1.05±0.04
		1 h	98.6±16.6	1.14±0.06
		1 h + aphidicolin	88.7±10.5	0.99±0.04
UV-B/A (KJ/m^2)	AA8	None	4.64±4.73	0.90±0.05
		1 h	2.36±1.15	0.86±0.06
		1 h + aphidicolin	1.74±0.38	0.87±0.04
	UV5	None	7.14±3.59	1.05±0.08
		1 h	23.45±9.22	1.01±0.04
		1 h + aphidicolin	17.40±9.81	1.02±0.05
Sun (KJ/m^2)	AA8	None	3.57±0.66	1.01±0.06
		1 h	4.54±1.03	0.87±0.03
		1 h + aphidicolin	3.87±0.79	0.91±0.04

Quantitative assessment of DNA damage belongs without any doubt to the most important subjects in radio - and photobiology, especially when induction and repair of damage has to be correlated with certain biological endpoints. The FADU technique, which is based on time-dependent alkaline denaturation of DNA under moderate denaturating conditions, detects DNA strand breaks in a convenient X-ray dose range. Two different approaches could be verified after radiation exposure: (i) the induction of primary single strand breaks and their disapperance after repair incubation of X-irradiated cells and (ii) the occurrence of secondary single-strand breaks caused by incision enzymes during excision repair (II) after UV exposure.

CHO cells proficient (AA8) or deficient (UV5) in the early steps of NER have similar capacities for X-ray induced DNA strand breaks, showing that for removal of such lesions the NER pathway is not required. For UV-C exposure, only few single strand breaks were induced for both cell lines. After repair incubation, especially in presence of the DNA polymerase inhibitor aphidicolin, the assay clearly permits a distinction between the repair deficient and proficient cell line. For UV5 cells, strand breaks are not detectable, neither in absence nor in presence of aphidicolin. For AA8 cells the yield of secondary strand breaks increases with dose, in presence of aphidicolin the amount of strand breaks is about twice as high as in absence of aphidicolin, reflecting the evolution of secondary strand break induction and rejoining. These results are in accordance with data obtained for human fibroblasts [5] and lymphocytes [6]. For 1 h repair incubation after exposure with artificial UV-B/A light the repair proficient cell line showed a certain amount of additional secondary strand breaks. The repair deficient cell line showed no response at all, indicating that a functional ERCC2 gene product is required for processing of UV-B/A-induced DNA lesions. Natural sun induced primary strand breaks in comparable amounts to artificial UV-B/A light in AA8 cells. After a repair incubation of 1 hour, however, the amounts of strand breaks for repair proficient cells was not significantly changed. This picture of a steady state implies, that either NER is disabled after solar exposure or repair of primary strand breaks progressed with the same rate as NER-induced secondary strand breaks appeared. Further studies are needed to solve this problem.

5. References

1. Goodhead D.T., Thacker J., Cox R. (1993) Weiss Lecture. Effects of radiations of different qualities on cells: molecular mechanisms of damage and repair. *Int. J. Radiat. Biol.* **63**, 543-556
2. Eisenstark A. (1987). Mutagenic and lethal effects of near-ultraviolet radiation (290-400 nm) on bacteria and phage. *Environ. Mol. Mutagen.* **10**, 317-337
3. Baumstark-Khan, C. (1994), Alkaline elution versus fluorescence analysis of DNA unwinding. *Methods in Enzymology* **234**, 88-102
4. Baumstark-Khan, C. (1992) Effect of aphidicolin on DNA synthesis, PLD-recovery and DNA repair of human diploid fibroblasts. *Int. J. Radiat. Biol.* **61**, 191-197
5. Alapetite C., Wachter T., Sage E., Moustacchi E. (1996). Use of the alkaline comet assay to detect DNA repair deficiencies in human fibroblasts exposed to UV-C, UV-B, UV-A and gamma-rays. *Int. J. Radiat. Biol.* **69**, 359-369
6. Green M.H., Lowe J.E., Harcourt S.A., Akinluyi P., Rowe T., Cole J., Anstey A.V., Arlett C.F. (1992). UV-C sensitivity of unstimulated and stimulated human lymphocytes from normal and xeroderma pigmentosum donors in the comet assay: a potential diagnostic technique. *Mutat. Res.* **273**, 137-144

A BIOSENSOR FOR ENVIRONMENTAL GENOTOXINS, CHEMICAL MUTAGENS AND RADIATION FACTORS, BASED ON THE SOS RESPONSE FUNCTION OF *ESCHERICHIA COLI* CELLS

L. R. PTITSYN

State Research Centre of RF: GNII Genetica
1, 1-st Dorozhny pr., 113545, Moscow, Russia

Abstract

A luminescent bacterial reporter assay, the SOS *lux* test, was developed for rapid detection of environmental genotoxins. The bioassay is based on recombinant plasmid carrying the promoterless *lux* operon of *Photobacterium leiognathi* fused with strong SOS inducible promoter. The DNA damage potency of chemicals or physical factors can be monitored by measuring the visible light emitted by growing culture. The sensitivity of the SOS *lux* test is similar to other bioassays, such as SOS Chromotest, *umu* test and Ames mutatest. The role of enzymes of DNA damage-repair system in SOS response of cells can be evaluated by using the appropriate mutant cells as a host cells in this test-system. The cell cultures as well as immobilized cells can be used for detection of genotoxins in SOS *lux* test.

1. Introduction

The biological test systems have been developed complementary to chemical and physical detection methods [1] to assess the environmental pollutants and genotoxins. Genotoxic agents (e.g. chemicals or radiation) induce a variety of DNA injuries. Depending on their nature, these lesions may be excised or reverted with no future consequences for the cell, may induce the SOS response which includes SOS mutagenesis, or remains unrepaired and leads to cell death.

The Ames test based on reversion to histidine prototrophy upon expose to mutagens is most extensively used and validated in studies using several hundreds of chemicals [2].

SOS dependent bacterial test systems make use of the fact that a cascade of cell functions known as the "SOS response" is induced in response to DNA damaging agents. The most commonly used bacterial tests based on the SOS system are the SOS Chromotest [3] and *umu* test [4]. In the SOS Chromotest, *E. coli* PQ37 cells with the structural genes for -galactosidase, *lacZ*, under the control of a SOS dependent gene *sulA* are used as "reporter". The *umu* test makes use of a recombinant *Salmonella ty-*

C. Baumstark-Khan et al. (eds.),
Fundamentals for the Assessment of Risks from Environmental Radiation, 161-166.

phimurium TA1535(pSK1002) strain with a fused *umuC::lacZ* operon carrying on the plasmid. In both systems the SOS inducing potency is determined from a colorimetric assay for -galactosidase synthesized in response to genotoxin. As shown, for 82% of the chemical compounds the genotoxic potency (determined by SOS Chromotest) is comparing with mutagenic potency (determined by Ames test) [5].

The bioluminescence reaction involves the oxidation of a long-chain fatty aldehyde and reduced flavin mononucleotide and leads to the emission of blue-green light with a peak at 490 nm:

$$FMNH_2 + RCHO + O_2 \rightarrow FMN + H_2O + RCOOH + h\upsilon \quad (1)$$

The main advantage for *lux* genes system is the availability of a real-time, *in vivo* and non-disruptive monitoring of it's expression [6].

We have constructed a biosensor with specificity for genotoxins by combining the SOS system regulatory component with the bioluminescence system as a rapid reporter component. The recombinant *E.coli* C600(pPLS-1) carries a plasmid with the promoterless *lux* operon (*luxCDABFE*) of *Photobacterium leiognathi* under control of a strong SOS promoter that originates from part of *cda* gene of the plasmid ColD [7]. This SOS *lux* test responds sensitively to a variety of genotoxins (chemicals and physical factors) and it's sensitivity is comparable to that of known assay systems, such as the Ames test, SOS Chromotest and the *umu* test.

2. Material and Methods

2.1. CONSTRUCTION OF pPLS-1 PLASMID AND BIOSENSOR

E. coli C600 (*F⁻, thi-1, thr-1, leuB6, lacY1, tonA21, supE44*), *AB1157 (F⁻, thi-1, thr-1, leuB6, proA2, hisG4, argE3, lacY1, galK2, ara-14, xyl-5, mtl-1, tsx-33, rpsL3, supE44*), and JC9239 (*recF143*; a mutant of AB1157) were used as a host strains.

DNA manipulations, including plasmid isolation, molecular cloning and DNA sequencing were performed by standard methods as described [8-9].

The pPLS-1 construction was described in detail in [10]. Briefly, the 6.7 kbp fragment of pPL-2 plasmid [11] (carrying promoterless operon of *Ph. leiognathi*) was fused with 1.7 kbp fragment of ColD plasmid (carrying SOS promoter of *cda* gene) and cloned into pBR322 derivative vector. The resulting plasmid pPLS-1 carries the *luxCDABFE* genes downstream of the SOS promoter and thereby has all functions necessary for SOS inducible bioluminescence. pPLS-1 was used to transform *E. coli* C600, as well as AB1157 and JC9239 cells.

2.2. THE SOS LUX ASSAY

E. coli C600(pPLS-1), as well as AB1157(pPLS-1) or JC9239(pPLS-1) cells were grown overnight at 37 °C in L-medium supplemented with 50 mkg/ml ampicillin. After dilution (1:50) in fresh L-medium, the culture was incubated at 37°C until the optical density at 560 nm (OD_{560}) reached about 0.2-0.3. Aliquots of this culture (0.18 ml) were transferred into polypropylene luminometric tubes, and 0.02 ml of appropriate concentrations of the test chemicals in aqueous solution was added. The distilled water (0.02

ml) was added to the control tubes. After incubation with aeration at 37°C for 90 min and then at room temperature for 10 min, the light emission (mV) of the cultures was measured (LKB Luminometer 1250, Pharmacia Biotech, Uppsala, Sweden), and simultaneously the cell density (OD_{560}) was determined. Light measurements were done within 1 sec. The modified procedure for γ radiation testing was described [10].

The following chemicals were used in the assay: mitomycin C (MMC) (PL Biochemicals), N-methyl-N'-nitro-N-nitrosoguanidine (MNNG) (Sigma), dimethylsulphate (DMS) (Sigma). *Cis*-diammine-dichloroplatinum(II) (*cis*-DDP) and *trans*-diammine-dichloroplatinum(II) (*trans*-DDP) were kindly provided by Dr. V. B. Ivanov (Moscow).

The factor of SOS induction F_i is equal to the relative light emission Lux_i/Lux_o of treated and control cultures divided by the relative OD_{560} values OD_i/OD_o:

$$F_i = \frac{Lux_i \cdot OD_0}{Lux_0 \cdot OD_i} \qquad (2)$$

This correction for cell concentration is necessary, because some genotoxins delay or impair cell growth and thereby influence the total light emission of the culture.

2.3. CELL IMMOBILIZATION

Cells were immobilized into agarose granules [12]. The size of granules formed was 0.1-0.3 mm. The 1 ml column with this matrix was used as the on-line registration cell. M9 medium with 0.5.% casamino acids was used as eluent. The flow rate of medium was 5 ml/h and the temperature was 28 °C. To induce the SOS response, 10^{-6} M MMC was added to flow medium during 30 min and farther the column was washed by medium without MMC. The level of light emission was measured during 8 h.

3. Results and Discussion

The MMC, MMNG and DMS are the chemicals with known mutagenic potential [13, 5]. These substances have a different mutagenic mechanisms: MMC predominantly induces DNA intrastrand crosslinking, MNNG and DMS induce alkylation of DNA bases. The γ radiation preferentially induce DNA base damage and strand breaks. All of these genotoxins induce light emission in C600(pPLS-1) progressively by increasing the substrate concentration in the culture medium. The lower limit of detection which is given by that dose at which F_i reaches a value of 2 was derived from a linear plot of F_i versus dose. The concentrations of chemicals and dose γ radiation which induced light expression 2-fold over the background level in SOS *lux* test are comparable to that of other SOS bioassays and the data for mutagenic activity of these genotoxins as determined by the Ames test (Table 1). These results indicate that the SOS *lux* test is an effective method for screening of genotoxins and monitoring the genotoxic potency of radiation.

The SOS *lux* test as compared with other bioassay has some practical advantages: the registration time is very short (1 s), the measurement is non-destructive for the test bacteria and can be repeated several times from the same culture during the whole incu-

TABLE 1. A comparison of the sensitivity of the SOS *lux* test with that of other mutagenicity or genotoxicity tests

Genotoxin	*Lower limit of detection (mol/l) for chemicals, (Gy) for γ rays*			*Revertants (per nmol/l) or lower limit of detection (Gy)*
	SOS lux test	*SOS Chromotest* [a) b)]	*umu test* [c)]	*Ames Test* [a) d)]
MMC	$5.0\cdot10^{-9}$	$1.7\cdot10^{-8}$	$1.5\cdot10^{-7}$	1200
MNNG	$7.1\cdot10^{-7}$	$5.0\cdot10^{-7}$	$2.0\cdot10^{-6}$	44
DMC	$7.5\cdot10^{-6}$	$6.7\cdot10^{-6}$	$3.0\cdot10^{-4}$	0.1
γ rays	2.56[e)]	<5	n.d.	3-4

a) from [3] for chemicals;
b) from [5] for γ rays;
c) from [13];
d) from [14] and [15] for γ rays;
e) data were kindly provided by O. Komova and E. Krasavin (JINR, Dubna, Moscow Region);
n.d. = not determined

bation period. Depending on the aims of experiment and the type of genotoxin, the sensitivity of the SOS *lux* test can be increased by using the appropriate mutant host strains for the pPLS-1 plasmid. Also, the role of mutations in genes of repair or recombination systems on the SOS response can be evaluated by using the mutant strains.

The AB1157(pPLS-1) and JC9239(pPLS-1) (*recF* mutant of AB1157) cells were used for studies the genotoxicity of *cis*- and *trans*-DDP. *Cis*-DDP is a widely used antitumor drug that binds to DNA by preferentially forming intrastrand crosslinks between the N7 atoms of adjacent purines (the main adduct: *cis*-DDP{d(GpG)} - 65% of total lesions). The *trans*-DDP produces interstrand DNA adducts and is therapeutically inactive [16].

These complexes show significant differences in SOS inducing potency and regions of active concentrations. The both compounds posses the cytotoxic activity on the tester cells that leads to decreasing of light emission at it's high concentrations (Table 2.). The cytotoxicity of *cis*-DDP to *recF* mutant is significantly increasing, that leads to decreasing the concentration region for SOS response and reduction of F_i value. *Trans*-DDP does not change the cytotoxicity to *recF* mutant. So, the type of DNA-platinum adducts determines the level of SOS response of the cells, and mutations in recombination pathways can change the SOS response induced by certain types of adducts.

The possibility of *in vivo* monitoring of SOS response in SOS *lux* test allows to use the reporter strain in immobilized form in on-line mode system. We use the procedure of C600(pPLS-1) cells immobilization in agarose granules to obtain the matrix for SOS

TABLE 2. The SOS response of *E.coli* AB1157(pPLS-1) and JC9239(pPLS-1) to *cis*-DDP and *trans*-DDP

	cis-DDP		*trans-DDP*	
	AB1157 (pPLS-1)	*JC9239 (pPLS-1)*	*AB1157 (pPLS-1)*	*JC9239 (pPLS-1)*
Minimal SOS inducing conc., mol/l	$3 \cdot 10^{-6}$	$0.7 \cdot 10^{-6}$	$6 \cdot 10^{-5}$	$3 \cdot 10^{-5}$
Conc. for maximal SOS response, mol/l	$3 \cdot 10^{-4}$	$2 \cdot 10^{-5}$	$3 \cdot 10^{-4}$	$3 \cdot 10^{-4}$
Maximal Fi	62.5	6.4	3.8	2.5

lux test. The 1 ml column with this matrix was used as a registration cell. The addition of 10^{-6} M MMC in medium flow leads to significant increase the light emission of matrix (after 3 h - more than 100-fold as comparing to non-induced matrix). So, this bioassay can be easily adopted to immobilized mode for on-line monitoring of pollutants.

Acknowledgement. The author thank Dr. G. Horneck for helpful advice and support, Dr. O. Komova and Prof. E. Krasavin for valuable comments on the work. This work is a part of studies which was supported by a COPERNICUS Grant CIPA CT-94 0122 from the European Commission.

4. References

1. Scheller, F., and Schmid, R. D. (1991) *Biosensors: Fundamentals, Technologies and Applications*, VCH Verlagsgesellschaft, Weinheim.
2. Mersch-Sundermann, V., Schneider, U., Klopman, G., and Rosenkranz, H. S. (1994) SOS induction in *Escherichia coli* and *Salmonella* mutagenicity: a comparison using 330 compounds, *Mutagenesis* **9**, 205-224.
3. Quillardet, P., Huisman, O., Ari, R. D., and Hofnung, M. (1982) SOS chromotest, a direct assay of induction of an SOS function in *Escherichia coli* K12 to measure genotoxicity, *Proc. Natl. Acad. Sci. USA* **79**, 5971-5975.
4. Oda, Y., Nakamura, S., Oki, I., Kato, T., and Shinagawa, H. (1985) Evaluation of the new system (umu-test) for the detection of environmental mutagens and carcinogens, *Mut. Res.* **147**, 219-229.
5. Quillardet, P. and Hofnung, M. (1993) The SOS chromotest: a review, *Mut. Res.* **297**, 235-279.
6. Chatterjee, J. and Meighen, E. A. (1995) Biotechnical applications of bacterial bioluminescence (*lux*) genes, *Photochem. Photobiol.* **62**, 641-650.
7. Frey, J., Ghersa, P., Palacios, P. G., and Belet, M. (1986) Physical and genetic analysis of the ColD plasmid, *J. Bacteriol.* **166**, 15-19.
8. Maniatis, T., Fritsch, E. F., and Sambrook, J. (1982) *Molecular Cloning: a Laboratory Manual*, Cold Spring Harbor Laboratory, N.Y.
9. Sanger, F., Micklen, S., and Coulson, A. R. (1977) DNA sequencing with chain-terminating inhibitors, *Proc. Natl. Acad. Sci. USA* **74**, 5463-5475.
10. Ptitsyn, L.R., Horneck, G., Komova, O., Kozubek, S., Krasavin, E.A., Bonev, M., and Pettberg, P. (1997) A biosensor for Environmental Genotoxin Screening Based on a SOS *lux* Assay in Recombinant *E.coli* Cells. *Applied and Environmental Microbiology* **63**, 4377-4384.
11. Ptitsyn, L. R., Fatova, M. A., and Stepanov, A. I. (1990) Expression of the bioluminescence system of *Photobacterium leiognathi* in *E. coli* cells, *Molecul. Genetika (Russ.)*, **16**, 26-29.

12. Birnbaum, S. and Bulow, L. (1988) Production and release of human proinsulin by recombinant *Escherichia coli* immobilized in agarose microbeads, *Enzyme Microb. Technol.* **10**, 601-605.
13. Nakamura, S., Oda, Y., Shimida, T., Oki, I., and Sugimoto, K. (1987) SOS-inducing activity of chemical carcinogens and mutagens in *Salmonella typhimurium* TA1535/pSK1002: examination with 151 chemicals, *Mut. Res.* **192**, 239-246.
14. Isildar M, Bakale G., (1984), Radiation-induced mutagenicity and lethality in Ames tester strains of Salmonella. *Radiat. Res.* **100**, 396-411
15. Kozubek S., Krasavin E.A., Amirtayev K.G., Tokarova B., Soska I., Drasil V., Bonev M., (1989), The induction of revertants by heavy particles and gamma-rays in Salmonella tester strains. *Mutat. Res.* **210**, 221-6
16. Corda, Y., Job, C., Anin, M.F., Leng, M., and Job, D.(1993) Spectrum of DNA-platinum adduct recognition by prokaryotic and eukaryotic DNA-dependent RNA polymerases, *Biochemistry* **32**, 8582-8588.

THE NATURE OF THE SOS-INDUCING LESIONS CAUSED BY UV LIGHT IN *E. COLI* CELLS DEFICIENT IN UMU C GENE PRODUCT

O. V. KOMOVA [1], E. S. CANDIANO [1], E. A. KRASAVIN [1],
M. N. BONEV [2], G. HORNECK [3]

[1] *Joint Institute for Nuclear Research, Dubna, Russia*
[2] *Institute for Nuclear Research and Nuclear Energy, Sofia, Bulgaria*
[3] *Institute of Aerospace Medicine, Cologne, Germany*

1. Introduction

A major repair mechanism in *E.coli* cells is nucleotide excision repair. This process is initiated by the Uvr ABC nuclease which provides damage recognition and incision. The rate of incision is limited by the number of UvrC molecules per cell (approximately 20). Products of *polA* and *uvrD* genes are required to complete repair, release Uvr complex from DNA and thus stimulate turnover of enzymes [1, 2].

Unexcised DNA lesions caused by UV light require the intervention of the SOS inducible repair system to be bypassed. Both *in vitro* and *in vivo* evidence suggest that the signal for SOS induction consists of regions of single-stranded DNA that are generated when the cell attempts to replicate a damaged template or when DNA replication complex proceeds past lesions and forms gaps in daughter strand [3]. In consequence of the SOS signal generation, about 20 genes are derepressed. Among them are *umuC* and *umuD* encoding the products essential for translesion synthesis accompanied by a significant elevation of mutation frequency. It was suggested that in SOS-induced cells polymerase III in an altered low-fidelity mode mediated by UmuC, UmuD' and RecA is able to replicate over lesions in such a way that an incorrect nucleotide is inserted opposite the lesion [4, 5]. In *umuC* mutant cells this mechanism does not function and the lesions impeding DNA replication will contribute to the SOS induction.

In the present study we have attempted to clarify the nature of these lesions, by use of the SOS *lux* test for quantitative estimation of the SOS induction [6]. This method is based on the recombinant plasmid pPLS-1 containing a promoterless luciferase operon of *Photobacterium leiognathi* (reporter component) under control of a strong SOS promoter of cda gene of ColD plasmid (receptor component). *E.coli* cells containing this construction are inducible to high level of light production (as a result of luciferase reaction) in the presence of DNA lesions.

C. Baumstark-Khan et al. (eds.),
Fundamentals for the Assessment of Risks from Environmental Radiation, 167-172.

2. Material and Methods

The strains of E.coli K-12 used were: AB1157/*pPLS1(WT)*; AB1886/*pPLS1(uvrA6)*; GW2110/*pPLS1(umuC122:Tn5)*; SK7772/*pPLS1(ΔuvrD291)*.

E. coli cells were grown overnight at 37 °C in L-medium [7] supplemented with 50 μg/ml ampicillin. After dilution (1:50) in fresh L-medium, the culture was incubated at 37 °C until the optical density at 560 nm reached about 0.35. Then cells were harvested by centrifugation, resuspended in M9 [7] buffer and irradiated with UV light. In some cases photoreactivating treatment was applied immediately after exposure or after 30 min starvation. Aliquots of 1 ml of the irradiated suspension were transferred into 20 ml L-medium and incubated at 30 °C with shaking. At the required time 0.5 ml were transferred into luminometric tubes and light emission (MIPHI luminometer) and cell density (spectrophotometer SPECOL 11) were measured. Light measurements (imp/s) were carried out at 30 °C during 1s. To prevent photoreactivation, all experiments were done under yellow light.

UV radiation at 254 nm (low pressure Hg lamp at a dose rate 0.1 W/m^2) was used in this assay. Photoreactivating treatment was done for 30 min at 30 °C using a luminescent lamp (40 W).

The factor of SOS induction F_i was calculated as the relative light emission of irradiated suspension and unexposed sample Lux_i/Lux_0 divided by the relative optical density values OD_i/OD_0:

$$F_i = \frac{Lux_i \cdot OD_0}{Lux_0 \cdot OD_i} \quad (1)$$

3. Results and Discussion

Effect of the *umuC* mutation on the induction of the SOS response was tested for two UV doses 5 J/m^2 and 20 J/m^2. In both cases induction factor increased more rapidly and reached higher maximum in *umuC* deficient cells compared to wild type cells (Figure 1). Therefore we suggest that UmuC protein (perhaps on the constitutive level) participates in the processing of some SOS-inducing structures caused by UV or prevents formation of some of them.

Figure 2a (curves 2 and 3) shows that a bulk of the lesions giving rise to the high SOS response in mutant cells exposed to 5 J/m^2 are eliminated by the photoreactivating treatment. It allows to conclude that they are pyrimidine dimers. In view of this, so considerable effect of the *umuC* mutation on the SOS induction seems to be unexpected because unexcised dimers inducing SOS system are mainly removed by postreplication recombinational repair without participation of the UmuC protein [8]. It could be suggested that overlapping dimers require UmuC to be repaired, but the dose 5 J/m^2 is not sufficient to produce such configurations. Therefore, we can speculate that dimers inducing SOS response in *umuC* deficient strain are resistant to the excision or excision repair is stimulated when this protein presents in cell.

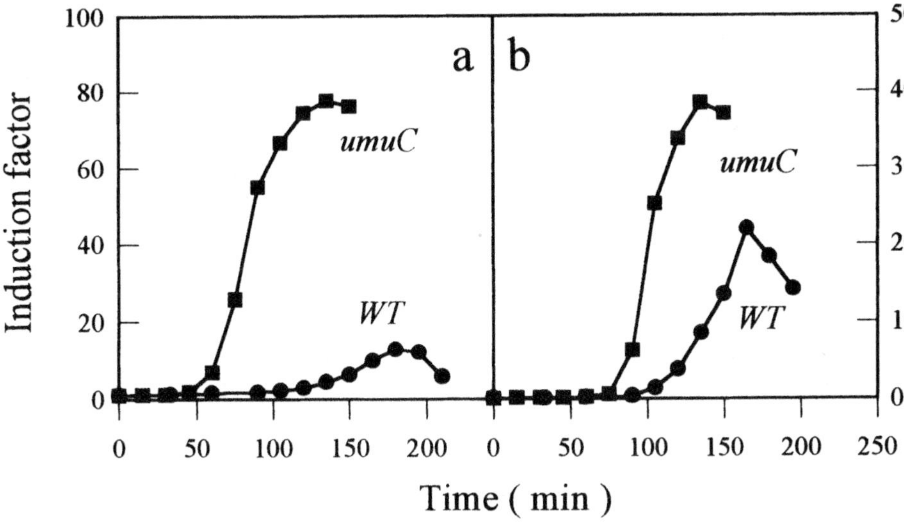

Figure 1. Effect of the umuC mutation on the induction of the SOS response by exposure to 5 J/m^2 UV (a) and to 20 J/m^2 UV (b).

In order to investigate the efficiency of excision repair in mutant strain we applied the procedure of starvation of irradiated cells in M9 buffer (condition under which SOS function is repressed but excision repair occurs normally). Figure 2a (curves 1 and 2) shows that after 30 minutes starvation of the cells irradiated with dose 5 J/m^2, maximum of response is reduced and it reaches only half of that observed after immediate postirradiation incubation in rich medium. Thus, excision repair can remove pyrimidine dimers inducing SOS signal in *umuC* mutant strain in the absent of the replication forks in force. When SOS-induction is permitted immediately after UV exposure replication forks probably reach them earlier then their excision occurs, and they will contribute to the SOS response due to generation of gaps in daughter strand.

The comparison of these curves reveals also the increase of the initial rate of the SOS induction in the cells held in buffer after irradiation (Figure 2a insert, curves 1 and 2) This effect becomes more significant with prolongation of starvation (data not shown) and increasing dose of irradiation (Figure 2b). Presumably some SOS-inducing structures appear or accumulate during starvation (may be in some cells) which lead to the enhancement of the SOS response. The signal generated by them is non-photoreversible (Figures 2a insert and 2b, curves 2 and 3). It allows to suggest that they may be produced at 6-4 photoproducts. which are unrepairable by photolyase.

Because single-stranded DNA is considered to be the critical inducing signal [9], the main candidates on the role of these structures are unfilled excision gaps. Such gaps can be formed when DNA polymerase I can not carry out repair synthesis. The lack of its activity can be caused by the generation of unsuitable primers for polymerase I in

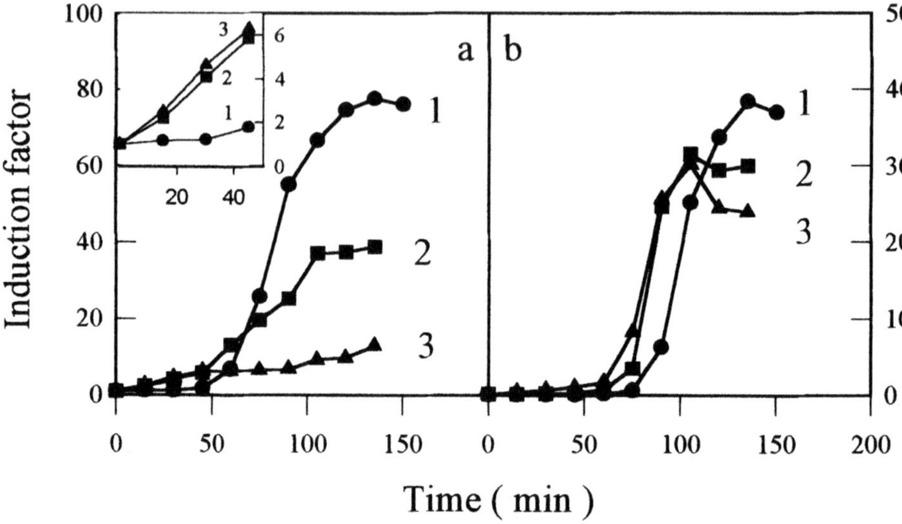

Figure 2. Effect of starvation and visible light treatment on the induction of the SOS response in *umuC* mutant cells exposed to 5 J/m^2 UV (a) and to 20 J/m^2 UV (b). The cells were incubated in rich medium in the dark immediately after exposure (1) or held in M9 buffer for 30 min after exposure and then incubated in rich medium in the dark (2) or illuminated by visible light for 30 min after exposure and then incubated in rich medium in the dark (3). The insert shows initial part of SOS response kinetic curves.

consequence of incision. In the cells deficient in incision (*uvrA*$^-$) generation of such structures does not occur. Actually, we have not obtained any changes in SOS response kinetics after starvation in this strain (data not shown).

Another explanation of data obtained can be done. As it was mentioned above DNA polymerase I is involved in removal of the uvrABC nuclease complexes from the DNA at the damaged site as well as in polymerisation step of excision repair. The lack of this ability can result in accumulation of some molecular structures during starvation. They represent regions of DNA containing incised pyrimidine dimers associated with uvrABC nuclease complexes.

This explanation seems to be more acceptable. Firstly, according to the last experimental data *in vitro,* incised oligonucleotides containing photoproducts can not be removed by uvrABC nuclease and require polymerase I to be released in the process of the repair synthesis [1]. Though, at present no common viewpoint on this problem exists, the availability of exposed gaps in DNA seems to be improbable.

We observed still more pronounced effect of starvation on the initial rate of the SOS induction in *uvrD*$^-$ cells exposed to 5 J/m^2 UV (Figure 3). It is known that this mutant is just deficient in excision of olygonucleotides containing damage [10]. But *uvrD* gene product (helicase II) as well as polymerase I is required for the removal of uvrABC complexes from the DNA. In this case the structures mentioned above will also accumulate in the DNA.

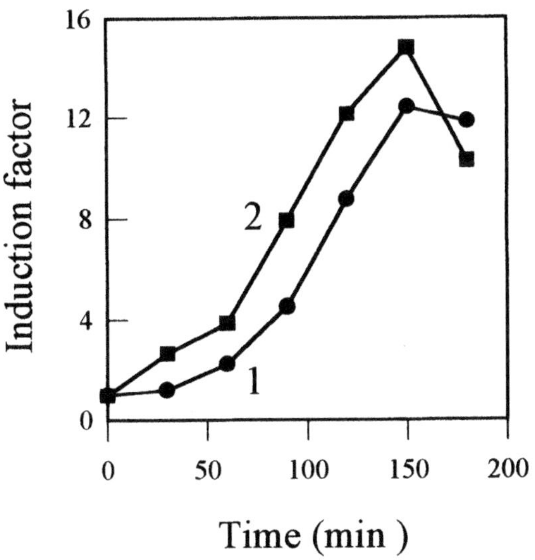

Figure 3. Effect of starvation on the induction of the SOS response in uvrD mutant cells exposed to 5 J/m^2 UV. The cells were incubated in rich medium in the dark immediately after irradiation (1) or held in M9 buffer for 30 min after exposure and then incubated in rich medium in the dark (2).

Thus, results presented here allow to conclude that the nature of the SOS-inducing lesions in *umuC* mutant strain is different at low and high doses. At low doses SOS response is mainly caused by pyrimidine dimers unexcised to the moment when replication fork reaches them. In this case gaps in daughter strand of the DNA opposite the lesions can be left by replication machinery and give rise to the SOS response. These dimers can be removed by photoreactivating treatment or by excision repair during prolonged period of starvation in the absence of active replication forks. In addition, there is small part of photoproducts resistant to visible light and excision which cause SOS induction very quickly after the incubation starts. These lesions completely define the SOS response after high UV doses. We have suggested that they are photoproducts (more probably 6-4 photoproducts), incision of which results in generation of unsuitable primers for polymerase I. As a result, structures representing incised photoproduct associated with uvrABC complexes are accumulated in the DNA because polymerase I function is necessary to release these enzyme molecules. Such structures as prevent replication as reduce efficiency of excision repair. We have not observed the effect of starvation on the initial rate of the SOS induction in wild type cells (data not shown) .Therefore, we have supposed that polymerase III in complex with UmuC and UmuD` can substitute polymerase I in this process. It allows to complete replication (in error-prone mode) and provide turnover of incision enzymes to stimulate excision repair.

Prolonged SOS signal in *umuC* mutant strain may be necessary in order to synthesize some inducible protein factor. This factor can be required to stimulate alternative mechanism of replication bypass (error-free) without participation of UmuC protein. As it was shown, the reiniciation of DNA replication downstream from a blocked replisome took place in *E.coli* cells irradiated by UV [11]. This process termed "induced replisome reactivation" (IRR), requires RecA and addition protein - IRR factor but does not require product of *umuC* gene. It was suggested that this factor is the inhibitor of RNaseH. We can speculate that alternative mechanism of translesion synthesis occurring in *umuC* mutant is IRR.

References

1. Caron, P.R., Kushner, S.R.and Grossman, L. (1985) Involvement of helicase II (*uvrD* gene product) and DNA polymerase I in excision mediated the uvrABC protein complex, *Proceedings of the National Academy of Sciences of USA* **82**, 4925-4929
2. Husain, I., Houten, B.V., Thomas, D.C. and Abdel-Monem, M (1985) Effect of DNA polymerase I and DNA helicase II on the turnover rate of uvrABC excision nuclease, *Proceedings of the National Academy of Sciences of USA* **82**, 6774-6778
3. Rupp, W.D. and Howard-Flanders, P.(1968) Discontinuities in DNA synthesized in an excision-defective strain of *Escherichia coli* following ultraviolet irradiation, *J.Molecular Biology* **31**, 291-304
4. Walker, G.C. (1995) SOS-regulated proteins in translesion DNA synthesis and mutagenesis, *Trends in Biochemical Science* **20**, 416-420
5. Echols, H. and Goodman, M.F. (1990) Mutation induced by DNA damage: a many protein affair, *Mutation Research* **236**, 301-311
6. Ptitsyn, L.R., Horneck, G., Komova, O., Kozubek, S., Krasavin, E.A., Bonev, M., and Pettberg, P. (1997) A biosensor for Environmental Genotoxin Screening Based on a SOS *lux* Assay in Recombinant *E.coli* Cells. *Applied and Environmental Microbiology* **63**, 4377-4384.
7. Miller, J (1972) *Experiments in molecular genetics.* Cold Spring Harbor Laboratory Press, Cold Spring Harbor, NY
8. Smith, K.C., Wang, T.V. and Sharma, R.C. (1987) *RecA*-dependent DNA repair in UV-irradiated *Escherichia coli, J.Photochemistry and Photobiology, B: Biology* **1**, 1-11
9. Sassanfar, M. and Roberts, J.W. (1990) Nature of the SOS-inducing signal in *Escherichia coli, J. Molecular Biology* **212**, 79-96
10. Kuemmerle, N.B. and Masker, W.E. (1980) Effect of the *uvrD* mutation on excision repair, *J.Bacteriology* **142**, 535-546
11. Khidir, M.A., Casaregola, S. and Holland, I.B. (1985) Mechanism of transient inhibition of DNA synthesis in ultraviolet irradiated *E.coli*: Inhibition is independent of *recA* whilst recovery requires RecA protein itself and an additional inducible SOS function, *Molecular and General Genetics* **199**, 133-140

DNA INJURIES, DAMAGE INDUCTION AND REMOVAL.

Recommendations for Research Needed

B.D. MICHAEL [1], G. RONTO [2] AND
C. BAUMSTARK-KHAN [3], J. CADET [4], M. FRANKENBERG-SCHWAGER [5], O.V. KOMOVA [6], W.-U. MÜLLER [7],
M.R. QUASTEL [8], E. SAGE [9]

[1] *Gray Laboratory Cancer Research Trust, Mount Vernon Hospital, Northwood Middlesex, UK*
[2] *Institute of Biophysics, Semmelweis University of Medicine, Budapest, Hungary*
[3] *DLR, Inst. of Aerospace Medicine, Radiobiology Unit, Köln, Germany*
[4] *Département de Recherche Fondamentale sur la Matière Condensée, SCIB/Laboratoire "Lesions des Acides Nucléiques", CEA/Grenoble, France*
[5] *Universität Göttingen, Klin.Strahlenbiologie und Klin.Strahlenphysik, Goettingen, Germany.*
[6] *Joint Institute for Nuclear Research, Division of Radiation and Radiobiological Research, Dubna, Russia*
[7] *Institut für Medizinische Strahlenbiologie, Universitätsklinikum Essen, Germany*
[8] *Department of Epidemiology and Institute of Nuclear Medicine, Ben Gurion University of the Negev, Beer Sheva, Israel*
[9] *CNRS, Institut Curie, Section de Recherche, Paris, France*

1. Introduction

Most of the physical and chemical processes involved in DNA damage induction have been characterized in general terms, both for UV and for ionizing radiations. The relative contributions of direct energy absorption and of indirect damage via reactive intermediates are broadly understood. Thus we have a general, although incomplete, picture of the influence of radiation type (wavelength for UV; quality or linear energy transfer (LET) for ionizing radiations) on biological outcome. Some aspects of the dose-effect relationships for DNA damage induction and of the kinetics of its repair are understood. These relationships provide key information for the development of mechanistic models of human risk/exposure relationships.

C. Baumstark-Khan et al. (eds.),
Fundamentals for the Assessment of Risks from Environmental Radiation, 173-178.

2. State of Knowledge

2.1. INDUCED DAMAGE

- We have good knowledge of UV-C-induced damages (most photoproducts are identified), except for the CC-photoproduct.
- Action spectra have been characterized for some end-points (bacteriophage killing, spore inactivation, 6-4-photoproducts, cyclobutane dimers, 8-hydroxyguanine, single-strand breakage).
- For ionizing radiation, some progress has been made towards the development of models of radiation damage to DNA and chromatin which can predict spectra of lesions and their complexity.
- We have only fairly good knowledge for low LET induced lesions (strand breaks, base damages), although some damages are still black boxes (limited information for regional multiple damaged sites, alkali labile lesions, clustered damages). More is known about the damages induced by low LET radiation than for high LET radiation.

2.2. DNA REPAIR

- Several of the repair pathways for UV and for ionizing radiation have been characterized. These include photoreversal of cyclobutane dimers and 6-4-photoproducts, excision repair of bulky damage, transcription-coupled repair, base excison of oxidative damage, mismatch repair, homologous repair, non-homologous repair (end joining, illegitimate rejoining).
- Although there are similarities in the DNA repair pathways of *E. coli* and man, not all of the genes involved have been identified and even less information is available about the gene products involved in repair.
- A little is known about links of repair processes to replication and more is known about transcription and specific cellular regulation and the cell cycle. Inherited defects in repair are also linked to immunodeficiency syndromes and to premature ageing.
- There is some information about cancer-proneness and repair from Xeroderma pigmentosum and Ataxia telangiectasia and some other hereditary diseases. Some indications on DNA repair gene defects come from human tumours (repair of adducts) although the sequence of events is still unclear.
- There is some evidence for the involvement of signal transduction in the induction of repair systems.
- There is evidence that individuals vary widely in their repair capacities.
- It is known that not all damages are repaired successfully. Some damages are misrepaired and others remain unrepaired. These persisting damages may have deleterious genetic and somatic consequences.
- While mutation may bring detriment in respect of heriditable and somatic diseases, it is also a source of benefit in the evolutionary process. A better understanding of

mutation is likely to contribute to its applications, for example in plant and animal breeding.

3. Problems to be Solved

3.1. DOSIMETRIC CONSIDERATIONS

- Improved standardisation is needed for lamps, filters and dosimetry, for both, UV-A and UV-B studies. Standardisation of dose in terms of physical units rather than biologically based units is recommended.

3.2. INDUCED DAMAGE

- A priority is the identification of UV-A-damage, its precursors, nature and interrelations with other damage. It is also necessary to determine to what extent the yield of UV-A products depends on local physiological conditions. Potential systems to protect against UV-A damage by chemical means should be investigated. The UV-A chromophore has not yet been identified. It is necessary to clarify the role of any radical reactions involved.
- Very little is known about cytosine photoproducts although these clearly have an important role in UV-B mutagenesis.
- Improved methodology is required for quantification of tandem cytosine photoproducts.
- The influence of DNA sequence, conformation and association with proteins on the nature and yields of photoproducts needs to be determined.
- The role of UV-A needs to be related to that of UV-B and integrated to sunlight.
- With ionizing radiation, the critical lesions still need to be identified. The main candidates are considered to be double-strand breaks (dsb) and clustered lesions induced by ionizing radiations of all types. In addition it is considered that structurally derived fragments, or „regionally damaged sites“ (RDS), may also have an important role. Assays for these lesions need to be developed. Among these assays, it will be important to determine the relative contributions of base damages and strand breaks and their involvement in lesion complexity.
- In addition to radiation induced products, the background levels of damage attributable to oxidative metabolism need to be determined accurately.
- Efforts should be made to identify and quantify the yields of DNA-protein crosslinks induced by UV and by ionizing radiations of various qualities.

3.3. DNA REPAIR

- Rates and efficiencies of repair need to be evaluated for the different classes of lesions. Interactions between different classes of lesions need to be included in these studies. Induction and repair of damage need to be related to final outcomes, for example, mutation and carcinogenesis. The influence of repair fidelity needs to be included.

- Further work is required to quantify mutational spectra as markers of UV-A and UV-B exposure. Similar approaches may also provide „signatures“ to distinguish exposures to high versus low LET ionizing radiation.
- Repair and misrepair of clustered lesions and of base damage after ionizing radiation are at present only poorly understood and need detailed studies. At a very basic level, the influence of proximal lesions on each other's repair need to be determined using model systems.
- Further work is needed to elucidate the relationship between the initial damages and their cellular processing and the eventual endpoints, such as chromosomal damage and mutation, both for UV and ionizing radiation.
- Little is known about somatic mutations. Rapid methods need to be developed to assay for these changes in exposed individuals.
- With both UV and ionizing radiation induced lesions, the ways that they initiate signalling responses, apoptosis, adaptive responses and gene induction and act on cell cycle checkpoints need to be clarified. This may be important in understanding the actions of mixed fields, for example ionizing radiation and UV.
- In addition to DNA damage as an initial signalling event membrane targets have to be considered.
- Variability of individual responses to radiation has an important bearing on susceptibility to risk and on treatment strategy. At present fast and robust assay systems are not generally available and need to be developed.

4. Justification for Further Research Needed

- DNA damage and its processing are key factors in the deleterious effects of the rising environmental pressures on human populations. Damage and repair also play important roles in the treatment of malignancies. Cancer is a disease which, with certain exceptions, arises predominantly in the older age groups. Life expectancy is likely to increase due to improvements in medical care. The burden of cancer treatment, both in terms of patient numbers and the need to control the disease for longer periods of residual life span can only be expected to increase. Prevention of cancer therefore becomes a topic of even greater importance. Also the development of treatment strategies which achieve higher cure rates and longer periods free from disease will become more important. Therefore, improved understanding of DNA damage in cancer treatment will have an increasingly important role in the management of the disease.
- There is a growing awareness that individual variability of response to DNA damaging agents plays an important role in respect of cancer susceptibility and avoidance strategies. Individual variability is also important in respect of optimisation of treatment since present failure to identify in advance the small fraction of patients who will over-respond means that the majority of patients will receive suboptimal treatment doses. The knowledge of individual patient responsiveness would allow their treatment to be tailored optimally in each case and the development of tests at the DNA level are likely to enable this to be achieved. Research in the area of

evaluating the responsiveness of individual patients would have significant applications in cancer therapy.

- Skin cancer is a high priority for prevention because of the present rapid rise in its incidence. It is one of the few cancers for which the major causative agent has been clearly identified and for which there are good prospects for effective prevention.
- DNA damage induced by various ionizing radiations is believed to have a significant role in the induction of heritable defects.

5. Recommended Approaches (Minimal/Optimal Requirements)

5.1. EFFECTS OF ULTRAVIOLET RADIATION

5.1.1. Minimal Requirements

- The photolesions induced by UV-A and UV-B need to be identified and quantified in vitro and in vivo.
- The yields and the spectra of mutations induced by UV-A and UV-B in vitro and in vivo need to be determined and correlated with biological endpoints including carcinogenesis and immunosuppression.
- The mechanisms and extent of repair of lesions induced by UV-A in vitro need to be determined.
- For the different photoproducts improved determinations of the action spactra have to be made and correlated with defined biological endpoints.

5.1.2. Optimal Requirements

- Actions of mixed wavelengths in relation to the endpoint mentioned above need to be determined in relation to simulated and natural sunlight.
- Repair kinetics after exposure to simulated and natural sunlight need to be determined.
- Action spectra for the different photoproducts need to be determined using polychromatic light.
- The chromophores involved in the induction of UV-A damage need to be identified.
- The induction and repair of photoproducts in vivo need to be studied.

5.2. EFFECTS OF IONIZING RADIATION

5.2.1. Minimal Requirements

- Techniques need to be developed to measure clustered lesions. Techniques to measure fragmentation of DNA in the range from 10 bp to 10 Mbp need to be improved.
- The yields of base damages induced in cellular DNA need to be re-assessed.
- The effects of LET on the yields of single and clustered lesions need to be evaluated.

- The cellular processing of single and complex lesions needs to be evaluated in relation to endpoints relative to risk.
- The link between initial DNA damages and the eventual formation of chromosomal aberrations needs to be explored.

Optimal Requirements

- DNA-protein crosslinks need to be identified and quantified both in terms of induction and repair.
- The genetic changes involved in oncogenic transformation need to be understood.
- The putative roles of DNA damage and repair in relation to genomic instability need to be evaluated.

5.3. EFFECTS GENERALLY RELATED TO DNA DAMAGE

- A very important aspect are genes involved in repair of UV- and ionizing radiation-induced damage. Principal approaches involve comparative studies in E. coli, yeast and mammalian cells and address gene sequences and and the corresponding structure and function of the proteins.
- The roles of DNA repair deficiencies in relation to specific human syndromes and cancer predisposition have to be evaluated. In this respect human and rodent cell lines defective in specific gene products will provide important mechanistic information. Mice carrying the corresponding knock-out genes are of extreme value.
- For cell signalling, the role of DNA lesion as a trigger needs to be determined. Membrane and DNA triggered responses need to be evaluated in relation to the processing of DNA damage.
- DNA-based predictive assays of radiosensitivity need to be inproven and further developed for individualization of protocols in therapy and diagnostics.
- Assays for somatic cell muatation need to be developed and applied as a biological indicator of radiation exposure.
- Mechanistic studies of low-dose mutations of germ lines by radiation need to be carried out.
- The actions of clastogenic factors in relation to DNA damage induction need to be studied.
- The involvement of DNA damage and repair in cellular senescence need to be studied.
- Assays need to be developed to determine the spectra of point mutation induced in mammalian cells.

BIOLOGICAL DOSIMETRY OF ABSORBED RADIATION DOSE BASED ON THE FREQUENCIES OF CHROMOSOMAL ABERRATIONS IN HUMAN LYMPHOCYTES

A.T. NATARAJAN [1], G. OBE [2]

[1] *MGC Dept. of Radiation Genetics and Chemical Mutagenesis, State University of Leiden*
P.O. Box 9503, NL-2300 RA Leiden, The Netherlands
[2] *Dept. of Genetics, University of Essen*
P.O. Box 45117, D-45117 Essen, Germany

1. Introduction

In the investigation of radiation accidents, it is important to estimate the dose absorbed by exposed persons in order to plan their therapy. For these tasks informations such as magnitude of dose received, as external or internal radiation, or as partial or whole body irradiation are necessary. For example following exposure to high doses (> 5 Gy) of low LET radiation the victim may need bone marrow transplantation and in case of partial body exposure to high doses the victim may be isolated and kept under very hygienic conditions in order to allow repopulation from surviving stem cells in the bone marrow.

2. Methods To Estimate Radiation Doses

Usually in accidents the victims do not carry a physical dosemeter unless the accident happens in a nuclear facility. In the absence of any physical dosimetry one has to resort to biological methods. Even if physical dosimetric information is available, it is better to confirm this estimate with a biological dosimetry. Currently the fully developed biological indicator for exposure to ionizing radiation is the study of chromosomal aberrations in peripheral blood lymphocytes. Other techniques such as ESR measurements of tooth enamel are available but not yet well validated.

3. Chromosomal Aberrations

Most of the circulating lymphocytes are in a presynthetic stage of the cell cycle (G0) in the body. Immediately following radiation exposure the chromosomal aberrations are formed. Lymphocytes with aberrations can circulate in the body for months to many years. Lymphocytes are usually stimulated in culture for about 48 h with a mitogen, whereby they enter the cell cycle and undergo mitoses. The mitotic cells are arrested by

C. Baumstark-Khan et al. (eds.),
Fundamentals for the Assessment of Risks from Environmental Radiation, 179-186.

colcemid and fixed. Air dried preparations are stained and chromosomal aberrations are analyzed in the first mitosis in culture [1].

4. Types Of Chromosomal Aberrations

Exposure of blood lymphocytes at the G0 stage leads to chromosome type of aberrations such as dicentrics, rings, translocations, inversions and acentric fragments. Except for translocations and inversions all other aberration types are unstable and will be gradually eliminated during cell division.

5. Dose-Response Curves

In general the frequencies of dicentrics are used to estimate absorbed doses. The frequencies of dicentrics induced by radiation *in vitro* and *in vivo* appears to be very similar for a given dose and therefore, dose-response curves generated from *in vitro* studies can be used with confidence to estimate doses of *in vivo* exposure. The frequency of exchange aberrations (dicentrics and translocations) increase in a linear quadratic manner following acute low LET radiation [$Y=k+\alpha D+\beta D^2$]. With high LET radiation a linear dose response is observed [$Y=k+\alpha D$]. For chronic low LET exposure the yield is mostly linear with the dose (Fig. 1).

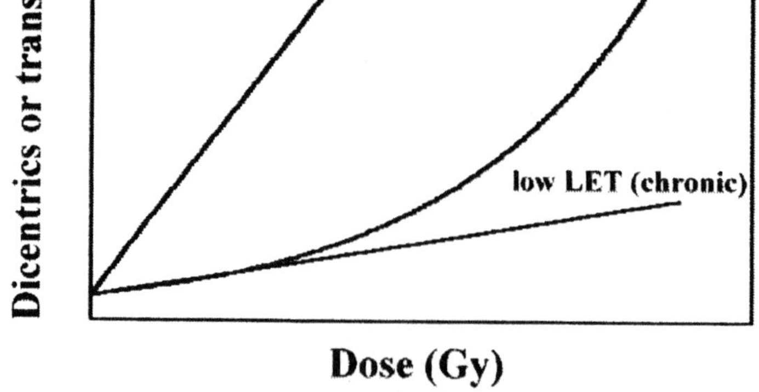

Figure 1. Dose effect relationships for dicentrics or translocations following exposure of lymphocytes to high LET, low LET acute or low LET chronic irradiation

6. Dose Estimates

The aberrations among lymphocytes after low LET radiation follow a Poisson distribution, while aberrations induced by high LET radiation as well as partial body radiation are overdispersed among the lymphocytes. From the distribution of aberrations among the cells and the nature of the accident one can use an appropriate calibration curve to estimate the dose. Thus, in principle it is possible to estimate doses following exposure to acute as well as chronic low LET radiation, high LET radiation, partial body radiation and a mixture of these. The calibration curves should be based on scoring large numbers of cells for each dose point, especially at the low dose range (up to 1 Gy) using at least 2 different blood donors, preferably a male and a female. The spontaneous frequency of dicentrics in lymphocytes is about 1 in 1000 cells [1]. In a suspected exposure case, a minimum of 500 to 1000 cells are to be scored and if the frequencies are less than the spontaneous one the person can be considered to be unexposed. There are several factors which can influence the accuracy of the dose estimate. These include (a) the calibration curve, (b) the number of cells scored and (c) the number of aberrations observed and the distribution of aberrations. Based on these factors a dose estimate can be made by using the maximum likelyhood method [2]. The sensitivity of this technique is limited to >0.1-0.2 Gy and very much depends on the number of cells scored in each case.

For making dose-estimates the blood sample should be drawn within a day following exposure. For delayed sampling a correction factor can be applied assuming that the average lifetime of lymphocytes is about 3 years [1]. However, the estimates for the average lifespan of lymphocytes vary considerably. In a recent study on the radiation victims of the Goiânia accident, the average half-life based on the disappearance of lymphocytes carrying dicentrics was estimated to be between 110 and 470 days depending on the radiation dose received [3]. Cytogenetic studies of atom bomb victims from Hiroshima and Nagasaki indicate that a long lived sub-population of lymphocytes can have a lifespan of more than 50 years. This makes the dose estimates based on dicentric frequencies from delayed sampling, using a correction factor for the average half-life of lymphocytes, not very reliable.

Unlike dicentric chromosomes which are lost in the body with time, balanced translocations are expected to persist. Thus, one can use the frequencies of translocations for dose estimates immediately after irradiation as well as for past exposure. Translocations can be detected by using chromosome banding techniques but these are very laborious. Recently the Fluorescence *In Situ* Hybridization Technique (FISH) using chromosome specific DNA libraries has been successfully employed to detect translocations [4, 5]. Though the frequencies of translocations and dicentrics are expected to be induced at equal frequencies, *in vitro* radiation studies show that translocations are induced at a much higher rate than dicentrics. This is mainly due to the contribution of incomplete and complex translocations in addition to balanced reciprocal translocations [6, 7].

In view of the persistant nature of translocations it has been suggested that their frequencies can be used to estimate doses of past exposures. For example, translocation frequencies in atom bomb survivers appear to be similar to the frequencies induced *in vitro* by different doses of radiation [8]. However, other studies on radiation victims from Goiânia and Chernobyl indicate that the frequencies of translocations at later sampling times may not reflect the frequencies induced at the time of the accident [9, 10, 11, 12].

7. The Goiânia Accident

In this accident more than 100 individuals were contaminated with ^{137}Cs in September, 1987. The ^{137}Cs was from a radiotherapy unit which was stolen, broken into pieces and distributed to several individuals. Immediately after discovery of the accident, blood samples were collected and processed for chromosome analysis at the Institute of Radiation Protection and Dosimetry, Rio de Janeiro. The frequencies of dicentrics and their distribution were determined. Since the exposure in this accident was chronic in nature, a dose-response curve which had been generated earlier based on low dose rate γ-rays was used to estimate the doses. The doses of 129 individuals were estimated which varied from 0-7 Gy. From the distribution of aberrations, individuals with whole body and partial body irradiation were identified [3]. Later on a new dose response curve was generated with protracted irradition and based on that curve the dose estimates were recalculated [13]. The comparison of the individual dose estimates and the

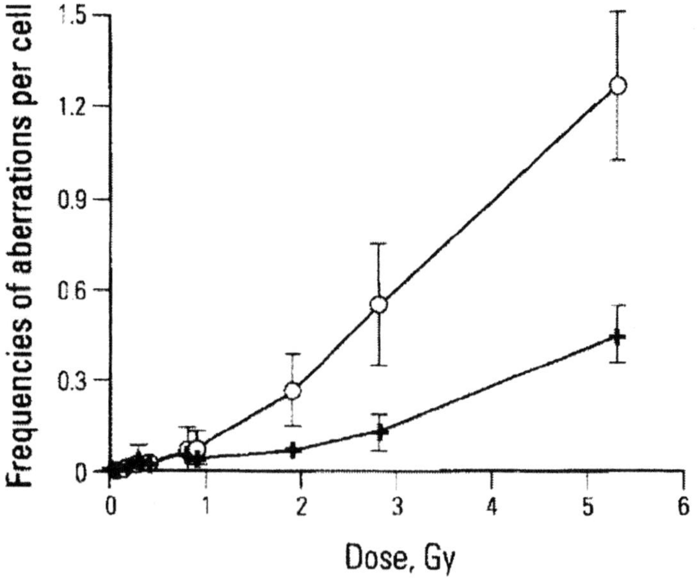

Figure 2. Comparison of the frequencies of dicentrics analyzed in Goiânia victims in 1987 (O) and of translocations (+) analyzed in Goiânia victims in 1993 (figure taken from [18]).

history of the accident indicated a very close correlation between highly exposed individuals and high doses [14], demonstrating the validity of using frequencies of dicentrics to estimate absorbed radiation dose in accidents.

8. Retrospective Dosimetry Based On Translocation Frequencies

Earlier studies using banding techniques indicated that the translocation frequencies remain stable over many years following therapeutic radiation of patients with ankylosing spondylitis [15]. However, the initial translocation frequencies were not determined in these patients. In addition to the studies on the atomic bomb victims, retrospective dosimetry has been used for a radiation worker accidentally exposed to tritium [16], cancer patients undergoing radiation therapy [17] and Chernobyl victims [11, 12]. These data suggest that with few exceptions the frequencies of translocations remain constant for several years. In all these cases the initial frequencies of dicentrics or translocations were not determined. Victims of the Goiânia accident form an unique cohort of a radiation exposed population for which data on initial frequencies of translocations are available [10]. In a follow-up study the frequencies of translocations in 24 of

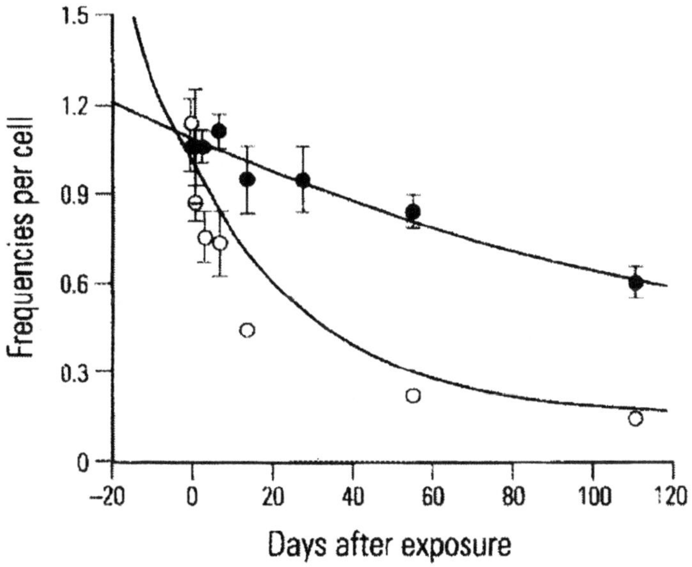

Figure 3. Frequencies of dicentrics (O) and translocations (●) over time following irradiation with X-rays and analyzing splenocytes after sacrificing mice at different times following irradiation. Dicentrics were analyzed in Giemsa stained preparations and translocations by means of FISH with DNA libraries of chromosomes 1, 11 and 13 (figure taken from [18]).

these victims who recieved estimated doses between 0.1 to 5.3 Gy were determined using FISH technique. Several coctails of chromosomes covering about 80 % of the genome were used and a total of 71,245 metaphases were analyzed. The frequencies of translocations in the group of individuals who received more than 1 Gy were lower than the initial dicentric frequencies. The difference was about a factor of 2 to 3. In individuals who received doses less than 1 Gy the translocation frequencies were also lower than the initial dicentric frequencies and in individuals who received 0.4 Gy or less the frequencies of translocations were similar or more than the initial frequencies of dicentrics. At these low doses to get reliable dose estimates, it is obvious that one should score 10,000 to 20,000 cells per individual, making it a very difficult task to perform (Fig. 2). A similar conclusion was also reached in the follow up studies of the Estonian clean up workers in the Chernobyl accident [9]. In such cases at best one can discriminate between the exposed and control group without attempting to estimate individual doses.

In a mouse model we studied the question as to whether and with what characteristics dicentrics and translocations decline with time. Mice were irradiated with 2 Gy X-rays and the frequencies of dicentrics and translocations were determined in splenocytes 0 to 112 days after irradiation [18]. As can be seen from Fig. 3, the translocations disappear much slower than dicentrics with time [19].

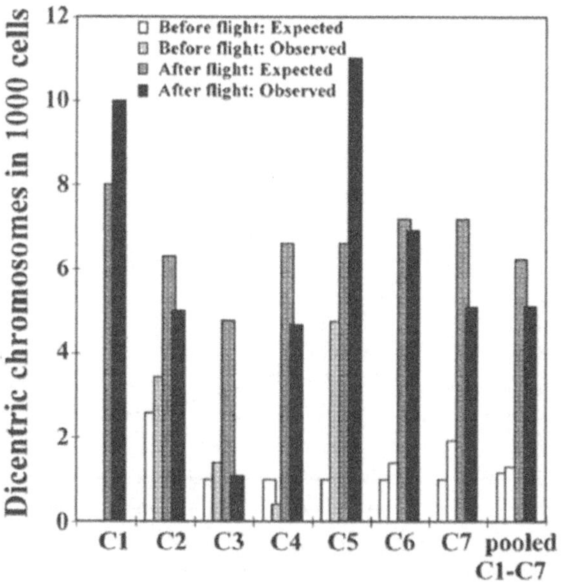

Figure 4. Observed and expected frequencies of dicentric chromosomes in seven cosmonauts (C1-C7) from MIR missions (120-198 days). The expected frequencies were calculated using data on physical dosimetry and assuming a dicentric yield of 0.03/cell/Gy for low LET radiation, 0.5/cell/Gy for high LET radiation, a background level of 0.001/cell and half-life of 3 years for lymphocytes (data taken from [20]).

9. Cosmonauts

Astronauts and cosmonauts are exposed in the orbit to low and high LET radiations. Analyses of lymphocytes of cosmonauts following MIR-missions indicate an increase in the frequency of cells carrying chromosomal aberrations. Based on these frequencies attempts have been made to estimate the dose accumulated by the cosmonauts [20]. A good correlation was found between the observed and expected frequencies of aberrations (Fig. 4), indicating the possibility to use this methodology for future biological dosimetric studies and radiation risk estimates on astronauts and cosmonauts.

10. References

1. IAEA (1986) *Biological Dosimetry: Chromosomal Aberration Analysis for Dose Assessment*, Technical Reports Series No. 260, International Atomic Energy Agency, Vienna.
2. Papworth, D.G. (1975) Curve fitting by maximum likelihood, *Radiat. Bot.*, **15**, 127-140.
3. Ramalho, A.T., Nascimento, A.C.H., and Natarajan, A.T. (1988) Dose assessments by cytogenetic analysis in the Goiânia (Brazil) radiation accident, *Radiat. Prot. Dosim.*, **25**, 97-100.
4. Natarajan, A.T., Vyas, R.C., Wiegant, J., and Curado, M.P. (1991) A cytogenetic follow-up study of the victims of a radiation accident in Goiânia (Brazil), *Mutation Res.*, **247**, 103-111.
5. Pinkel, D., Straume, T., and Gray, J.W. (1986) Cytogenetic analysis using quantitative, high sensitivity, fluorescence hybridization, *Proc. Natl. Acad. Sci. (USA)*, **83**, 2934-2938.
6. Bauchinger; M., Schmid; E., Zitzelsberger; H., Braselmann; H., and Nahrstedt, U. (1993) Radiation-induced chromosomal aberrations analysed by two colour fluorescence in situ hybridization with composite whole chromosome-specific DNA probes and a pancentromeric DNA probe, *Int. J. Radiat. Biol.*, **64**, 179-184.
7. Natarajan, A.T., Vyas, R.C., Darroudi, F., and Vermeulen, S. (1992) Frequencies of X-ray induced chromosome translocations in human peripheral lymphocytes as detected by in situ hybridization using chromosome-specific DNA libraries, *Int. J. Radiat. Biol.*, **61**, 199-203.
8. Lucas, J.N., Awa, A., Straume, T., Poggensee, M., Kodama, Y., Nakano, M., Ohtaki, K., Weiner, H.-U.G., Pinkel, D., Gray, J., and Littlefield, G. (1992) Rapid translocation frequency analysis in humans decades after exposure to ionizing radiation, *Int. J. Radiat. Biol.*, **62**, 53-63.
9. Granath, F., Darroudi, F., Auvinen, A., Ehrenberg, L., Hakulinen, T., Natarajan, A.T., Rahu, M., Rytomaa, T., Tekkel, M., and Veidebaum, T. (1996) Retrospective dose estimates in Estonian Chernobyl clean-up workers by means of FISH, *Mutation Res.*, **369**, 7-12.
10. Natarajan, A.T., Santos, S.J., Darroudi, F., Hadjidikova, V., Vermeulen, S., Chatterjee, S., van der Berg, M., Grigorova, M., Sakamoto-Hojo, E.T., Granath, F., Ramalho, A.T., and Curado, M.P. (1998) ^{137}Cesium-induced chromosome aberrations analyzed by Fluorescence In Situ Hybridization: Eight years follow up of the Goiânia radiation accident victims, *Mutation Res.* **400**, 299-312
11. Salassidis, K., Schmid, E., Peter, R.U., Braselmann, H., and Bauchinger, M. (1995) Dicentric and translocation analysis for retrospective dose estimation in humans exposed to ionizing radiation during Chernobyl nuclear power plant accident, *Mutation Res.*, **311**, 39-48.
12. Salassidis, K., Georgiadou-Schumacher, V., Braselmann, H., Müller, P., Peter, R.U., and Bauchinger, M. (1995) Chromosome painting in highly irradiated Chernobyl victims: a follow-up study to evaluate the stability of symmetrical translocations and the influence of clonal aberrations for retrospective dose estimation, *Int. J. Radiat. Biol.*, **68**, 257-262.
13. Ramalho, A.T. (1992) PhD Thesis, "Subsidios a tecnica de dosimetria citogenetica gerados a partir da analise de resultados obtidos com o acidente radiologico de Goiânia". Universidade Federal do Rio de Janeiro, Rio de Janeiro, Brazil.
14. IAEA (1988) *The Radiological Accident in Goiânia*, International Atomic Energy Agency STI/PUB/815, Vienna.

15. Buckton, K.E. (1983) Chromosome aberrations in patients treated with X-irradiation for ankylosing spondylitis, in T. Ishihara and M.S. Sasaki (eds.) *Radiation-Induced Chromosome Damage in Man,* Alan R. Liss, New York, pp. 491-511.
16. Lucas, J.N., Poggensee, M., and Straume, T. (1992) The persistence of chromosome translocations in a radiation worker accidentally exposed to tritium, *Cytogenet. Cell Genet.*, **60**, 255-256.
17. Gebhart, E., Neubauer, S., Schmitt, G., Birkenhake, S., and Dunst, J. (1996) Use of three-color chromosome in situ suppression technique for detection of past radiation exposure, *Radiat. Res.*, **145**, 47-52.
18. Natarajan, A.T., Boei, J.J.W.A., Darroudi, F., Van Diemen, P.C.M., Dulout, F., Hande, M.P., and Ramalho, A.T. (1996) Current cytogenetic methods for detecting exposure and effects of mutagens and carcinogens, *Environ. Health Persp.*, **104**, 445-448.
19. Hande, M.P., Boei, J.J.W.A., Granath, F., and Natarajan, A.T. (1996) Induction and persistence of cytogenetic damage in mouse splenocytes following X-irradiation analysed by flourescence in situ hybridization. I. Dicentrics and translocations, *Int. J. Radiat. Biol.*, **69**, 437-444.
20. Obe, G., Johannes, I., Johannes, C., Hallmann, K., Reitz, G., and Facius, R. (1997) Chromosomal aberrations in blood lymphocytes of astronauts after long-term space flights, *Int. J. Radiat. Biol.* **72**, 727-734

CHROMOSOME DAMAGES IN COSMONAUTS' BLOOD LYMPHOCYTES AS A MEASURE OF RADIATION EFFECT.

Problems of Modeling and Calibration

B. S. FEDORENKO, V. M. PETROV, S. V. DRUZHININ

State Scientific Centre of the RF - Institute for Biomedical Problems
Khoroshovskoye Shosse, 76-a, 123007, Moscow, Russia

Abstract

The frequency and types of chromosome aberrations in lymphocyte culture of peripheral blood of 24 cosmonauts have been studied. It has been ascertained that the yields of aberrant cells and of chromosome-type aberrations increase after long-term space missions. Proceeding from the linear-quadratic relationship between the effect on dose of high-energy protons, a calibration curve has been obtained after exposure of blood samples in vitro. There was a good agreement between individual doses of space radiation and appropriate doses of 9-GeV protons. It has been concluded that relatively low doses of space radiation result in instability of genome, which, in its turn, may be a cause of an increase in risks of development of the long-term consequences.

1. Introduction

An increase in yields of cells with chromosome aberrations, a change of correlation between yields of different types of chromosome aberrations are known to be a peculiarity of radiation influence on hereditary apparatus of mammalian cells, whereas chromatid-type aberration yield remains unchanged [3, 5]. One of the basic and most studied technique of biological dosimetry is based on the analysis of unstable chromosomal aberrations in human peripheral blood lymphocyte culture [4].

Numerous investigations in different countries show that chromosomal aberration yield is an objective showing of individual or mean populational dose of ionizing radiation. However, it is necessary to follow the two principle conditions: radiation effects should be assessed after single short-term radiation exposure and cytogenetic examination should be undertaken in as short as possible periods after irradiation. According to some literature data, during first 5-20 weeks after irradiation the level of unstable chromosomal aberrations does not change and it allows to assess the dose of acute radiation exposure [4]. Provided the above conditions are not met, we ought to talk about indication of radiation exposure. At the same time there are approaches of retrospective dose

C. Baumstark-Khan et al. (eds.),
Fundamentals for the Assessment of Risks from Environmental Radiation, 187-194.

assessment in cases of chronic or prolonged irradiation, based on the rate of elimination of aberrant lymphocytes from circulating blood, on the regularities of aberration distribution in lymphocytes and others [10]. Anyway cytogenetic analysis of cosmonauts' blood lymphocytes enables obtaining some important additional information on space radiation exposure, which can not be obtained in another way.

The purpose of the investigations conducted by us was studying qualitative and quantitative regularities of cytogenetical damage development in lymphocyte culture of cosmonauts' peripheral blood, resulted from space radiation exposure. Taking into account that space radiation comprises mainly high-energy protons (~80%), it seemed expedient to obtain the calibration curves. For this purpose we used the frequency of dicentrics and centric rings, which had been obtained earlier after irradiation of donor blood samples with 9-GeV, 645-MeV and 50-MeV protons [7, 8]. In this case we proceeded from the linear-quadratic dependence of the effect on dose.

2. Materials and Methods

The investigations of chromosome damages in cosmonauts' blood lymphocytes envisaged studying the spontaneous level of chromosome aberrations during the pre-flight period (1-2 months prior to first space missions) as well as scoring the frequency of structural chromosomal lesions a day after the completion of a space mission. A total of 24 cosmonauts were examined. Pre-flight examinations were made of 14 cosmonauts (Table 1). The cosmonauts aged from 33 to 54 years. Heparinized blood was taken from the cosmonauts and cultured according to the usual technique [11]. Blood samples were placed in culture tubes containing RPMI-1640 medium, 20 % serum, phytohaemagglutinin and antibiotics, and incubated at 37°C. Metaphases were harvested after 48 h of culture, including a 2-h colchicin block. Metaphases spreads were then treated for Giemsa staining, and 200-1500 mitoses were analysed for each cosmonaut. All chromosomal aberrations were taken into account without caryotyping. The data received were statistically processed.

Dose measurements onboard the space ships were done with the use of a radiometer R-16 as well as individual thermoluminescent dosimeters ID-3M.

TABLE 1. A schedule of performing the cytogenetical analyses in cosmonauts' lymphocyte culture

Number of cosmonauts	*Schedule*
14	before the 1st flight
11	after the 1st flight
8	before the 2nd flight
6	after the 2nd flight
4	before the 3rd flight
3	after the 3rd flight

3. Results and Speculation

Analysis of chromosome damages in cosmonauts' blood lymphocytes during the pre-flight period showed that the frequency of chromosome-type aberrations varied from 0.23 to 2.5%. Table 2 illustrates the frequency of aberrant cells, chromosome and chromatid-type aberrations (per 100 cells) and their mean values before the space flights.

As it is shown, the total number of aberrant cells and chromosome aberrations was equal to 1.29 and 0.71% respectively. The number of dicentrics and centric rings was as high as 0.08%. Therefore, it can be considered that during the pre-flight period the frequency of chromosome-type aberrations in cosmonauts' blood lymphocytes corresponds to the assumed values for relatively young healthy persons. The number of aberrant cells in practically healthy people varies from 0 to 5%, being equal to 1% on the average for people aged 20-40 years, and this number increases up to 1.5% by the age of 40-50 years [13]. 3 of the 14 cosmonauts exhibited dicentrics and centric rings whose mean frequency was as high as 0.08%. 5 of the cosmonauts had acentric rings. A cell with a translocation was revealed in one cosmonaut. The cosmonauts 3 and 19 displayed a higher yield of single fragments, which was as high as 2.0 and 2.6% respectively.

TABLE 2. Background yields of chromosome aberrations in cosmonauts' peripheral blood lymphocytes

Cases	*No. of cells scored*	*Aberrant cells, %*	*Chromosome-type aberrations, %*			*Chromatid-type aberrations, %*
			excess acentrics	*dicentrics and centric rings*	*total*	
3	200	2.5	0.5	0	0.5	2.0
5	200	2.0	1.0	0	1.0	1.0
6	200	2.0	2.0	0	2.0	0
10	300	2.0	2.0	0	2.0	0
11	200	2.5	2.5	0	2.5	0
12	200	1.0	0	0	0	1.0
13	150	0.67	0.67	0	0.67	0
14	110	0.9	1.8	0	1.8	0
18	1415	1.34	0.71	0.21	0.92(1t)	0.42
19	500	3.4	0.8	0.2	1.0	2.6
20	2100	0.61	0.14	0.09	0.23	0.38
21	500	2.0	0.8	0	0.8	1.2
22	500	0.8	0.4	0	0.4	0.4
23	500	0.4	0.4	0	0.4	0.2
mean	7075	1.29±0.14	0.62±0.09	0.08±0.03	0.71±0.10	0.62±0.09

Quantitative analysis of spontaneous chromosomal changes attests that single and twin fragments make up 95% of the total number of chromosome aberrations. On the average there was 0.01-0.02 spontaneous aberrations per a cell. In fact, each aberrant cell contained no more than one aberration. The number of aberrations per one aberrant cell is an important showing which describes the radiation exposure. The experiments showed that exposure to sparsely ionizing radiation at a dose of 50 cGy induced no more than one aberration per one cell. As the dose increases, the number of chromosome aberrations per one aberrant cell exponentially increases with dose [6].

The yield of dicentrics in lymphocyte culture is usually used for dose assessment. It is stipulated by their low background level (1-10 dicentrics per 1000 cells), easy identification and the dose-dependent increasing of their number after exposure to radiation [13].

The cosmonauts underwent X-ray diagnostic procedures. The X-ray doses under local irradiation usually does not exceed 0.5-2.0 cGy [14]. However, in practice some cosmonauts may undergo repeated X-ray procedures which, naturally, lead to an increase in the individual dose. Though X-ray doses reached a few tens of cGy (0.4 to 57.8 cGy), we did not reveal a significant increase in the background level of chromosome aberrations in the cosmonauts' blood during the pre-flight period. Although the tree cosmonauts [1] 18, 19 and 20 exhibited dicentrics and centric rings, the X-ray doses they received were minimal. By the way, a considerable increase in chromosomal aberration yield is possible at a dose of 0.7 cGy and higher, as cited in [9].

The total doses of space radiation over the space flights were 1.0 to 30.9 cSv and did not depend upon their duration. For most of the cosmonauts, their absorbed doses remained at a low level and did not exceed the permissible dose limit. However, the biological effectiveness of space radiation can be significantly higher. The quality factor (QF) of space rays, measured by the use of proportional counters such as NAUSICAA, CIRCE (France) or TEPC (USA) onboard the station MIR, varies from 1.8 to 2.2 [2, 12]. These values correspond to the RBE coefficients for accelerated charged particles with high energies (protons, helium ions and carbon ions), calculated on the base of aberrant cell yields and total number of chromosome aberrations in human lymphocyte culture after in vitro exposure [7, 8].

Table 3 illustrates yields and types of chromosome aberrations after long space missions. Aberrant cell yields and yields of dicentrics and centric rings increase after each flight and decrease during the inter-flight period which lasts no less than a year.

Aberrant cell yield increases mainly at the expense of dicentrics and centric rings (Figure 1) being the markers of radiation exposure. It should be noted that only one of three cosmonauts after his third flight showed an increase in yield of dicentrics and centric rings. It could be explain by the insufficient number of metaphases scored.

The mean numbers of chromosome aberrations (Table 4) also point out the distinguished increase in chromosome-type aberration yield after each space flight.

It should be also mentioned that an increase in chromatid fragment yield in some cosmonauts as well as a rare appearance of cells with translocations take place after the first and second flights.

TABLE 3. Post-flight yields of chromosome aberrations in cosmonauts' peripheral blood lymphocytes

Cases	*No. of cells scored*	*Aberrant cells, %*	*Chromosome-type aberrations, %*			*Chromatid-type aberrations, %*
			excess acentrics	*dicentrics and centric rings*	*total*	
after the 1st flight						
3	200	5.5	4.0	0.5	4.5	2.5
5	930	1.3	0.86	0.75	1.61	0
6	100	2.0	2.0	0	2.0	0
10	900	1.76	1.1	0.66	1.76	0
11	220	3.63	3.61	0.45	4.06	0
12	352	1.4	0.85	0.56	1.41	0.14
13	400	1.75	0.5	0.75	1.25	0.44
14	520	1.15	1.14	0.38	1.52	0
18	1387	1.68	0.22	0.67	0.89	0.79
20	1000	2.2	0.7	0.6	1.3(+1t)	1.0
21	1000	2.3	0.9	0.2	1.1(+2t)	1.3
23	1000	2.6	0.8	0.1	0.9(+1t)	2.0
before the 2nd flight						
3	500	2.4	0.4	0	0.4	2.0
5	850	1.05	0.35	0.38	0.73	0.35
7	1001	1.4	0	0.7	0.7	0.7
9	200	1.5	1.0	0	1.0	0.5
10	500	1.6	0.8	0	0.8	1.0
11	500	1.0	0.4	0	0.4	0.6
13	500	1.2	0	0	0	1.2
17	1560	0.76	0.64	0.06	0.7	0.12
after the 2nd flight						
3	1000	3.0	0.8"	0.1	0.9	1.7
5	536	2.93	0.37	0	0.37(3t)	2.0
7	1000	2.5	0	2.5	2.5	1.1
9	100	3.0	0	1.0	1.0	2.0
10	758	1.85	0.26	0.13	0.39	1.45
16	983	1.92	1.41	0.81	2.22	0.3
17	585	0.86	0.5	0	0.5	0.36
before the 3rd flight						
1	200	2.5	2.5	0	2.5	0
2	100	3.0	2.0	0	2.0	1.0
4	200	2.5	2.5	0	2.5	0
7	500	3.6	1.4	0.2	1.6	2.6
after the 3rd flight						
1	80	3.75	3.75	0	3.75	0
2	100	4.0	1.0	0	1.0	3.0
4	1020	1.09	0.4	0.69	1.09	0

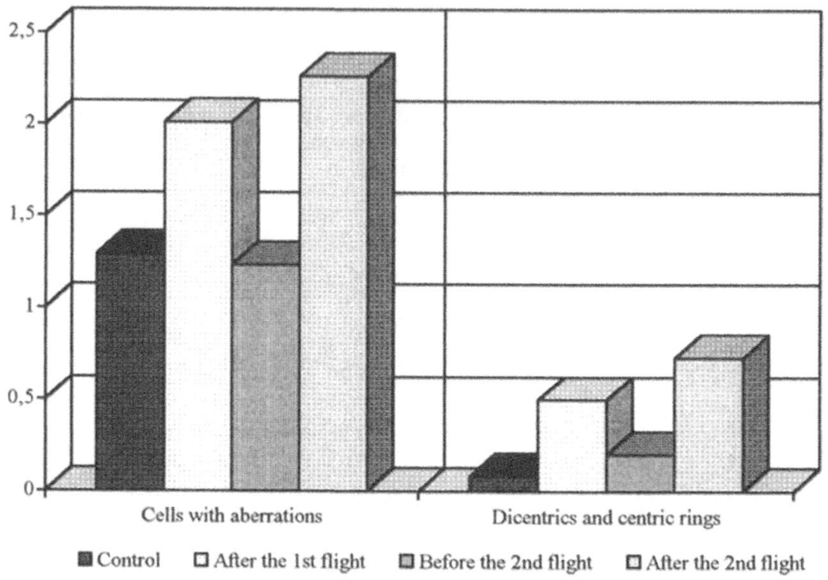

Figure 1. Chromosome aberration frequency in cosmonauts blood lymphocytes (per 100 cells)

TABLE 4. Chromosome aberration yield in cosmonauts' peripheral blood lymphocytes before and after space flights (per 100 cells)

Case	*No of cells scored*	*Aberrant cells*	*Chromosome-type aberrations*			*Chromatid-type*
			excess of acentrics	*dicentrics and centric rings*	*total*	*aberrations*
control	7075	1.29±0.14	0.62±0.09	0.08±0.03	0.71±0.10 (+1t)	0.62±0.09
after the 1st flight	8009	2.01±0.16	0.94±0.11	0.50±0.08	1.42±0.13 (+4t)	0.77±0.10
before the 2nd flight	5611	1.23±0.15	0.41±0.09	0.20±0.06	0.61±0.10	0.66±0.11
after the 2nd flight	4962	2.26±0.21	0.58±0.11	0.73±0.12	1.31±0.16 (+3t)	1.15±0.15
before the 3rd flight	1000	3.10±0.56	1.90±0.44	0.1±0.1	2.00±0.45	1.4±0.37
after the 3rd flight	1200	1.50±0.35	0.67±0.24	0.58±0.22	1.25±0.32	0.25±0.14

TABLE 5. Chromosome aberration yield and proton dose assessments

Case	Flight duration	Indiv. dose, cSv	*Results*					*Proton dose, cSv*		
			before flight		*after flight*		*Excess of D&R*	*9 GeV*	*645 MeV*	*50 MeV*
			No. of cells	*D&R*	*No. of cells*	*D&R*				
3	146	4.4	200	0	200	0.5	0.5	5	10	22
5	189	6.6	200	0	930	0	0	-	-	-
18	115	5.2	1415	0.21	1387	0.79	0.58	6	11	24
20	90	9.0	2100	0.09	1000	1.0	0.91	9	17	36

Thus, the cytogenetical study of cosmonauts' blood lymphocytes after long space flights has established the influence of a factor leading to an increase mainly in yields or arising of chromosome-type aberrations. This factor could be exposure to space radiation. Taking into account a higher biological effectiveness of accelerated charged particles as compared to standard radiation, heavy nuclei could lead to the increase in chromosomal aberration yield. Hereditary apparatus damages of the same type took place in plant objects exposed to space radiation [1].

The obtaining of calibration curves is a necessary element of biological dosimetry. For this purpose we used the yield of dicentric and centric rings, obtained after exposure of donor blood samples to 50-MeV, 645-MeV and 9-GeV protons [8]. In this case we applied the linear-quadratic relationship between dose and dicentric yields. Table 5 shows the results of biodosimetrical measurements for one space mission on MIR station. A good agreement between the individual doses of cosmic rays and the doses of 9-GeV protons, estimated by using the calibration curve, was revealed under the condition that the QF for 9-GeV protons was supposed to be equal to 1. At the same time the in vitro experiments showed a higher RBE of 9-GeV protons as compared to standard radiation. For protons of other energies the differences between the doses are significant.

4. Conclusions

The cytogenetic analysis of cosmonauts' peripheral blood lymphocytes during the preflight period did not reveal a significant deviation of chromosomal aberration yield from the assumed values for relatively young healthy persons.

The influence of relatively low doses of space radiation during long space missions leads to an increase in yield or appearance of chromosome-type aberrations (dicentrics and centric rings). This, in its turn, can lead to instability of the genome and also to an increase in risk of development of the long-term effects of cosmic irradiation. To solve the problems of biological dosimetry under long space missions it is necessary to conduct special studies of human blood lymphocytes exposed to accelerated charged particles, which would be directed at the development of the model of radiation influence on the human organism.

Acknowledgement. This work is supported in part by contract NAS-15-10110

References

1. Akatov U. A., Nevzgodina L.V., Sanovich V.A. et al. (1986) Radiation studies in space. Results of medical investigations onboard the station "Salute-6"-"Soyuz", Nauka, Moscow, p.335-348.
2. Badhwar G.D., Konradi A., Atwell W. et al. (1996) Measurements of the Linear Energy Transfer Spectra on the Mir orbital station and Comparison with Radiation Tpansport Models, *Radiat. Meas.* **26**, 2, 147-158.
3. Bauchinger M, Schmid E., Dresp J. (1979) Calculation of the dose-rate dependence of the dicentric yield after Co-γ-irradiation of human lymphocytes, *Int. J. Radiat. Biol.* **35**, 229-233.
4. Biological dosimetry: Chromosomal aberration analysis for dose assessment, IAEA, Vienna, Techn. Repts Ser., 260 (1986), 69p.
5. Bochkov N.P. (1971) Human chromosomes and irradiation, Atomizdat, Moscow. (Rus).
6. Bochkov N.P. (1993) A review of cytogenetical studies after Chernobyl accident, Vesnik RAMS, 6, 51-56. (Rus).
7. Fedorenko B.S., Savchenko N.Ya., Vorozhtsova S.V. et al. (1987) Biological effectiveness of helium ions and relativistic-energy protons, *Radiobiology* **27**, 3, 339-343. (Rus).
8. Gerasimenko V.N., Portman A.I., Bulanova M. et al. (1987) The influence of charged particles with relativistic energies on chromosomal aberration frequency in human blood lymphocytes (The dose-effect dependence and RBE values for deitrons and helium ions), *Radiobiology* **27**, 6, 743-747. (Rus).
9. Kudritsky Yu.K., Bikkulov R.I., Grigoryeva L.P. et al. (1974) Regularities of cytogenetical effects of ionizing radiations at low doses. *Rept. Staatl. Zentz. Strahlenschutz*, **157**, 89-92.
10. Lloyd D.C., Edwards A.A. (1983) Radiation-induced chromosome damage in man, Alan R. Liss Inc., New York, 23-49.
11. Moorhead P.S. Novell P.S., Wellimann W.J. et al. (1960) Chromosome preparation of leucocytes cultured from peripheral blood, *Exptl. Cell. Res.* **20**, 3, 613-616.
12. Nguen V.D. et al. (1991) Real time quality factor and dose equivalent meter "Circe" and its use on board the soviet orbital station "Mir", *Acta Astronautica* **23**, 217-226.
13. Sevan'kaev A.V. (1987) Radiosensitivity of human lymphocyte chromosomes during mitotic cycle. Energoatomizdat, Moscow. (Rus.).
14. Stavitsky R.V., Ermakov I.A., Lebedev L.A. et al. (1989) Equivalent doses in human organs and tissues under X-ray procedures. Energoatomizdat, Moscow. (Rus.).

ABERRATIONS INDUCED IN CHROMOSOME 1, 2 AND Y OF HUMAN LYMPHOCYTES BY THREE TYPES OF RADIATION OF DIFFERENT LET VALUE AS DETECTED BY FISH

E. LUKÁŠOVÁ [1], S. KOZUBEK [1], R. GOVORUN [2], M. REPIN [2], L. RÝZNAR [1], E. KRASAVIN [2], M. KOZUBEK [3] AND V. KROHA [4]

[1] *Institute of Biophysics, Academy of Sciences, Královopolská 135, 61265 Brno, Czech Republic*
[2] *Joint Institute for Nuclear Research, Dubna, Russia*
[3] *Faculty of Informatics, Masaryk University, Botanická 68a, 60200 Brno, Czech Republic*
[4] *Institute for Nuclear Physics, 250 68 Rez at Prague, Czech Republic*

Abstract

Fluorescence *in situ* hybridisation was used to determine the efficiency of the induction of aberrations in chromosomes 1, 2 and Y of human lymphocytes by γ-rays, fast neutrons of mean energy of 7 MeV and accelerated nitrogen ions. The frequency and the spectrum of aberration types induced by these three types of radiation were different. Nitrogen ions induced the highest number of aberrations. The most frequent were chromosome fragments followed by centric rings. Neutrons and γ-rays induced most frequently translocations followed by dicentric chromosomes. Efficiency of neutrons in the induction of aberrations was higher in comparison with γ-rays.

Induction of aberrations by both γ-rays and fast neutrons in Y chromosome was much less effective in comparison with chromosome 1. The low efficiency of these types of radiation in the induction of aberrations in Y chromosome is probably related to the lower content of DNA in this chromosome and to its higher content of heterochromatin.

2. Introduction

Exposure of mammalian cells to ionising radiation produces a variety of chemical lesions in DNA among them double-strand breaks (DSB) which are most important for the induction of chromosomal aberrations [1-5]. Chromosomal aberrations have been investigated for a long time by different techniques for either research purposes or for biological dosimetry. Conventional analysis of metaphase chromosomes used in biological dosimetry is usually limited to asymmetrical exchanges. In the past, banding analysis was necessary for the recognition of symmetrical exchanges but this procedure

C. Baumstark-Khan et al. (eds.),
Fundamentals for the Assessment of Risks from Environmental Radiation, 195-202.

is too time-consuming to be applied routinely for this purpose. Pinkel *et al.* [6] demonstrated that fluorescence *in situ* hybridisation (FISH) with chromosome-specific DNA probes provides a new technique for a rapid and accurate detection of chromosomal rearrangement and is especially useful for analysis of symmetrical translocations.

We have used the FISH method to evaluate different types of aberrations of the largest human chromosomes 1, 2 and one of the smallest human chromosomes, Y, in lymphocytes after irradiation with ^{14}N ions, fast neutrons and γ-rays. Attention was paid to the proportion of non-repaired fragments to exchange aberrations and symmetrical to asymmetrical exchange aberrations following exposure to ionising radiation of different LET. In addition, the sensitivity of Y chromosome was compared with that of chromosome 1.

3. Materials and Methods

3.1. HUMAN PERIPHERAL BLOOD LYMPHOCYTES

Lymphocytes were isolated from whole heparinised blood (16-20 U/ml) of healthy men. Plasma containing lymphocytes was obtained as supernatant after 30 min of blood sedimentation in the presence of 1% of gelatine. It was diluted (1:3) with RPMI-1640 medium containing 0.1 ml of phytohaemagglutinin (Murex, England) per 5 ml of suspension, 1% glucose, penicillin (100 U/ml) and streptomycin (100 μg/ml). Incubation was performed at 37°C for 48 h, including exposure to colcemid.

3.2. CYTOLOGICAL PREPARATIONS

Preparations were made by routine air-dry procedure, which included harvesting by centrifugation, hypotonic shock and fixation in acetic acid/methanol (1:3). Slides for *in situ* hybridisation were stored at -20°C.

3.3. IRRADIATION

Lymphocytes were irradiated in G_0 phase of the cell cycle at room temperature by ^{137}Cs γ-rays (0.66 MeV; dose rate 3.6 Gy/min). The exposure to 50 MeV of ^{14}N ions (mean value of LET =77 keV/μm) was performed at a synchrocyclotron U 400 M (Joint Institutes of Nuclear Research, Dubna, Russia). Fast neutrons of mean energy of 7 MeV were generated by a cyclotron U-120 M in the Institute of Nuclear Physics of the Academy of Sciences of the Czech Republic in Rez. The generator is based on the $^{9}Be(d,n)$ reaction. Irradiated lymphocytes were incubated for 50-56 h before the harvesting.

3.4. *IN SITU* HYBRIDISATION

Cambio biotinylated chromosome 1 and Oncor digoxigenated chromosome 2 paints were used to visualise these chromosomes on the same slide, whereas Cambio biotinylated Y chromosome paint was used separately. The *in situ* hybridisation was performed according to the recommendation of the manufacturers, as described by Lichter *et al.* [7]. Slides were observed under a Zeiss fluorescence microscope equipped with a triple band-pass filter (AHF, Germany) specific for FITC, rhodamine and DAPI. The images were acquired by a cooled colour CCD camera (C5310, Hamamatsu) and stored in computer memory using a frame-grabber (MuTech Corp., USA)

3.5. CHROMOSOME ABERRATION ANALYSIS

Chromosome aberrations were classified according to Savage [8] in 100 - 200 mitosis per radiation dose. In the case of a dicentric chromosome and a centric ring one acentric fragment was subtracted from the number of fragments presented in the cell. Y chromosome aberrations at high doses of γ–rays and neutrons were evaluated in interphase nuclei. Nuclei were scored as normal if they showed only one yellow signal corresponding to Y chromosome stained by FITC. If the nucleus contained two or more clearly separated signals, the Y chromosome was classified as damaged.

Results

4.1. INDUCTION OF ABERRATIONS BY THREE DIFFERENT TYPES OF RADIATION

Frequency and the spectrum of aberration types were evaluated in the first mitosis after irradiation. The nitrogen ions induced the highest number of aberrations per cell in comparison with neutrons and γ-rays for all doses investigated. The most frequent aberration type in lymphocytes irradiated with nitrogen ions were chromosome fragments followed by centric rings and translocations (Table 1). On the contrary, neutrons and γ-rays induced most frequently translocations; fragments and centric rings were less frequent. Approximately the same number and the same types of aberrations found in chromosome 1 were also found in chromosome 2 (Table 1). Gamma-rays and neutrons did not induce aberrations in Y chromosome up to a radiation dose of 7 and 5 Gy, respectively. Such high radiation doses induce a high number of aberrations in chromosome 1.

4. 2. PROPORTION OF SYMMETRICAL TO ASYMMETRICAL ABERRATIONS

Two major classes of chromosomal aberrations induced by ionising radiation in G_0 lymphocytes were symmetrical translocations and asymmetrical exchanges (dicentrics, centric rings and fragments). The ratio of symmetrical translocations to asymmetrical exchanges was highest in γ-irradiated lymphocytes and lowest in those irradiated with nitrogen ions (Table 2). The effectiveness of neutrons to induce exchange aberrations was lower than that of nitrogen ions and higher than that of γ-rays (Table 2). The asymmetrical exchanges represent about 75 % of all aberrations in lymphocytes irradiated by different doses of nitrogen ions while in those irradiated with γ-rays the asymmetrical aberrations comprise only about 50 %. These aberrations represent about 60 % in lymphocytes irradiated with neutron doses of 3 and 5 Gy.

TABLE 1. Aberrations in chromosome 1 and 2 induced by different doses of ^{14}N ions, fast neutrons and γ-rays. R: radiation type, CH: chromosome, D: dose, AC :aberrant cells, A/100: number of aberrations per 100 cells, CF: chromosome and chromatid fragments, T: translocations, DIC: dicentrics, CR: centric rings, I: insertions, n: neutrons

R	CH	D[Gy]	AC [%]	A / 100	Aberration type [%]				
					CF	T	DIC	CR	I
		0.50	40.0±4.5	51.5±5.1	27.0±3.7	7.5±1.9	1.5±1.1	15.0±2.7	0.5±0.5
		0.75	35.0±5.9	53.0±6.6	21.0±3.3	13.0±3.6	4.0±2.0	11.0±3.3	4.0±2.0
^{14}N	1	1.00	32.8±5.2	45.9±4.6	30.4±4.2	5.7±1.7	1.6±1.1	4.1±0.8	3.3±1.0
		1.50	52.9±7.1	73.1±8.4	32.5±5.3	19.2±4.3	10.5±3.2	12.5±3.5	1.0±1.0
		2.00	53. 0±7.2	89.0±9.4	40.0±6.0	22.0±4.7	8.0±3.0	22.0±4.7	1.0±1.0
		3.00	58.4±8.2	154±13.4	86.0±1.0	40.7±6.9	5.8±2.6	11.6±3.7	10.5±3.5
		0.50	40.5±4.5	58.5±6.4	30.0±4.0	11.0±3.2	5.0±2.5	9.0±2.8	2.0±1.5
		0.75	46.0±4.9	72.0±8.2	33.0±4.1	21.0±4.0	6.0±2.7	9.0±2.8	4.0±2.0
^{14}N	2	1.00	35.4±5.2	58.2±6.4	42.1±5.9	6.3±1.8	4.7±2.1	2.4±1.0	2.3±0.9
		1.50	51.5±6.9	79.0±8.6	41.0±5.7	13.3±3.6	11.4±3.4	7.6±1.9	3.8±1.0
		2.00	51.9±6.9	75.0±8.5	26.0±3.5	22.0±4.7	8.7±3.0	17.3±3.7	1.0±1.0
		3.00	58.0±7.0	96.0±9.6	64.1±8.2	26.1±5.2	3.3±1.8	15.2±2.7	2.2±0.8
		1	6.4±1.8	6.4±1.8	1.9±1.0	2.5±1.1	0.5±0.5	0	1.5±0.9
γ-rays	1	3	19.4±2.6	19.4±2.6	4.4±1.3	10.5±1.9	2.0±0.8	0	2.5±0.9
		5	45.4±5.0	48.6±5.2	14.5±2.9	21.0±3.5	7.0±2.0	1.3±0.9	2.5±1.2
		7	77.0±8.6	82.0±9.1	22.0±5.2	40.0±6.3	10.0±3.2	2.0±1.4	4.0±2.0
n	1	3	54.0±5.6	71.0±9.6	15.0±3.0	26.0±6.2	19.0±5.6	6.0±2.4	5.0±2.2
		5	73.0±8.2	105±9.8	26.6±2.0	40.0±7.2	22.6±6.2	8.3±2.8	7.4±2.6

4. 3. PROPORTION OF FRAGMENTS TO EXCHANGE ABERRATIONS

Differences in the induction of chromosomal damage by sparsely and densely ionising radiation is more evident when the frequency of fragments to exchange aberrations is compared. The ratio of these frequencies is highest in cells irradiated with nitrogen ions and lowest in those irradiated with γ-rays. The effectiveness of neutrons to induce fragments is lower than the effectiveness of nitrogen ions and higher than that of γ-rays (Table 2).

TABLE 2. Ratios of symmetrical (SE) to asymmetrical exchanges (AE) and fragments (F) to total exchanges (TE) in chromosome 1 of human lymphocytes irradiated with γ-rays, nitrogen ions, and neutrons

Radiation	Dose[Gy]	TE [%]	AE [%]	SE [%]	F [%]	SE / AE	F / TE
	1	6.4	4.0	2.2	1.9	0.60	0.29
γ-rays	3	19.4	10.9	8.9	4.4	0.81	0.22
	5	48.6	21.0	27.6	14.5	1.31	0.29
	7	82.0	40.0	42.0	22.0	1.05	0.26
	0.50	51.5	44.0	7.5	27.0	0.71	0.52
	0.75	53.0	40.0	13.0	21.0	0.30	0.39
^{14}N	1.00	45.9	40.2	5.7	30.4	0.14	0.66
	1.50	75.7	56.5	19.2	32.5	0.33	0.42
	2.00	93.0	71.0	22.0	40.0	0.30	0.43
	3.00	154.6	118.9	35.7	86.6	0.30	0.55
Neutrons	5	71.0	45.0	26.0	15.0	0.57	0.30
	7	104.9	64.9	40.0	26.6	0.61	0.32

TABLE 3. Aberrations in the Y chromosome induced by γ-rays and fast neutrons in interphase nuclei of human lymphocytes

Type of radiation	Dose [Gy]	Cells scored	Nuclei with 1 signal	Nuclei with 2 and 3 signals	Y chromosome damaged [%]
0	0	206	206	0	0
γ-rays	7	100	98	2	2.00±1.4
γ-rays	30	183	137	46	25.0±3.7
γ-rays	45	94	59	35	37.2±6.3
Neutrons	15	100	96	4	4.00±2.0
Neutrons	30	104	72	32	30.8±5.4

4.4. ABERRATIONS OF Y CHROMOSOME INDUCED BY γ-RAYS AND NEUTRONS

Aberrations in the Y chromosome were observed only after high doses of γ-rays (30 Gy) and neutrons (15 Gy) (Table 3). Only small number of mitoses could be analysed after these doses and therefore the damage to Y chromosome was evaluated in interphase nuclei. The frequency of damaged nuclei increased with the dose of radiation and was comparable for the same doses of fast neutrons and γ–rays (Table 3).

5. Discussion

Nitrogen ions were most efficient in the induction of exchange aberrations as compared to neutrons and γ-rays. The most frequent aberration types induced by nitrogen ions in chromosome 1 are asymmetrical exchanges; these aberrations represent a substantially smaller fraction of all aberrations in cells irradiated with neutrons and γ-rays. Also the proportion of fragments in relation to exchange aberrations is higher in cells irradiated with nitrogen ions than in those irradiated with neutrons and γ-rays. The observed differences indicate that the initial damage induced by high LET radiation of nitrogen ions is qualitatively different from that induced by fast neutrons and low LET radiation. The effectiveness of neutrons to induce aberrations is lower than the effectiveness of nitrogen ions and higher than that of γ-rays. The differences are probably related to the spatial energy distribution of these types of radiation differing in LET. When cells are exposed to sparsely ionising radiation, the quanta of deposited energy are distributed randomly in cell nuclei; thus, for radiation doses used in our experiments all cells are potentially exposed to a risk of radiation damage. However, in the case of particles of densely ionising radiation the dose is non-homogeneously deposited both in terms of the dose distribution inside each particle track and in the number of the particles per cell nucleus [9]. Cell nuclei that do not lie in the path of particle will not have aberrations induced by radiation in contrast to those that are traversed. According to Poisson statistics, the mean number of particles of nitrogen ions with LET of 77 keV/μm that traverse a nucleus of G_0 lymphocytes (cross section of 75 μm^2) at the dose of 0.5 Gy is 3.04. About 60 % of nuclei will be hit by >2 particles at this dose of radiation and about 5 % of nuclei will not be hit. At a dose of 3 Gy 60 % of cells will be hit by > 8 particles. We can assume that cells with more hits per nucleus enter mitosis much later and they do not occur among the mitoses exposed to 1 and 2 hits at the time taken for aberration scoring after irradiation. The frequency of cells with higher number of hits increase with the dose of nitrogen ions; this is manifested by a decreasing number of registered damaged cells at higher doses at the time used for scoring of radiation damage. This effect is not so pronounced at higher doses of neutrons.

The observed higher ratio of fragments to exchange aberrations indicates that the initial molecular damage from high-LET radiation of nitrogen ions is qualitatively different from the damage induced by γ-rays and neutrons. The lesions produced by particle tracks of higher energy seem to be more complex and therefore less reparable, as it was shown by other investigators [10-14].

The frequencies of chromosome aberrations induced by γ-rays and neutrons was much higher in chromosome 1 than in Y chromosome. While there were aberrations in chromosome 1 in about 80 % of cells after 7 Gy of γ-rays and 5 Gy of neutrons, respectively, there were no aberrations in the Y chromosome. The DNA content of the Y chromosome is about 4.6 times lower than that of chromosome 1 [15], which is obviously one of the reasons for its lower sensitivity to radiation. Moreover, the presence of only one copy of this chromosome in male nuclei should decrease the probability of its radiation damage by a factor of 2. The total reduction of radiation sensitivity expected at the level of one order of magnitude (4.6x2=9.2) corresponds to our observations. The probability of aberration induction by fast neutrons should be also considered on the basis of the LET distribution. The cross-section for induction of aberrations by particles with higher LET should be close to the geometrical cross-section of the sensitive site. This is about 1.3 μm^2 for the Y chromosome and 67 μm^2 for both chromosomes 1 as it results from our measurements of fluorescence signals of these chromosomes in interphase nuclei. The ratio of the cross-sections of these chromosomes is about 1:50 and, consequently, the induction of aberrations by recoil particles produced by neutrons should be 50 times lower for the Y chromosome when compared to chromosome 1.

Another factor that may influence the frequencies of lesions induced by ionising radiation in chromosomes 1 and Y is the structure of these chromosomes. Chromosome 1 is mainly euchromatic, whereas the predominant part of the Y chromo-some is composed of highly condensed heterochromatin. The condensed chromatin appears to be more resistant to radiation damage in comparison to euchromatin. This has been demonstrated for different cell types [16-19].

Acknowledgement. This work was supported by the Grant Agency of Czech Republic (Grant No. 202/ 96/0602 and 202/96/1718) and the Grant Agency of the Ministry of Health of the Czech Republic (Grant No. NM-15-3)

References

1. Natarajan, A.T. and Obe, G. (1978) Molecular mechanisms involved in the production of chromosomal aberrations. I. Utilisation of *Neurospora* endonuclease for the study of aberration production in G_2 stage of the cell cycle, *Mutat. Res.* **52**, 137-149.
2. Natarajan, AT., Darroudi, F., Mullenders, L.H.F., and Meijers, M. (1986) The nature and repair of DNA lesions that lead to chromosome aberrations induced by ionizing radiation, *Mutat. Res.* **160**, 231-236.
3. Schmid, E., Bauchinger, M., and Mergenthaler M. (1976) Analysis of the time relationship for the interaction of X-ray induced primary breaks in the formation of dicentric chromosomes, *Int. J. Radiat. Res.* **30**, 339-346.
4. Bedford, J.S. and Cornforth, M.N. (1987) Relationship between the recovery from sublethal X-ray damage and the rejoining of chromosome breaks in normal human fibroblasts, *Radiat. Res.* **11**, 406-423.
5. Ward, J.F. (1988) DNA damage produced by ionizing radiation in mammalian cells: Identities, mechanisms of formation and reparability, *Prog. Nucl. Acid. Res. Mol. Biol.* **35**, 95-125.
6. Pinkel, D., Straume, T., and Gray, J. (1986) Cytogenetic analysis using quantitative, high sensitivity, fluorescence hybridisation, *Proc. Natl. Acad. Sci. USA* **83**, 2934-2938.
7. Lichter, P., Cremer, T., Borden, J., Manuelidis, I., and Ward, D.C. (1988) Delineation of individual human chromosomes in metaphase and interphase cells by *in situ* suppression hybridisation using recombinant DNA libraries, *Human Genet.* **80**, 224-234.
8. Savage, J.R.K. (1975) Classification and relationships of induced chromosomal structural changes, *J. Med. Genet.* **12**, 103-122.
9. Kraft, G., Krämer, M., and Scholz, M. (1992) LET, track structure and models: a review. *Rad. Environm. Biophys,* **31**, 161-180.
10. Weber, K.J. and Flentje, M. (1993) Lethality of heavy ion-induced DNA double-strand breaks in mammalian cells, *Int. J. Rad. Biol.* **64**, 169-178.
11. Heilmann, J., Rink, H., Taucher-Scholz, G., and Kraft, G. (1993) DNA strand break induction and rejoining and cellular recovery in mammalian cells after heavy ion radiation, *Rad. Res.* **135**, 46-55.
12. Prise, K.M. (1994) Use of radiation quality as a probe for DNA lesion complexity. *Int. J. Rad. Biol.* **65**, 43-48.

13. Prise, K.M, Folkard, M., Newman, H.C., and Michael, B.D. (1994) The effect of radiation quality on lesion complexity in cellular DNA. *Int. J. Rad. Biol.* **66,** 537-542.
14. Jenner ,T.J., Belli, M., Goodhead, D.T., Ianzini, F., Simone, G., and Tabocchini, M.A. (1992) Direct comparison of biological effectiveness of protons and alpha-particles of the same LET. III. Initial yield of DNA double-strand breaks in V79 cells, *Int. J. Rad. Biol.* **61,** 631-637.
15. Mendelson, M.L., Mayall, B.H., Bogart, E., Moore D.H., and Perry, B.H. (1973) DNA content and DNA-based centromeric index of the 24 human chromosomes, *Science* **179,** 1126-1129.
16. Natarajan, A.T. and Ahnstrom, G. (1973) Induced chromosomal aberrations and heterochromatic.In: *Modern Aspects of Cytogenetics: Constitutive Heterochromatin in Man.* (R. A. Pfeifer ed.) Stuttgart, Schatauer, pp.215-223
17. Bedford, J.S. and Muhlman-Diaz, M. (1991) Damage selectivity in chromosomes In: *Proceedings of the 9ht ICRR Congress*, Vol **II**. (J.D. Chapman, W.C. Dewey and G.F. Whitmore eds.) N.Y. Academic pp. 212-217.
18. Vyas, R.X., Darroudi, F., and Natarajan, A.T. (1991) Radiation induced chromosomal breakage and rejoining in interphase-metaphase chromosomes of human lymphocytes, *Mutat. Res.* **249**, 29-35.
19. Natarajan, A.T., Balajee, A.S., Boei, J.J.W.A., Chatterjee, S., Darroudi, F., Grigorova, M., Noditi, M., Oh, H.J., Slijepcevic, P., and Vermeulen, S. (1994) Recent developments in the assessment of chromosomal damage, *Int. J. Rad. Biol.* **66**, 615-623.

CHROMOSOMAL ABERRATIONS AS A BIOMARKER FOR COSMIC RADIATION

L. V. NEVZGODINA

State Scientific Center of Russian Federation - Institute of Biomedical Problems
Khoroshevskoye Shosse 76a, Moscow 123007, Russia

Abstract

Higher-plant seeds were flown in spacecraft experiments for periods from a few days to 1.5 years and chromosomal aberrations were analyzed. Results of simulation experiments with accelerated heavy-ion beams are discussed. Our data show that chromosomal aberrations can be used as dosimeters. The data are compared with results obtained with human peripheral lymphocytes of in vitro. The experimental data indicate that the number of aberrant cells (multiple-aberrant cells included) rises, depending on space flight duration, HZE particle fluence, HZE particle-induced absorbed dose, and on the topography of the HZE particle passage through the cells. Methodological aspects to estimate biological effects of cosmic radiation and the associated radiation risk during long-term space flights is discussed.

1. Introduction

Physical dosimetry methods do not accurately estimate radiation impact on biological systems because they fail to fully characterize the complicated mechanisms of radiobiological processes. This is particularly true for cosmic radiation, which is of complex composition and shows a broad range of radiation characteristics. It should be noted that cosmic radiation includes the HZE particles of high ionizing power. Even low doses of cosmic radiation are known to damage somatic and generative tissues. Thus, there is a need to quantitate genomic effects induced by cosmic radiation.

In view of this, physical dosimetry methods have to be supplemented with biological methods of recording effects of cosmic radiation. Chromosomal aberrations are an essential criterion associated with DNA damages in biological objects. With objects suitable for chromosomal analyses the radiation-induced chromosomal aberrations are preferable to other biological reactions because they can be quantitated. Chromosomal aberrations are of different types which give information on the mechanism of their origin.

C. Baumstark-Khan et al. (eds.),
Fundamentals for the Assessment of Risks from Environmental Radiation, 203-208.

Chromosomal aberrations have been used to identify cellular damage when studying the kinetics of cell populations in irradiated tissues and malignant tumors. In space research, this particular biological marker is used to estimate the cosmic radiation hazard for space crew members [1, 2].

2. Method

Seeds of higher-plants are convenient as biomarker of cosmic radiation. Radiobiological syndromes of plants and animals have some traits in common, such as the existence of critical tissues and organs, identical types of cytogenetic damages. The radiobiological reactions of plant meristem cells are much alike the reactions of animal stem cells and involve loss of multiplication ability, proliferate death, and occurrence of chromosomal damages. At the same time, the seeds are advantageous over actively-metabolizing systems because of the feasibility of long-term exposure on board of spacecrafts (up to a few years). They can be used to analyse the correlation between the localization of heavy-particle hits and biological effects. In addition plant seeds are easy to handle.

Fig. 1 shows a seed section of *Lactuca sativa* indicating the sensitive targets of the apical stem and root meristems. The dimensions and locations of the sensitive structure elements of the seed are shown. Chromosomal aberrations were taken to be the radiation damage criterion for space research. Location of heavy ion hits in the seeds were determined by solid-state track detectors.

The experiments were made with sample groups of seeds stored on Earth, in-flight heavy ion-hit seeds, and in-flight heavy ion-nonhit seeds.

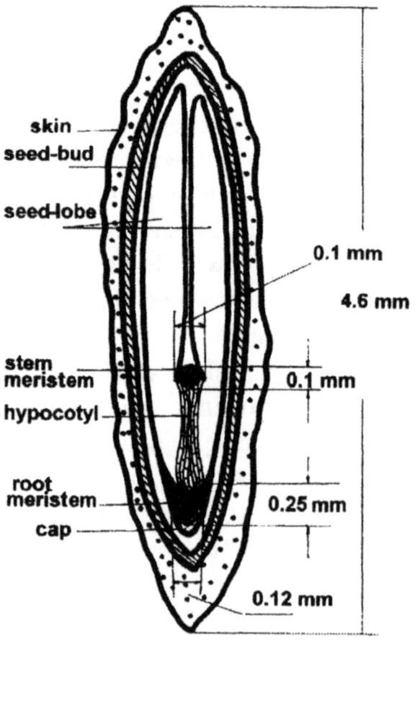

Figure 1. Histological section of *Lactuca sativa* seed.

Chromosomal aberrations were analysed in air-dry seeds on different spacecrafts from a few days to 457 days. Some experiments used solid-state track detectors in order to identify those objects which were direct affected by heavy ions.

3. Results and Discussion

Pioneer seed experiments carried out without detecting heavy ions demonstrated an increased frequency of aberrant cells in the germinating seeds exposed during all in-flight experiments. Cells with multiple aberrations were observed (Fig. 2). The high variability of occurance of aberrant cells as well as the repeated occurence of multiple-aberration cells was suggested to be due to heavy ion passages. A clear correlation between heavy-ion hit events and the yield of cells with multiple chromosomal aberrations was confirmed by topographic detection of particle passages through *Lactuca sativa* seeds [3-5]. The locations of heavy-ion hits in different areas of the seeds proved also to be of great importance. The highest effect was observed when heavy ions hit the apical root meristem [6].

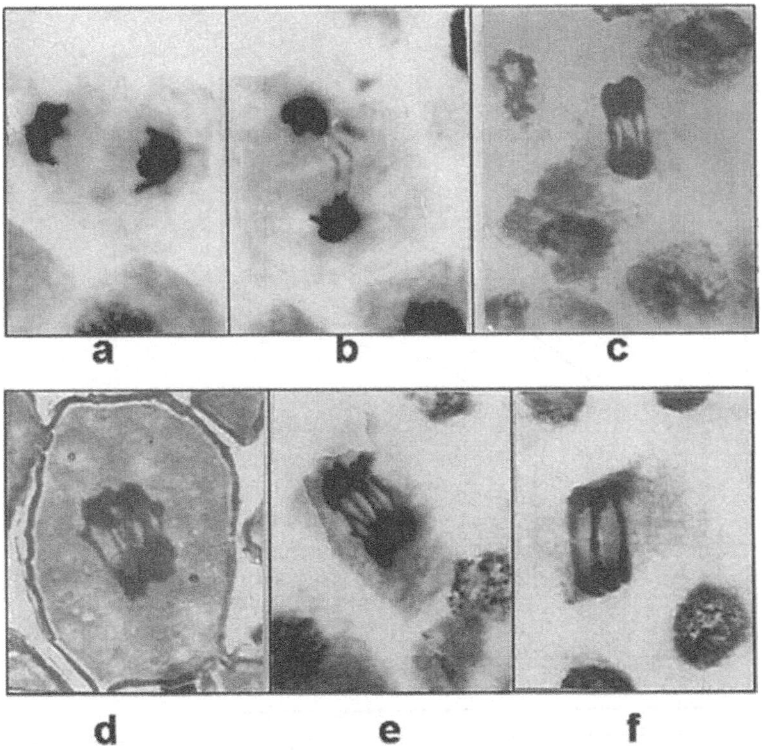

Figure 2. Aberrant cells in telophases of germinated seeds exposed in the Salyut space station. (a) Normal telophase. (b) Chromosomal bridge with fragment. (c, d, e, f) Telophases with multiple chromosomal aberrations.

The experiments carried out on Salyut-6 orbital station during its 457-day flight have shown that the radiobiological effects are functions of absorbed dose and heavy-ion fluence (Fig. 3). Aberrant cells were most frequently seen in seeds which were hit in-flight by heavy ions ($Z \geq 6$) with LET≥200 μm for the dose of 63.4 mGy (fluence 44.2 cm^{-2}). The same level of aberrant cells was obtained [7] in seeds exposed to gamma-quanta and to 630 MeV protons with doses of 10 Gy and 5 Gy, respectively. This result shows the extremely high biological effectiveness of heavy-ion impact on cells.

Important information can be obtained by analyzing the number of aberrations per aberrant cell (Fig. 4).

Multiple damaged chromosomes were observed most frequently in the heavy-ion hit seeds during 127-201 day long space flights. As the exposure time increased to 457 days, the number of multiple damages did not rise, probably because of cell death. The number of chromosome-type aberrations with fragmentation increased in the heavy-ion hit seeds exposed for 201 days and in the hit and nonhit seeds exposed for 457 days. Chromatid-type aberrations tended to increase with flight duration, but at a lower rate compared with the chromosome-type aberrations.

The data show that the biological effectiveness of heavy ions is much higher than the effectiveness level that could have been expected from the relatively low absorbed doses and heavy-ion fluences.

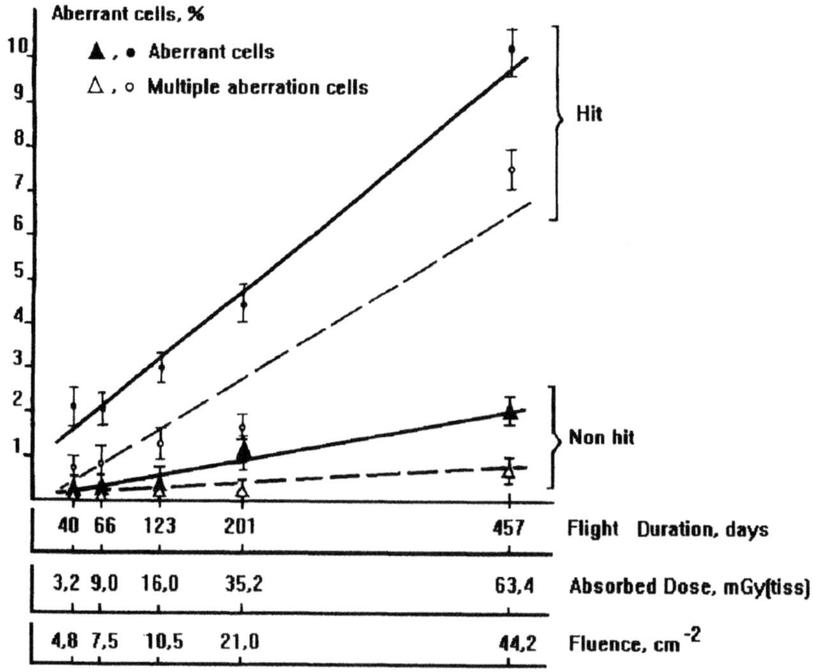

Figure 3. Yield of aberrant cells and of cells with multiple aberrations in *Lactuca sativa* seeds exposed in the Salyut space station

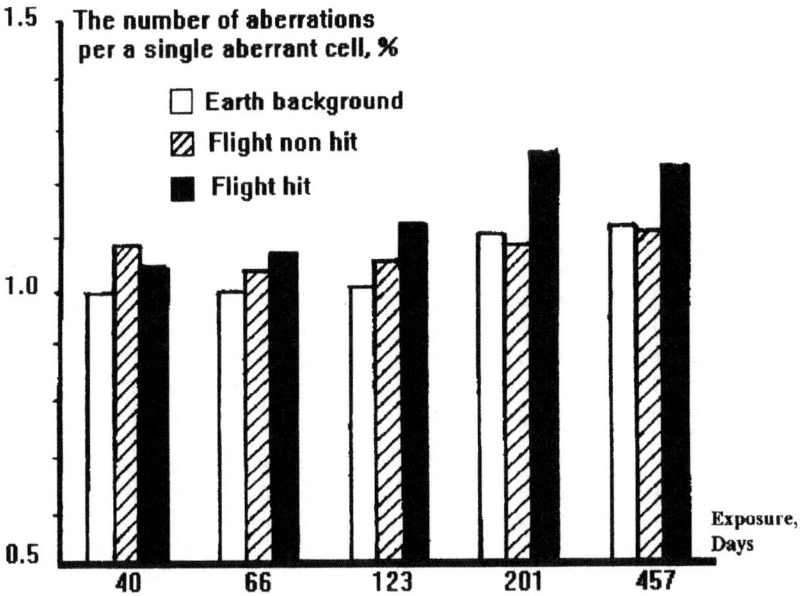

Figure 4. Number of chromosomal aberrations per aberrant cell in *Lactuca sativa* seeds exposed in the Salyut space station.

It is of great interest to compare data obtained for seeds and for human blood lymphocytes (in vivo). The cytogenetic analysis of blood lymphocytes of cosmonauts based on the Sabatier et al. [1] showed that the chromosomal aberrations were significantly elevated following space fights of 177 and 194 days. Short-term flights (14 days) did not lead to aberrant cells. At the same time, an increase was noted in the number of multiple chromosomal damags per aberrant cell. For example, blood of different cosmonauts included cells with 3-4 and more damages, while up to 10 chromosomal damages per single cell were observed in some cases.

An increased frequency of the seed chromosome damage in single cells was observed even during short-term flights, starting from 14 days. As the flight duration increased, the damage also increased, mainly in the heavy ion-hit seeds. This indicates that multiple-damaged cells correlate with heavy ion impacts. It should be noted that cells with multiple chromosomal aberrations were observed in the peripheral blood of ameliorators after Chernobyl disaster (more than 10 acentric fragments per cell). This effect was attributed to the incorporated α-emitting radionuclides of high ionizing power [8].

Thus, the comparison of the results of cytological analyses of lymphocytes of cosmonaut and the apical seed meristem in space experiments indicates that characteristics of chromosomal damages are comparable. This is also indicated by the experiments with dandelion species (*Taraxacum officinale*) gathered in two years after Chernobyl disaster [9] and with the pine needles from the Chernobyl zone [10]. The multiple chromosomal aberrations found in the meristemal cells of those plants were attributed to the impact of high-ionizing α-radiation.

Chromosomal aberrations can be used as universal biological markers of radiation. In long-term manned space flights, plant seeds are easy to use, and can, additional to the analysis of lymphocytes from cosmonaut, serve as a biological marker for radiation action.

4. References

1. Sabatier, L., Fedorenko, B.S., Gerasimenko, V.N., Ditrillaux, B., Hoffscir F. *et al.* (1995) Cromosome abnormalities in lymphocytes of cosmonauts' peripheral blood after long-term space flight, *Aerospace and Environmental Medicine*, **29**, No.1, 26.
2. Fedorenko, B.S. (1995) Biological effects of heavy charged particles. Main results and prospective investigations in the context of interplanetary missions. *Aerospace and Environmental Medicine*, **29**, No.2, 16-21.
3. Grigoryev, Yu.G. and L.V.Nevzgodina, L.V. (1978) Cosmic Radiation Heavy Ion Impact on Lattice Seeds Exposed on Board Cosmos-782 (Bioblock Experiment), in *Advances in Space Biophysics*, Acad. Sci. USSR Publishing House, pp.52-57, Pushchino.
4. Nevzgodina, L.V., Grigoriev, Yu.G., and Marenny, A.M. (1990) In: *Heavy Ion Impact on Biological Objects*, Energoatomizdat, Moscow, p.110.
5. Facius, R., Scherer, K., Reitz, G., Bücker, H., Nevzgodina, L.V., Maximova, E.N., (1994) Particle trajectories in seeds of *Lactuca sativa* and chromosome aberrations after exposure to cosmic heavy ions on Cosmos Biosatellite 9 and 9. *Advances in Space Research* **14**, (10)93-(10)103
6. Maximova, E.N. (1985) Effect of heavy charged particles of galactic cosmic radiation on seeds, *Space Biology and Aerospace Medicine*, No.3, 71-74.
7. Nevzgodina, L.V., Maksimova, E.N., and Kaminskaya, E.V. (1989) Impact of single galactic cosmic ray heavy ion hits on Lactuca sativa seeds exposed on board Salyut-6 and 7 manned space stations *Cosmic Biology and Aerospace Medicine*, No.6, 66-70.
8. Semov, A.B., Iofa, E.L. Akaeva, E.A., and Shevchenko, V.A. (1994) Dose dependence of chromosomal aberration induction in Chernobyl disaster ameliorators, *Radiation Biology, Radioecology*, **34**, 865.
9. Taskaev, A.I., Shevchenko, V.A., Popova, O.N., Abramov, V.I., Shevchenko, V.V. *et al.* (1988) Ecological and genetic consequences of Chernobyl disaster for fauna, *Scientific Report Series of USSR Academy of Sciences*, Syktyvkar.
10. Sidorov, V.P. (1994) Cytogenetic effect induced by Chernobyl disaster in pine needle cells, *Radiation Biology, Radioecology*, **34**, 847.

BIOPHYSICAL MODELS FOR THE FORMATION OF RADIATION-INDUCED CHROMOSOME ABERRATIONS BASED UPON THE EXPERIMENTAL OBSERVATIONS MADE UNDER PCC AND MITOTIC CONDITIONS

R. GREINERT, O. BOGUHN, E. DETZLER, C. THIEKE AND D. HARDER

Institute of Medical Physics and Biophysics, University of Göttingen Goßlerstr. 10f, 37073 Göttingen, Germany

Abstract

Radiation-induced chromosome rearrangements are key events for cytolethality and somatic mutations. In order to study the formation kinetics of exchange-type chromosome aberrations in the time period following irradiation, but still within the same phase of the cell cycle, we applied the PCC (premature chromosome condensation) technique on human T-lymphocytes in G_0 phase, using variable delay times between irradiation and cell fusion. PCC was combined with C-banding to identify dicentric chromosomes and with FISH painting to identify chromosome translocations. Besides the acentric chromosome fragments seen in PCC immediately after irradiation and subsequently gradually disappearing according to DNA double-strand break repair, a first intermediate product of the exchange-type aberrations, manifested as incomplete translocations with a yield proportional to dose (αD), is formed in less than 30 min. This mechanically unstable intermediate is transformed into stable exchanges (complete translocations) in the course of a few hours. On the other hand, yield component βD^2 requires about 10 hours to be fully developed in the form of PCC-detectable complete chromosome translocations.

The detection of this kinetic difference has been the last step in a chain of observations which led to differing biophysical models for the formation of the linear and quadratic yield components. For the *β component*, the results of fractionated irradiation and of cell fusion experiments consistently confirm Lea's model of *pairwise interaction between chromatin lesions* induced upon *two* chromosomes or chromosome arms, and lesion lifetimes indicate that they are DNA double-strand breaks in the course of enzymatic repair. But, according to the aberration yields resulting from irradiations with ultrasoft X rays and from RBE comparisons with the clustering of ionizations and excitations along the particle tracks, the *α component* appears to originate from a *single severe radiation-induced molecular lesion* in the chromatin of only *one* chromosome, able to initiate molecular interaction with another, intact chromatin thread. This lesion could be a cluster of DNA double-strand breaks, causing DNA deletions, possibly of the

C. Baumstark-Khan et al. (eds.),
Fundamentals for the Assessment of Risks from Environmental Radiation, 209-218.

size of one or more nucleosomal DNA windings or of chromatin fibre loops, setting free several DNA strands able to invade the intact chromatin of other chromosomes. Temperature jump experiments indicate that this lesion could be formed during thermalization of a cluster of ionizations and excitations, causing local melting of the DNA. Such "thermal spike" model would be consistent with the accepted theory of bubble nucleation in bubble chambers by the secondary electrons of X-rays and high-energy electrons.

The difference between the mechanisms forming the α and β components, described here for human T-lymphocytes, may have various applications for the understanding of high-LET particle bioeffects and even for radiotherapy, but needs confirmation for other cell types.

1. Introduction

Lea's [1] and Neary's [2] proposed mechanism for the production of radiation-induced structural chromosome aberrations of the "exchange type", i.e. of rearrangements involving irregular joints between different chromosomes or chromosome arms, can be called the mechanism of "pairwise lesion interaction". Its historical fundament is the observation that the dose dependence of the yield of these aberrations,

$$y = \alpha \cdot D + \beta \cdot D^2 \tag{1}$$

contains the quadratic component βD^2, indicative of a second-order reaction between radiation-induced chromatin lesions situated on different chromosomes or chromosome arms, and that this component is reduced by protraction and fractionation, explainable by lesion repair proceeding in the time between lesion induction and pairwise interaction. The mechanism of "pairwise lesion interaction" has been beautifully confirmed by the cell-fusion experiment of Cornforth [3] who showed that inter-cell chromosome exchanges in the first metaphase of a hybrid, made by fusion between two cells, were decisively enhanced if *both* of the fused cells had been irradiated before fusion. Identification of the pairwise interactive lesions with DNA double-strand breaks in the course of enzymatic repair has been possible by comparing their repair kinetics (accurately observable in split-dose experiments) with that of DNA double strand breaks as measured in molecular repair experiments [4].

In attributing the dose-proportional component αD of the exchange-type aberration yield as well to "pairwise lesion interaction", Lea and Neary assumed that the pair of radiation-induced molecular lesions on two different chromosomes or chromosome arms was resulting from the passage of *one* ionizing particle through or near the chromatin of the *two* chromosomes, thus rendering a dose-proportional yield. Experimental support for this model came from the observed proportionality of coefficient α with the LET of the investigated particles, electrons, protons and heavier ions, corresponding to an increase of the aberration yield per particle traversal with the *square* of LET. This doubtlessly indicated the "double hit" character of the mechanism, but not necessarily the location of the two hits on two different chromosomes. So, this model became challenged when it was observed that ultrasoft X-rays of only about 250 eV produced considerable α values although the "ranges" (track lengths) of their photoelectrons of ca. 5 nm could hardly be believed to be able to bridge the distance between the two chromatin threads [5]. This meant that *another* mechanism - still implying energy deposition in the form of a double hit - had to be looked for.

Our research, based upon the hope that this obvious problem might lead to further mechanistic insight, was guided by the working hypothesis that the process underlying the α component had a "fast, short-range" character, whereas the mechanism giving rise to the β component appeared to be of "slow, long-range" nature [6]. The argument for this suggestion had been the gradual disappearance of the β component with increasing LET which - in view of the already known slow kinetics of the mechanism leading to the β component [7, 8] - signalized a "kinetic advantage" of the α component. Delayed PCC (premature chromosome condensation) experiments with X-irradiated human lymphocytes appeared as an approach suitable to look into the formation kinetics of the exchange-type aberrations within the G_0 phase, long before they could be made visible in the first metaphase of an artificially stimulated cell cycle [9]. In the conduction of this program it was observed that the α component of dicentric chromosomes was actually formed in a "fast" mode, taking no longer than the minimum time required for cell fusion and chromosome condensation (30 min), while it took about 10 hours until the β component reached its full magnitude [10]. The attempt to further consolidate the insight gained, namely that the α component is actually coined by a fast production process with essential differences compared to the origin of the β component, lead to two further experiments whose results will be reported and discussed in the present paper. Dicentrics observable in delayed PCC were now produced by alpha particles, and FISH staining was combined with delayed PCC in order to get mechanistic information about the fate of individual chromosomes.

2. Materials and Methods

2.1. CELL PREPARATION

Human lymphocytes were separated by the Ficoll-Paque method from fresh blood samples obtained from healthy male donors. Chinese hamster ovary cells (CHO-K1), routinely grown in α-MEM containing 5% FCS, were used as mitotic cells to induce PCC by fusion with the lymphocytes.

2.2. IRRADIATION AND DELAY TIME

2.2.1. α-Particle Irradiation

The device used for α-particle irradiation was an Am-241 source equipped with a He-flooded moving collimator, kindly provided by D. Frankenberg (Göttingen). The most probable energy of the α particles reaching the cell nucleus is 3.45 MeV. One milliliter of α-MEM containing approximately $2 \cdot 10^6$ lymphocytes was deposited upon the bottom of an exposure dish without touching the walls to avoid side-effects based on adhesion. The bottom consisted of a mylar foil (thickness 1.5 μm). The dose-rate behind this foil is 1.3 Gy/min, and the remaining range of the α particles in the medium is about 20 μm. To allow all lymphocytes to settle at the bottom, the dish was left standing for at least 1 hour in the irradiation device before the shutter was opened. The doses applied ranged from 1 to 5 Gy. The irradiation device was installed in a room cooled down to 4 °C, so that the experimental conditions were the same as for the X-ray irradiation.

2.2.2. X-Ray Irradiation

The lymphocytes were irradiated at 4 °C with 150 kV X-rays, filtered with 0.5 mm Al, at a dose rate of 0.99 Gy/min.

2.2.3. Delay and Fusion Times (t_D and t_F)

After irradiation cells were either fused immediately (t_D=0 h) or incubated up to 24 h (t_D=0-24 h) at 37 °C to allow repair and misrepair to proceed before cell fusion. In all cases the following fusion time was t_F=55 min.

2.3. PCC AND STAINING OF THE MICROSCOPIC SLIDES

Mitotic cells and lymphoctes were mixed in a round-bottom culture tube and centrifuged at 150g for 3 min. For cell fusion the culture medium was discarded, and 100 µl of 55% (w/v) polyethylene glycol (PEG) was added. This suspension was mixed carefully with 1 ml PBS and then centrifuged. The pellet was resuspended in 0.7 ml α-MEM (5% FCS) containing 0.2 µg/ml Colcemid. The cells were then incubated at 37 °C for t_F=55 min to allow cell fusion and condensation of the lymphocyte chromosomes. After this time the cells were treated with 0.56% KCl, centrifuged, fixed in methanol:acetic acid (3:1) and dropped onto wet slides.

For the evaluation of fragments, dicentric chromosomes and other chromosomal aberrations, the preparations were C-banded with $Ba(OH)_2$ and 5% Giemsa or painted with FISH, respectively. We used a whole chromosome painting probe for human # 4 (Vysis, Spectrum Orange, red fluorescence) mixed with an all human centromere probe (Oncor, FITC-labelled, green fluorescence) and a # 4-specific centromere probe (Vysis, Spectrum Orange, red fluorescence). The chromosomes were counterstained with DAPI (blue fluorescence).

2.4. SCORING AND IMAGE ANALYSIS

For evaluations using C-banding, approximately 60 cells per data point were scored for dicentric chromosomes by using a light microscope equipped with a CCD color camera (CF15 MCC, Kappa, Germany). Camera images were stored into a frame grabber and analysed by digital image processing using the program OPTIMAS (BioScan Inc., USA). This procedure allowed contrast enhancement and background correction to support the identification even of poorly stained C-bands (see fig. 1).

For the analysis of FISH-painted cells, approximately 100 cells per data point were scored using a fluorescence microscope (Zeiss) coupled with a CCD monochrome camera (Hamamatsu C-4880). The camera images (see fig. 2) were processed using the program HiPic.

Figure 1 shows the chromosome figures seen when C-banding is used to make dicentric chromosomes detectable. The number of excess acentric fragments is obtained by subtracting 46 from the total number of pieces of chromosomal material, as usual in metaphase cells. Figure 2 shows the staining pattern when chromosome #4 is marked by

FISH and the centromeres are indicated by the all-human centromere marker. This method permits to score all dicentric figures in which #4 participates. It can be distinguished whether the asymmetric translocation is *complete*, i.e. associated with a bicoloured acentric fragment, or *incomplete*, i.e. not associated with a bicoloured fragment. In the case of incomplete dicentrics the chromosome pattern generally contains two not-joined acentric fragments per dicentric with one of the two fragments coloured. In about 5% of the incomplete dicentrics one acentric piece is joint with the two centromeric pieces, so that a three-coloured form with two centromeres appears. This three-coloured form is only seen at delay times less than about 3 h.

3. Results

A synoptic presentation of the scoring results of dicentrics is given in fig. 3a-c for the three modifications of the delayed PCC experiment. With C-banding and 4 Gy of 150 kV X-rays we obtained the result of fig. 3a already mentioned in the introduction [10], an initial plateau exclusively attributable to the α component and a subsequent rise to a higher plateau due to delayed appearance of the β component. Fig. 3b shows the recent results with C-banding and irradiation of the lymphocytes with 2 Gy of 3.45 MeV alpha particles. The full magnitude of the dicentrics yield occured already at t_D=0, and the yield did not significantly change when t_D was further increased. The observation that the α component of exchange-type aberrations is formed in a fast mode, first made in

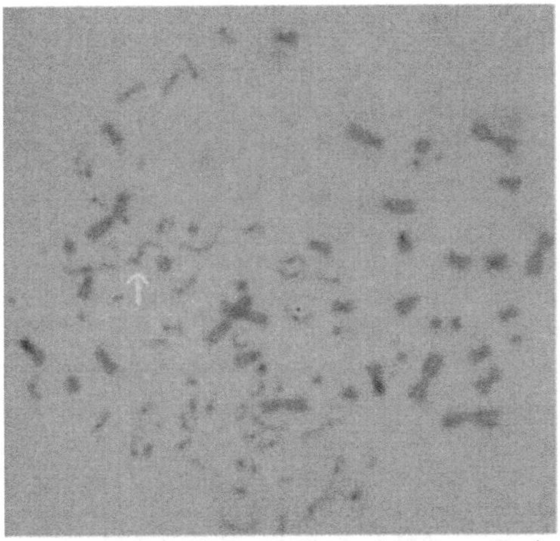

Figure 1. Typical micrograph of dicentrics scoring in human T-lymphocytes after 4 Gy of X-rays in PCC+C-banding technique. Arrow: Dicentric interphase chromosome.

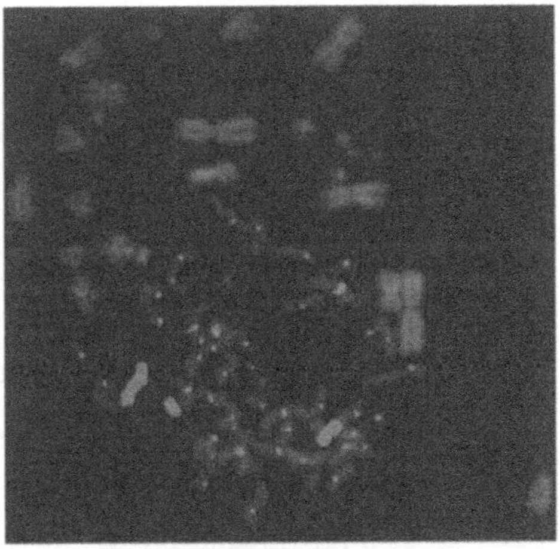

Figure 2. Typical micrograph of PCC+FISH+all human centromere staining of human T-lymphocytes after 4 Gy of X-rays. Red: One intact chromosome 4 and one incomplete dicentric involving chromosome 4 (note unattached red fragment).

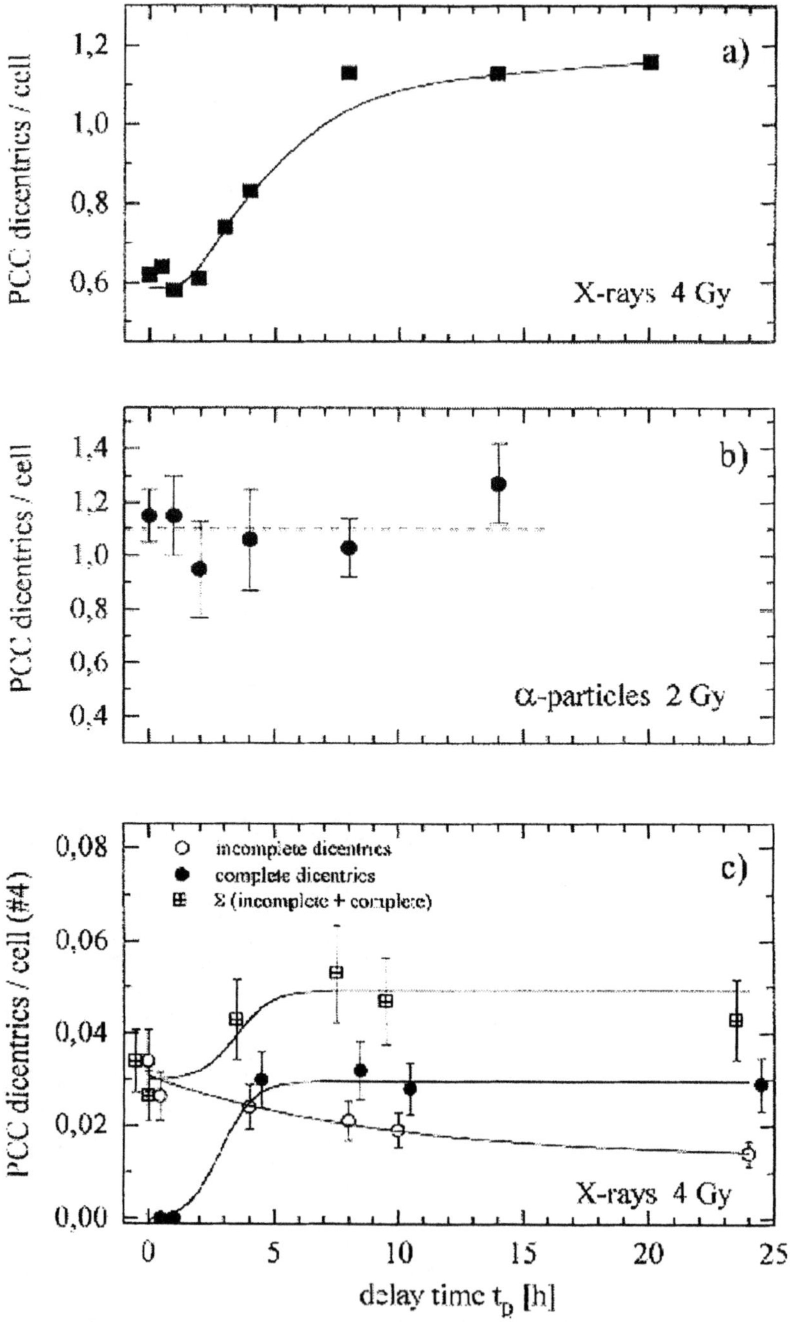

Figure 3. Cellular yield of dicentric chromosomes in human T-lymphocytes, observed by C-banding under conditions of delayed PCC, as a function of delay time tD. a.) After 4 Gy of 150 kV X-rays [10], b.) after 2 Gy of 3.45 MeV alpha particles, c.) after 4 Gy of 150 kV X-rays with distinction of incomplete, complete and sum of all dicentrics.

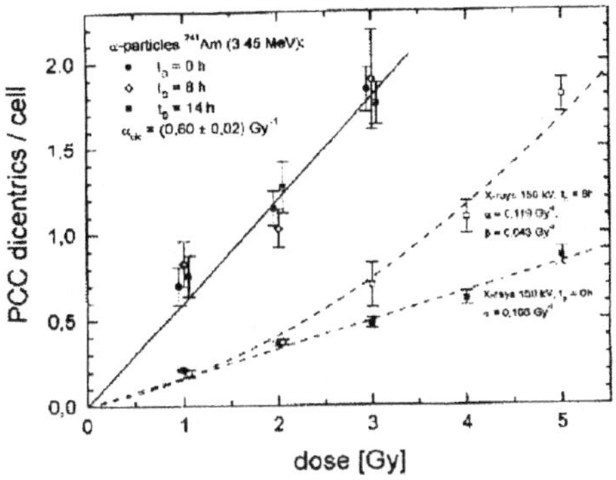

Figure 4. Dose-dependent cellular yield of dicentric chromosomes in human T-lymphocytes, observed by C-banding under conditions of delayed PCC after 3.45 MeV alpha particle irradiation (full curve), respectively 150 kV X-irradiation (dashed curves).

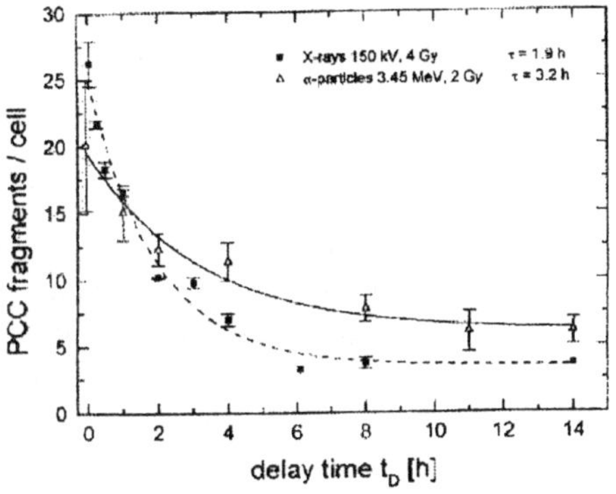

Figure 5. Cellular yield of excess acentric fragments in human T-lymphocytes, observed by C-banding under conditions of delayed PCC, after 3.45 MeV alpha particle irradiation (full curve), respectively 150 kV X-irradiation (dashed curve) as a function of delay time tD.

the previous X-ray experiment, was hereby confirmed for the case of a prevailing α component.

Additional information about the fast-appearing form of the dicentrics was obtained by the combination of centromere detection with FISH staining, as shown in fig. 3c for an experiment in which 4 Gy of 150 kV X-rays were applied. It turned out that all of the early-appearing dicentrics were *incomplete,* i.e. not associated with a bicoloured acentric fragment. These incomplete forms were subsequently gradually replaced by *complete* forms without a change in the sum of the dicentrics yield up to about t_D=2 h, the onset point of the contribution from the β component. Such fast formation after X-irradiation is also seen for the *sum* of all exchanges (all chromosome figures with colour change), but all early forms are incomplete, i.e. not appearing as pairs of bicoloured chromosomes, and are gradually reduced while the complete reciprocal translocations and dicentrics appear with some delay [11]. An incomplete form often seen at the early stage is a combination of two acentric fragments, not yet accompanied by the combination of two joint centric fragments. The sum of all incomplete forms as well as the subgroup of incomplete dicentrics have strictly dose-proportional dose-effect relationships, which we measured at t_D=0 and t_D=8 h. For the β-component no incomplete form has been observed.

The dose-effect relationship for dicentric chromosomes measured with C-banding at three times (t_D=0 h, 8 h and 14 h) was linear, and the RBE compared to the α component of the X-ray experiment, measured as the ratio between the α values, was approximately 4 (fig. 4). We also observed the time-dependent reduction of the number of excess acentric fragments (fig. 5). The irreparable excess fragments remaining after long delay times showed an RBE value of about 4 between the alpha-particle experiment and the X-irradiation (fig. 5). The difference in the repair time constants (1.9 and 3.2 h) was due to an individual variation between blood donors.

4. Discussion

The results given in figure 3a-c indicate that the fast appearance of the α component of the dicentric chromosomes is a general phenomenon, irrespective of changes in radiation quality. Even after the shortest fusion plus condensation time which we applied (30 min), the α component of the dicentrics showed its full initial plateau value as in fig. 3a. Although it is necessary to confirm this for more types of radiation and for other cell lines, we can so far summarize that three experiments have consistently pointed to the conclusion that there must be a mechanistic difference compared to the β component of the exchange figures, characterized by the faster appearance of the dose-proportional component in delayed PCC experiments.

Figure 3c as well as our paper [11], which refers to all types of interchromosomal exchange, give the important additional information that the fast appearing exchanges are *incomplete*. What is seen in the first two hours is a bi-coloured entity, consisting either of two centric pieces, of one centric and one acentric piece, of two acentric pieces or sometimes of three pieces, but the full pattern of a complete reciprocal translocation or a complete dicentric aberration, which comprises *two* chromosomal entities made up from the two participating chromosomes, has not yet been developed at this time. This incomplete pattern is the manifestation of the ongoing molecular interaction process forming the complete exchange, since this pattern is gradually replaced by the complete exchanges. We assume that the production process involves *an intermediate product,* not yet stable with respect to the forces exerted by the artificial chromosome condensation, and that the bi-coloured entity seen at this stage is the manifestation of this intermediate product. The yield of the incomplete exchanges is proportional to dose at all times, thereby constituting another difference compared to the β component, which latter exclusively appears in the complete form.

Another characteristic feature of the exchanges contributing to the α component has to be seen in the efficiency of their production as a function of the LET of the radiation applied. As fig. 4 shows, the RBE of the 3.45 MeV alpha particles compared to the α component of 150 kV X-rays is about 4, which corresponds to the proportionality of the RBE with dose-mean restricted LET, a regularity already well-known from metaphase observations of the dicentrics [12]. As mentioned in the introduction, this proportionality indicates that *two or more statistically independent energy deposition events,* occuring subsequently along the particle track, are necessary to induce the exchange-type aberration. But approximately the same RBE value, namely about 4, is observed for the production of the irreparable excess acentric fragments remaining after the disappearance of the reparable dissociations (fig. 5). Irreparable chromosome fragmentation therefore requires the same clustering of energy deposition events as the induction of an interchromosomal exchange, a most striking result.

Theoretically, there are two mechanisms which would result in this parallelity: 1) The two or more energy deposition events along the particle track correspond to radiation damage on *two different* chromatin fibres (Lea's and Neary's pairwise lesion interaction model). In this case the irreparable excess acentric fragments would be regarded as by-products of interchromosomal exchange reactions when the exchange remains incomplete or the reaction fails to produce a stable exchange but leads to fragmentation of the two participating chromosomes. But this explanation suffers from two imperfections: a) The relatively large number of the irreparable excess acentric fragments (comparison of fig. 5 with fig. 3 shows a ratio of about 5:1) does not qualify them as by-products of the more rare interchromosomal exchanges; b) Given this model, the irreparable excess acentric fragments would frequently be produced *in pairs* which means that their number per cell should be clearly overdispersed, i.e. with a variance-to-mean-value ratio exceeding unity [13, 14]. But the mitotic experiments only indicate a slight overdispersion of the excess acentric fragments [15].

Therefore we give more credit to mechanism 2: Both the irreparable fragments and the exchanges originate from clusters of energy depositions in *one* chromosome. This model does not imply inconsistencies with regard to the number and overdispersion of the irreparable fragments, as discussed before. In addition, there is accumulating Monte Carlo and experimental evidence that such clusters of energy deposition and consequently of molecular damage in *single* chromosomes actually occur [16, 17]. We have already shown that for such clusters the RBE is proportional to dose-mean restricted LET, as required if these clusters were to cause irreparable fragments and interchromosomal exchanges [11]. Mechanism 2 also offers a solution for the problem with the short range of the photoelectrons of ultrasoft X-rays mentioned in the introduction, because clustered damage is now required in only *one* chromosome.

In discussing the results of the delayed PCC experiments we have thereby arrived at a sum of arguments which bear the tendency to partly (i.e. in matters of the α component) step away from Lea's and Neary's pairwise interaction model of the interchromosomal exchanges, which would mean a great change compared to what has been believed since 50 years. *The α component of the interchromosomal exchanges would now have to be attributed to the interaction of one chromosome, affected by clustered damage, with another, intact chromosome.* The nature of the interaction would have to be basically different from that of the β component, which latter involves the enzymatic system for double-strand break repair. The new model for the α component would perhaps imply that clustered damage to a chromosome leads to free DNA strand ends and/or to opened loops of chromatin threads able to invade certain vulnerable sites in other, intact chromosomes. As judged from the knowledge of the mechanism of bubble generation in overheated liquids, it cannot be excluded that clustered damage to chromatin involves reaction steps occuring in the initial, high-temperature phase of the thermalization of the radiation-induced energy deposit, a “thermal spike“ which could lead to local melting of the DNA. The relatively high speed of the reaction leading to irregular interchromosomal joints can perhaps be explained by its dependence on relatively few reaction steps such as in the case of enzyme-supported fast ligation. The initial incompleteness of the exchange-type aberrations and the well-known temperature dependence of the α component [18] could thereby find their appropriate explanation. The mechanistic difference between the α and β components would perhaps offer the possibility to influence them separately, e.g. for the purposes of radiotherapy. Of course these are more questions than answers, but the evidence from the PCC experiments clearly suggests that the reaction model for the α component be revised.

5. References

1. Lea, D.E. (1946) Actions of radiations on living cells. Cambridge University Press, Cambridge
2. Neary, G.J. (1965) Chromosome aberrations and the theory of RBE: 1. General considerations, *Int. J. Radiat. Biol.* **9**, 477-.502
3. Cornforth, M. N. (1989) On the nature of interactions leading to radiation-induced chromosomal exchanges. *Int J. Radiat. Biol.* **56**, 635-643
4. Greinert, R., Volkmer, B., Virsik-Peuckert, R.P. and Harder, D. (1996) Comparative study of the repair kinetics of chromosomal aberrations and DNA strand breaks in proliferating and quiescent CHO cells. *Int. J. Radiat. Biol.* **70**, 33-43
5. Virsik, R.P., Schäfer, Ch., Harder, D., Goodhead, D.T., Cox, R. and Thacker, J.(1980) Chromosome aberrations induced in human lymphocytes by ultrasoft Al_k and C_k X-rays. *Int. J. Radiat. Biol.* **85**, 545-557
6. Virsik, R.P., Blohm, R., Hermann, K.-H. and Harder, D. (1981) Fast, short-ranged and slow, distant-ranged interaction processes involved in chromosome aberration formartion. In: *Seventh Symposium on Microdosimetry*, Oxford (H.G. Ebert, Ed.). Harwood Academic Publishers, London
7. Virsik, R.P. and Harder, D. (1980) Recovery kinetics of radiation-induced Chromosome aberrations in human G_0 lymphocytes. *Radiat. Environ. Biophys.* **18**, 221-238
8. Virsik-Peuckert, R.P., Volkmer, B. and Mehmel, D. (1995) Influence of ara A on the radiation-induced repair process in Chinese hamster cells studied at the level of chromosome aberrations. *Int. J. Radiat. Biol.* **58**, 1009-1023.
9. Pantelias, G.E., Iliakis, G.E., Sambani, C.D. and Politis, G. (1993) Biological dosimetry of absorbed radiation by C-banding of interphase chromosomes in peripheral blood lymphocytes. *Int. J. Radiat. Biol.* **63**, 349-354
10. Greinert, R., Detzler, E., Volkmer, B. and Harder, D. (1995) Kinetics of chromosome aberration formation in X-irradiated human lymphocytes: Analysis by delay-time PCC experiments. *Radiat. Res.* **144**, 190-197
11. Greinert, R., Detzler. E., Bartels, E., Schulte, K., Boguhn, O., Thieke, C., Virsik-Peuckert, R.P. and Harder, D. (1997) Further evidence for the association of chromosome aberration yield coefficient α with "fast, short range" and of coefficient β with "slow, long-range" pairwise interaction between DNA lesions. In: *Microdosimetry. An interdisciplinary approach* (D.T. Goodhead, P. O'Neill and H.G. Menzel, eds.) 176-179.
12. Bartels, E.R. and Harder, D. (1990) The micodosimetric regularities of nanometer regions. *Rad. Prot. Dosim.* **31**, 211-215
13. Virsik, R.P. and Harder, D, (1981) Statistical interpretation of the overdispersed distribution of radiation-induced dicentric chromosome aberrations at high LET. *Radiat. Res.* **85**, 13-23
14. Greinert, R., Harder, D. (1997) Biophysical analysis of the dose-dependent overdispersion and the restricted linear energy transfer dependence expressed in dicentric chromosome data from alpha-irradiated human lymphocytes. *Radiat Environ.Biophys.* **36**, 89-96
15. Virsik, R.P. and Harder, D. (1991) Analysis of radiation-induced acentric fragments in human G0 lymphocytes. *Radiat. Environ. Biophys.* **19**, 29-40
16. Ottolenghi, A. and Merzagora, M. (1997) Theoretical and experimental bases for mechanistic models of radiation-induced DNA damage. In: *Microdosimetry. An interdisciplinary approach* (D.T. Goodhead, P. O'Neill and H.G. Menzel, eds.) 103-110
17. Rydberg, B.(1996) Clusters of DNA damage induced by ionizing radiation: Formation of short DNA fragments. II. Experimental detection. *Radiat.Res.* **145**, 200-209
18. Gumrich, K., Virsik-Peuckert, R.P. and Harder, D. (1986) Temperature and the formation of radiation-induced chromosome aberrations. I. The effect of irradiation temperature. *Int. J. Radiat Biol.* **49**, 665-672

ADVANCED PRECISION LIGHT MICROSCOPY FOR THE ANALYSIS OF 3D-NANOSTRUCTURES OF CHROMATIN BREAKPOINT REGIONS.

Towards a Structure-Function Relationship of the BCR-ABL Region

M. HAUSMANN [1], A. ESA [1], P. EDELMANN [1],
L. TRAKHTENBROT [2], H. BORNFLETH [1], B. SCHNEIDER [1],
J. BRADL [1], I. BEN-BASSAT [2], G. RECHAVI [2], C. CREMER [1]

[1] *Applied Optics and Information Processing, Institute of Applied Physics, University of Heidelberg, Albert-Überle-Str. 3-5, D-69120 Heidelberg, Germany*
[2] *Institute of Hematology, Chaim Sheba Medical Center, Tel Hashomer, and Sackler School of Medicine, Tel-Aviv University, Tel-Aviv, Israel*

Abstract

The microscopic analysis of the 3D-nanostructure of chromatin in intact cell nuclei can contribute to a better understanding of the organization of breakpoint regions and thus of the formation of chromosome aberrations induced by ionizing radiation. Specific labelling of very small chromatin targets with dyes of different spectral signatures is feasible by fluorescence in situ hybridization. The use of DNA clones mapping breakpoint regions differently, allows precise measurements of the 3D-topology of these regions. However, 3D-fluorescence microscopy beyond the classical resolution limit is required. This can be fulfilled using a confocal laser scanning microscope in combination with principles of spectral precision distance microscopy. As an example, results obtained by digital image analysis are shown for the *bcr-abl* region involved in the formation of the Philadelphia chromosome [t(9;22)] of leukemia patients. These results support cytogenetic and molecular biology studies which have indicated that always one parental allele only may be involved in the rearrangement. For further studies with improved precision, spatially modulated excitation fluorescence microscopy can be applied, which is a promising technique to allow an almost molecular distance resolution equivalent in 3D-fluorescence light microscopy of intact cell nuclei.

1. Introduction

For the induction of radiation induced chromosome aberrations, for instance, translocations or dicentric chromosomes, various models suggest a fundamental role of the 3D-architecture and of related organization principles of the genome in the interphase cell nucleus. In recent years, highly specific delineations by fluorescence in situ hybridiza-

C. Baumstark-Khan et al. (eds.),
Fundamentals for the Assessment of Risks from Environmental Radiation, 219-230.

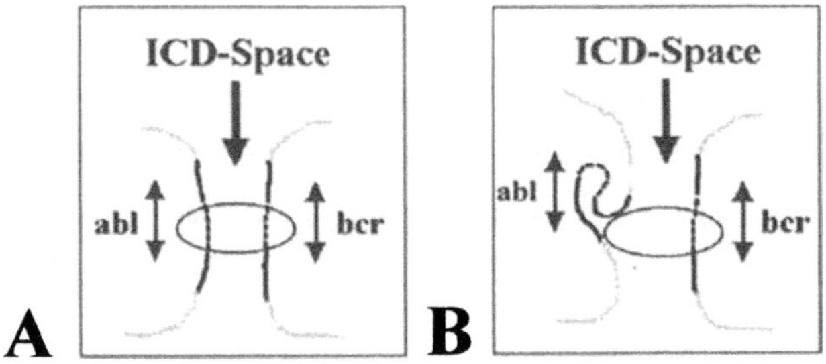

Figure 1: Schematic representation of two possible configurations (according to the ICD-model assumptions) of the abl-region of chromosome 9 and the bcr region of chromosome 22. In the "open" configuration (A) both breakpoint regions contact the ICD-space in such a way that one repair complex performs the illegitimate DNA-coupling. In the "closed" configuration (B) one breakpoint region (here e.g. abl) is protected against simultaneous binding of the same DNA repair complex to both the first and the second (= bcr) breakpoint region. (- - - - breakpoint cluster; ——— associated DNA clones, ellipsis repair complex)

tion (FISH) of chromosomes and chromosomal subregions have indicated that the cell nucleus is subjected to a high degree of compartmentalization, containing distinct, mutually exclusive chromosome territories subdivided into chromosome domains [for review: 1]. These chromosome territories may have characteristic enveloping surfaces. For a better understanding of the link between the initial damage to the DNA induced by radiation and its consequences on genome structure and function, an improved knowledge of the 3D-genome architecture is required. A model based hypothesis suggests that the formation of radiation induced interchromosome exchanges occurs in a small volume at risk between the surfaces of adjacent chromosome territorries [2]. In the recently postulated "ICD-model" this volume of risk has been identified to represent the interchromosomal domain (ICD) space of dimensions of some ten nanometers [1, 3, 4]. Thus, intraterritory chromatin arrangements might favour or reduce the possibility of illegitimate recombinations of certain double strand breaks (Figure 1) [5].

Therefore, it appears to be useful to accurately study the nano-topology of typical breakpoint cluster regions in detail, as for instance, the *bcr* and *abl* regions of chromosome 22 and 9, respectively. These regions are known to be involved in the formation of the Philadelphia chromosome of patients suffering from chronic myeloid leukemia, where it is found in more than 90% of the cases [6].

2. Requirements for the Analysis of the Genome Topology

To study the topology of breakpoint regions in three dimensionally conserved cell nuclei, some requirements concerning labelling of DNA and microscopy have to be considered:

a) Short DNA sequences have to be labelled specifically with fluorescent dyes: FISH [7, 8] has the potential to label the DNA of chromatin in cell nuclei from whole chromosome territories down to individual genes with fluorochromes of different spectral signatures (e.g. different fluorescence excitation or emission spectra).

b) Appropriate DNA clones which map the breakpoint region and relevant neighbouring regions have to be characterized and amplified by molecular biological techniques to be suitable as FISH probes.
c) Three-dimensional, high-resolution fluorescence microscopy has to be applied to image small FISH-labelling sites in three-dimensionally conserved cell nuclei with high sensitivity: So far, the confocal laser scanning fluorescence microscope [9-11, for review: 12] is the optical system with the best 3D-resolution to detect fluorescent micro-objects which is commercially available [13]. Here, in this report, some practical limitations and techniques to overcome these shortcomings for the analysis of the genome nanostructure will be presented.
d) Digital image analysis has to be used to quantify the distances and spatial distribution of the bary centers of FISH labelling regions. Here, it will be shown how digital image analysis contributes to the technique of "spectral precision distance microscopy" which allows precise distance measurements far beyond the classical resolution of the diffraction limited microscopy optics [14, 15].

3. Confocal Laser Scanning Microscopy

3.1. PRINCIPLE

To study the 3D-structure of the genome in intact cell nuclei (typical diameter in the order of 10 μm), 3D-fluorescence images have to be acquired. This can be performed by optical sectioning of a cell nucleus into a complete series of 2D-images. For this purpose, microscope optics performing an appropriate depth discrimination of each 2D-section is required. In a confocal laser scanning microscope (CLSM) this can be obtained by the confocal arrangement of a pointlike excitation light source and a pointlike detector [12]. By means of digital image reconstruction the cell nucleus can be visualized in a pseudo 3D-image in false colors (see Figure 2). For this purpose, highly sophisticated 3D visualization and image analysis computer programs are available.

In practice, this can be performed by a monomode laser beam coupled into a high quality microscope via a glass fibre. This laser beam can for instance be focused on an excitation pinhole. The light emerging from this pinhole is made parallel by an appropriate lens and scanned in x and y directions (lateral beam movement) by (a) galvanometer driven mirror(s). The light is focused diffraction limited into a spotlike volume of the specimen by an objective of high numerical aperture. This excitation spot can be moved through the specimen in the focal plane (= one optical section). Moving the microscope stage step by step in z-direction (axial direction) allows to move the objective's focal plane through the specimen and to record a series of optical sections.

The fluorescence light excited by the illumination light spot is collected by the same objective and focussed on a confocally arranged second pinhole to ensure that only fluorescence light from the illuminated region of the actual focal plane is detected. Out of focus contributions are suppressed. Behind the pinhole dichroic mirrors and spectral filters separate the fluorescence light into independent optical channels according to its spectral signature, assuming the ideal case of negligible spectral overlap of the filters. Several photomultipliers can be used to record the fluorescence signals.

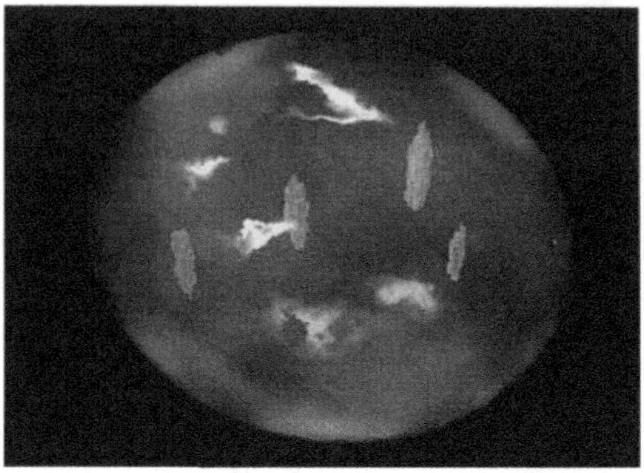

Figure 2: 3D-image of a Ph^+-cell nucleus of a CML patient. The DNA sequences shown in Figure 3 were labelled by FISH. The cell nucleus with the labelling sites was recorded by the CLSM. From the 2D-image sections the labelling sites were segmented by thresholding using the Cavalieri estimator. A 3D-data stack of twentyeight 2D-images was obtained and the voxel image was visualized. The labelling site with three colors (red, green, blue) is located on the Ph-chromosome. The green labelling site is located on the conserved chromosome 9 and the red-blue one on the conserved chromosome 22. From the der9 chromosome only a red labelling site is visible. The false colors of the labelling sites correspond to Figure 3.The cell nucleus shape was obtained from the fluorescence background.

3.2. OPTICAL RESOLUTION OF A CONFOCAL LASER SCANNING MICROSCOPE

The spatial resolution of a CLSM is usually expressed by the full width at half maximum (FWHM) of the point spread function [13] which is the diffraction limited image of a fluorescent pointlike object. Under ideal conditions, the FWHM is about 150 nm lateral and due to physical reasons about 350 nm axial. However, biological specimen such as cell nuclei after FISH labelling are far from these ideal conditons. In situ measurements of the spatial resolution under typical, i.e., practical conditions revealed even higher values although the relative difference between axial and lateral resolution was maintained [16, 17]. As a consequence, the error of distance measurements between two points of interest depends on the spatial orientation of these points [16]. To overcome these shortcomings of resolution anisotropy, micro axialtomography was introduced into fluorescence microscopy [16, 18, 19]. In this device, the cell nuclei are mounted onto the surface of a glass fibre which can be tilted perpendicularly to the optical axis of the microscope. The cell nuclei can always be tilted into such a perspective that two points of interest are in the same focal plane. All distance measurements can be performed with the optimum resolution and the smallest error values [17, 20].

4. Precision Distance Microscopy

In contrast to the resolution which is given by the FWHM of the point spread function, the localization accuracy is given by the error of the coordinates for the intensity maximum. For well corrected microscope objectives with nearly symmetrical lateral and axial point spread functions, the intensity maximum corresponds to the intensity bary center. Thus, the localization accuracy of an object can be expressed, for instance, by the standard deviation of the intensity bary center coordinates of a series of measurements. Under practical conditions using a biological specimen, it was found that the localization accuracy may be estimated to about a tenth of the value of the corresponding FWHM of the point spread function. The application of axialtomography in combination with a CLSM results at least in a localization precision of about ±20 nm in all three directions [20]. With the concept of precise localization, it was possible to determine distances >1 FWHM of objects of the same spectral signature with an accuracy considerable below the optical resolution [19].

Moreover, this concept offers possibilities to measure distances <1 FWHM with high accuracy if the pointlike objects carry a different spectral signature. The idea of this so called "spectral precision distance microscopy" [15] is explained in the following example:

Let us assume two DNA sequences located in the cell nucleus within a volume with dimensions below one FWHM. These two DNA sequences can be FISH labelled each with a different spectral signature. The diffraction limited images can the be recorded independently for each labelling site, and the intensity bary centers can also be determined independently from each other with the localization accuracy valid for targets of the spectral signature considered. From the intensity bary centers determined by appropriate algorithms for digital image analysis the euclidean distances between the target loci bary centers can be easily calculated. The precision with which these distances below the optical resolution can be measured determines the resolution equivalent:

The resolution equivalent (i.e. the smallest distance between targets of different spectral signature which can reliably be detected) strongly depends on the localization error which is influenced by the optical resolution, the signal to noise ratio, the detector sensitivity and the digitization accuracy [14, 15, 21, 22]. Most prominent are effects of chromatic shifts between the two targets of different spectral signatures. These chromatic shift effects are due to insufficient chromatic corrections of the microscope objectives for high precision measurements under the optical microconditions used. Especially, the depth localization of targets (z-localization) is influenced by these shift effects. However, chromatic shifts can precisely be measured using appropriate calibration objects [14, 23]. By means of digital image analysis the calibration values can be used to correct the measured data of the cell nuclei.

5. Application of Spectral Precision Distance Microscopy

As an example for spectral precision distance microscopy, preliminary results of the analysis of the topology of the *bcr-abl* breakpoint region in cells from a patient with chronic myeloid leukemia are presented. These data can be seen as a first step towards a structure-function model of this region which is involved in the formation of the translocation t(9;22) (q34;q11), the Philadelphia (Ph) chromosome. The Ph-chromosome is a cytogenetic hallmark in chronic myeloid leukemia and acute lymphocytic leukemia.

On the molecular level, the translocation t(9;22) (q34;q11) results in the fusion of a part of the *abl* proto-oncogene and a part of the *bcr* gene. Cytogenetic and molecular studies of the *bcr-abl* fusion gene suggested an exclusive maternal [24] or paternal [25] origin. These results do not necessarily contradict each other since both studies were performed on a limited number of cases selected according to different criteria. Other complex recombinations may be involved in the Ph-chromosome formation. To better understand possible mechanisms contributing to the *bcr-abl* rearrangement in the genome, a precise study of the chromatin organization around these regions in normal and pathologically aberrant cell nuclei is required. In the following, first preliminary data concerning one CML-case will be presented.

5.1. SPECIFIC DNA-LABELLING

Bone marrow cells from a patient with CML were fixed in methanol/acetic acid (3:1; v:v) without hypotonic treatment. For FISH, a combination of the following DNA-clones was used:

- a cosmid clone (cos-abl 8) containing a 40 kb fragment flanking the *abl*-gene on chromosome 9 [26];
- a YAC-clone (B99E11) containing a 115 kb fragment spanning the *bcr*-breakpoint cluster on chromosome 22 [27];
- a BAC-clone (52C3) containing a ca. 90 kb fragment spanning the *bcr*-breakpoint cluster on chromosome 22 [28].

In Figure 3, a schematic representation of the mapping of the DNA-clones and the expected arrangement in Ph+ and Ph– cells is shown. The DNA-clones were purified and the human insert was amplified by PCR. 300 ng aliquots of the PCR amplified DNA probes were differentally labelled:

- the 40 kb-probe with FITC-12-dUTP;
- the 115 kb-probe with digoxigenin-11-dUTP (visualization by TRITC-anti-digoxigenin);
- the 90 kb-probe with biotin-14-dCTP (visualization by CY5-avidin).

In situ hybridization with the three DNA probes was performed simultaneously after preannealing with Cot1-DNA and salmon sperm DNA. The target DNA was denatured in 70 % formamide, 2xSSC at 76 °C for 4 min followed by dehydration by an ethanol series. The denatured and preannealed probe mixture was added to the target for hybridization at 37 °C overnight. After several washing steps, the avidin and anti-digoxigenin detection kit was added and the cell nuclei were counterstained with DAPI. Further details of the protocol will be published elsewhere.

5.2. MICROSCOPY

3 D-image acquisition was performed using a Leica TCS NT confocal laser scanning microscope equipped with an Argon-Krypton laser for fluorescence excitation at 488 nm, 568 nm, and 647 nm. For each nucleus, the three dyes were detected simultaneously using three photomultiplier detection channels with appropriate optical filters for FITC-, TRITC-, and CY5-fluorescence. Optical sectioning was performed using an oil objective 63x/NA1.4. For image analysis, the labelling sites were segmented for each color independently using the Cavalieri estimator [17, 20]. For a series of relevant thresholds, the bary centers of the segmented intensity distributions were calculated. The z-coordinates of the loci were corrected for the chromatic shift according to Born-

fleth et al. [14]. These chromatic shift values were compatible with the results obtained by the Cavalieri estimator:

a) FITC-CY5 310 nm
b) FITC-TRITC 240 nm
c) TRITC-CY5 80 nm

From the corrected localization values the euclidean distances of the intensity bary centers of the DNA-labelling sites were calculated.

5.3. RESULTS

Distances between labelled DNA targets were measured both in Ph⁺ and Ph⁻ cells. In Figure 4, cumulative frequency distributions of the distances obtained from the presumptive Ph chromosome characterized by three associated, differentially colored labelling sites are shown. In a part of the nuclei distances larger than the effective resolution limit (FWHM) were found (up to 950 nm). In the *bcr-abl* fused region about 50 % of the distances between the labelling sites derived from the 40 kb (chromosome 9) and115 kb (chromosome 22) fragments (see Figure 3) were below 300 nm. The labelling sites derived from the 40 kb (chromosome 9) and the 90 kb (chromosome 22) fragments tended to have larger distances (50% distance: ca. 600 nm).

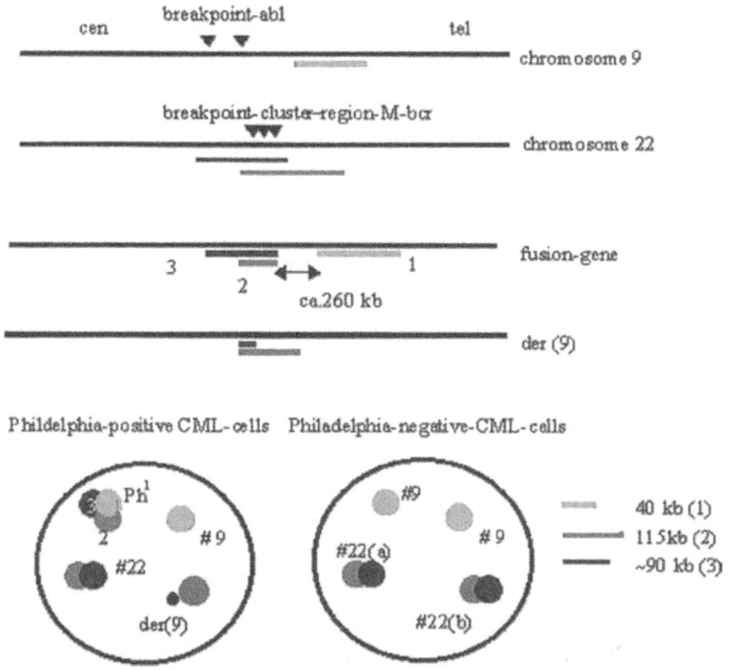

Figure 3: Schematic representation of the relative localization of the three DNA-clones used for FISH labelling and their expected arrangement in Ph⁺ and Ph⁻ cells

In Figure 5, cumulative frequency distributions of distances between the 115 kb and 90 kb fragments on the presumably intact chromosomes 22 characterized by the two associated differentially colored labelling sites are shown. In each Ph^- cell nucleus, the shorter distance was assigned to 22(a) and the larger one to 22(b). The two alleles of 22(a) and 22(b), respectively, in Ph^- cell nuclei differed significantly in their distances (applying the two sided Kolmogorov-Smirnov test). The intact chromosome 22 of Ph^+ cell nuclei clearly correlated with chromosome 22(a) in Ph^- cell nuclei showing shorter distances.

5.4 BIOLOGICAL IMPLICATIONS

The results reported in 5.3 suggest that for the CML case analyzed, always the same "more compact" (shorter distances) allele of chromosome 22 may be preserved in Ph^+ cells. Consequently, the "more open" (larger distances) chromosome 22 may be involved in the *bcr-abl* fused chromatin region. These suggestions are compatible with molecular cytogenetic studies of the parental origin of the *bcr-abl* fusion [24, 25]. If so, small variations in local 3D-nanostructures of chromatin breakpoint regions might play a fundamental role in the induction of radiation induced leukemia. To test this hypothesis more cases have to be analyzed with a statistically relevant number of cell nuclei. However, the results so far obtained indicate the feasibility of such an approach of spectral precision distance microscopy to analyze small FISH labelling sites in 3D-conserved cell nuclei. Using other DNA-clones for FISH probes, the technique may also contribute to answer the question whether the fusion gene is of maternal or paternal origin.

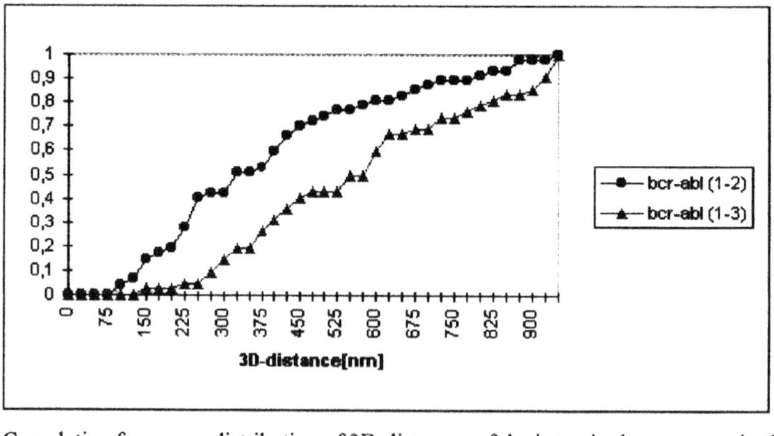

Figure 4: Cumulative frequency distribution of 3D-distances of the intensity bary centers in the *bcr-abl* fused (three color) region in Ph^+-cells (see also Figure 3). *bcr-abl* (1-2) shows the distances (FITC-TRITC distance) of the labelling sites derived from the 40 kb and 115 kb fragments (data obtained from 47 nuclei). *bcr-abl* (1-3) shows the distances (FITC-CY5 distance) of the labelling sites derived from the 40 kb and 90 kb fragments (data obtained from 42 nuclei).

6. Further Improvements - the SME Microscope

The principle of spectral precision distance microscopy may considerably contribute to analyze the 3D-nanostructure of the genome. However, with smaller DNA-clones for FISH probes and for a more refined analysis, the resolution and the effective localization accuracy of the CLSM may become insufficient, especially for small fluorescent labelling sites of low intensity. To overcome those shortcomings, an epifluorescent laser microscope with spatially modulated excitation (SME) was developed [29, 30], which in principle is similar to the standing-wavefield fluorescence microscope [31-34]. In the SME-fluorescence microscope [29], coherent light of an Argon ion laser (λ=488 nm, 514 nm) or a Krypton laser (λ=568 nm, 647 nm) can be coupled into the setup mounted on a thermally stabilized Invar bread board via a single mode, polarization conserving glass fibre. The light is expanded by means of a collimator and split into two beams of equal intensity by means of a beamsplitter. The parallel light of the two beams is focussed into the back focal plane of two opposite objectives (100x/NA0.7-1.4 Planapo) using an achromatic lens in front of each objective. Since the objectives work according to infinite optics, this results in parallel light beams of planar wave fronts in the object space between the two objectives. Interference of these two parallel light beams results in a standing wave field. The interference angle between the two light beams and the optical axis of the setup can be varied in order to vary the distances of the interference maxima.

The specimen mounted on a standard glass slide can be moved piezoelectrically along the direction of the optical axis (z-direction = perpendicular to the wave fronts) with a minimum stepsize of 20 nm. The spatially modulated wave field excites the

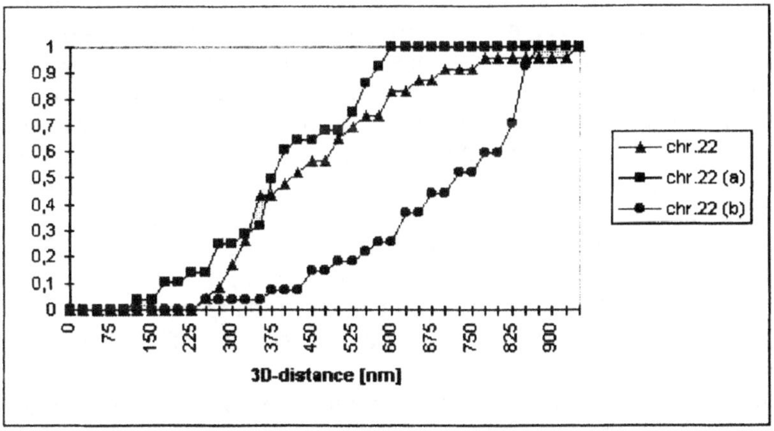

Figure 5: Cumulative frequency distribution of the 3D-distances of the intensity bary centers of the 115 kb and 90 kb labelling sites (TRITC-CY5 distance) in intact chromosomes 22 obtained from 27 Ph– cell nuclei (22(a), 22(b)) and 23 Ph+ cell nuclei. The two parental alleles 22(a) and 22(b) differed significantly in their distance distribution. The chromosome 22 in Ph+ cell nuclei correlated to the shorter distances of 22 (a).

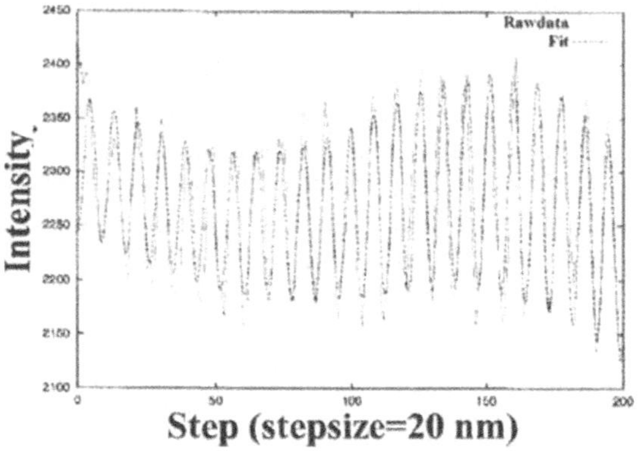

Figure 6: Example of precise axial (z-) localization of a fluorescent labelling site (ca. 17-18 kb in the "Prader-Willi" region of chromosome 15) in a cell nucleus. Along an axis parallel to the optical axis the relative fluorescence intensity versus the z-axis in steps of 20 nm is shown (rawdata). The dashed line (fit) represents a non-linear least-square fit from which the localization of the intensity maximum (intensity bary center) and the localization accuracy (here 1 nm) were calculated.

fluorochrome label with a fluctuating intensity during specimen movement. Using one objective and appropriate optical filters, 2D-images are recorded by a cooled digital CCD-camera for each focus setting. Thus, a complete 3D-data stack of up to 400 images can be obtained. Controlling of the microscope, image acquisition and image analysis can be performed on a personal computer running under the LINUX operating system.

After image analysis, a pointlike fluorescence object is represented by a fringe pattern along the z-direction. A typical intensity distribution along an axis parallel to the z-axis through a chromatin labelling site in a cell nucleus is shown in Figure 6. By means of a non-linear least-squares fit algorithm such intensity distributions can be evaluated to localize the main maximum (= site of the intensity bary center) accurately. For test particles of one spectral signature, this procedure resulted in a localization accuracy for the z-direction of 1 to 2 nm. High-precision distance measurements in the range of 300 nm up to about 2.5 μm resulted in an average 3d-distance accuracy of 27 nm [30]. The results of localization accuracy were confirmed by data obtained in conserved cell nuclei. Fluorescent targets of FITC labelled DNA-clones (ca. 17-18 kb) of chromosome 15 were recorded by the SME fluorescence microscope (one spectral signature). The intensity bary centers were again localized with an accuracy of 1 to 2 nm for the z-component.

From this localization accuracy, the axial optical resolution equivalent of the SME-fluorescence microscope can be estimated to be one of the best values obtainable for 3D-microscopes. However, the lateral localization accuracy is still compatible with a conventional high quality epifluorescence microscope. To overcome these shortcomings, micro axialtomography [16] may be applied so that two target points can be tilted into a plane parallel to the optical axis of the SME-microscope where the optimum localization accuracy can be obtained. Moreover, multi wavelength excitation has been be implemented. This will allow distance measurements considerably below the optical resolution limit, if the DNA labelling sites carry different spectral signatures. For such measurements the principles of "spectral precision distance microscopy" described above for confocal imaging may be applied so that the combination of all the techniques described in this article may lead to a resolution equivalent appropriate to analyze the 3D-architecture of individual nucleosome arrangements in a given breakpoint region.

Acknowledgement. The authors gratefully acknowledge the financial support by DFG, the BMBF and the EU. A. Esa receives a scholarship of the Friedrich Ebert Stiftung. The authors are indebted to Dr. U. Weier (Lawrence Berkeley National Laboratory), Dr. N. Heisterkamp (Los Angeles Children's Hospital) and Dr. U.-J. Kim (California Institute of Technology, Pasadena) for providing the DNA clones. The DNA probes were prepared in the laboratory of Dr. U. Weier. The authors also thank Dr. I. Solovei (Institute of Anthropology and Human Genetics, Munich) for the FISH labelling used for Figure 5. Precision distance microscopy is a patent application submitted by C. Cremer, M. Hausmann, J. Bradl, and B. Rinke.

7. References

1. Cremer, C., Münkel, C., Granzow, M., Jauch, A., Dietzel, S., Eils, R. Guan, X.Y., Meltzer, P.S., Trent, J.M., Langowski, J., Cremer, T. (1996) Nuclear architecture and the induction of chromosomal aberrations. *Mut. Res.* **366**, 97-116
2. Savage, J.R.K., Papworth, D.G. (1973) The relationship of radiation-induced dicentric yield to chromosome arm number. *Mut. Res.* **19**, 139 - 143
3. Zirbel, R.M., Mathieu, U., Kurz, A.; Cremer; T., Lichter, P. (1993) Evidence for a nuclear compartment of transcription and splicing located at chromosome domain boundaries. *Chromos. Res.* **1**, 92 - 106
4. Cremer, T., Kurz, A., Zirbel, R., Dietzel, S., Rinke, B., Schröck, E., Speicher, M.R., Mathieu, U., Jauch, A., Emmerich, P., Scherthan, H., Ried, T., Cremer, C., Lichter, P. (1993) Role of chromosome territories in the functional compartmentalization of the cell nucleus. *Cold Spring Harb. Symp. Quant. Biol.* **58**, 777-792
5. Cremer, C., Bornfleth, H., Bradl, J., Esa A., Kreth, G., Rinke, B., Eils, R., Münkel, C., Dietzel, S., Granzow, M., Jauch, A., Guan, X-J., Meltzer, P.S., Trent, J.M., Trakhtenbrot, L., Zink, D., Langowski, J., Hausmann, M., Cremer, T. (1998) Nuclear architecture and the formation of chromosome aberrations. *Radiat. Res.* **148**, 513-515 (in press)
6. Mitelman, F. (1991) Catalog of chromosome aberrations in cancer. Alan R. Liss, New York
7. Lichter, P., Cremer T. (1992) Chromosome analysis by non-isotopic in situ hybridization. In: D.E. Rooney, B.H. Czepulkowski (eds.) Human Cytogenetics-a Practical Approach. IRL-Press, Oxford, pp. 157 - 192
8. Gray, J.W., Pinkel, D., Brown J.M. (1994) Fluorescence in situ hybridization in cancer and radiation biology. *Radiat. Res.* **137**, 275 - 289
9. Cremer C., Cremer, T. (1978) Considerations on a laser-scanning-microscope with high resolution and depth of field. *Microsc. Acta* **81**, 31 - 44
10. Brakenhoff, G.J., van der Voort, H.T.M., van Spornsen, E.A., Linnemans, W.A.M., Nanninga, N. (1985) Three dimensional chromatin distribution in neuroplaszoma nuclei shown by confocal laser scanning microscopy. *Nature* **317**, 748 - 749
11. Stelzer, E.H.K., Marsman, H.J.B., Wijnaendts van Resandt, R.W. (1986) A setup for a confocal scanning laser interference microscope. *Optik* **73**, 30 - 33
12. Pawley, J.B. (1995) Handbook of biological confocal microscopy. Plenum Press, New York

13. Gu, M. (1996) Principles of three-dimensional imaging in confocal microscope. World Scientific Publishing, Singapore
14. Bornfleth, H., Sätzler, K., Eils, R., Cremer C. (1998) High-precision distance measurements and volume-conserving segmentation of objects near and below the resolution limit in three-dimensional confocal fluorescence microscopy. *J. Microsc.* **189**, 118-136
15. Cremer, C., Edelmann, P., Esa, A., Bornfleth, H., Schneider, B., Bradl, J., Rinke, B., Trakhtenbrot, L., Dietzel, S., Hausmann, M., Cremer, T. (1998) Spektrale Präzisionsmikroskopie in der Genomforschung. *Z. Med. Phys.* In press
16. Bradl, J., Rinke, B., Hausmann, M., Schneider, B., Cremer, C. (1996) Improved resolution in „practical" light microscopy by means of a glass fibre 2π-tilting device. *Proc. SPIE* **2628**, 140-146
17. Rinke, B., Bradl, J., Edelmann, P., Schneider, B., Hausmann, M., Cremer, C. (1996) Image acquisition and calibration methods in quantitative confocal laser scanning microscopy. *Proc. SPIE* **2926**, 190 - 200
18. Bradl, J., Hausmann, M., Schneider, B., Rinke, B., Cremer, C. (1994) A versatile 2π-tilting device for fluorescence microscopes. *J. Microsc.* **176**, 211-221
19. Bradl, J., Rinke, B., Edelmann, P., Krieger, H., Hausmann, M., Cremer, C. (1996) Resolution improvement in 3D-microscopy by object tilting. *Microsc. Anal.* **11/96**, 9-11
20. Bradl, J., Rinke, B., Esa, A., Edelmann, P., Krieger, H., Schneider, B., Hausmann, M., Cremer, C. (1996) Comparative study of three-dimensional localization accuracy in conventional, confocal-laser-scanning and axial tomographic fluorescence light microscopy. *Proc. SPIE* **2926**, 201-206
21. Manders, E.E.M., Verbeek, F.J., Aten, J.A. (1993) Measurement of co-localization of objects in dual-colour confocal images. *J. Microsc.* **169**, 375 - 382
22. Manders, E.E.M., Hoebe, R., Strackee, J., Vossepoel, A.M., Aten, J.A. (1996) Largest contour segmentation: a tool for the localization of spots in confocal images. *Cytometry* **23**, 15 - 21
23. Manders, E.E.M. (1997) Chromatic shift in multicolour confocal microscopy. *J. Microsc.* **185**, 321-328
24. Haas, O.A., Argyriou-Tirita, A., Lion, T. (1992) Parental origin of chromosomes involved in the translocation t(9;22). *Nature* **359**, 414 - 416
25. Litz, C.E., Copenhaver, C.M. (1994) Paternal origin of the rearranged major breakpoint cluster region in chronic myeloid leukemia. *Blood* **83**, 3445 - 3448
26. Heisterkamp, N., Groffen, J., Stephenson, J.R. (1983) The human v-*abl* cellular homologue. *J. Mol. Appl. Genet.* **2**, 57 - 68
27. Lengauer, C., Riethman, H.C., Speicher, M.R., Taniwaki, M., Konecki, D., Green, E.D., Becher, R., Olson, M.V., Cremer T. (1992) Metaphase and interphase cytogenetics with alu-PCR yeast artificial chromosome clones containing the *bcr* gene and the protooncogenes c-raf-1,c-fms, and c-erbB-2. *Cancer Res.* **52**, 2590 - 2596
28. Kim, U.-J., Birren, B.W., Slepak, T., Mancino, V., Boysen, C., Kang, H.-L., Siman, M.I., Shizuya, H. (1996) Construction and characterization of a bacterial artificial chromosome library. *Genomics* **34**, 213 – 218
29. Schneider, B., Bradl, J., Kirsten, I., Hausmann, M., Cremer, C. (1998) High precision localization of fluorescent targets in the nanometer range by spatially modulated excitation fluorescence microscopy, In: J. Slavik *(Ed.) Fluorescence Microscopy and Fluorescent Probes, Vol. II,* Plenum Press, pp 71-76
30. Hausmann, M., Schneider, B., Bradl, J., Cremer, C. (1997) High-precision distance microscopy of 3D-nanostructures by a spatially modulated exciation fluorescence microscope. *Proceedings SPIE* **3197**, 217-222
31. Lanni, F., Waggoner, A.S., Taylor, D.L. (1986) Standing wave luminescence microscopy. US Patent No. 4 621 911
32. Bailey, B., Farkas, D.L., Taylor, D.L., Lanni, F. (1993) Enhancement of axial resolution in fluorescence microscopy by standing-wave excitation. *Nature* **366**, 44-48
33. Gustafsson, M.G.L., Agard, D.A., Sedat, J.W. (1995) Sevenfold improvement of axial resolution in 3D widefield microscopy using two objective lenses. *Proc. SPIE* **2412**, 147-156
34. Freimann, R., Pentz, S., Hörler H. (1997) Development of a standing-wave fluorescence microscope with high nodal planem flatness. *J. Microsc.* **187**, 193 – 200

CHROMATIN STRUCTURE AND ITS CELL CYCLE KINETICS IN INTACT AND IRRADIATED CELL NUCLEI.

Structural Aspects of Radiation Induced Leukemia

S. KOZUBEK [1], E. LUKÁŠOVÁ [2], L. RÝZNAR [1], M. KOZUBEK [2], A. LIŠKOVÁ [1], V. KROHA [3]

[1] *Institute of Biophysics, Academy of Sciences, Královopolská 135, 61265 Brno, Czech Republic*
[2] *Faculty of Informatics, Masaryk University, Botanická 68a, 60200 Brno, Czech Republic*
[3] *Institute for Nuclear Physics, 250 68 Rez at Prague, Czech Republic*

Abstract

The prototype of the genetic aberration which is found in leukemias is the t(9;22) translocation involving *ABL* and *BCR* genes typical of chronic myeloid leukemia (CML) which is frequently induced by radiation. In this study, the structural characteristics of these chromosomes and genes were investigated. We showed that both in interphase and prometaphase the average distance between *ABL* and *BCR* genes is substantially shorter than that of randomly distributed genes. Therefore, one of the reasons of high incidence of CML may be the physical closeness of *ABL* and *BCR* genes in cell nuclei. The chromatin structure undergoes substantial changes during the cell cycle. We have found that, after the stimulation of lymphocytes, the *ABL* and *BCR* genes move towards the membrane, their mutual distances increase, and the minimum distance between heterologous ABL and BCR genes increases. From experiments in which sister cell nuclei were analysed we can conclude that the cell cycle dependent movement of individual genes is predetermined to some extent at the moment of cell division. The so called intermingling or random walk of chromatin are rather limited. The original chromatin structure and its cell cycle dependent evolution are disturbed in irradiated cells. We found that the positions of *ABL* and *BCR* genes in irradiated and stimulated lymphocytes are shifted to the centre of the nucleus, they are closer to each other and also the distances between heterologous *ABL* and *BCR* genes are shorter. Therefore, radiation increases the probability of *ABL-BCR* interaction during the first cell cycle after irradiation. Average values of the topological parameters of *ABL* and *BCR* genes are similar for different cell lines and different individuals. On the other hand, we found statistically significant variations in bone-marrow cells of healthy individuals (and CML patients). We suggest that individuals with shorter *ABL-BCR* distances are more likely to suffer from leukemia in future.

C. Baumstark-Khan et al. (eds.),
Fundamentals for the Assessment of Risks from Environmental Radiation, 231-242.

1. Introduction

At the present time it is well known that the induction of malignant diseases is connected with a deregulation of cell division in the so-called restriction point (R-point) occurring in the G_1 phase of the cell cycle, in which the subsequent development of the cell is determined [1]. The regulation mechanism of cell proliferation includes a number of genes of two types: proliferation activating genes (oncogenes) and proliferation suppressing genes (tumour suppressors). Deregulation of these genes can be caused by various defects; e.g. mutations, chromosomal aberrations (mainly translocations), deletions, integration of oncoviruses, etc. It has been found that many of translocations are specific for a defined type of malignancy and that they represent often single chromosomal change in malignant cells [2]. Thus, t(9;22), giving rise to the so-called Ph chromosome, is present in 95% of patients with chronic myeloid leukemia (CML) [3]. Some other types of translocations in haemoblastic malignancies are, e.g., t(15;17) in promyeloid leukemia, t(9;22), t(4;11) or t(8;14) in lymphoblastic leukemia, t(14;18), t(8;14), and t(11;14) in non-Hodgkin's lymphomas NHL [2,4,5,6]. It is well known that CML (and also other mentioned malignancies) are frequently induced by ionising radiation [7].

We have shown that the physical distance can play an important role in the processes leading to the translocations that are responsible for oncogenic transformation of cells [8, 9]. For example, the positions of critical genes involved in the formation of Ph-chromosome are not random in the interphase cell nucleus; both the *ABL* and *BCR* genes are located close to the centre of the nucleus, which obviously leads to substantial enhancement of the induction probability of translocation.

In this contribution we summarise our results of the investigations of the topology of chromosomes (9 and 22) and oncogenes (ABL and BCR) in human cell nuclei. We present evidence that neither the positions of the above mentioned chromosomes and genes nor the cell cycle changes of these positions are random in cell nuclei. The mutual positioning of these chromosomes and genes increases the probability of exchange processes between them. It will be further shown that the regularities of the cell cycle dependent changes of chromatin structure can be strongly affected by ionising radiation.

2. Materials and Methods

2.1. CULTIVATION AND IRRADIATION OF CELL CULTURES

Human lymphocytes were isolated from the whole heparinized blood (16-20 U/ml) of healthy individuals. Plasma containing lymphocytes was obtained after 2 h of blood sedimentation at room temperature. The cell culture was divided into two 3 ml aliquots and irradiated by 0 and 5 Gy of ^{137}Cs γ-rays (0.66 MeV; dose rate 3.6 Gy/min), respectively. Control and irradiated samples were diluted (1:3) with RPMI-1640 medium containing 0.1 ml of phytohaemagglutinin (PHA) (Murex, England) per 5 ml of the suspension, 1% glucose, penicillin (100 U/ml) and streptomycin (100 mg/ml). The incubation was performed at 37°C for 50 h. Colcemid was added 2 h before the end of incubation.

HL-60 cell cultures were incubated in similar way in RPMI medium containing serum and antibiotics, at 37°C and 5% CO_2. Human lung fibroblasts *IMR 90* (American Type Culture Collection) were grown as monolayers on microscope slides in RPMI-1640 medium containing 10% foetal bovine serum and antibiotics. Some cultures were treated with nocodazole (40 ng/ml, SIGMA) for 4 -12 h to arrest the cells at prometaphase.

2.2. FIXATION OF CELLS

2.2.1. Fixation by Standard Cytogenetic Procedure

After hypotonic treatment (0.075 M KCl) for 20 min the cell culture was fixed 3 times by cold (-20°C) methanol/acetic acid (3:1). After the last centrifugation the concentration of the cell suspension was adjusted by fixation solution and dropped on slides.

2.2.2. Fixation by Paraformaldehyde

The cells were rinsed in phosphate-buffered saline(PBS) and fixed for 5 min with buffered paraformaldehyde (4% solution in PBS, Sigma). The slides were washed twice in PBS and permeabilized in 0.1 N HCl with 0.01% Triton X-100 for 15 min followed by two washings in 0.1 M Tris-HCl, pH 7.2 and stored in 50% glycerol in PBS until used. Prior to *in situ* hybridisation, cells were rinsed with PBS and digested with RNase (700 mg/ml in 10 mM Tris, 1 mM NaCl, and 0.1% Triton X-100) at 37°C for 30 min, then washed 3 times in PBS. Then the slides were incubated in 50% formamide with 2xSSC, pH 7.0, overnight. A growing population of *HL-60* cells was sedimented by centrifugation, washed twice by PBS, and spread on a slide. *HL-60* cells were fixed according to the same procedure as the fibroblasts, however, they were not digested by RNase before hybridisation.

2.3. DNA PROBES, IN SITU HYBRIDIZATION, AND PROBE DETECTION

2.3.1. Cells Fixed by Methanol/Acetic Acid

M*bcr-abl* (digoxigenated *bcr*, biotinylated *abl*) as well as c-*myc* and telomere 14 probes were purchased from ONCOR, hybridised, and detected according to earlier published recommendations [8] with small modifications. The chromosomes were denatured at 70°C in 70% formamide with 2xSSC, pH 7.0 for 2 min, dehydrated in cold (-20°C) 70%, 80%, 95% ethanol (2 min each), then air dried. The probe was prewarmed at 37°C for 5 min and vortexed. The M*bcr-abl* DNA probe (10 μl) was applied to the denatured and dried slide, covered by a glass coverslip of 25x25mm, sealed with rubber cement and incubated overnight at 37°C in a humidified chamber.

Whole chromosome-specific DNA probes modified by digoxigenin (Oncor, Gaithesburg, USA) and those modified by biotin (Cambio, Cambridge, UK) were used. The procedure of hybridisation was performed according to the instructions of the manufacturers.

2.3.2. Cells Fixed by Paraformaldehyde

Slides were rinsed twice with PBS, drained well and the remaining liquid was eliminated in air. Immediately 10 μl of DNA probe was applied on the slide which was covered with a coverslip and sealed with a rubber cement. The cellular and probe DNA were denatured simultaneously at 72 °C for 10 min. Hybridisation was carried out for 16 h at 37 °C as in previous case.

2.3.3. Post Hybridisation Wash

The post-hybridisation wash was performed in 50% formamide with 2xSSC, pH 7.0 at 43°C for 15 min followed by two 5 min washes in 2xSSC containing 0.1% Tween 20, pH 7.0 at 37°C. Slides were then transferred to 1xPBD (ONCOR) and incubated at room temperature for 3 min. Rhodamine-antidigoxigenin and FITC-avidin (60 ml) was applied to a slide and incubated for detection under a plastic coverslip at 37°C for 5 min. The washing was performed in 3 changes of 4xSSC containing 0.1% Tween 20, pH 7.0 at 43°C for 5 min. The temperature of the slides was equilibrated to room temperature in the 4xSSC containing 0.1% Tween 20, and the cells were conterstained with DAPI (0.02 mg/ml) in the Vectashield.

2.4. DATA ACQUISITION AND STATISTICAL ANALYSIS

The green and red fluorescent signals for *ABL* and *BCR* genes were detected on blue background. Digital images were generated using a single-chip cooled color camera (Hamamatsu C5310). The camera was attached to a LEICA microscope equipped with filters for FITC, Rhodamine, and DAPI. The images were digitised (Matrix Vision frame-grabber) with 24-bit resolution (8 bits per colour) and processed using a Pentium 120 MHz computer. A special software, FISH 2.0, was developed for the automated analysis of the FISH-painted nuclei. FISH 2.0 runs under the MS-Windows environment and contains a number of procedures for image analysis.

The nuclei were detected using simple thresholding in blue colour. The optimal threshold was found by an automatic histogram analysis. The green and red signals

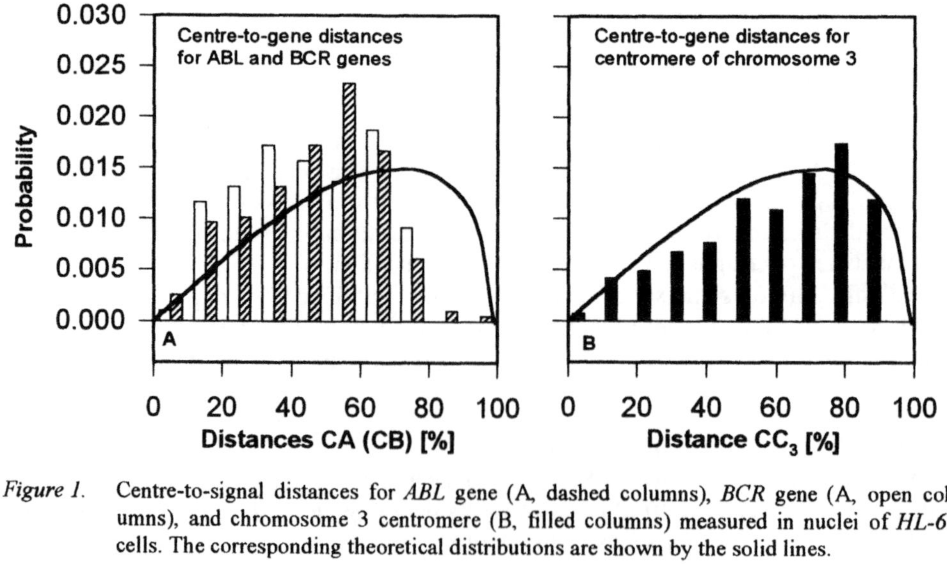

Figure 1. Centre-to-signal distances for *ABL* gene (A, dashed columns), *BCR* gene (A, open columns), and chromosome 3 centromere (B, filled columns) measured in nuclei of *HL-60* cells. The corresponding theoretical distributions are shown by the solid lines.

inside nuclei were detected by means of a modified watershed algorithm. The co-ordinates of the signals, the distance of the signals from the centre of nucleus (in percent of local radius), their intensity, and their area were computed. The significance of the differences between various distributions was calculated using the Student's t-test at $p=0.05$. All distances are expressed as the percentage of nuclear radius.

The following notation will be used throughout the remainder of this paper: distance between homologous *ABL* genes, AA; distance between homologous *BCR* genes, BB; minimum of four distances *ABL-BCR* in the nucleus, AB_m; centre of nucleus-to-*ABL* distance, CA; centre of nucleus-to-*BCR* distance, CB. All distances between genes were normalised to the average nuclear radius; centre-to-gene distances were normalised to the local radius (local radius is the distance between the nuclear centre and the membrane in the direction of the signal). The average values of the above-mentioned distances will be $\overline{AA}$, $\overline{BB}$, $\overline{CA}$, $\overline{CB}$, and $\overline{AB_m}$. Similar notation will be used also for c-*MYC* and telomere 14 genes (MM, TT, CM, CT, and MT_m). In the case of centromeres similar notation will be used: e.g. the centre-to-centromere of chromosome 3 distance will be CC_3.

3. Results

3.1. THE POSITIONS OF PROTOONCOGENES *ABL* AND C-*MYC* IN INTERPHASE NUCLEI

The positions of critical genes participating in oncogenic translocations in interphase nuclei of various human cell lines are not random. Theoretically, the average distance of a point, randomly placed in a sphere, from the centre of the sphere in the 2D projection is 58.8% of the radius. The average centre-to-gene distances of the *ABL* or *BCR* genes (CA, CB) determined for human lymphocytes, fibroblasts, bone marrow cells and *HL-60* cells are 35-50% of nuclear radius. On the other hand in the case of chromosome 3 centromere the centre-to-signal distances are in agreement with random distribution (Fig.1). Mutual distances between each *ABL* gene to one or another *BCR* gene were determined. The minimum value of the four measured distances (AB_m) per the cell nucleus was determined. The distribution of minimum distances between non-homologous *ABL* and *BCR* genes (Fig.2) does not correspond to Monte-Carlo calculations (solid line). According to the theory, the average value should be about $\overline{AB_m}=42\%$ of radius, meanwhile experimental average values for this type of distributions were $\overline{AB_m}=24\text{-}36\%$ for different cell lines used in our experiments. Another protooncogene investigated in our experiments was c-*MYC* which is involved in the translocation t(8;14) in Burkitt's lymphoma and other non-Hodgkin lymphomas. The position of the corresponding target gene *IGH* was determined using telomere 14q sequence situated in close proximity of the *IGH* gene. The average centre-to-gene distances for c-*MYC* and *IGH* in stimulated human lymphocytes are $\overline{CM}=52\text{-}57\%$ and $\overline{CT}=42\text{-}48\%$. The average minimum distance between heterologous genes was $\overline{TM}_m=33.5\pm2.1\%$. For human fibroblasts we observed substantially higher values for $\overline{CM}=54\text{-}61\%$, $\overline{CT}=58\text{-}59\%$, and $\overline{TM}_m=40.3\pm3.5$.

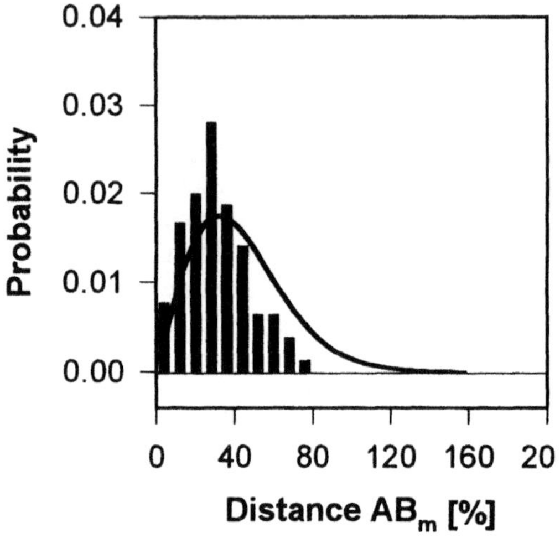

Figure 2. Distribution of minimum *ABL-BCR* distances (AB_m) for *HL-60* cells. Theoretical distribution calculated by Monte-Carlo simulation is shown by the solid line (2 points simulating *ABL* genes and 2 points simulating *BCR* genes were randomly placed in a sphere; 4 distances between heterologous genes were calculated in 2D projection and minimum distance was taken; the calculation was repeated 10^4 times and corresponding distribution was constructed).

3.2. POSITIONS OF CHROMOSOMES 9 AND 22 IN THE PROMETAPHASE ROSETTES

The condensation of the chromatin is observed already in the prometaphase where the rosettes of chromosomes can be found in cells fixed by paraformaldehyde. Using dual colour painting we were able to detect chromosomes 9 and 22 in the rosettes and determine their positions. Chromosomes 9 and 22 were found mostly on the opposite sides of the rosettes. Chromosomes 22 are predominantly nearer to the centre of cell nucleus as compared to chromosomes 9. In most prometaphase rosettes chromosomes 9 and 22 are near to each other. Owing to the high degree of condensation of chromosomes in prometaphase, the *ABL* and *BCR* genes are inside the painted areas and consequently also close to each other. We can expect that also in the subsequent stages of the mitosis the chromosomes 9 and 22 remain close to each other.

3.3. CHANGES OF THE TOPOLOGICAL PARAMETERS OF THE *ABL* AND *BCR* GENES DURING THE CELL CYCLE

Changes of the positions of *ABL* and *BCR* genes during the cell cycle were investigated in stimulated human lymphocytes. Human T-lymphocytes are naturally synchronised in G_0-phase of the cell cycle. After stimulation by PHA they proceed to G_1 and then to S and G_2 phases of the cell cycle. The positions of *ABL* and *BCR* genes were investigated in dependence on the time of stimulation by PHA. The time kinetics of different

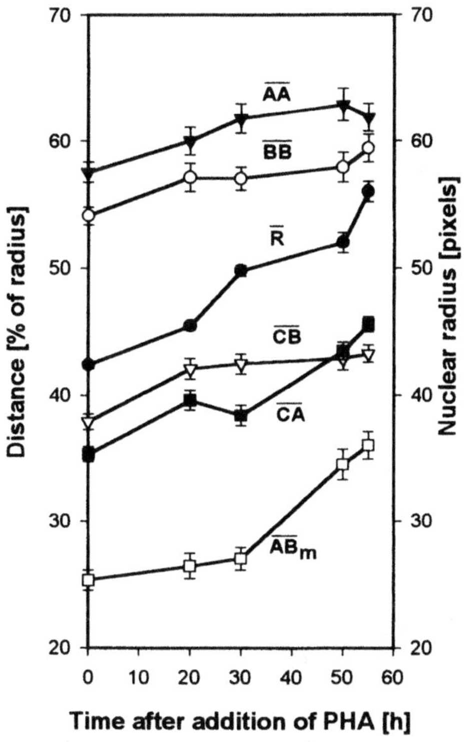

Figure 3. Changes of the topological parameters of the *ABL* and *BCR* genes in stimulated human lymphocytes in dependence on the time after the addition of PHA. R - nuclear radius [pixels]. The meaning of the symbols is explained in Materials and Methods.

parameters measured in hybridised nuclei are shown in Fig.3. Obviously, after the stimulation of lymphocytes the *ABL* and *BCR* genes move towards the membrane, their mutual distances increase, and the minimum distance between heterologous *ABL* and *BCR* genes increases. The mentioned parameters are normalised to the nuclear radius; the changes of the absolute values are even more pronounced.

3.4. THE POSITIONS OF CHROMOSOMES 9 AND 22 IN SISTER CELLS

The cells growing attached onto the slides can be regularly observed and their movement can be traced. In this way the sister cells (cells arising from the same division) can be determined and the location of their chromosomes in interphase nuclei can be investigated. The comparison shows how regular is the movement of individual genes during the cell cycle and to what extent the positions of genes are influenced by random motion. Thus, we measured the distance between chromosome 3 centromeres in the nuclei of sister cells. We observed different distances in the range between 20 and 150% of nuclear radius. However, there was good correlation between the distances of homologous centromeres in sister cells (Fig.4). In the second series of experiments we investigated the location of chromosomes 9 and 22 in human fibroblasts. We found that the positions of chromosomes in the nuclei of sister cells are symmetrical with similar

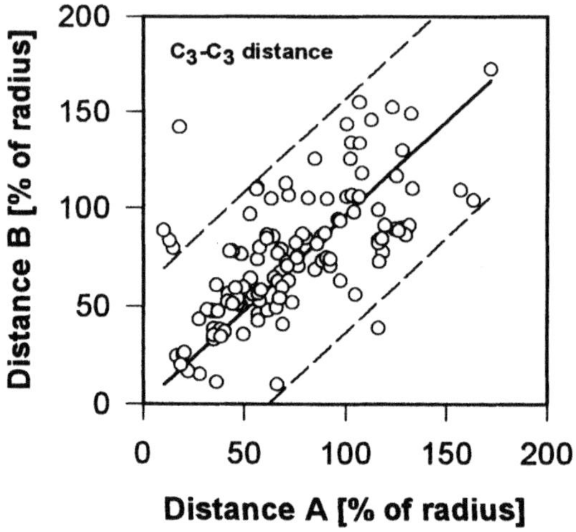

Figure 4. 2D distances between the chromosome 3 centromeres (C_3C_3) in the nuclei of human fibroblasts were measured in cell A and its sister, B. The distances are given in % of nuclear radius.

motifs in both cell nuclei. It seems from these experiments that the kinetics of the chromatin motion during the cell cycle is determined at the moment when the cell is born. The initial kinetics of chromatin movement is the same in both sister cells.

3.5. THE POSITION OF THE *ABL* AND *BCR* GENES IN IRRADIATED CELL NUCLEI

In lymphocytes stimulated after irradiation with 5 Gy of γ-rays the *ABL* and *BCR* genes are shifted to the centre of the nucleus. The distances between homologous genes (AA, BB) as well as those between heterologous genes (AB_m) are shorter [8]. It was shown in chapter 4.3 that the distances CA, CB, AA, BB, and AB_m increase with the time after the addition of PHA to human lymphocytes. Therefore it seems that the changes observed after irradiation might reflect just inhibition of chromatin changes caused by the stimulation of cells. It follows from these experiments that radiation, beside the induction of DNA lesions, decreases the distances between homologous (AA, BB) as well as heterologous (AB_m) genes during the first cell cycle of stimulated lymphocytes and consequently increases the probability of the induction of translocations.

3.6. TOPOLOGICAL CHARACTERISTICS OF THE *ABL* AND *BCR* GENES IN DIFFERENT CELL LINES AND DIFFERENT INDIVIDUALS

The values of average centre-to-gene distances, gene-to-gene distance and minimal gene-to-gene distances for heterologous genes are similar for different cell lines and different individuals. However, precise measurements showed statistically significant variations. The results (Table 1) indicate that the topological characteristics may differ for various individuals. As can be seen the distances between homologous genes in lymphocytes are smaller than those in bone-marrow cells; the distance AB_m is smaller in bone-marrow cells and even smaller in CML patients where also the centre-to-gene distances are the shortest. It is interesting to note that the centre-to-gene distances for the *ABL* gene are always similar to those for the *BCR* genes for different cell lines and also for different individuals.

TABLE 1. Geometrical parameters of chromatin structure in human cell lines derived from healthy individuals and patients suffering from chronic myeloid leukemia (CML). The values for different individuals are consecutively numbered.

Cell type	Number of measurement	$\overline{CA}$ [% of radius]	$\overline{CB}$ [% of radius]	$\overline{AA}$ [% of radius]	$\overline{BB}$ [% of radius]	$\overline{AB}_m$ [% of radius]
Bone marrow cells	1	47.8±1.5	46.8±1.5	70.5±0.26	75.5±1.29	27.40±1.81
	2	39.7±1.8	45.7±1.9	69.3±2.43	64.9±1.50	25.23±2.19
	3	43.9±1.8	40.8±1.7	60.2±1.24	64.6±1.18	29.97±1.92
	Average	43.8	44.4	66.8	68.3	27.52
Lymphocytes	1	45.1±1.8	44.9±1.8	66.3±3.8	65.2±4.8	30.6±1.8
	2	47.9±1.2	44.2±1.4	69.1±5.8	61.8±4.5	34.1±2.8
	3	36.9±1.0	43.4±1.3	63.5±2.7	58.4±2.5	35.2±1.6
	4	45.2±1.2	43.4±1.4	60.2±2.5	57.7±2.3	37.0±1.7
	5	44.7±1.2	40.9±1.3	58.4±2.4	62.7±2.4	31.7±1.7
	Average	43.9	43.4	63.5	61.2	33.7
CML patients	1	40.5±0.9	43.5±1.1	66.5±2.1	62.9±1.8	9.6±0.7
	2	51.0±4.2	45.6±3.5	75.4±7.8	68.7±5.5	15.4±3.9
	3	41.9±1.2	43.9±1.3	65.9±2.4	67.5±2.3	9.6±0.9
	4	42.3±1.4	43.3±1.3	66.2±2.5	67.0±2.3	7.7±0.9
	5	36.8±1.6	39.9±1.2	55.5±2.3	56.7±2.6	6.4±0.8
	6	42.4±1.2	39.6±1.3	61.2±2.6	61.3±2.7	8.6±1.5
	7	38.8±1.7	40.6±1.6	56.6±3.9	59.5±3.4	4.3±0.9
	8	42.4±1.6	41.6±1.8	59.0±2.9	59.8±3.2	5.5±0.7
	9	39.1±1.6	38.7±1.6	56.9±3.4	58.0±3.4	5.7±0.8
	10	36.9±1.7	37.6±1.7	54.7±4.3	53.1±3.9	8.1±2.3
	11	42.4±1.4	40.2±1.5	57.4±3.0	64.8±3.1	5.6±1.0
	Average	41.3	41.3	61.4	61.8	7.9

4. Discussion

This paper presents information on the location of two chromosomal loci (*ABL* and *BCR*) in the nuclei of different cell types and on the changes of these locations after irradiation of cells with γ-rays. The *ABL* and *BCR* loci represent critical regions for translocations giving rise to oncogenic transformation of blood cells leading to chronic myeloid and acute lymphoblastic leukemia. The casual relationship between ionising radiation and the induction of leukemia is evident: (i) the induction of leukemia after irradiation is much more frequent as compared to the induction of other malignancies (for example, the frequency of leukemias is almost doubled in Japan A-bomb survivors meanwhile the other malignancies are increased only by 7% [10]); (ii) *ABL/BCR* translocation is frequently found in leukemia [2]; (iii) ionising radiation induces *ABL/BCR* translocation [11]. Our investigations showed non-random distribution of the *ABL* and *BCR* genes which are frequently found in a very small part of the nuclear volume (about 35% of the nuclear volume). The distributions of the *BCR* gene are surprisingly very similar to those of *ABL* gene. It leads to non-random distribution of *ABL-BCR* distances in which shorter distances are very frequent even in normal (Ph negative) cells. This is obviously an important factor in the induction of *ABL-BCR* translocations. Further we observed that chromosomes 9 and 22 which involve *ABL* and *BCR* genes are located in close vicinity in prometaphase plates of human fibroblasts. In terms of territorial structure of interphase nucleus it means that the proximity of chromosomes 9 and 22 enables contact between *ABL* and *BCR* genes through the inter-domain space.

The investigation of the cell cycle kinetics of the positions of *ABL* and *BCR* genes, which was performed with stimulated lymphocytes, showed movement of both genes towards the membrane after stimulation, increase in the gene-to-gene distances for both *ABL* and *BCR* genes and also increase of the gene-to-gene distances between heterologous genes. This observation is in agreement with findings of Weimer [12] who observed similar movement of centromeres in stimulated lymphocytes. However in spite of the movement towards the membrane the *ABL* and *BCR* genes remain in the inner part of the nucleus (in stimulated lymphocytes the $\overline{\text{CA}}$ and $\overline{\text{CB}}$ distances approach to the value 45-50% of nuclear radius).

Another interesting experiment which is related to the mechanism and kinetics of the chromatin movement is the comparison of sister cell nuclei (the nuclei of two cells arising from the same cell division). Similar location of various genes was found in sister human fibroblasts. Obviously the cell cycle dependent movement of individual genes is predetermined to some extent at the moment of cell division. The random changes of the chromatin structure are either limited (below 1 μm) or slow (kinetics in the range of hours). What is the fate of the *ABL* and *BCR* genes during this "programmed movement" in the cell cycle is not known but owing to the fact that the *ABL-BCR* distance substantially increases (Fig.3), we can speculate that the "program" involves partial separation of the two sites during such an important event in the cell life like the replication of DNA.

It was shown earlier that the chromatin structure can be influenced by ionising radiation [13]. We observed the inhibition of the cell cycle dependent changes of the chromatin structure in irradiated and subsequently stimulated lymphocytes in both the first G_1 and $S(G_2)$ stages of the cell cycle. In irradiated cells both the *ABL* and *BCR* genes are frequently found near the centre of the cell nucleus, which increases the probability of their possible interaction. It seems, therefore, that radiation not only induces DNA lesions that are necessary prerequisites for the induction of translocations but it also changes the structure of chromatin and its kinetics in such a way that it increases the probability of interaction between the critical genes.

If the chromatin structure is correlated to the incidence of leukemia then the individual variations of the structure might reflect the individual sensitivity to the induction of leukemia. We have compared bone-marrow cells, lymphocytes, and bone-marrow cells in CML patients in different individuals (Table 1). The results show that the positioning of *ABL* and *BCR* genes in different cell lines is similar, however, precise measurements indicated statistically significant variations. These differences might play an important role in different dispositions of these individuals to the induction of *ABL/BCR* translocation and consequently also to leukemias.

Acknowledgement. This work was supported by the Grant Agency of the Czech Republic (Grants No. 202/96/1718 and 202/97/0874) and the Ministry of Education of the Czech Republic (Grant No. VS-97031).

5. References

1. Strauss, M., Lukáš, J., and Bartek, J.(1995) Unrestricted cell cycling and cancer, *Nature medicine* **1**, 1245-1246.
2. Rabbitts ,T.H. (1994) Chromosomal translocations in human cancer, *Nature* **372**, 143-150.
3. Kurzrock, R., Gutterman, J.L., and Talkpaz, M. (1988) The molecular genetics of Philadelphia chromosome-positive leukemias. N. *Engl. J. Med.* **319**, 990-998.
4. Lafage-Pochitaloff, M., Alcalay, M., Brunel, V., Longo, L., Sainty, D., Simonetti, J., Birg, F., and Pelicci, P.G. (1995) Acute promyelocytic leukemia cases with nonreciporcal PML/RARa or RARs/PML fusion genes. *Blood* **85**, 1169-1174.
5. de Boer, C.J., Schuuring, E., Dreef, E., Peters, G., Bartek, J., Kluin, P.M., and van Kreken, J.H.J. (1995) Cyclin D1 protein analysis in the diagnosis of mantle cell lymphoma, *Blood* **86**, 2715-2723.
6. Dyer, M.J.S., Lillington, D.M., Bastard, C., Tilly, H., Lens, D., Heward, J.M., Stranks, G., Morilla, R., Monard, S., Guglielmi, P., Kluin, Nelemans J.C., Hagemeijer, A., Young, B.D., and Catovsky, D. (1996) Concurrent activation of MYC and BCL2 in B cell non- Hodgkin lymphoma cell lines by translocation of both oncogenes to the same immunoglobulin heavy chain locus, *Leukemia* **10**, 1198-1208.
7. Preston, D.L., Kusumi, S., Tomonage, M., Izumi, S., Ron, E., Kuramoto, A., Kamada, N., Dohy, H., Matui, T., Nonaka, H., Thompson, E.E., Soda, M., and Mabuchi, K. (1994) Cancer incidence in atomic bomb survivors. Part III: Leukemia, Lymphoma and multiple myeloma, 1950-1987, *Rad. Res.* **137**, S68-S97.
8. Kozubek, S., Lukášová, E., Rýznar, L., Kozubek, M., Lišková, A., Govorun, R.D., Krasavin, E.A., and Horneck, G. (1997) Distribution of ABL and BCR genes in cell nuslei of normal and irradiated lymphocytes, *Blood* **89**, 4537-4545.
9. Lukášová, E., Kozubek, S., Kozubek, M., Kjeronská, J., Rýznar, L., Horáková, J., Krahulcová, E., and Horneck, G. (1997). Localisation and distance between ABL and BCR genes in interphase nuclei of bone marrow cells of control donors and patients with chronic myeloid leukemia, *Human Genet.* **100**, 525-535.
10. Mabuchi, K., Preston, D.L., Pierce, D.A., Shimizu, Y., and Vaeth, M. (1995) Fifty years after the A-bombs and the current cancer risk. In *Proceedings Radiation Research* 1895-1995, Würzburg, pp. 1109-1114

11. Ito, T. (1995) Induction of ABL-BCR fusion genes by in vitro X-irradiation. *Jpn. J. Cancer Res.* **84**, 105-112.
12. Where, R., Haaf, T., Kruger, J., Poot, M., and Schmid, M. (1992) Characterization of centromere arrangements and for random distribution in G0, G1, S, G2, G1, and early S, phase in human lymphocytes, *Hum. Genet.* **88**, 673-682.
13. Kozubek, M., Lukášová, E., Filková, H., Jarošová, M., Krahulcová, E., Horáková, J., Horneck, G., and Kozubek, S. (1995) Computer analysis of FISH-painted cellular genoms - spatial distribution of chromosomes in human cells. In Proceeding of the Symposium *Radiation Biology and its Application in Space Research* (Eds S. Kozubek, G. Horneck), Kiramo, Brno.

ADVANCES IN COMPUTER ANALYSIS OF RADIATION-INDUCED CHANGES IN THE HUMAN GENOME

M. KOZUBEK [1], S. KOZUBEK [2], E. LUKÁŠOVÁ [2], L. RÝZNAR [2], A. LIŠKOVÁ [2], J. AMRICHOVÁ [3] AND G. HORNECK [4]

[1] *Faculty of Informatics, Masaryk University, Botanická 68a, 60200 Brno, Czech Republic*

[2] *Institute of Biophysics, Acad. of Sciences, Královopolská 135, 61265 Brno, Czech Republic*

[3] *Faculty of Science, Masaryk University, Kotlárská 2, 60200 Brno, Czech Republic*

[4] *Deutsches Zentrum für Luft- und Raumfahrt, Institut für Luft- und Raumfahrtmedizin, Abt. Strahlenbiologie, 51140 Köln, Germany*

Abstract

Radiation-induced chromosomal aberrations can be detected using fluorescence *in situ* hybridization (FISH). Additional to numerical and structural aberrations changes in spatial arrangement of chromosomes can be detected by appropriate DNA-probes. Using FISH, both metaphase spreads as well as interphase nuclei can be analyzed, the latter being important for non-dividing cells.

The analysis of FISH-stained microscopic slides can be done either manually or by means of a computer. The second approach is obviously preferable as the analysis is quicker, more precise, not biased and can be automated. Moreover, the computer can work overnight and process large amounts of data. The possibility to analyze a lot of images (i.e. numerous metaphase spreads or interphase nuclei) is of great value for studies scoring rare events such as aberrations induced by environmental radiation. If, as for low doses, the percentage of damaged cells is very low a lot of data is needed for proper statistical scoring.

The computer-aided analysis of FISH-stained slides can be divided into three main parts: (i) search for the object (e.g. finding the metaphase spreads or interphase nuclei on the slide), (ii) capture of the image (the image must be recorded using a suitable CCD camera, digitized and saved into the computer memory), (iii) image analysis (computer must find the metaphase spreads or interphase nuclei, FISH-signals on them and evaluate the data).

This paper describes the automation of these three tasks and, subsequently, the whole process of FISH-stained slide analysis. The paper concentrates on the system developed in Brno based on a motorized Leica DMRXA microscope, a cooled CCD camera, a PC computer which drives both the microscope and the camera, and own

C. Baumstark-Khan et al. (eds.),
Fundamentals for the Assessment of Risks from Environmental Radiation, 243-248.

software for image acquisition and analysis. The system was already successfully used in several studies of radiation-induced changes in the human genome.

1. Introduction

Fully automatic acquisition and processing of FISH-stained slides is a very complex and difficult task. Several commercially available systems exist which carry out automated acquisition and classification of metaphase spreads painted by conventional cytogenetical methods (e.g. Giemsa staining), but no system is commercially available for automatic fluorescence imaging.

The first successful attempt to build such an automatic analyzer for FISH-stained image analysis was made at the Delft University by Netten *et. al.* [1, 2] who managed to build a system for automatic dot counting on FISH-painted interphase nuclei. The system worked, however, only for slides painted with one stain (counterstain and one centromeric or specific gene DNA-probe).

Recently, we have developed a similar system at the Institute of Biophysics in Brno. It analyses interphase nuclei and scores aberrations by detailed hybridization dot analysis. It was further improved in order to analyze also multi-color fluorescence images. Although it is optimized for analysis of centromeric and specific gene DNA-probes, it can help to examine nuclei painted with whole-chromosome probes or even FISH-painted metaphases (e.g. translocations studies).

The automation of the whole process of acquisition and analysis has been done step by step for about four years. First, the off-line analysis of the images was programmed [3], then image capture and storage to the hard disk was automated [4] and finally the localization and focusing of the nuclei was automated. The complete software was developed using the C programming language under Microsoft Windows. This system (or its earlier developed parts) was used in most of our studies [5, 6, 7].

2. System Description

The system consists of a motorized microscope, a cooled CCD camera and a computer which controls both the microscope and the camera. The computer is connected to the microscope via a serial interface and to the camera controller via a parallel port. The image is acquired using an appropriate frame-grabber plugged into the computer (see Table 1 for details).

The computer can easily control the exchange of objectives and filter-cubes, and the movements of shutter and of the microscope stage in x-, y- and z- directions. It can also modify exposure time of the camera and grab image from the camera into computer memory. The image can then be processed in two steps: during on-line and during off-line analysis.

The on-line analysis is performed immediately after acquisition of the image. The task of this quick process is to check the image quality. Poor quality images are not saved to the hard disk, whereas acceptable quality images (such which have passed the quality check) are stored for further off-line processing.

TABLE 1. Hardware and Software Specifications of the Analysis System #

Microscope	DMRXA, Leica
Stage x-y control	Motorized
z-axis control	Motorized
Nuclei localization	Manual or automatic
Nuclei focusing	Automatic
Image acquisition and storage	Automatic
Camera	Color cooled CCD (Hamamatsu C5310)
Digitizer (frame-grabber)	PC Image SRGB, Matrix Vision
Computer	PC Pentium (120 MHz, 40 MB RAM)
Operating system	Microsoft Windows 95
Software used	FISH 2.0 (our own software)
On-line analysis	Image quality control
Off-line analysis (stack of images)	Automatic FISH-signal analysis (e.g. overnight)
Speed of analysis (time per image)	Image localization: 5-10 seconds image focusing: 5–10 seconds image acquisition: 20-40 seconds on-line analysis: 0.5-1 seconds off-line analysis: 10-40 seconds

the system is regularly updated, the presented configuration is of October 1997

After the microscopic slide is processed, a stack of images (usually several hundreds) is located in a specific directory on the hard disk. Each image can now be shifted to another computer, where it can be processed off-line (while a new stack of images can be acquired from the microscope). Thus, the acquisition and on-line analysis of one stack of images can be performed in parallel with the off-line processing of another stack.

Both acquisition with on-line processing and off-line processing are fully automated and can run overnight. All results of the analysis are stored to the hard disk. The final analysis of the data is performed in a spreadsheet program.

3. Automatic Acquisition and On-line Analysis

The acquisition procedure consists of the following main steps: (i) location of the objects to be acquired, (ii) focusing on the counterstained image and (iii) acquisition of images for all stains including the counterstain using narrow-band filters. The microscopic slide is sequentially scanned until the complete rectangle area containing objects is processed. Focusing is performed using standard differential techniques (see Table 2). The whole algorithm works as follows:

1. Initialization of the microscope, the frame-grabber and the CCD camera controller
2. Moving the object stage to the first/next position to be examined
3. Setting the filter-cube for the selected counterstain
4. Setting proper camera exposure time for the counterstain
5. Opening of the shutter
6. Performing software-controlled focusing of the counterstained image
7. Acquisition of the focused counterstained image
8. Closing of the shutter
9. Checking the quality of the counterstained image using histogram analysis
10. For poor quality images continue with step 19
11. Save the counterstained image to the hard disk
12. For each stain used for hybridization do steps 13 to 18
    13. Setting the filter-cube for the given stain
    14. Setting proper exposure time for the given stain
    15. Opening of the shutter
    16. Acquisition of the image corresponding to the given stain
    17. Closing of the shutter
    18. Save the acquired image to the hard disk
19. If there are more microscope stage positions to move to, go to step 2
20. Shut-down of the microscope, the frame-grabber and the CCD camera controller

4. Automatic Off-line Analysis

During off-line analysis the images were reloaded one after the other from the hard disk and analyzed. The main tasks of the off-line analysis are (i) finding of the objects (interphase nuclei) using thresholding followed by binary image refinement and (ii) detection of the FISH-signals on the objects using a modified watershed algorithm (see Table 2 for details). For each image a corresponding text file containing the results of the analysis is written to the hard disk. The analysis ends when the whole stack of images (several hundreds) is processed. The algorithm can be summarized as follows:

1. Loading of the first/next counterstained image from the hard disk
2. Loading of all corresponding images for all stains used for hybridization
3. Subtraction of the CCD camera noise from all loaded images
4. Analyze the counterstained image as follows:
    5. Analyze the intensity histogram and calculate
        6. Mean background noise intensity and standard deviations
        7. Mean objects intensity and standard deviations
        8. Optimal threshold for consequent segmentation
        9. Perform segmentation of objects using the thresholding approach
        10. Smooth boundaries of detected objects
        11. Separate touching and overlapping objects using boundary analysis
        12. Compute attributes for all detected objects
        13. Save counterstained image analysis results to the hard disk

14. For each object found in the counterstained image do
    15. For each stain used for hybridization do
        16. Selection of the image corresponding to the stain
        17. Selection of the area of the corresponding counterstained object
        18. Find FISH-signals using a modified watershed algorithm
        19. Compute attributes for the detected FISH-signals
        20. Save results to the hard disk (append to previous results)
21. If there are any unanalyzed images left, go to step 1
22. End of the analysis

TABLE 2. Image Analysis Techniques used for the Analysis

focusing	differential approach - searching maximum of focus function given as a sum of differences in 8-neighbourhood - using parabolic curve fitting (least squares algorithm)
image quality check	intensity histogram analysis - determining: - mean background noise intensity with standard deviation - mean objects intensity with standard deviation - optimal threshold (local minimum between the two maxima)
noise removal	removal of CCD camera stripes produced by long exposure time - using typical stripes characteristics or - using median filter
linear filters	Gaussian smoothing (necessary for the consequent modified watershed algorithm)
basic segmentation	thresholding approach - using optimal threshold computed during image quality check
binary image processing	nuclei boundary smoothing - using modified Jarvis' march [8, 3] with suitable smoothing range touching and overlapping nuclei separation - using break pairs method [9, 3] or - using erosion followed by dilation without touching [9, 3]
segmentation of FISH-signals	modified (anti-)watershed algorithm [10, 3] - flood simulation by gradual thresholding
measuring	computation of standard parameters for all objects: - mass center (x-and y- coordinates) - size and perimeter - mean intensity and maximal/minimal intensity - roundness and radius for round objects - other parameters when necessary
classification	scoring of aberrations based on measured parameters - numerical scoring (in case of numerical aberrations such as trisomy) - spatial arrangement analysis (superimposed FISH-signals in case of trans-locations)
statistical analysis	standard statistical evaluation in a separate spreadsheet system (based on all data evaluated from one microscopic slide)

5. Discussion

We have built a unique system capable of acquiring and analyzing images from FISH-stained microscopic slides in a fully automatic mode. Although the system is quite new and still in the test phase it is of great value in our studies of radiation-induced changes in the human genome. It enables the processing of large quantities of image data with high accuracy and thus gives rise to good statistics even for rare events studies, especially when the percentage of aberrant cells is very low.

We have shown that combined usage of motorized microscopes and modern computers together with suitable image processing techniques makes the FISH-studies much easier, quicker, more precise, not biased and automatic. Moreover, using computer processing it is possible to perform also very complex studies otherwise impossible (e.g. complex statistical evaluations of enormous data quantities). The described approach to combine microscopy and image processing can help not only in studies of radiation-induced aberrations but also in other human genome projects.

Acknowledgment. This research was supported by the Grant Agency of the Czech Republic (Grants No. 202/96/1718 and 202/97/0874) and by the Ministry of Education of the Czech Republic (Grant No. VS-97031).

6. References

1. Netten, H., L.J. Vliet, H. Vrolijk, W.C.R. Sloos, H.J. Tanke, I.A. Young (1996) Fluorescent dot counting in interphase cell nuclei, *Bioimaging*, **4**, 93-106.
2. Netten, H. (1997) *Automated Image Analysis of FISH-Stained Cell Nuclei*, PhD thesis, Technische Universiteit Delft.
3. Kozubek, M. (1995) Analysis of images obtained by fluorescence in situ hybridization, MSc thesis, Masaryk University, Brno.
4. Kozubek, M., S. Kozubek, E. Lukášová, J. Amrichová, L. Rýznar, A. Lišková (1998) Semi-automated FISH-stained interphase nuclei analysis in human genome studies, In: *Fluorescence Microscopy and Fluorescent Probes*, vol. 2 (ed. J. Slavík), Plenum Press, New York, 183-189.
5. Kozubek, S., E. Lukášová, L. Rýznar, M. Kozubek, A. Lišková, R.D. Govorun, E.A. Krasavin, G. Horneck (1997) Distribution of *ABL* and *BCR* genes in cell nuclei of normal and irradiated lymphocytes, *Blood* **89**, 4537-4545.
6. Lukášová, E., S. Kozubek, M. Kozubek, J. Kjeronská, L. Rýznar, J. Horáková, E. Krahulcová, G. Horneck (1997) Localisation and distance between *ABL* and *BCR* genes in interphase nuclei of bone marrow cells of control donors and patients with chronic myeloid leukemia, *Human Genetics* **100**, 525-535.
7. Kozubek, S., E. Lukášová, J. Amrichová, M. Kozubek, L. Rýznar, A. Lišková, V. Kroha, J. Šlotová, G. Horneck: Comparison of three-dimensional and two-dimensional image acquisition and analysis of non-dehydrated and dehydrated nuclei, submitted
8. Preparata, F.P., M.I. Shamos (1985) *Computational Geometry*, Springer.
9. Joyce-Loebl (1985) *Image Analysis: Principles and Practice*, Joyce-Loebl.
10. Serra, J. (1982) *Image Analysis and Mathematical Morphology*, Academic Press.

CHROMOSOMAL DAMAGES AND MUTAGENESIS IN MAMMALIAN AND HUMAN CELLS INDUCED BY IONIZING RADIATION WITH DIFFERENT LET

R. D. GOVORUN

Joint Institute for Nuclear Research, Division of Radiation and Radio-biological Research
141980 Dubna, Russia

Abstract

On the basis of literature and proper data the inference was made about the essential role of structural chromosomal (and gene) damages in spontaneous and radiation-induced mutagenesis of mammalian and human cells on HPRT-loci. The evidence of the increasing role of these damages in the mutagenesis after the influence of ionizing radiation with high LET are considered. The consequences of HPRT-gene damages have been examined hypothetically. The heterogeneity of mutant subclones on their cytogenetical properties were revealed experimentally. The data reflect a phenomenon of the reproductive chromosomal instability in many generations of mutant cells. The mutagenesis of mammalian cells is also accompanied by the impairment of chromosome integrity with high probability as a stage of appropriate genome reorganization because of changed vital conditions.

1. Introduction

The problem of chromosomal damages and mutagenesis of irradiated mammalian and human cells attracts special attention in connection with environmental pollution and radioecological situation in the world.

At this time data about mutations in genes like HPRT, APRT, TK and others, which cause cell resistance to the action of some toxic analogues of the purines and the pyrimidines, are obtained. We are investigating the mutants of the HPRT-locus and data concerning this gene are presented below. The human gene is located at the X-chromosome, it includes 44 kb of genomic DNA and has 9 exons. The HPRT-gene codes for hypoxanthine guanine phosphoribosyltransferase, which catalyzes the condensation of 5'-phosphoribosyl-1-pyrophosphate and the purine bases of hypoxanthine and guanine during the synthesis of the mononucleotides, utilized by the cells during DNA synthesis.

C. Baumstark-Khan et al. (eds.),
Fundamentals for the Assessment of Risks from Environmental Radiation, 249-258.

As known the mammalian cells can use exogenic purines or synthesize them *de novo*. They can also incorporate exogenic purine analogues as 6-thioguanine (6-TG), which are toxic for them. In selective medium containing 6-TG only such cells survive and form colonies, which have an impaired or absent HPRT activity as a consequence of mutations in the HPRT-locus. This fact allows the isolation of individual surviving colonies and the investigation of HPRT-mutant clones.

2. Results and Discussion

As our investigations have shown [1] the types of dose effect curves for mutation induction and chromosomal aberrations, as well as survival curve of Chinese hamster cells depend on the radiation's LETs. After irradiation with γ-rays and heavy ions with LET up to about 80 keV/μm induction curves are nonlinear and approximate to linearity at higher LET values. The RBE–LET dependence is described by a curve having a maximum at 80–100 keV/μm (Figure 1). The close values of RBE factors for mutation induction and chromosomal aberration frequency lead to the speculation, that the same primary damage, namely double-strand breaks of DNA cause these changes. As a result of radiation exposure both chromosomal aberrations and microstructural damages of DNA leading to gene mutations are formed.

Literature data indicate the role of structural DNA damages in the induction of gene mutations in mammalian and human cells [2-14]. The data in Table 1 allow the assessment of the nature of HPRT-mutations in these cells. The changes were not revealed by blot-analysis for the majority of spontaneous mutants (up to 85%). For X- and γ-irradiation the fraction of altered mutants becomes prevailing (up to 55–100%). Partial or total gene deletions prevail for both types of mutants, but after irradiation the fraction of total deletion is raised. Data obtained by sequencing or PCR methods show a high degree of diversity for the gene changes.

Thus, the loss of one to two exons or parts of it, frameshifts, small deletions, splice-mutations and base-pairs substitutions were observed among spontaneous mutants from human T-lymphocytes [15]. It was estimated that deletions up to 22 kb from the 5'-end and 29 kb from the 3'-end of the HPRT-gene did not infringe the viability of spontaneous mutants [7]. There is also evidence that breaks in the HPRT-gene are of non-random distribution for spontaneous and radiation-induced mutants [5, 13]. Radiation-induced mutants of Chinese hamster cells display a duplication of a part of the 5'-end of the gene and gene deletions in different locations [7, 13].

For isolation of HPRT-mutated clones it is necessary to express the premutational damage to a mutation. This is achieved by growing the irradiated cells for several division cycles (3–5 up to 10 and more: the „expression time") under normal physiological conditions before the selection of mutated cells is achieved by cultivation in medium containing 6-TG. It is known, that the structural chromosomal aberrations form during the first postirradiation mitosis. Moreover, gene deletions having sizes less than the microscope resolution allows to visualize will develop. The minimal size to be visualized is about 1.2–1.3 kb, which comes close to the appropriate size necessary for a successful blot-analysis (1.5–14 kb) [4, 13]. It is shown earlier, that for the majority of

TABLE 1. Spontaneous and radiation-induced HPRT-mutations in mammalian and human cells.

Types of mutants	Cell line	mutants #	Normal Southern patterns %	Altered Southern patterns: Total number N	Total number %	Mutation types: Total gene deletion	Partial gene deletion	Exon deletion	References
	Human lym-	14	64	5	36	-	5	-	[3]
Spontaneous	phoid cell line,	35	43	20	57	1	19	-	[4]
mutants	TK-6	85	61	33	39	10	21	-	[5][1)]
	Human T-	21	43	12	57	-	12	-	[6]
	lymphocytes	326	85	48	15	8	40	-	[7]
		105	87	14	13	-	14	-	[8]
	Chinese ham-	52	86	7	14	1	6	-	[9]
	ster cells, V79	44	82	8	18	6	2	-	[10]
X-rays: 1.5 Gy	Human	28	46	15	54	15		-	[3]
X-rays, 1 Gy	lymphoid cell	33	48	17	52	10	7	-	[4]
^{125}I dUrd: 0.5	line,	68	16	57	84	28	29[2)]	-	[4]
and 0.1 µCi/ml [3)]	TK-6	21	38	13	62	10	3[2)]	-	[4]
X-rays	Human T-	33	48	17	52	3	14	-	[11]
X-rays	lymphocytes	24	25	18	75	-	-	10	[12]
γ-, α-radiations	Chinese ham-	15	-	15	100	-	14[2)]	-	[13][1)]
X-rays: 2–4 Gy	ster cells, V79	113	31	78	69	46	32	-	[9]
	line CHO	89	11	79	89	54	25[4)]	-	[14]

1) single partial duplications are observed also; 2) and rearrangements (deletion sizes – 1,5–14 kb)
3) it is similar to 1 Gy of X-rays; 4) deletion sizes – 5–13 kb

HPRT-mutants the partial or total gene deletions were important. Data from metaphase analysis of such unstable chromosomal aberrations, as fragments and especially interstitial deletions indirectly confirm the role of structural gene damages in mammalian and human cell mutagenesis. Cytogenetic size analysis of interstitial deletions attracts attention: very small up to large circular or oval structures, acentrical rings. Structural gene deletions with different sizes are also registered among interstitial chromosomal deletions. Obviously, the probability of multiple chromosome breaks, which are a prerequisite for formation of such structures, increases after irradiation with heavy ions having high LETs. In our investigation [16] the number of interstitial deletions and acentric fragments increased extremely after irradiation with heavy ions reaching a maximum at about 100 MeV/µm (Figure 2).

The role of structural chromosomal damages in mutagenic processes in mammalian and human cells can indirectly be confirmed by the formation of translocations. In this case genetic matter is not lost and these aberrations are kept for a long time in the cell

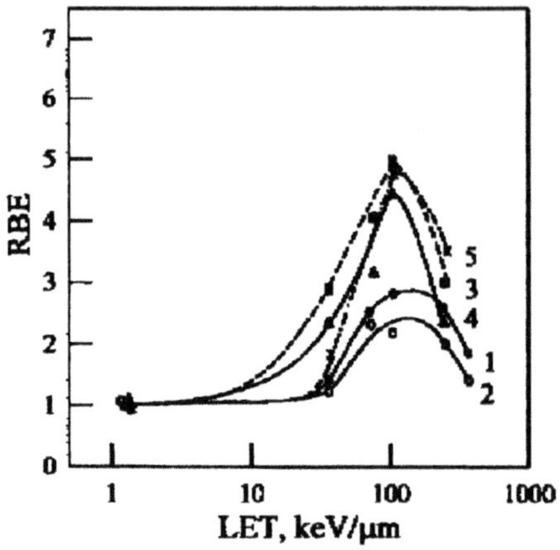

Figure 1. RBE of heavy ions as a function of LET on survival (1 – 10 % level of survival, 2 – D_0), on induction of mutations (3 – mutation frequency 5×10^{-5}, 4 – mutation frequency 10×10^{-5}), on induction of chromosomal aberrations (5). (Chinese hamster cells, V79).

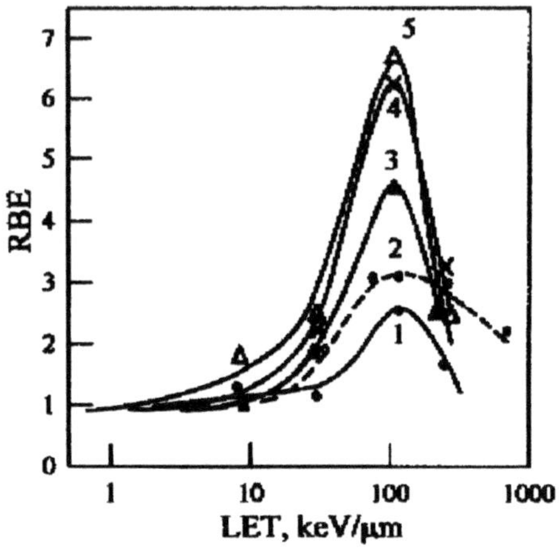

Figure 2. RBE of heavy ions as a function of LET on cytogenetical criterions: 1 – dicentrics and acentric rings, 2 – survival, 3 – total chromosomal aberrations, 4 – interstitial deletions, 5 – double acentric fragments. (Chinese hamster cells, V79).

population. Subsequently such cells can give rise to the development of mutational processes in the organisms and to cancerogenesis. Our investigations of such aberrations in irradiated human lymphocytes using the FISH-method [17, 18] varify the high frequency of chromosome-1 translocations (Figure 3). Their number increased nonlinear with γ-ray doses. After ^{14}N-ion irradiation (RBE of ^{14}N-ions with LET ~80 keV/μm exceeds 3) the dose dependency approximates to linearity.

Mutant clones which show no visible changes in the blot-analysis are supposed to display true point gene mutations or microdeletions. Radiation-induced gene mutations are essentially identical to spontaneous ones. As a result of the destruction of pyrimidine and imidoazole rings and the loss of bases due to broken glycoside bonds, different single transversions and transitions in DNA codons of genes must arise. As a result of base modification only single transitions arise: as a consequence of desamination the adenine can transform to hypoxanthine, which codes for guanine, and the cytosine can transform to uracil, which codes for thymine. Accordingly base-pair substitutions of AT to GC and GC to AT will occur during DNA replication. The majority of transitions and also some transvertions have no influence on matrix activity of DNA, as they do not interrupt transcription and replication processes. Phenotypically any transitions and transversions result in a amino acids replacement in the enzyme. For mammalian and human cells it is not always possible to distinguish small chromosomal damages (for example, interstitial and terminal microdeletions, insertions) from point mutations,

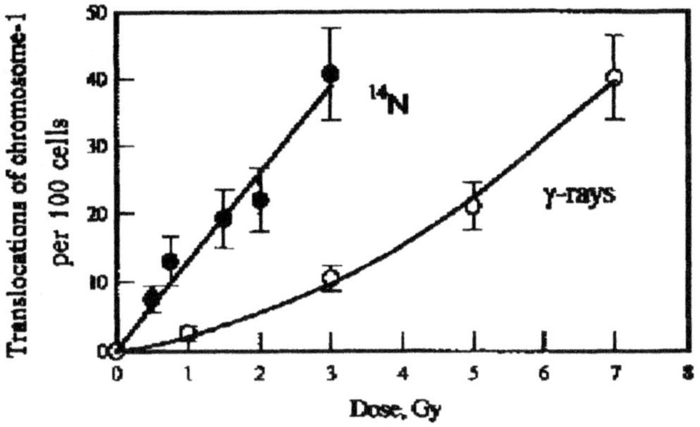

Figure 3. The frequency of translocations of chromosome -1 in human lymphocytes as a function of γ-rays and heavy ion doses.

which inactivate this gene. Phenotypically such structural mutations are displayed in a loss of individual amino acids or, on the contrary, their additional occurrence in enzyme molecules.

As has been marked earlier, the realization of a DNA damage as a mutation leading to 6-TG resistance requires several cell division cycles in normal physiological growth conditions, before the cells will be passed to selective medium. The damaged base, which remains unrepared during the first replication cycle, will cause a base substitution in the newly synthesized DNA (transition or transvertion) which in turn will lead to the occurrence of a potentially mutated cell (Figure 4), additionally a normal daughter-cell will be born simultaneously. After the second division cycle the cell having a changed base pair in its DNA, will give rise to one potentially mutant and one mutant cell, the normal cell will give rise to two normal cells. After seven cell divisions (provided that all cells will be viable) 1-potential mutant, 63 mutant and 64 normal cells will appear in the colony under normal growth conditions. During cultivating in medium containing 6-TG only the cells carrying mutations of the HPRT-gene which results in impairment or absence of the ability to incorporate this toxic analogue have the chance to survive and to form colony. The appearance of the cells with different kinds of HPRT-gene mutations (dependent on the types of the base damage) in growing populations is not excluded. If bases are modified, all cells will have the same mutation, but if bases were degraded or a break of the glycoside bond resulted in a loss of the bases, different transvertions and transitions are made possible and daughter-cells with different HPRT-gene mutations will arise from the same ancestor (potentially mutant) cell. Thus non-uniform cell populations having different mutagenic properties will emerge.

Mutagenic events in mammalian and human cells are obviously connected with chromosomal changes. These changes may either concern chemical organization or impair structural integrity of chromosomes. On a cellular level, the expression of gene damages is complex, possible consequences which arise from HPRT-gene damages are

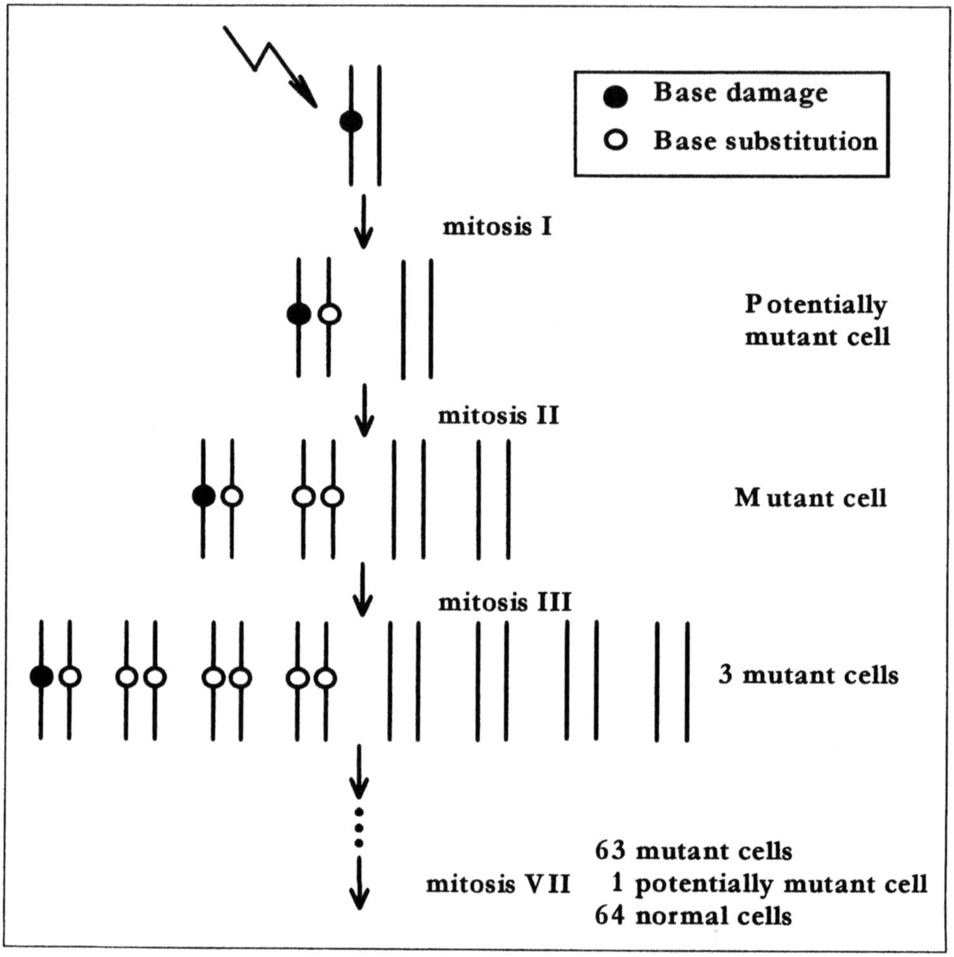

Figure 4. The formation of mutant cells.

demonstrated in figure 5. HPRT-mutations can be connected (i) to intragenic changes ("point" mutations, microdeletions, inversions, addition of a gene sites) and (ii) with deletions (total gene loss or partial gene loss, if structural X-chromosomal deletions occur). In case of base damages the majority of emerging transitions or transvertions do not interrupt the processes of transcription and translation. As a result the synthesis of normal enzyme according to the degenerative genetic code (1) can happen. Chain termination can occur, as a consequence of appearing terminate codons and accordingly enzyme synthesis will be suppressed (2); substitution of amino acid in the enzyme may cause total loss (3) or reduction (4) of the enzyme activity. Three nucleotides covering gene microdeletions of terminal or interstitial locations will result in a loss of one codon (terminal or within the m-RNA) and thus enzyme activity can be absent (5) or impaired

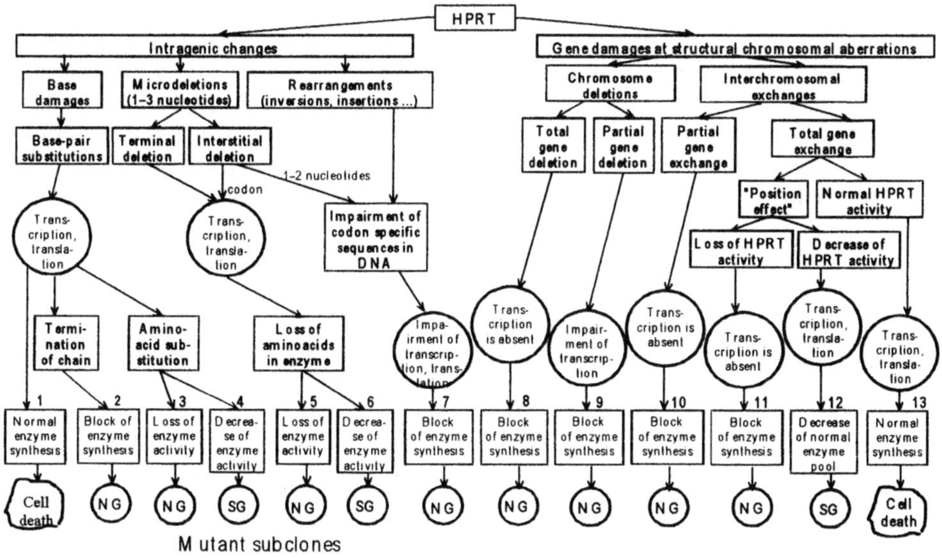

Figure 5. The consequences of HPRT gene damages.

(6) as well. Interstitial deletion with sizes of 1-2 nucleotides and gene microreorganization (like inversions, insertions and others) will impair codon specific sequences on large parts of the DNA resulting in "frameshifts" at transcription level and impairment of translation. In such cases, enzyme synthesis is not realized (7). The majority of X-chromosomal structural aberrations in the HPRT-gene region which are connected with total or partial deletions result in the absence of enzyme synthesis (8, 9). The same is true for interchromosomal partial exchange of the gene (10). Chromosomal exchanges which do not lead to a damage of the HPRT-locus itselfes can influence enzyme synthesis as well, and HPRT-activity may either be totally absent due to "the position effect" (11) or reduced due to a decreased pool of intact enzymes (12). If the HPRT-enzyme synthesis is not impaired, the HPRT-activity will survive (13) and as a consequence, the cell will die.

The expression of gene damages requires a number of cell generations to be born, as the impairment of gene activity and as a consequence the deficit in the appropriate enzyme activity has to be complete for cellular survival. The loss of the active enzyme can be regulated on the level of transcription, translation and posttranslation processes. The fate of irradiated cells during cultivating on selective medium will be different. Cells which are free of damages, and also mutant cells possessing normal enzyme activity will die (1, 13). Others will be capable to form colonies (2–12), but their phenotypical properties will vary. It may be assumed, that those cells which synthesize modified enzymes having a reduced activity or which synthesize less intact enzymes (4, 6, 12), will form colonies slower, than others. Such mutant population will be composed of cells, which can not incorporate sufficient purine analogues during the cell cycle. We observed such a delay of HPRT-mutants' growth for Chinese hamster V79-cells. The fraction of slowly growing mutants increased from 25 up to 85% with increasing the dose from 2 to 7 Gy. Spontaneous mutants exhibiting a slow growth [19] and a partial HPRT activity [4] have been seen for human lymphoid TK-6 cells.

Moreover, our experiments have shown that a further consequence of mutations is an increase of chromosome instability in mutant subclone populations [20]. The mutants were highly heterogeneous concerning the frequency of aberrant cells and the numbers of chromosomal aberrations (Table 2). Classifying diploid mutans according to these parameters we distinguished several groups of mutants: group I did not differ essentially from the control. The single radiation-induced mutants of this group were found with the aberration frequency 2 to 4 times less. For mutants of the group II the level of aberrant cells was about 2 to 4 times higher than for the control. The mutants of the group III have very high level of chromosomal damages (>30% of aberrant cells). Percentage of radiation-induced mutants of the group I decreases from 70 to 33% but the group II increases from 14 to 50% when doses increased from 2 to 7 Gy (Figure 6). The mutants of the group III in fact were found only for spontaneous ones. There was only one radiation-induced mutant (7 Gy) which differed from all others according to its chromosomal aberration spectrum: the stable translocation was found in 88% of its descendant cells, the level of other aberrations did not differ from the mutants of group II.

Moreover, 15-20% tetraploid mutants were additionally observed. They were characterized by the increased level of chromosomal aberrations compared to diploid controls. The frequency of chromosomal aberrations in the descendants of radiation-induced tetraploid mutants (2 and 3 Gy), was a few times higher. The only tetraploid mutant, induced by a dose of 7 Gy differs from all others: the stable translocation was observed in almost 90% of the cells, the level of other aberrations was high as well. Extremely high levels of chromosomal aberrations are observed for spontaneous tetraploid mutants (about 50% of aberrant cells).

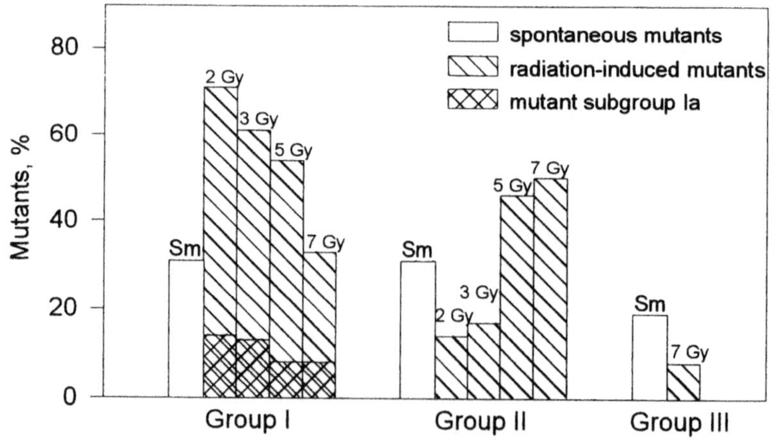

Figure 6. The percentage of diploid mutants in groups.

TABLE 2. Chromosomal aberrations in cells of diploid mutants.

Types of mutants	Dose, Gy	Group	Number of mutants	Cells with aberrations, %	Total aberrations per 100 cells	Fraction of mutants, %
Control	--	--	14*	8.7 ± 0.8	9.9 ± 0.8	78
Spontaneous mutants	--	I b	5	8.6 ± 1.3	10.4 ± 1.4	31
		II	5	16.8 ± 1.8	21.2 ± 2.1	31
		III	3	39.7 ± 3.7	63.4 ± 4.6	19
Radiation-induced mutants	2	I a	1	4.0 ± 2.0	5.0 ± 2.2	14
		I b	4	9.8 ± 1.6	10.3 ± 1.6	57
		II	1	29.0 ± 5.4	44.0 ± 6.6	14
	3	I a	3	3.0 ± 1.0	3.0 ± 1.0	13
		I b	11	9.0 ± 0.9	10.2 ± 1.0	48
		II	4	19.8 ± 2.2	28.9 ± 2.3	17
	5	I a	1	4.0 ± 2.0	4.0 ± 2.0	8
		I b	6	7.8 ± 1.1	9.5 ± 1.3	46
		II	6	18.7 ± 1.8	22.3 ± 1.9	46
	7	I a	1	2.0 ± 1.4	2.0 ± 1.4	8
		I b	3	7.0 ± 1.5	7.3 ± 1.6	25
		II	6	17.5 ± 1.7	23.7 ± 1.9	50
		III**	1	88[18] ± 9.4	110[22]±10.5	8

* Normal subclones;
** 88 % of cells have interchromosomal symmetrical exchange; [ ] other types of aberrations

The heterogeneity of the cytogenetical properties and the increase of chromosomal instability of mutants were a result of HPRT-mutations or other genetical changes which additional occur in mammalian cells. The data reflect the phenomenon of reproductive instability of chromosomes for many generations of mutant cell. The increased level of chromosomal aberrations in many mutants cells shows that the impairment probability for chromosomal integrity raises in mutants inevitably, possibly as a consequence of genome reorganization, necessary for the changed vital conditions.

This analysis confirms an essential role of structural chromosomal damages (and accordingly structural gene damages) for radiation-induced mutagenesis of mammalian and human cells. At the same time mutagenesis is accompanied by chromosomal instability of mutant cell populations. This property is a suitable criterion for the assessment of risks from environmental radiation.

References.

1. Shmakova, N.L., Krasavin, E.A., Govorun, R.D., Fadeeva, T.A., Koshlan', I.V. (1997) Lethal and mutagenic effects of irradiation of various LET on the mammalian cells, *Radiation Biology and Radioecology*, **37**, 85–91 (Russ).

2. Sankaranarayanan, K. (1991) Ionizing radiation and genetic risks. III. Nature of spontaneous and radiation-induced mutations in mammalian *in vitro* systems and mechanisms of induction of mutations by radiation, *Mutation Research*, **258**, 75–97.
3. Liber, H.L., Call, K.M., Little, J.B. (1987) Molecular and biochemical analysis of spontaneous and X-ray-induced mutations in human lymphoblastoid cells. *Mutation Research*, **178**, 143–153.
4. Whaley, J.M., Little, J.B. (1990) Molecular characterization of hprt mutants induced by low- and high -LET radiation in human cells, *Mutation Research*, **243**, 34–35.
5. Gennett, J.N., Thilly, W.G. (1988) Mapping large spontaneous deletion endpoints in the human HPRT gene. *Mutation Research*, **201**, 149–160.
6. Turner, D.R., Morley, A.A., Haliandros, M., Kutlaca, R., Sunderson, B.J. (1985) *In vivo* somatic mutants in human lymphocytes frequently result from major gene alterations, *Nature*, **315**, 343–345.
7. Niclas, J.A., Hunter, T.C., O'Neill, J.P., Albertini, R.J. (1989) Molecular analysis of *in vivo* hprt mutations in human T-lymphocytes. III. Longitudinal study of hprt gene structural alterations and T-cells clonal origins, *Mutation Research*, **215**, 147–160.
8. Bradley, W.E.C., Garean, J.L.P., Seifert, A.M., Messing, K. (1987) Molecular characterization of 15 rearrangements among 90 human in vivo somatic mutants shows that deletions predominate, *Mol. Cell Biol.*, 7, 956–960.
9. Breimer, L.H. (1988) Ionizing radiation-induced mutagenesis. *Br. J. Cancer*, **57**, 6–18.
10. Thacker, J, Ganesh, A. (1989) Molecular analysis of spontaneous and EMS-induced mutations of the *hprt* gene in hamster cells. *Mutation Research*, **210**, 103–112.
11. Skulimowski, A.W., Turner, D.R., Morley, A.A., Sanderson, B.J.S., Haliandros, M. (1986) Molecular basis of X-ray-induced mutations at the HPRT locus in human lymphocytes *Mutation Research*, **162**, 105–112.
12. O'Neill, J.P., Hunter, T.C., Sullivan, L.M., Niclas, J.A., Albertini, R.J. (1990) Southern blot analysis of human T lymphocyte mutants induced in vitro by gamma irradiation. *Mutation Research*, **240**, 143–149.
13. Thacker, J., Fleck, E.W., Morris, T., Rossiter, B.J.F., Morgan, T.L. (1990) Localization of deletion break-points in radiation-induced mutants of the hprt gene in hamster cells, *Mutation Research*, **232**, 163–170.
14. Morgan, T.L., Fleck, E.W., Poston, K.A., Denovan, B.A., Newman, C.N., Rossiter, B.J.F., Miller, J.H. (1990) Molecular characterization of X-ray-induced mutations at the HPRT locus in plateau-phase Chinese hamster ovary cells, *Mutation Research*, **232**, 171–182.
15. Rossi, A.M., Thijissen, J.C.P., Tates, A.D., Vrieling, H., Natarajan, A.T., Lohman, P.H.M., Van Zeeland, A.A. (1990) Mutation affecting RNA splicing in man are detected more frequently in somatic than in germ cells, *Mutation Research*, **244**, 353–357.
16. Nasonova, E.A., Govorun, R.D., Krasavin, E.A. (1989) Induction of structural chromosome mutations in mammalian cells by accelerated heavy ions. In: Proc. of the Workshop on Genet. effects of charged particles, Dubna. 4–6 Oct., 1988. JINR D19–89–143, 169–182 (Russ.).
17. Repin, M.V., Govorun, R.D., Krasavin, E.A., Lukášová, E., Kozubek, S. (1996) Stable and unstable chromosome aberrations in human lymphocytes induced in vitro by γ-rays, *Radiation Biology and Radioecology*, **36**, 848–851 (Russ.).
18. Repin, M.V., Govorun, R.D., Lukášová, E., Kozubek, S., Krasavin, E.A. (1997) FISH-analysis of stable and unstable chromosome-1-aberrations induced by γ-rays and accelerated heavy ions in human blood lymphocytes. In: Proc. Sissakian Memor. Sympos. "Problems of Biochemistry, Radiat. and Space Biol." 22–25 January 1997, Moscow, Dubna, 132.
19. Yandell, D.W., Dryja, T.P., Little, J.B. (1986) Somatic mutations at a heterozygous autosomal locus in human cells occur more frequently by intragenic structural alterations, *Somat. Cell Mol. Genet.*, **12**, 255–263.
20. Govorun, R.D., Koshlan', I.V., Krasavin, E.A., Shmakova, N.L. (1996) Mutagenic influence of γ-radiation on Chinese hamster cells. Cytogenetic characterization of HPRT mutants, *Radiation Biology and Radioecology*, **36**, 852–859 (Russ).

RADIATION-INDUCED MUTATION AT MAMMALIAN MINISATELLITE LOCI

Y.E. DUBROVA [1,2], J. BROWN [1], V.N. NESTEROV [3], N.G. KROUCHINSKY [3], V.A. OSTAPENKO [3], M. PLUMB [4], P. BOIS [1] AND A.J. JEFFREYS [1]

[1] *Department of Genetics, University of Leicester, Leicester LE1 7RH, UK*
[2] *Vavilov Institute of General Genetics, Moscow B-333, Russia*
[3] *Research Institute for Radiation Medicine, Mogilev, 212004, Belarus.*
[4] *MRC Radiation and Genome Stability Unit, Harwell, Oxon, OX11 0RD, UK*

Abstract

Germline mutation at human minisatellite loci has been studied among children born in areas of the Mogilev district of Belarus heavily polluted after the Chernobyl accident. The overall mutation rate was twice as high in the exposed group as in control families. An elevated rate was seen at all three independent sets of minisatellites (detected separately by multi-locus probes 33.15, 33.6 and six single-locus probes), indicating a generalised increase in minisatellite germline mutation rate in the Belarus families. Individual radiation doses for chronic external and internal exposure to ^{137}Cs were estimated for each parent in the exposed group. Mutation rate was significantly higher in the more exposed Belarus families, providing direct evidence for radiation induction of germline mutation. Minisatellite mutations were also scored in the progeny of acute γ-irradiated and non-irradiated CBA/H male mice. Mutation induction was only attributable to the spermatogonia stage of spermatogenesis and the frequency of mutations increased linearly with radiation dosage from 0.5 Gy to 1Gy. Our data suggest that minisatellite loci provide the efficient system for the monitoring of germline mutation in humans.

1. Introduction

Estimating the genetic hazards of radiation and other mutagens in humans depends on an appropriate experimental system for mutation scoring. The shortcoming of current approaches for genetic monitoring is the necessity to use very large numbers of individuals (often more that 100,000) to detect increases in mutation rate [1]. Tandem repeat minisatellite loci have several potential advantages for monitoring germline mutation in man. Firstly, the frequently very high rate of spontaneous mutation altering allele

C. Baumstark-Khan et al. (eds.),
Fundamentals for the Assessment of Risks from Environmental Radiation, 259-264.

length (repeat copy number) [2, 3] provides a system capable in principle of detecting induced mutations in relatively small population samples. Secondly, the probable selective neutrality of minisatellites ensures that mutants will not be lost prenatally, as can occur at protein loci [1]. Finally, it is possible to extend mutation screening to germinal and somatic DNA using single molecule PCR approaches to detect and quantify the incidence of mutants in an individual [4]. We have previously obtained evidence that germline mutation at minisatellites is remarkably sensitive to ionising radiation, both in mice and in humans [5, 6]. To obtain detailed information on the effects of radiation on the mammalian germline, we have extended this analysis to more exposed human families and experimentally irradiated mice.

2. Population Groups and Mutation Scoring

Blood samples were collected from 48 additional families (father, mother, child) which, together with the 79 families studied previously [6], gave in total 127 families inhabiting the heavily polluted rural areas of the Mogilev district of Belarus (Bychovskii region, 19 families; Cherikovskii region, 40 families; Krasnopol'skii region, 38 families; Slavgordskii region, 23 families; Krichevskii region, 4 families and Mogilevskii regions; 3 families). The whole cohort is composed of children born between February and September 1994 for whom both parents were continuously resident in the Mogilev district from the time of the Chernobyl accident. For all parents the local level of surface contamination by ^{137}Cs exceeded 1 Ci/km^2. The control sample consisted of 120 non-irradiated Caucasian families from the United Kingdom.

DNA fingerprints were produced using two multi-locus minisatellite probes 33.6 and 33.15 [7] and eight hypervariable single-locus minisatellite probes MS1, MS31, MS32, CEB1, CEB15, CEB25, CEB36 (loci *D1S7, D7S21, D1S8, D2S90, D1S172, D10S180, D10S473*, respectively) and B6.7 (located on chromosome 20q13), chosen for their relatively high spontaneous mutation rate ([2, 3, 8, 9] and G. Vergnaud, unpublished). These probes provided sufficient information to verify the parentage of all children analysed, even in the presence of mutation [3].

3. Results

3.1. MUTATION RATE

To determine the extent to which multi-locus and single-locus minisatellite probes detect overlapping sets of bands, the same blots were hybridised with all ten probes. In most cases minisatellite probes detected sets of completely independent bands, whereas some bands detected by probe 33.15 are derived from minisatellites MS1 and MS31. We therefore present three different estimates of mutation rate based on independent sets of minisatellites (Table 1).

For each of the ten probes, the spontaneous mutation rate determined from the control families (Table 1) was similar to that previously measured in Caucasian populations ([2, 3, 8, 9] and G. Vergnaud, unpublished). In contrast, the frequency of mutant bands was elevated in the offspring of irradiated parents.

TABLE 1. Mutation rates in control and exposed populations

Probe	Control group, 120 families			Exposed group, 127 families			Ratio	P^a
	No. bands	No. mutations	Mutation rate per band	No. bands	No. Mutations	Mutation rate per band		
33.15	1903	19 (7+2)[b]	0.0100	2221	44 (8+5)[b]	0.0198	1.98	0.0073
33.6	2492	15	0.0060	2977	31	0.0104	1.73	0.0499
B6.7	218	5	0.0229	250	11	0.0440	1.92	0.1917
CEB1	218	16	0.0734	250	28	0.1120	1.53	0.0927
CEB15	218	4	0.0184	250	7	0.0280	1.53	0.3880
CEB25	218	3	0.0138	250	12	0.0480	3.49	0.0470
CEB36	218	2	0.0092	250	4	0.0160	1.74	0.5319
MS32	218	3	0.0138	250	2	0.0080	0.58	0.5790
Total	1308	33	0.0252	1500	64	0.0427	1.69	0.0067
Grand total	5099	56	0.0110	6616	136	0.0206	1.87	0.00003

[a] Probability using Fisher's exact test of independence (two-tailed).
[b] Number of mutants scored by MS1 and MS31 are shown in brackets.

Using multi-locus probe 33.15, a statistically-significant two-fold increase in mutation frequency was found in the offspring of irradiated parents. A very similar increase in mutation rate was also found by profiling with multi-locus probe 33.6. Another estimate of mutation rate was based on scoring mutants at six independent single-locus minisatellites. For this set of loci the mutation rate in the exposed families was 1.7 times higher than in the control group. We also estimated the total frequency of mutant bands in offspring using all eight independent probes (33.15, 33.6, B6.7, CEB1, CEB15, CEB25, CEB36 and MS32). A highly statistically significant 1.9 fold increase in the mutation rate (P=2.53x10^{-5}) was found in the exposed group.

3.2. CHRONIC RADIATION AND MUTATION RATE

The individual radiation doses for chronic exposure were estimated for 126 families in the exposed group, using published data on the annual external and internal exposure to ^{137}Cs in soil, milk and vegetables in different localities of the Mogilev district [10] and family histories after the Chernobyl accident. The parental radiation dose for each family was taken as the mean value of paternal and maternal doses obtained from the time of the Chernobyl accident to the conception of the child.

The mean dose over all families for parental external and internal chronic exposure to ^{137}Cs was 27.6±3.3mSv (s.d.=36.8mSv, median=19.7mSv). Families within the exposed group were therefore divided according to the median of the distribution (Fig. 1a)

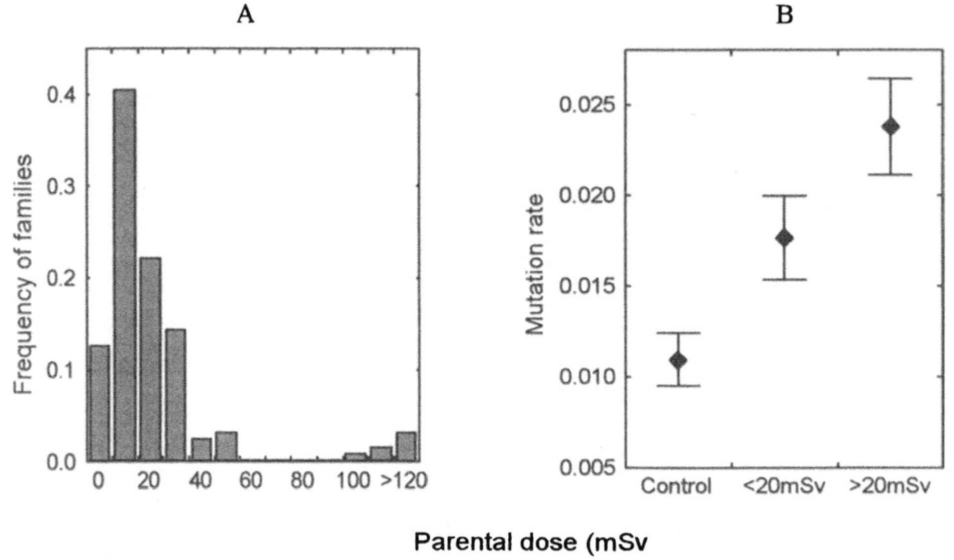

Figure 1. Parental radiation dose for external and internal chronic exposure to ^{137}Cs and mutation rate. A, Distribution of parental dose. B, Frequency of minisatellite mutation (±s.e.m., data pooled for probes 33.15, 33.6, B6.7, CEB1, CEB15, CEB25, CEB36 and MS32) in offspring grouped according to parental dose.

into less exposed (<20mSv) and more exposed (>20mSv) families, and the total mutation rate (probes 33.15, 33.6, B6.7, CEB1, CEB15, CEB25, CEB36 and MS32) was estimated for both sets of families. The mutation rate in the more exposed families was 1.35 times higher than in the less exposed (P=0.0154), and both were higher than in the unexposed population ($P=3.03\times10^{-6}$ and 0.0030, respectively; Fig. 1b). Moreover, a marginally-significant positive correlation between the parental dose and the number of mutant bands in offspring was found within the exposed group (Kendall's τ=0.1111, P=0.0672), consistent with a steady increase of mutation rate with radiation dose.

4. Discussion

The main result of the present study is that, despite increasing the number of families in the exposed group by more than 50%, and profiling of all families with an additional five minisatellites, the difference in the mutation rate between the two groups remains unchanged from that previously reported [6]. The new evidence refutes the hypothesis that the increased mutation rate in the Mogilev group might reflect the existence in this population of minisatellite loci with unusually high spontaneous mutation rates. Scoring of mutations by three independent sets of minisatellite loci detected separately by 33.15, 33.6 and six different single-locus probes revealed an increased mutation rate in the exposed group for most of the minisatellite systems, indicating that increased mutability cannot be attributed to a single locus which has accumulated unusually unstable alleles in the Mogilev population. Thus, the most probable cause of increased mutation rate in the exposed group is the influence of environmental mutagens. The correlation between parental exposure and minisatellite mutation rate found within the exposed group suggests that the observed increase in mutation rate may be directly caused by ionising radiation. To the best of our knowledge, these data and our previous results [6] provide the first experimental evidence for radiation-induction of human germline mutation. In contrast, data collected during the past 40 years on children of the Hiroshima and Nagasaki atomic bomb survivors using standard mutation monitoring systems have not provided evidence for any statistically significant differences in mutation rate between exposed and control families [11].

Nevertheless, major questions remain concerning dose-response parameters for mutation induction. It should be stressed that all studies on the induction of minisatellite mutation in mice have failed to produce any reliable relationships between radiation dose and mutation frequency [5, 12]. We therefore have recently extended mouse mutation analysis to evaluate in detail the relative effects of γ-radiation on mutation induction (Dubrova *et al.*, in preparation).

Male CBA/H mice were given an acute exposure to 0.5 or 1Gy γ-radiation and mated to untreated CBA/H females 3, 6 and 10 weeks post-irradiation. All parents and offspring were profiled by two multi-locus minisatellite probes MMS10 (Bois *et al.*, in preparation) and 33.15 [7] and two hypervariable single-locus minisatellite probes Hm2 and Ms6-hm [13, 14]. Our data suggest that mutation induction is only attributable to the spermatogonia stage of spermatogenesis, providing direct evidence for meiotic origin of minisatellite germline mutation in mice. The frequency of mutations at the spermatogonia stages increased linearly with radiation dosage from 0.5Gy to 1Gy with the doubling dose approximately 0.5Gy, a value close to that obtained in mice using other monitoring systems [15]. These recent results provide support for our studies on human populations exposed to radiation following the Chernobyl accident, indicating that enhanced minisatellite mutation rate may reflect radiation-induced damage to DNA accumulated in early spermatogonial stem cells.

Acknowledgements. This work was supported by grants from the Wellcome Trust (to Y.E.D. and A.J.J.), the Medical Research Council and the Royal Society (to A.J.J.), the Ministry of Health of Belarus (to V.N.N., N.G.K and V.A.O). A.J.J. was also supported by an International Research Scholars Award from the Howard Hughes Medical Institute.

References

1. Dubrova, Y.E. (1992) Principles and approaches for the determination of gene mutation rate in human populations, *Genetika (Russia)* **28**, 5-12.
2. Jeffreys, A.J., Royle, N.J., Wilson, V. and Wong, Z. (1988) Spontaneous mutation rate to new length alleles at tandem-repeat hypervariable loci in human DNA. *Nature* **332**, 278-291.
3. Jeffreys, A.J., Turner, M. and Debenham, P. (1990) The efficiency of multi-locus DNA fingerprint probes for individualization and establishment of family relationships, determined from extensive casework, *American Journal of Human Genetics* **48**, 824-840.

4. Jeffreys A.J., Tamaki, K., MacLeod, A., Monckton, D.G., Neil, D.L. and Armour, J. (1994) Complex gene conversion events in germline mutation at human minisatellites, *Nature Genetics* **6**, 136-145.
5. Dubrova, Y.E., Jeffreys, A.J. and Malashenko, A.M. (1993) Mouse minisatellite mutations induced by ionizing radiation, *Nature Geneics* **5**, 92-94
6. Dubrova, Y.E., Nesterov, V.N., Krouchinsky, N.G., Ostapenko, V.A., Neumann, R., Neil, D.L. and Jeffreys, A.J. (1996) Human minisatellite mutation rate after the Chernobyl accident, *Nature* **380**, 683-686.
7. Jeffreys, A.J., Wilson, V. and Thein, S.L. (1985) Hypervariable "minisatellite" regions in human DNA, *Nature* **316**, 67-73.
8. Vergnaud, G., Mariat, D., Apiou, F., Aurias, A., Lathrop, M. and Lauthier, V. (1991) The use of synthetic tandem repeats to isolate new VNTR loci: cloning of a human hypermutable sequence, *Genomics* **11**, 135-144 (1991).
9. Olaisen, B., Bekkemoen, M., Hoff-Olsen, P and Gill, P. (1993) Human VNTR mutation and sex, in S.D.J, Pena, R. Chakraborty, J.T. Epplen and Jeffreys, A.J. (eds.), *DNA Fingerprinting: State of the Science,* Birkhauser, Basel, pp. 63-69.
10. Drosdovitch, V.V., Minenko, V.F., Ulanovski, A.V. and Shemyakina, E.V. (1989) *Prognosis of Doses of Exposure for Population of BSSR from the Caesium Radionuclides,* Ministry of Health of Belarus, Minsk.
11. Neel, J.V., Schull, W.J., Awa, A.A., Satoh, C., Kato, H., Otake, M. and Yoshimoto, Y. (1990) The children of parents exposed to atomic bombs: estimates of the genetic doubling dose of radiation for humans, *American Journal of Human Genetics* **46**, 1053-1072.
12. Fan, Y.J., Wang, Z., Sadamoto, S., Ninomiya, Y., Kotomura, N., Kamiya, K., Dohi, K., Kominami, R. and Niwa, O. (1995) Dose-response of radiation induction of a germline mutation at a hypervariable mouse minisatellite locus, *International Journal of Radiation Biology* **68**, 177-183.
13. Kelly, R., Bulfield, G., Collick, A and Jeffreys, A.J. (1989) Characterization of a highly unstable mouse minisatellite locus: evidence for somatic mutations during early development, *Genomics* **5**, 844-856.
14. Gibbs, M., Collick, A., Kelly, R. and Jeffreys, A.J. (1993) A tetranucleotide repeat mouse minisatellite displaying substantial somatic instability during early preimplantation development, *Genomics* **17**, 121-128.
15. Luning, K.G. and Searle, A.G. (1971) Estimates of genetic risks from ionizing irradiation, *Mutation Research* **12**, 291-304.

CHROMOSOME DAMAGE.

Recommendations for Research Needed

G. OBE [1], B.S. FEDORENKO [2] AND
M. BENO [3], C. CREMER [4], Y.E. DUBROVA [5], R.D. GOVORUN [6],
R. GREINERT [7], D. HARDER [7], M. KOZUBEK [8], W.-U. MÜLLER [9],
A.T. NATARAJAN [10]

[1] *Department of Genetics, University of Essen, D-45117 Essen, Germany*
[2] *State Scientific Center of the Russian Federation, Institute for Biomedical Problems, 123007 Moscow, Russia*
[3] *Institute of Preventive and Clinical Medicine, SK-833 01 Bratislava, Slovak Republic*
[4] *Institute of Applied Physics, University of Heidelberg, D-69120 Heidelberg, Germany*
[5] *Vavilov Institute of General Genetics, Moscow B-333, Russia*
[6] *Division of Radiation and Radiobiological Research, Joint Institute for Nuclear Research, Dubna 141980, Moscow Region, Russia,*
[7] *Institute of Medical Physics and Biphysics, University of Göttingen, D-37073 Göttingen*
[8] *Faculty of Informatics, Masaryk University, 60200 Brno, Czech Republic*
[9] *Institut für Medizinische Strahlenbiologie, Universitätsklinikum Essen, D-45122 Essen, Germany*
[10] *Department. of Radiation Genetics and Chemical Mutagenesis, State University of Leiden, 2300 RA Leiden, The Netherlands*

1. Brief Statements of Knowledge

These recommendations focus on the action of ionizing radiation on eukaryotic chromosomes.

Eukaryotic chromosomes are complex associations of mainly DNA and various types of proteins such as histones and non-histone proteins. There is one DNA molecule in a presynthetic chromosome which is replicated during the S-phase of the cell cycle leading to a chromosome with two chromatids which are distributed to the daughter cells in mitosis. Agents which damage chromosomal DNA lead to chromosomal aberrations (CA). It has been shown that DNA double strand breaks (DSB) are ultimate lesions for the induction of CA. CA may be of chromosome type with both chromatids affected at the same site (dicentric chromosomes, translocations, ring chromosomes,

C. Baumstark-Khan et al. (eds.),
Fundamentals for the Assessment of Risks from Environmental Radiation, 265-268.

fragments) or of chromatid type with only one chromatid affected (chromatid interchanges, chromatid breaks). With low LET radiation, at low doses, a single hit event give rise to a chromosome exchange (alpha component of the dose-effect curve) and at higher doses in addition to this, breaks from two independent events can interact to form an exchange (beta component of the dose-effect curve). It has been presumed that at low doses of ionizing radiation clustered damage in a single chromosome might be sufficient for the formation of exchange aberrations. The efficiency of the production of lesions in the chromosomal DNA is dependent on LET. High LET particles on their traversal through the cell nucleus can produce a multiplicity of CA resulting in overdispersion of aberrant cells.

The frequencies of dicentric and ring chromosomes as well as of translocations in peripheral lymphocytes (PL) of persons exposed to radiation can be used as indicators of radiation dose (biological dosimetry). The reason for this is that PL irradiated *in vitro* or *in vivo* have comparable frequencies of CA. PL with dicentrics and/or translocations disappear with time and this is mainly correlated with the half-life of the cells. The half-life of PL is estimated to be about 3 years, though in some subpopulations of PL it seems to be much longer or shorter. The decline of the number of dicentrics and/or translocations with time is an important aspect in biological dosimetry.

There are well established dose-response-relations between radiation dose and the frequency of dicentrics for various radiation qualities. The dose that can be monitored for an acute low LET radiation exposure, without knowledge of the preirradiation spontaneous frequency of dicentrics, is above 0.1 Gy. In case the preirradiation frequencies are known the lowest limit of detection may be 0.05 Gy. Reciprocal translocations which can be detected with the method of fluorescence *in situ* hybridization (FISH) may improve biological dosimetry especially with regard to radiation exposures which occurred long time ago (retrospective dosimetry), because the population of PL with reciprocal translocations will be renewed by supply from the irradiated red bone marrow. In this field additional work has to be done before firm conclusions can be drawn. Micronuclei which are formed from radiation induced CA or whole lagging chromosomes may also be used as indicators of exposure.

There is a marked interindividual variability in radiation response. We do not know whether interindividual differences seen at the level of dicentric chromosomes reflect the radiation sensitivity of the whole body but, it is tempting to assume that there is some correlation between these phenomena. If this could be demonstrated it would contribute to the solution of the problem that for a person being exposed to radiation, physical dose alone does not describe radiation risk, because individual radiation sensitivity makes a major contribution to risks finally observed. Not only the physical dose is an important factor but also the response of the body to this dose.

Other types of alterations in chromosomes are sister chromatid exchanges (SCE). It has been shown that DSB can lead to SCE in prereplicative chromosomes. These SCE may result from inversions and translocations in single DNA molecules and may not be „real“ SCE. This type of chromosomal damage has not yet been routinely applied in radiation biology.

2. Unsolved Problems

The mechanisms leading to the formation of CA from various types of primary lesions in the DNA are not well known. Using the method of premature chromosome condensation (PCC) some information was obtained on reaction steps such as intermediate forms of aberrations. These intermediate forms need to become more clarified, especially since their study could lead to a better understanding of „potentially lethal damage". Another aspect of cellular reaction to radiation which is not yet explainable is genomic instability which has been shown *in vitro* as well as *in vivo*. Following exposure of cells to ionizing radiation, genetic changes may be expressed many cell generations later. Subtle changes in the chromatin such as radiation induced SCE and translocations could play a role in this phenomenon.

The influence of the arrangement of chromatin in the interphase nucleus on the formation of CA is not known sufficiently. With the FISH method it has been shown that in interphase the chromosomes are organized in distinct domains. In one recent model, the chromatin of single chromosomes is confined to distinct territories with an intranuclear channel system between them, in which cellular activities such as repair, transcription and splicing may take place. Other models of the structure of the interphase nucleus are under discussion. More knowledge on the structure of the cell nucleus would obviously be of great importance for a better understanding of the formation of CA.

More research is necessary to unravel the correlation between CA and cancer. Some types of cancer are known to be associated with highly specific types of CA such as translocations. Prospective analyses indicate that people with high frequencies of spontaneous CA in their PL may have a higher risk to develop cancer. More prospective studies concerning this correlation have be to done.

It is known that there are considerable interindividual variabilities with respect to cellular repair. To which extent this correlates with cancer risk is not yet clear. Other important aspects which are worth analyzing in more detail are clastogenic factors in the serum of irradiated people, radiation induced programmed cell death (apoptosis), and adaptive response of cells to low doses of radiation.

3. Justification of Further Research

In the framework of environmental hazards, studies on the induction of chromosomal damage and cancer by ionizing radiation are of major significance:

1. to contribute to a realistic estimation of radiation risk
2. to identify the proportion of cancer risk due to ionizing radiation
3. to base radiation protection regulations on qualitative knowledge of cancer induction
4. to get more insight in the mechanism of cancer induction
5. to base radiation therapy on solid understanding of radiosensitivity of normal and tumour tissues of individual patients
6. to use radiation as a tool to analyse structural aspects of chromosomes and their arrangement in the interphase nucleus
7. to study repair of radiation induced DNA damage
8. to look at the mechanism of how radiation-induced SCE are formed

9. to explain the phenomenon of radiation-induced genetic instability
10. to inquire into radiation-induced apoptosis
11. to supply evidence as to whether cells can adapt to low doses of radiation
12. to investigate the origin and nature of radiation-induced clastogenic factors

4. Recommended Approaches

Like in other realms research progress will largely benefit by the introduction of new techniques such as methods of molecular biology and advanced methods of microscopy and biocomputing. More emphasis should be put into these methodological aspects.

The analysis of CA in PL of persons exposed to ionizing radiation is still the most feasible method to study the effects of such exposures on the human genome. Further developments to automatize the analysis of CA and micronuclei would be of great help in this type of research.

More studies are necessary on the effect of complex and variable radiation exposure of astronauts and cosmonauts during space flights to find out as to what extent data on physical dosimetry and on CA in PL of cosmonauts and astronauts are correlated. Data on the biological effects of high charge and high energy particles (HZE) are especially needful with this respect.

Important questions to be answered in the field of biological dosimetry are how long after an exposure dicentric chromosomes and/or translocations can be found in PL (half-life of PL with aberrations). Studies on the biology of PL using biochemical, physiological and cytogenetical methods are necessary to further analyse of this aspect.

A main field of interest is the possible correlation of chromosomal aberrations and cancer. Prospective studies are necessary concerning this aspect and these should be funded.

The three-dimensional structure of chromatin in normal and malignant cells, its dependence on the cell cycle and its possible intercellular and/or interindividual variability should be studied intensively. This could lead to a better understanding of the formation of aberrations in general and of cancer specific aberrations.

More research into the phenomenon of genomic instability is mandatory because up to now the reason for this phenomenon is entirely unclear. One should look to subtle changes in the genome such as small inversions, translocations, insertions, deletions and gene mutations.

There are other fields of research which are important for an understanding of the action of radiation on cells such as the induction of programmed cell death (apoptosis), possible adaptive responses of cells to low doses of radiation, leading to less damage when the cells are exposed to higher doses later on, or the formation of „clastogenic factors“ in the serum of exposed persons.

These types of studies may help to answer the question whether the variability observed not only in cytogenetic but in all bioindicator assays, tells us something about the individual radiosensitivity with regard to lethal, mutagenic, carcinogenic, and teratogenic effects.

RADIATION INDUCED MUTATIONS

J. KIEFER

Strahlenzentrum der Justus-Liebig-Universität
Leihgesterner Weg 217, D 35392 Giessen, Germany

Abstract

Mutations constitute an important risk to human health. If induced in the germline they may be transmitted to the progeny causing hereditary diseases. Somatic mutations contribute to the induction of cancer. The basic mechanisms can be studied at the cellular level in tissue culture systems. On the other hand can mutations also be investigated in human lymphocytes and be used as a kind of „Biodosimeter". This review will briefly outline the methods presently available and then concentrate on three aspects, the dependence of mutation induction on dose rate with gamma-rays, the influence of wavelength with UV and radiation quality with ionising radiation and the molecular structure of radiation induced mutation at the level of the gene. It will be shown that, contrary to common belief, a reduction in dose rate will not necessarily lead to a decreased effectiveness. The proliferative state of the investigated populations plays here a decisive role. The mutagenic action of heavy charged particles is another interesting case. If cells could be killed by a single traversal no mutants could be recovered. It is shown that this is not the case, even with ions of high effective charge which is of relevance for risk assessment of space radiation.

The study of molecular patterns of radiation induced mutants demonstrate that heavy charged particles do not necessarily lead to larger deletions than with sparsely ionising radiation and that „non contiguous deletions" seem to be a signature of the action of densely ionising radiations.

1. Introduction

X-rays was the first agent shown to be able to induce mutations, a discovery which was honoured by the Nobel Prise to H. J. Muller [1]. This marked not only the starting point of modern mutation research but was also a significant milestone to the development of quantitative cellular genetics and molecular biology as evidenced e. g. by the works of Delbrück, Zimmer and Timofeeff-Ressovsky [2]. Their "Trefferprinzip der biologischen Strahlenwirkung" ("Hit principle of biological radiation action") paved the way to modern quantitative analysis of genetic phenomena at the molecular level. Apart from fundamental interest mutations play a very important role in the assessment of natural and

C. Baumstark-Khan et al. (eds.),
Fundamentals for the Assessment of Risks from Environmental Radiation, 269-280.

man-made environmental hazards. If originating in the germ cells they give rise to possible adverse inherited defects in the progeny, thus constituting a "genetical risk". (for a recent summary see [3]). Somatic mutations, on the other hand, contribute to cancer induction [4-6]. Point mutations create enzymes of unwanted changed properties, silent protoncogenes may be activated and tumour suppressors switched off if they are lost by deletions.

There are also more practical aspects: Mutations can now also be studied in vivo in exposed individuals and serve as a "biodosimeter" in situations where it is difficult to reconstruct the extend of the original exposure [8]. This growing field is not restricted to radiation impact but has also an important significance with chemical pollutants.

Basic mechanisms whose understanding is a prerequisite to rational application can best be studied in in-vitro systems. This short review restricts itself to investigations on mammalian cells. It will discuss some of the techniques involved and review some recent results from the author's laboratory.

2. Approaches to Study Mutations in Mammalian Cells

Mutations are rare events and difficult to score directly. Most techniques involve, therefore, procedures of phenotypic selection which in mammalian cells (unlike in bacteria and simple unicellular eucaryotes) are always based on resistance to certain cytotoxic agents. Another problem is recessivity. Since the information for a gene is usually encoded on two homologous chromosomes a mutation will phenotypically show up only if it is dominant, unless it is located on a sex chromosome or the cell is either hetero- or hemizygous. The latter approach necessitates the use of special cell strains.

Another problem found with all selective procedures is "mutation expression". As the mutation takes time to be fully established cells have to be cultured several days in normal medium before they are subjected to selection. Theoretically the "expression curve", i. e. the number of mutants as a function of time in non-selective growth should rise gradually to a plateau and remain there but this is hardly the case in the real world. Commonly a maximum is passed followed by a decline suggesting some growth disadvantage of the mutants which is not yet understood. The time of scoring is thus critical and whenever possible full expression curves should be recorded.

All methods to identify mutants require cell multiplication, i. e. dead mutants cannot be found. In the dose-effect curves this can be compensated for by relating the number of mutants to the number of colonies in non-selective medium ("mutants per survivor" or "mutant frequency") but this implies that the colony forming ability is the same in mutants and non-mutants. This is not necessarily true, particularly if large deletions are involved encompassing vital genes. The commonly found decrease of mutant frequencies with higher doses is presumably related to this phenomenon.

2.1 THE HPRT SYSTEM

Hypoxanthine-guanine-phosphoribosyl-transferase (HPRT) is an enzyme which catalises the incorporation of purine bases into nucleic acids. The gene is located on the X-chromosome so that male cells behave as if they were haploid with respect to this gene. Mutation induction studies can also be performed with female cells as the second X-chromosome is usually inactive but care has to be taken with transformed cells which may have several of them being not always metabolically silent. Special problems can arise when the molecular structure is analysed. Inactivation of the HPRT-gene leads to resistance towards 6-thio-guanine and can be easily scored in selective medium [9-11]. The cells do not lose their viability since they use normally another pathway for nucleotide incorporation. Because of its relative simplicity the HPRT-system is quite popular and widely used but shortcomings must not be overlooked: As loss of activity can be brought about both by point mutations and larger deletions one has to face a wide spectrum of possible changes at the molecular level asking for a more detailed analysis.

2.2 THE OUABAIN SYSTEM

Ouabain is an East-African arrow poison (the word is Kissuaheli) binding to Na-K-ATPase but not at the active site. This can be prevented by a point mutation leaving the overall enzyme activity unchanged. This mutation is thus semi-dominant and circumvents the problem of recessivity but it cannot be caused by larger deletions. Oubain resistance thus earmarks true point mutations [12].

2.3 THE TK-SYSTEM

The problem of recessivity can be overcome by using cells where the second gene locus is either inactivated (heterozygosity) or completely missing (hemizygosity). The human TK6 line is heterozygous for thymidine kinase (TK) [13]. The loss of this enzyme leads to resistance towards fluorothymidine. It can be caused by inactivation of the unmutated gene but also by "loss of heterozygosity" (LOH). This effect may either reflect real loss of parts of or the complete intact chromosome (rare since mostly lethal) but also by recombinational events leading to the transfer of the inactive gene to the respective site of the homologous chromosome (gene conversion, mitotic recombination). These events are quite common in yeast where the repair of DNA double strand breaks proceeds mainly via mitotic recombination but was so far considered unlikely in mammalian where break-endjoining is the main pathway. A decision whether LOH occurs is only possible by molecular analysis. Recent reports indicate, contrary to expectations, that LOH is frequent in TK6 mutants [14].

2.4 HYBRID SYSTEMS

Large deletions causing mutations may span parts of a chromosome where genes essential for vitality are located. These mutant are hence not viable and will not be picked up by the usual methods leading to a severe underestimation of mutant frequencies. This problem was solved by Waldren et al. [15] by constructing a hamster cell which contains a human chromosome with selectable markers. These are genes for cell surface structures which can be recognised by specific antibodies whose binding kills the cell. The mutant frequencies found after X-ray exposure was orders of magnitude larger than for HPRT which was measured in the same cells. This suggests that loss of essential genes is of great importance although the human chromosome is also a much bigger target.

2.5 THE GLYCOPHORIN ASSAY

Although this is not an in vitro assay it should be mentioned here for completeness. Mutations at the glycophorinA locus can be found in human erythrocytes after exposure to mutagenising agents. The test is performed in vitro in human erythrozytes [16].

2.6: "DNA FINGERPRINTING"

Selective procedures involve commonly only single genes, and the recorded mutant frequencies are quite small. This is even true for the hybrid cell described above. This means that the assay requires large numbers of cells, artefacts due to the selection procedure cannot be eliminated. This all could be overcome if mutations in large parts of the genome could be scored so that the much bigger target would also lead to higher yields. Mammalian cells contain a large amount of non-coding DNA which is organised in the form of small repeated sequences, called "satellite DNA". Since they are non-coding these repeats accumulate "neutral" mutations so that the molecular pattern constitutes a kind of a "DNA fingerprint" because it is different from one individual to the other. For the analysis the DNA is radioactively labelled and broken down to smaller fragment by restriction nuclease action. The fragments are separated by gel electrophoresis and hybridised to satellite specific probes after blotting. The resultant bands can then be scored for mutations [17]. This method can be used both in vitro and also for exposed individuals. As far as radiation is concerned there so far only one in-vitro study on UV and protons [18] (but see the contribution by Dubrova et al., this issue).

2.7 MOLECULAR ANALYSES

Phenotypic selection can only yield the dependence of mutant frequencies on exposure parameters irrespective of the molecular changes involved. These are, however, of particular interest when comparing different types of radiations and it may be hoped to obtain characteristic “signatures” helpful in the analysis of unknown environmental impacts and in the understanding of underlying mechanisms.

The first method employed towards this goal was the investigation of restriction patterns (see [19] for a review with ionising radiations) which is now commonly replaced by gene amplification using the "Polymerase Chain Reaction" (PCR) [20]. Examples will be given below.

3. Mutation Induction by Radiation

3.1 RADIATION TYPES AND RADIATION QUALITY

3.1.1 Ultraviolet Llight: Action Spectra

An action spectrum displays the dependence of a certain biological effect on wavelength. It can be used to identify the responsible chromophor and give clues to mechanisms. It is also an important information to assess the risks associated with changes in environmental solar exposure as may be brought about by the reduction of the ozone layer. To determine a correct action spectrum is not a trivial task but this will not be further elaborated here, some relevant remarks will be found in [21] and in the paper by R. Tyrell in this volume.

Action spectra have played an important, although often overlooked, role in the history of modern biology. Already in 1930 Gates [22] published an action spectrum for the inactivation of E. coli which clearly showed that nucleic acids played the key role - more than a decade before the biochemical work by Avery et al [23] who earned for this the Nobel Prise. The first action spectrum for mutation induction - also in bacteria - was published by Knapp et al. [24].

We have studied action spectra for mutation induction in V79 Chinese hamster cells, both at the HPRT-locus and for ouabain resistance. Results are summarised in Fig. 1. One sees that the general shape follows DNA absorption although there appear to be deviations in the UV-B and UV-A region with ouabain. This points to the involvement of additional chromophores and may be of relevance with regard to the ozone depletion problem.

The International Radiation Protection Agency (IRPA) has suggested a "UV hazard weighting function" which is compared to our data in fig. 2. A good correlation is seen demonstrating the usefulness of action spectroscopy.

3.1.2 IonisingRadiations: Dependence on Ionisation Density.

It is well known that the effectiveness of ionising radiations depends on the spatial distribution of energy deposition events. As a quantitative measure very often the "Linear Energy Transfer" (LET) is used although it is clear now that radiation quality cannot be adequately be described by a single quantity. LET is defined as "energy imparted per unit pathlength" and neglects thus any fine-structure of particle tracks. Our laboratory has studied mutation induction by heavy charged particles over the last years. Results are in detail published elsewhere[28-30] and will only be briefly summarised.

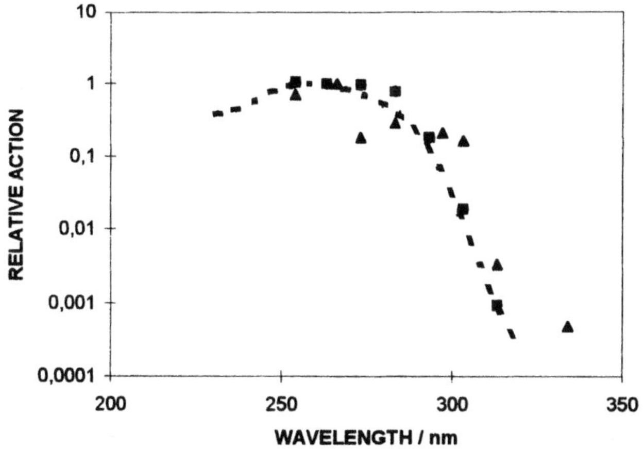

Figure 1: Action spectra for mutation at the HPRT-(squares) and the ouabain-locus (triangles). DNA absorption is shown for comparison in broken lines. Data from [25, 26].

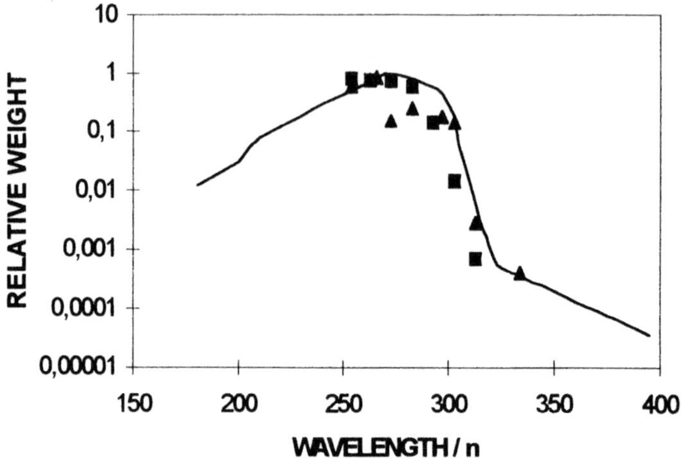

Figure 2: Comparison of the IRPA weighting function [27] with mutation induction at the HPRT-(squares) and the ouabain-locus (triangles)

A special problem with densely ionising particles must be addressed. They deposit large amounts of energy by single traversals, i. e. in a highly structured manner. In such a situation the quantity "dose" may loose its well-defined meaning, particularly at low values. It is the average of energies per mass element and thus includes also those which were not hit at all. This fraction can be quite substantial as demonstrated in fig. 3. With the environmentally relevant dose of 1 mGy even with X-rays only about 6% of nuclei are traversed by an electron, with alpha-particles the situation is more dramatic: here only 6 out of thousand nuclei are in fact traversed. On the other hand, those cells which are hit experience rather high doses: an alpha-particle deposits about 0.16 Gy, an Fe-ion nearly 1 Gy per single traversal. This has to be taken into account for an appropriate analysis and calls for experiments where the action of individual ions can be studied. Such "microbeam" studies have recently been initiated [31] and will be further pursued by a number of groups in spite of considerable technical problems. These kind of studies were pioneered by H. Bücker's BIOSTACK concept [32].

The different effectiveness of charged particles is commonly described by the "Relative Biological Effectiveness" (RBE) which is based on a comparison of doses and whose applicability is thus limited as just discussed. For radiation protection purposes "radiation weighting factors" and/or "quality factors" (QF) were introduced by the International Commission on Radiological Protection [33] which are related to RBE having thus similar limitations. Nevertheless a comparison of experimental RBE-values for mutation induction with the recently suggested relationship of QF on LET may be useful to check at least for obvious discrepancies. Fig. 4 compares our calculated RBE-values for low "doses" with the ICRP-model, and it is seen that they are generally not at variance. It seems, however, that the ICRP estimate is not over-conservative.

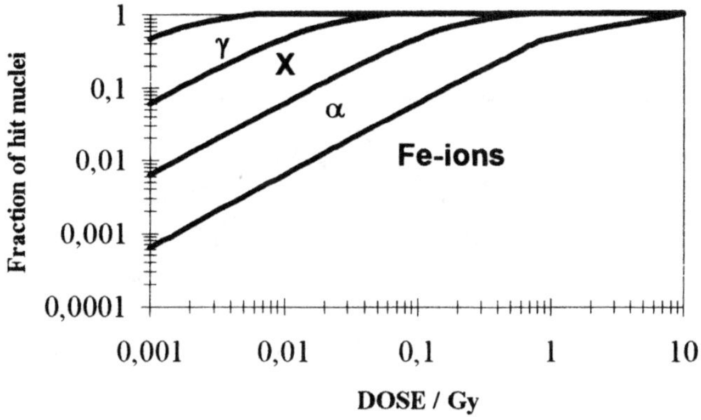

Figure 3: The fraction of cell nuclei which are not traversed by a charged particle as a function of dose.

3.2 DOSE RATE AND THE TIME FACTOR

Most mutation induction experiments are performed with high dose rates and short exposure times. This is, however, not the situation found in the real world where individuals are subjected to low dose rates over long time periods. It was commonly assumed that lowering the dose rate yields smaller effects which was based mostly on survival experiments where this kind of behaviour is well documented. There are considerably fewer experiments for mutation induction, none with ultraviolet light, a few with gamma-rays. Long-term investigations with cells in culture pose an experimental challenge. If they are grown in batch culture they reach confluence after a number of hours and stop proliferation. In order to keep them under steady state conditions the principle of "continuous culture" known from microbiology was adapted [34]. By regular replacing an aliquot of the cell suspension by fresh medium a state of exponential growth can be maintained over many weeks. The first results of mutation induction by continuous gamma-irradiation were performed with V79 Chinese hamster cells demonstrating for the first time an unexpected "inverse dose rate effect" [34, 35]. Since no dose rate influence was found with human TK6 cells [36] which could have been an indication that rodent cells behave in an atypical way further studies were carried out with the human P3 line, derived from a teratocarcinoma [37]. Here a similar result was obtained as with hamster cells although less pronounced: While with a dose rate of 50 mGy/hour a clear reduction in mutagenic action was seen no difference was found between acute exposure and that at a dose rate of 5 mGy/hour [39]. This surprising effect depends obviously on the proliferative status of the population [40]- it is absent in stationary cells where a decrease in dose rate always leads to a reduction in mutant frequencies. Fig. 5 summarises these findings which were in the meantime also confirmed by others [38]. There is presently no explanation for this surprising phenomenon which is clearly of importance for environmental risk assessment, e g. with regard to the recently introduced "dose and dose rate effectiveness factor" [33].

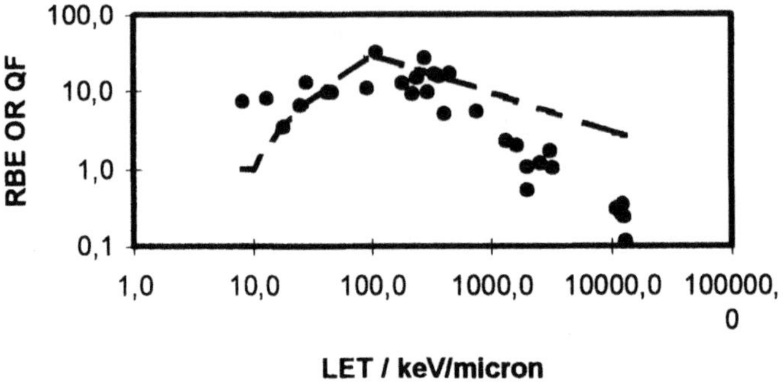

Figure 4: RBE values for mutation induction determined from the initial linear part of the dose-effect curves compared to the ICRP recommendation for the quality factor (broken line). Data from [28-30]

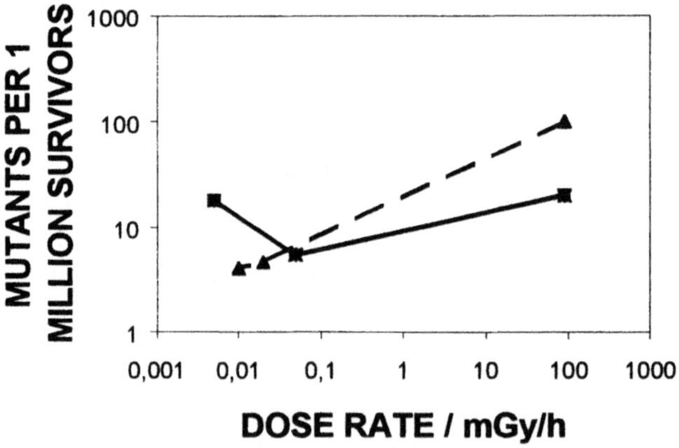

Figure 5: Mutant frequency after 4 Gy γ-rays with different dose rates. Solid curve: exponential cultures, broken line: plateau phase cells. Data from [39, 40]

3.3 MOLECULAR ANALYSIS

This paragraph will be restricted to the action of ionising radiations and concentrate on the HPRT gene since it will discuss the distribution of deletions. They are rare with UV where point mutations are dominant. They have to be studied by sequencing which is outside the scope of this small review.

The HPRT gene is (in hamster cells) about 34 kilobasepairs long and comprises 9 exons. The coding region spans only about 660 basepairs, i. e. most of the gene consists of non-coding DNA which has not yet been analysed.

In order to study the distributions of radiation induced deletions all exons are amplified with the aid of specific primers in a so-called "multiplex PCR" [41, 42]. The resulting DNA fragments are separated by gel electrophoresis. The different patterns are classified as "no" (no exon missing), "partial" (some but not all missing) and "total deletions" (no exon present). A special problem is the presence of background mutants in the population which may blur the result, particularly as they may expand clonally and hence suggest mutational hotspots which do not exist in reality. In order to account for them it is indispensable to analyse them always in parallel to obtain meaningful results [43].

While it was previously believed that ionising radiations induce mainly larger deletions [44] this statement has now to be modified. There is always a substantial fraction of so-called "wild type" mutants, i. e. where the changes are too small to be picked up by the technique. They represent presumably point mutations and can only be characterised further by sequencing. Nevertheless it is true that deletions - both partial and total - represent the majority with all types of ionising radiations. There is presently no general agreement whether their relative contribution changes with radiation quality. Most studies in this context are concerned with the comparison of X- or gamma-rays with alpha-particles. We find clearly an increase of total deletions with higher ionisation density (Fig. 6). The patterns were corrected for background mutants as described in detail elsewhere [44]. With alpha-particles (and also with carbon ions) a new pattern emerges. While with X- or gamma-rays deletions span always over neighbouring exons this is no longer the case with densely ionising radiations and we find "non contiguous" deletions which may be a signature for higher ionisation density. They have to the best of our knowledge not been reported previously in the literature.

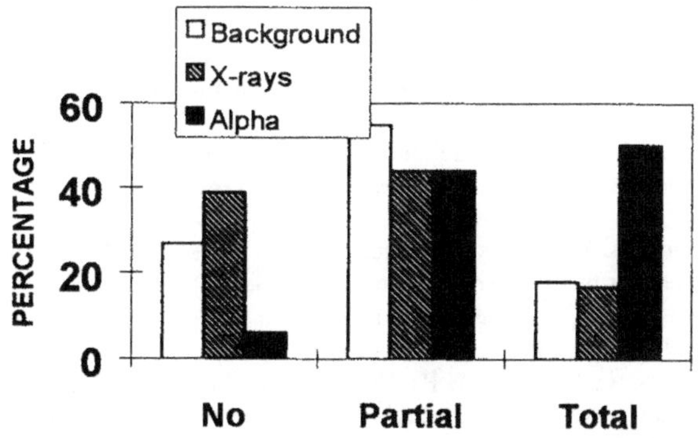

Figure 6: The distribution of mutations with no exon loss and with partial and total deletions in unirradiated background mutants and after X-ray and alpha particle exposure. Data from [43]

4. Concluding Remarks

Realistic risk assessment is concerned with low doses and low dose rates. Epidemiology cannot provide answers in this realm as very large populations had to be studied under well controlled conditions. This is not possible so that the only way out is to elucidate basic mechanisms which would allow an extrapolation to situations where measurements are not available and can also not be gained. Mutation studies figure prominently in this respect, mainly because of the proven connection to carcinogenesis [5-7]. Another aspect is the use of mutations for biodosimetry which has become possible by the development of in-vivo assays [17, 45, 46]. They are not only useful in radiation research but also for other mutagenic agents in our environment. For the HPRT gene a large databank has been estblished which is permanently updated. [47]

It could not be the objective of this chapter to give a comprehensive review of the field but it was the aim to highlight a few aspects of mutation research in the light of environmental risk assessment and to demonstrate as examples some results. The ultimate goal must be to understand the mechanism of radiation induced carcinogenesis. Experimental studies as described provide building blocks towards this ambitious undertaking. The way is undoubtedly still very long but a good theoretical understanding is the only means for a rational analysis according to the old well-known principle:

"*There is nothing more practical than a good theory*" (presumably Einstein)

Acknowledgements. The studies of our laboratory reported here were partly and at various times supported by the German Federal Ministry of the Environment (BMUNR), the German Space Agency DARA and "Gesellschaft für Schwerionenforschung" GSI. I thank my former and present co-workers who produced the results over the years: N. Crompton, F. Zölzer, K. Schenk-Meuser, C. Forster, F. König, M. Kohlpoth, M. Kuntze, T. Kranert, U. Stoll, P. Schmidt and E. Schneider.

References

1. Muller, H. J. Artificial transmutation of the gene. *Science* **66**, 84 (1927)
2. Timofeeff-Ressovsky, N. W., Zimmer, K. G., Delbrück, M. Über die Natur der Genmutation und der Genstruktur. *Nachr. Ges. Wiss. Göttingen Vl. N. F.* **1**, 189-245 (1935)
3. Sankaranaraayanan, K. Genetic effects of ionising radiation in man. *Ann. ICRP* **22 (1)**, 75-94 (1991)
4. Upton, A. C. Biological aspects of radiation carcinogenesis, in: *Radiation carcinogenesis: Epidemiology and biological significance* (Boice, J. D. and Fraumeni, J. F., eds.), Raven Press: New York (1984), p. 9
5. Barbacid, M. Mutations, oncogenes and cancer. *Trends in Genetics* **2**, 188-192 (1986)
6. Bishop, J. M. Molecular themes in oncogenesis. *Cell*, **64**, 235-248 (1991)
7. Perera, F. P., Whyatt, R. M. Biomarkers and molecular epidemiology in mutation/cancer research. *Mutat. Res.* **313**, 117-129 (1994)
8. Dacruz, A. D., Curry, J., Curado, M. P., Glickman, B. W. Monitoring HPRT mutant frequency over time in T-lymphocytes of people accidently exposed to high doses of ionizing radiation. *Env. Mol. Mutagen.* **27**, 165-175 (1996)
9. Hsie, A. W., Casciano, D. A., Couch, D. B., Krahn, D. F., O'Neill, J. P., Whitefield, B. L. The use of Chinese hamster ovary cells to quantify specific locus mutation and to determine mutagenicity of chemicals. *Mutat. Res.* **86**, 193-214 (1981)

10. Li, A. B., Carver, J. H., Choy, W. N., Hsie, A. W., Gupta, R. S., Loveday, K. S., O'Neill, J. P., Riddle, J. C., Stankowski, L. F. jr., Yang L. A guide for the performance of the Chinese hamster ovary Hypoxanthine-guanine-phosphoribosyltransferase gene mutation assay. *Mutat. Res.* **189**, 135-141 (1986)
11. Li, A. B., Gupta, R. S., Heflich, R. H., Wassom, J. S. A review and analysis of the CHO/HGPRT assay to determine the mutagenicity of chemical agents. *Mutat. Res.* **196**, 17-36 (1987)
12. Elmore, E., Barrett, J. C. Measurement of spontaneous mutation rates at the Na-K ATPase locus (ouabain resistance) of human fibroblasts using improved growth conditions. *Mutat. Res.* **97**, 393-404 (1982)
13. Liber, H. L., Thilly, W. G. Mutation assay at the thymidine kinase locus in diploid human lymphoblasts. *Mutat. Res.* **94**, 467-485 (1982)
14. Yandell, D. W., Dryla, T. P., Little, J. B. Somatic mutations at a hterozygous autosomal locus in human cells occur more frequently by allele loss than by intragenic structural alterations. *Somat. Cell Mol. Genet.* **12**, 255-263 (1986)
15. Waldren, C., Jones, C., Puck, T. T. Measurement of mutagenesis in mammalian cells. *Proc. Natl. Acad. Sci. US* **76**, 1358-1362 (1979)
16. Langlois, R. C., Bigbee, W. L., Jensen, R. H., Measurements of the frequency of human erthrocytes with gene expression loss phenotypes at the glycophorin A locus. *Human Genet.* **74**, 353-362 (1986)
17. Jeffreys, A. J., Wilson, V., Thein, S. L. Individual-specific "fingerprints" of human DNA. *Nature* **316**, 76-79 (1985)
18. Ogheri, S., Bruna, V., Cera, F., Favaretto, S., Cherubini, R., Celotti, L. Mutant frequency at the HPRT locus and in minisatellite sequences in Chinese hamster V79 cells irradiated with low-energy protons (31 keV/μm) and ultraviolet light (254 nm). *Radiat. Res.* **148**, 203-208, (1997)
19. Breimer, L. H.. Ionizing radiation-induced mutagenesis. *Brit. J. Cancer* **57**, 6-18 (1988)
20. Vosberg, H.-P.. The polymerase chain reaction: an improved method for the analysis of nucleic acids. *Human. Genet.* **83**, 1-15 (1989)
21. Kiefer, J., Forster, C., Nussbaum, A., Schenk-Meuser, K.: Ultraviolet action spectra of cellular effects in *Biologic Effects of Light 1995* (M. H. Holick, E. G. Jung, eds.), Walter de Gruyter Verlag, Berlin-New York (1996), p. 108-116
22. Gates, F. L.: A study of the bactericidal action of UV-light. III. The absorption of UV-light by bacteria. *J. Gen Physiol.* **14**, 31-42 (1930)
23. Avery, O. T., McLeod, C. M., MacCarthy, M. Studies on the chemical nature of the substance inducing transformation of pneumococcal types. *J. Exp. Med.* **79**, 137-158 (1944)
24. Knapp, E., Reuss, A., Risse, O., Schreiber, H.: Quantittative Analyse der mutationsauslösenden Wirkung monochromatischen UV-Lichtes. *Naturwiss.* **27**, 304 (1939)
25. Zoelzer, F., Kiefer, J. Wavelength dependence of inactivation and mutation induction to 6-thioguanine resistance in V79 Chinese hamster fibroblasts. *Photochem. Photobiol.* **40**, 49-53 (1984)
26. Forster, C. Action spectra for ultraviolet light induced cell inactivation and mutation to ouabain resistance in V79 Chinese hamster fibroblasts. Ph. D. thesis, University of Giessen (1993)
27. Moseley. H. Ultraviolet and laser safety. *Phys. Med. Biol.* **39**, 1765-1800 (1994)
28. Kranert, T., Schneider, E., Kiefer, J. Mutation induction in V79 Chinese hamster cells by very heavy ions. *Int. J. Radiat. Biol.* 58, 975-987 (1990)
29. Stoll, U., B. Barth, N. Scheerer, E. Schneider, J. Kiefer. HPRT-mutations in V79 Chinese hamster cells induced by accelerated Ni-, Au and Pb-ions. *Int. J. Radiat. Biol.*, **70**, 15-22 (1996)
30. Stoll, U., A. Schmidt, E. Schneider, J. Kiefer. Killing and Mutation of V79 Chinese Hamster Cells exposed to accelerated O- and Ne-Ions. *Radiat. Res.* **142**, 288-294 (1995)
31. Hei, T. K., Wu, Li-Jun, Liu, S.-X., Vannais, D., Waldren, C. A., Randers-Pehrson, G. Mutagenic effects of a single and an exact number of α particles in mammalian cells. *Proc. Natl. Acad. Sci US* **94**, 3765-3770 (1997)
32. Bücker H. Biostack, a study of the biological effects of HZE galactic cosmic radiation. in: *Biomedical Results of Apollo*. Ed. by Johnston RS, Dietlein LF & Berry CA. NASA SP-368, pp. 343-54, 1975.

33. ICRP 60. Recommendations of the International Commission on Radiological Protection. *Ann. ICRP* **21**, No. 1-3, 1991
34. Crompton, N. E. A., F. Zölzer, E. Schneider, J. Kiefer. Increased mutant induction by very low dose-rate-irradiation. *Naturwissenschaften* **72**, 439-440 (1985)
35. Crompton, N. E. A., B. Barth, J. Kiefer. Inverse dose rate effect for the induction of 6-thioguanine 60 resistant mutants in V79S Chinese hamster cells by Co-rays. *Radiat. Res.* **124**, 300-308 (1990)
36. König, F., J. Kiefer. Lack of dose-rate effect for mutation by -rays in human TK6 cells. *Int. J. Rad. Biol.* **54**, 891-898 (1988)
37. Zeuthen, J., Norgaard, J. O. R., Avner, P., Fellous, M., Wartiovaara, J., Vaheri, A., Rosen, A. and Giovanella, B. C. Characterization of a human ovarian tertocarcinoma-derived cell line. *Intern. J. Cancer*, **25**, 19-32. , (1980,)
38. Furuno-Fukushi, I., Aoki, K., Matsudaira, H. Mutation induction by different dose rates of gamma-rays in near-diploid mouse cells in plateau- and log-phase culture. *Radiat. Res.* **136**, 97-102 (1993)
39. Kohlpoth, M. Untersuchungen zur Mutationsinduktion durch ^{60} Co-Gamma-Strahlung in menschlichen Epithelzellen. Ph. D. thesis University of Giessen 1991
40. Kuntze, M. In-vitro Untersuchungen zur Untersuchungen zur Induktion und Expression strahleninduzierter Mutationen in menschlichen Zellen. Ph. D. thesis University of Giessen 1993
41. Rossiter, B. J. F., Fuscoe, J. C., Muzny, D. M., Fox, M., Caskey, C. T. The Chinese hamster HPRT gene: Restriction map, sequence analysis and multiplex PCR deletion screen. *Genomics* **9**, 247-256 (1991)
42. Xu, Z.D., Yu, Y.J., Gibbs, R.A., Caskey, C.T. and Hsie, A.W.. Multiplex DNA amplification and solid-phase direct sequencing for mutation anlysis at the HPRT lcus in Chinese hamster cells. *Mutat. Res.* **288**, 237-248 (1993).
43. Schmidt, P., Kiefer, J. Deletion-pattern analysis of alpha-particle and X-ray induced mutations at the HPRT locus of V79 Chinese hamster cells, submitted for publication
44. Thacker, J. The nature of muations induced by ionising radiation in cultured hamster cells. III. Molecular characterisation of hprt-deficient mutants induced by γ-rays orα-particles showing that the majority have deletions of all or part of the hprt gene. *Mutat. Res.* **160**, 267-275 (1986)
45. Cole, J., Skopek, J. R. Somatic mutant frequency, mutation rates and mutation spectra in the human population in vivo. *Mutat. Res.* **304**, 33-105 (1994)
46. Albertini, R. J., Nicklas, J. A., O'Neill, P., Robison, S. H. In vivo somatic mutations in humans: measurements and analysis. *Ann Rev. Genet.* **24**, 305-326 (1990)
47. Cariello, N. F., Craft, T. R., Vrieling, H., van Zeeland, A. A. Adams, T., Skopek, T. R. Human HPRT database - software for data entry and retrieval. *Env. Mol. Mutagen.* **20**, 81-83 (1992)

GENOMIC ALTERATIONS IN RADIOGENIC CELL TRANSFORMATION

T. C. YANG [1], M. DURANTE [2], K. A. GEORGE [3], AND S. YAMADA [1]

[1] *Radiation Biophysics Laboratory, NASA Johnson Space Center, Houston, TX 77058, USA*
[2] *Department of Physics, University "Federico II", Naples, Italy*
[3] *KRUG Life Sciences, Inc., Houston, TX 77058, USA*

The first author of this contribution, Dr. Tracy Yang, suddenly passed away on October 18th 1998. His co-authors wish to dedicate this contribution to his memory. His death is a great loss.

Abstract

For risk assessment, carcinogenic effects of radiation is very important to know, since most people are exposed to relatively low doses. Various experimental results indicate that cancer has cellular origin, and cultured mammalian cells have been used as a model system to study the radiation responses and the mechanism(s) of tumorigenesis at molecular and cellular level. Recent investigations focus on identifying the sensitive phase(s) of cell cycle and type(s) of genomic alteration which can be important in radiogenic cell transformation. Significant enhancement of sensitivity in G_2/M was observed for X- and γ–rays, but not for high-LET radiation. The role of chromosomal aberrations and cancer genes, including both oncogenes and tumor suppressor genes, in neoplastic transformation by ionizing radiation has been examined. Despite extensive efforts, there is no consistent results as to which oncogene(s) may be involved in radiation cell transformation. Cell fusion studies suggest that the transforming gene(s) is recessive. Some investigations have observed alterations in oncogene(s) in radiation transformed cells. These results, however, were not found in different cell systems. In addition to cancer gene studies, cytogenetic analyses of transformed cells were performed by many investigators and significant progress has been made. Common loss or translocation of specific chromosomes were detected in radiation transformed or tumor cells. Also genomic instability was observed in irradiated cells in culture. Overall the role of oncogenes in radiation cell transformation remains unclear. The tumor suppressor genes are likely to be involved in radiation carcinogenesis but are yet to be identified. Complex exchanges of chromosomes, which can be effectively induced by high-LET heavy ions, may be important genetic alterations in radiation cell transformation. Further studies on the relationship between chromosomal alterations and genomic instability and/or inactivation of tumor suppressor genes are warranted.

C. Baumstark-Khan et al. (eds.),
Fundamentals for the Assessment of Risks from Environmental Radiation, 281-288.

Introduction

The induction of cancer is one of the major potential hazards of radiation to human beings, and many efforts have been made in past decades to determine the quantitative relationship between radiation and tumor formation in experimental animals. The development of cell culture system has made it possible to probe the cellular and molecular mechanisms underlying the transformation of a normal cell to a malignant cell. In addition to the advantage of investigating malignant transformation directly at the cellular level without involving the complexity of various factors that are present in whole animals, the *in vitro* cell system also allows one to introduce other oncogenes to the same system, e.g., chemical carcinogens and oncogenic viruses, to examine the cocarcinogenic effects of radiation.

The model system used extensively for radiation cell transformation studies is an immortal cell line of mouse embryonic cells (C3H10T1/2) with strong density-dependent controls over cell division. When seeded into a tissue culture dish, C3H10T1/2 cells will grow to form a confluent monolayer of cells and then cease cell proliferation. An exposure of these cells to radiation is sufficient to transform some cells to a state where they exhibit altered growth control. As a result, these transformed cells will continue to divide after confluence has been reached and form distinct, densely-staining, multilayered foci. This marker makes the use of C3H10T1/2 cells for quantitative studies of radiation carcinogenesis at cellular level possible. Various investigators applied the focus formation assay to determine the dose-response relationship for low- and high-LET radiation, to study the dose rate effects, to investigate the RBE and LET relationship, and to examine interactions between radiation and chemical agents on cell transformation. Early studies showed that the dose-response curve changed from curvilinear to linear with an increase of LET, that the RBE increases with LET to a peak at about 100-200 keV/μm and than decreases [1, 2] and that the low dose rate of high-LET radiation, such as neutrons and heavy ions, can be more effective than high dose rates – an inverse dose rate effect [3, 4].

More recent studies were on oncogenic effects of low-energy alpha particles and on cell cycle effects. The interest in cell cycle studies is due to the finding of inverse dose-rate effects which has been suggested as a result of increase of radiosensitivity at certain cell cycle phases [5, 6, 7]. In addition, the advance of molecular biology has stimulated much interest in molecular mechanisms of oncogenic cell transformation by radiation. A brief review of recent advancement in radiation cell transformation is given below.

2. Oncogenic Effects of High-LET Radiation and Cell Cycle Dependence

Based on past success with the excellent quantitative properties of the C3H10T1/2 *in vitro* oncogenic transformation assay system, Miller et al [8] determined the relative biological effectiveness (RBE) for radiation-induced transformation for alpha particles of defined LET. They found that as the LET of the radiation was increased, the rate of induction of oncogenic transformation increased and the maximum RBE approached 20 at 90-100 keV/μm. At higher LETs, between 120 and 265 keV/μm, RBE dropped precipitously. Alpha particles with LET of 90-100 keV/μm were also found to be more effective than that with higher LETs (120-200 keV/μm) in primary explants of Syrian hamster embryo (SHE) cells [9]. These results with low energy alpha particles are consistent with early findings reported by Yang et al [1].

Stimulated by the suggestion that the inverse-dose-rate effect of high-LET radiation might be due to the variation of sensitivity of cell cycle, several investigators studied the cell cycle dependence for low- and high-LET radiation. Experimental results from these studies are summarized in Table 1.

For low-LET radiation, experimental results indicate that late G_2 and M are sensitive phases [10-12] although one study suggested Late S to G_2 [13].. Interestingly, no significant difference in radiosensitivity through the cell cycle was observed for high-LET radiation [14, 15]. Unlike to ionizing radiation, S was found to be more sensitive to UV lights [12, 16]. A worthwhile note is that the results seem to be consistent between human and mouse cells for both low- and high-LET radiation and for UV lights.

3. Genetic Alterations in Radiation Transformed Cells

The advancement of molecular biology provided new opportunity for radiation biologists to study the genetic alterations important in radiation-induced cell transformation at chromosome and gene level. Experimental results from various investigators are summarized in Table 2.

Since the discovery of oncogenes in mammalian cells, several studies were performed to determine possible role of these genes in radiation-induced cell transformation. Although alterations in some oncogenes were observed and reported by few investigators [18, 19], these results have not been confirmed by later studies [20-22].

TABLE 1. Neoplastic cell transformation by radiation: Cell cycle dependence

Radiation	Cell System	Sensitive Phase	Technique	References
γ rays	HeLa x Skin fibroblast	late G_2/M	shake off	[10]
γ rays	C3H10T1/2	G_2/M	roll off	[11]
X-rays	C3H10T1/2	late S to G_2	shake off	[13]
X-rays	C3H10T1/2	M	shake off	[12]
Fission neutron	HeLa x Skin fibroblast	(no enhancement in late M/early or late G_1)	shake off	[14]
Fission neutron	C3H10T1/2	(no enhancement in S/G_2 or G_2/M	shake off	[15]
αparticles	C3H10T1/2	(no enhancement; mid-G_1 slightly high)	shake off	[17]
UV-B	C3H10T1/2	S	shake off	[12]
UV	Human fibroblast	S	release from contact	[16]

TABLE 2. Radiation cell transformation: Cytogenetic and cancer gene alterations

System	Chromosome	Oncogene	Antioncogene	Reference
Human fibroblast (WI38; γ-rays)		k-ras amplification		[18]
3H10T1/2 (X-rays)		c-myc increased expression		[19]
Rat embryo cells (α-particle)		no change in H-, K-,N-ras, raf		[20]
C3H10T1/2		no alteration in c-myc, cHa-ras, c-Ki-ras		[21]
Human fibroblast (γ-ryas)	lost of 6, 14, 17			[23]
Human mammary cells (HZE)	alterations in 1, 4, 6, 8, 17	no detectable change in myc, ras	no detectable change in Rb, p53	[22]

Besides molecular biology methods, the development of new techniques to study chromosome aberrations, e.g., chromosome painting, promotes cytogenetic investigation of radiation cell transformation. Recently some investigations have been conducted to determine possible involvement of chromosome aberrations and tumor suppressor gene(s) in cell transformation. Yang et al [22] systematically examined the alterations in chromosomes and cancer genes. Among 16 autosomes examined for translocations, by using fluorescence in situ hybridization (FISH) technique, chromosome 3, 12, 13, 15, 16, and 18 appeared to be normal in transformed cells. Chromosome 1, 4, 6, 8, and 17 in transformed cells, however, showed patterns different from those in nontransformed cells. Southern and Western blot analyses indicated no detectable alterations in Rb, or p53 genes. Further studies of chromosome 17 by using in situ hybridization with unique sequence p53 gene probe and a centromere probe showed no loss of p53 gene in transformed cells. However, experimental results from cell fusion studies indicated that the transforming gene(s) is recessive. Another study also showed genomic alterations - loss of chromosome 6, 14, and 17 in human fibroblasts transformed by ionizing radiation [23].

Because radiation induced DNA damages may lead to genomic instability in mammalian cells, limited studies have been done to examine chromosome alterations in surviving clones from irradiated cells. Durante et al [24] studied the karyotypes of human mammary epithelial cells surviving X- and α-irradiation and found that both low- and high-LET radiation induced chromosomal instability in long-term cultures.

These results from cytogenetic and molecular studies with cultured cells are interestingly similar to that found in tumor cells of animals exposed to radiation. Table 3 shows experimental data from several radiation carcinogenesis studies with animals. No change in p53 was found in rat lung tumors induced by radiation from plutonium [25]. Also Rb gene in lung adenocarcinoma cells from mouse exposed to neutrons or γ-rays was the same as that in normal tissue [26]. Chromosomal alterations, however, were observed in a mouse acute myeloid leukemia study [27]. Genomic instability was also found in radiation induced mouse mammary tumor cells [28].

TABLE 3. Radiation carcinogenesis: Cytogenetic and cancer gene alterations

System	Chromosome	Oncogene	Antioncogene	Reference
Rat lung tumor (F344; plutonium)			p53 not involved	[25]
Mouse AML (CBA/H; n)	deletions in #2; trisomy #1, 6, Y			[27]
Mouse mammary tumor (Balb/c, C57BL/6; γ-rays)	instability			[28]
Mouse lung adenocarcinoma (B6CF1; n, γ-rays)			Rb not different; increase p53 deletion	[26]

4. Chromosomal Aberrations Induced by High-LET Radiation

For radiogenic cell transformation, various experimental evidences indicate that nuclear DNA is the primary target and that double strand DNA breaks are important lesions [29, 30]. Double strand breaks can form chromosomal aberrations, which in turn may cause genomic instability. An interesting question is which type(s) chromosomal aberrations are essential in neoplastic transformation. Recently we have conducted experiments with energetic iron particles (1 GeV/u; 150 keV/μm) and found that RBE for complex exchanges, i.e., more than three chromosome breaks involved, is higher than that for other types of aberrations. For complex exchanges, our preliminary data showed that RBE increases with LET. Presently we do not have any studies with LET greater than 150 keV/μm. It is expected that at LETs greater than 200 keV/μm, the RBE will decrease. These experiments, however, remain to be done.

For risk assessment and for mechanistic understanding of radiation effects, it is of interest to know how many particle hits are necessary to cause chromosome aberrations in mammalian cells. We have conducted experiments with human lymphocytes, using chromosome #4 and centromere FISH probes. The average chromosome #4 area, which localized in the cell nucleus and can be seen after being painted with FISH probes, is about 20 μm^2. A simple analyses showed that in average it takes 9-10 iron particle hits in chromosome #4 to induce 1 chromosomal aberration (Table 4) and that 4-5 iron particle hits in chromosome #4 to induce 1 chromosome exchange (Table 5). These results suggest that high-LET iron particles are effective in causing multiple chromosome damages once it hits the target.

Experimental data obtained in our laboratory with gamma rays, protons (250 MeV), and iron particles (1 GeV/u) indicate that these radiation are equally effective in causing PCC fragments for a given dose. In average, 5 Gy iron particles induce about 1 PCC fragment in chromosome #4 per cell, i.e., about 0.2 PCC fragment per chromosome #4 per cell per Gy, or, about 0.24 PCC fragment per iron particle hit in chromosome #4. Based on these information, it is found that about 2.2 PCC fragments are needed to produce one chromosomal aberrant cell and about 1.1 PCC to form chromosome exchange.

TABLE 4. Iron particle hits and the formation of chromosome aberrations.

Gy	# Particles/20 μm^2	Total # Spreads Analyzed	Total # Aberr. Observed	# Aberrants/ Particle/20 μm^2
0	0	1014	1	-
0.75	0.625	1005	48	0.0764
1.0	0.833	314	21	0.0802
2.0	1.666	480	66	0.0825
3.0	2.499	104	32	0.1231
4.0	3.333	109	50	0.1376
5.0	4.166	91	50	0.1318
6.0	4.999	59	37	0.1254
			Average:	0.1081

5. Future Research

Since the establishment of the rodent embryo cell systems, especially C3H10T1/2 cells, and the focus formation assay for oncogenic transformation studies, quantitative data on dose- response curves for various dose rates and RBEs for low- and high-LET radiation have been generated by investigators in different laboratories. In general, these results were in fairly good agreement. An inverse dose-rate effect for high-LET radiation was observed as that found in animal studies. For a better understanding of the mechanism(s) of the inverse dose-rate effect, investigations have been performed to determine the cell cycle dependence, and interesting results have been obtained. The advancement of molecular biology techniques and cytogenetic methods provides new opportunities for studying the basic mechanisms of radiogenic cell transformation. Up-to-date there has been very limited studies with these new techniques. There are many critical questions in radiation cell transformation, such as, which genomic alteration(s) are essential for the initiation of radiation cell transformation, how do these alteration(s) formed, and what are the critical pathways in oncogenic transformation by radiation? Only when sufficient information on basic mechanisms of initiation, promotion and progression become available, in addition to quantitative data on dose-response relationship for various radiation modality, can proper biophysical modeling of radiation cell transformation be developed. Biophysical modeling is essential for extrapolating animal cell data to human cell, from cell to organism, and then from animal to human. Mechanistic studies with both animal and human cells are thus urgently needed. In the meantime, biophysical modeling of specific chromosome aberrations for low- and high-LET radiation should be initiated, since chromosomal aberrations data from FISH studies can be accumulated very rapidly. The processes from the formation of DNA breaks to the completion of chromosomal aberrations in cell nucleus can be considered as the initiation and fixation stages in radiation carcinogenesis. With a solid understanding of these two stages, it is then possible to extend modeling to promotion and progression pro-

TABLE 5. Iron particle hits and the formation of chromosomal exchanges.

Gy	# Particles / 20 μm²	Total # Spreads Analyzed	Total # Aberr. Observed	# Aberrants / Particle / 20 μm²
0	0	1014	2	-
0.75	0.625	1005	87	0.1385
1.0	0.833	314	39	0.1491
2.0	1.666	480	141	0.1763
3.0	2.499	104	67	0.2577
4.0	3.333	109	104	0.2862
5.0	4.166	91	120	0.3165
6.0	4.999	59	94	0.3187
			Average:	0.2347

cesses. Genomic instability can be an important mechanism in progression, and research in this field just begin. How many cancer genes are important in radiation transformation of animal and human cells, what lesions or chromosomal aberrations are critical in altering the function of these genes, and can damages in these genes lead to genomic instability and how? These questions remain to be answered.

6. References

1. Yang, T. C., Craise, L. M., Mei, M., and Tobias, C. A. (1985) Neoplastic cell transformation by heavy charged particles. *Radiat. Res.* **104**, S-177-S-187
2. Suzuki, M., Watanabe, M., Suzuki, K., Nakano, K., and Kaneko, I. (1989) Neoplastic cell transformation by heavy ions. *Radiat. Res.* **120,** 468-476
3. Hill, C. K., Buonaguro, F. M., Myers, C. P., Han, A., and Elkind, M. M. (1982) Fission-spectrum neutrons at reduced dose rates enhance neoplastic transformation. *Nature* **298**, 67-69
4. Yang, T. C., Craise, L. M., Mei, M., and Tobias, C. A. (1986) Dose protraction studies with low- and high-LET radiation on neoplastic cell transformation in vitro. *Adv. Space Res.* **6,** (11)137-(11)147
5. Rossi, H. H. and Kellerer, A. M. (1986) The dose rate dependence of oncogenic transformation by neutrons may be due to variation of response during the cell cycle. *Int. J. Radiat. Biol.* **50**, 353-361
6. Brenner, D. J., and Hall, E. J. (1990) The inverse dose-rate effect for oncogenic transformation by neutrons and charged particles: a plausible interpretation consistent with published data. *Int. J. Radiat. Biol.* **58**, 745-758
7. Elkind, M. M. (1991) Enhanced neoplastic transformation due to protracted exposure of fission-spectrum neutrons: biophysical model. *Int. J. Radiat. Biol.* **59,** 1467-1475
8. Miller, R. C., Marino, S. A., Brenner, D. J., Martin, S. G., Richards, M., Randers-Pehrson, G., and Hall, E. J. (1995) The biological effectiveness of radon-progeny alpha particles. II. Oncogenic transformation as a function of linear energy transfer. *Radiat. Res.* **142**,54-60
9. Martin, S. G., Miller, R. C., Geard, C. R., and Hall, E. H. (1995) The biological effectiveness of radon-progeny alpha particles. IV. Morphological transformation of Syrian hamster embryo cells at low doses. *Radiat. Res.* **142,** 70-77

10. Redpath, J. L. and Sun, C. (1990) Sensitivity of a human hybrid cell line (HeLa x skin fibroblast) to radiation-induced neoplastic transformation in G_2, M and mid-G_1 phase of the cell cycle. *Radiat. Res.* **121,** 206-211
11. Cao, J., Wells, R. L., and Elkind, M. M. (1992) Enhanced sensitivity to neoplastic transformation by ^{137}Cs gamma rays of cells in the G_2/M-phase age interval. *Int. J. Radiat. Biol.* **62,** 191-199
12. Cao, J., Wells, R. L., and Elkind, M. M. (1993) Neoplastic cell transformation of C3H mouse embryo cells, 10T1/2: cell-cycle dependence for 50 kV X-rays and UV-B light. *Int. J. Radiat. Biol.* **64,** 83-92
13. Miller, R. C., Geard, C. R., Geard, M. J., and Hall, E. J. (1992) Cell-cycle-dependent radiation-induced oncogenic transformation of C3H10T1/2 cells. *Radiat. Res.* **130**, 129-133
14. Redpath, J. L., Antoniono, R. J., Sun, C., Gerstenberg, H. M., and Blakely, W. F. (1995) Late mitosis/early G_1 phase and mid-G_1 phase are not hypersensitive cell cycle phases for neoplastic transformation of HeLa x Skin fibroblast human hybrid cells induced by fission-spectrum neutrons. *Radiat. Res.* **141,** 37-43
15. Saran, A., Pazzaglia, S., Pariset, L., Rebessi, S., Coppola, M., Di Majo, V., and Covelli, V. (1993) Cell cycle dependence of C3H10T1/2 cell transformation by fission spectrum neutrons and X-rays. (Abstract) Forty-First Annual Meeting of the Radiation Research Society, Dallas, TX
16. Maher, V. M., Rowan, L. A., Silinskas, K. C., Kateley, S. A., and McCormick, J. J.(1982) Frequency of UV-induced neoplastic transformation of diploid human fibroblasts is higher in xeroderma pigmentosum cells than in normal cells. *PNAS, USA* **79,** 1119-1136
17. Bettega, D., Calzolari, P., Costa, A., Noris Chiorda, G., and Tallone, L. (1995) Oncogenic transformation of C3H10T1/2 cells exposed to alpha particles: Sensitivity through the cell cycle. *Radiat. Res.* **142,** 276-280
18. Mizuki, K., Nose, K., Okamoto, H., Tsuchida, N., Hayashi, K. (1985) Amplification of cKi-ras gene and aberrant expression of c-myc in WI-38 cells transformed in vitro by gamma irradiation. *Biochem Biophys Res Com* **128,** 1037-1043
19. Billings, P. C., Shuin, T., Lillehaug, J., Miura, T., Roy-Burman, P., and Landoph, J. R. (1987) Enhanced expression and state of the c-myc in chemically and x-ray-transformed C3H10T1/2 C18 mouse embryo fibroblasts. *Cancer Res.* **47,**3643-3649
20. Endlich, B., Salavati, R., Zhang, J., Weiss, H., and Ling, C. C. (1993) Molecular analysis of rat embryo cell transformants induced by α-particles. *Int. J. Radiat. Biol.* **64,** 715-726
21. Krolewski, B. and Little, J. B. (1994) Search for oncogene mutations in X-ray-transformed mouse 10T1/2 cells by denaturing gradient gel electrophoresis blotting. *Int. J. Radiat. Biol.* **65,** 147-156
22. Yang, T. C., George, K., Tavakoli, A., Craise, L. M., and Durante, M. (1996) Radiogenic transformation of human mammary epithelial cells in vitro. *Radiat. Oncol. Investigations* **3,** 412-419
23. Reinhold, D. S., Walicka, M., Elkassaby, M., Milam, L. D., Kohler, S. K., Dunstan, R. W., and McCormick, J. J. (1996) Malignant transformation of human fibroblasts by ionizing radiation. Int. *J. Radiat. Biol.* **69,** 707-716
24. Durante, M., Grossi, G. F., and Yang, T. C. (1996) Radiation-induced chromosomal instability in human mammary epithelial cells. *Adv. Space Res.* **18,** (1/2)99-(1/2)108
25. Kelly, G., Stegelmeier, B. L., and Hahn, F. F. (1995) p53 alteration in plutonium-induced F344 rat lung tumor. *Radiat. Res.* **142**, 263-269
26. Zhang, Y. and Woloschak, G. E. (1997) Rb and p53 gene deletions in lung adenocarcinomas from irradiated and control mice. *Radiat. Res.* **148**, 81-89
27. Bouffler, S. D., Meijne, E. I. M., Huiskamp, R. and Cox, R. (1996) Chromosomal abnormalities in neutron-induced acute myeloid leukemia in CBA/H mice. *Radiat. Res.* **146**, 349-352
28. Ponnaiya, B., Cornforth, M. N., and Ullrich, R. L. (1997) Radiation-induced chromosomal instability in Balb/c and C57BL/6 mice: The difference is as clear as black and white. *Radiat. Res.* **147,** 121-125
29. Yang, T. C., Craise, L. M., Mei, M., and Tobias, C. A. (1989) Neoplastic cell transformation by high-LET radiation: Molecular mechanisms. *Adv. Space Res.* **9,** (10)131-(10)140
30. Yang, T. C., Mei, M., George, K. A., and Craise, L. M. (1996) DNA damage and repair in oncogenic transformation by heavy ion radiation. *Adv. Space Res.* **18**, (1/2)149-(1/2)158

INDUCIBLE RESPONSES AND PROTECTIVE FUNCTIONS OF MAMMALIAN CELLS UPON EXPOSURE TO UV LIGHT AND IONIZING RADIATION

B. KAINA [1], S. HAAS [1], S. GRÖSCH [1], T. GROMBACHER [1],
J. DOSCH [1], T. BISWAS [2], I. BOLDOGH [2], S. MITRA [2],
AND G. FRITZ [1]

[1] *Division of Applied Toxicology, Institute of Toxicology, Medical Faculty, University of Mainz*
Obere Zahlbacher Straße 67, D - 55131 Mainz, Germany
[2] *Searly Center of Molecular Science and 3Department of Microbiology and Immunology, University of Texas Medical Branch*
Galveston, TX 77555, USA

Abstract

In mammalian cells, ultraviolet (UV) light as well as ionizing radiation (IR) transcriptionally activate the early-responsive genes *c-fos*, *c-jun*, *junB* and *junD*. The induction of *fos* and *jun* by UV-C is currently understood to occur via activation of the EGF receptor and the Ras, Raf, ERK and JNK cascade leading ultimately to phosphorylation of transcription factors such as Fos and Jun (AP-1). This, finally, gives rise to transcriptional activation of AP-1 dependent target genes. Another gene we have recently demonstrated to be immediate-early inducible upon UV-irradiation encodes the Ras-related small GTPase RhoB. The pathway of *rhoB* induction appears to be different from *fos/jun* because (i) *rhoB* does not respond to treatment with TPA and (ii) UV stimulation of *rhoB* appears to be independent of ERK and JNK. The physiological function of RhoB is still poorly understood. Whether genes involved in cellular defense against UV-C and IR are targeted by Fos/Jun is an intriguing question. The fact that Fos/Jun are induced by genotoxic exposures indicates that this very early genotoxic response plays a physiological role in cellular defense against induced DNA damage. We have experimentally addressed this by utilizing cells for genotoxicity studies which are unable to induce c-Fos because of homozygous mutation of the gene. c-Fos deficient cells were shown to be hypersensitive to the cytotoxic effect (both necrotic and apoptotic death) of UV-C. Also, they responded with a markedly elevated level of UV-induced chromosomal aberrations. The extent of overall DNA repair synthesis was not significantly reduced, indicating that DNA repair deficiency is not the cause of the UV hypersensitivity observed. Interstingly, c-Fos deficient cells were cross-sensitive to a variety of chemical DNA damaging agents, but not to IR. The recovery from UV and chemical agent-induced immediate block to DNA replication was impaired in c-Fos deficient cells. From this we hypothesized that induction of Fos/Jun is required to activate gene(s) the

C. Baumstark-Khan et al. (eds.),
Fundamentals for the Assessment of Risks from Environmental Radiation, 289-300.

products of which are involved in abolition of replication blockage upon the induction of DNA damage. Induction of Fos/Jun as well as p53 may also be directly involved in regulation of DNA repair genes, such as O^6-methylguanine-DNA methyltransferase which is inducible upon genotoxic treatment. Another gene induced especially by oxidative stress is apurinic endonuclease which appears to be involved in adaptive response. Overexpression of RhoB in NIH3T3 cells gives rise to an altered survival response upon genotoxic exposure indicating this inducible function also to play a role in cellular defense.

1. Introduction

Mammalian cells respond on exposure to ultraviolet (UV) light, ionizing radiation (IR) or chemical DNA damaging agents with modification of a diversity of preexisting proteins and activation of genes [1]. The responses can be grouped, according to the time sequence of its appearance, into early, delayed and late responses. The immediate-early responses include both activation of preexisting proteins (such as EGF receptor, Ras, Raf, stress-activated protein kinases, Jun) as well as transcriptional activation of immediate-early genes, e.g. *c-fos*, *c-jun* and other members of the *fos/jun* family [2, 3]. These immediate responses give rise to increase in the activity of various genes whose biological significance is only partially understood. To the delayed responses belong the activation of repair functions leading to removal of damage from DNA and bypass of lesions during replication. The induction of genotoxic effects as well as reproductive cell death can be considered as late response. Interphase death observed with high radiation doses might be an early response. Whether apoptosis and necrosis induced by genotoxic treatments occurs early, i.e. within the cell cycle of treatment, appears to depend on the experimental system used. Our data obtained with rodent cells treated with UV-C and chemical genotoxic agents revealed apoptosis to be a late response. Here we wish to address notably the issue of induced gene activation by genotoxic stress, with special emphasis of UV-C and ionizing radiation (X-rays, γ–rays) and its possible role for cellular defense against the harmful effects of radiation.

2. Immediate-early Genes Induced by Radiation

The majority of data on gene activation was gained from studies with cells cultivated *in vitro* and irradiated with UV-C light (notably 260 nm). Although human individuals are usually not exposed to this wavelength, the responses observed may be considered to be representative, being a general phenomenon observed after treatment with many, if not all, genotoxic agents. For UV-C, a very early trigger of gene induction consists in modification of the epidermal growth factor (EGF) receptor by phosphorylation. This in turn gives rise to activation of GTP-binding proteins (e.g. Ras and Rac), the activation of a cascade of protein kinases and finally phosphorylation of preexisting c-Fos and c-Jun. This phosphorylation is thought to be decisive for transactivation of *c-fos* and *c-jun* itself.

The increase in the expression of *c-fos* and *c-jun* mRNA by UV-C is dose dependent (Fig. 1A). We would like to stress the point that the induction of *c-fos* and *c-jun* mRNA can be observed already with quite a low dose range of UV-C (below 10 J/m^2) which is not yet significantly toxic for the cells.

Recently, we have identified a novel type of immediate-early inducible gene, represented by *rhoB* [4]. *rhoB* does not code for a transcription factor. It encodes the Ras-related small GTPase RhoB which belongs to the family of Rho proteins covering RhoA, RhoB, RhoC, Rac and Cdc42. As shown in Fig. 1B, *rhoB* mRNA level but not the amount of *rac* mRNA is largely enhanced 45 min after UV-C treatment of NIH3T3 fibroblasts. As detected by Westen blot analysis, RhoB proteins generated upon UV-C irradiation immediately translocate to the membrane (G. Fritz and B. Kaina, in preparation).

As a general phenomenon, both UV-C, ionizing radiation and radiomimetical chemical agents evoke the induction of the immediate-early genes noted above as well as genes which respond at a slightly later time (several hours after the genotoxic treatment). Table 1 summarizes various of these inducible genes which are supposed to be

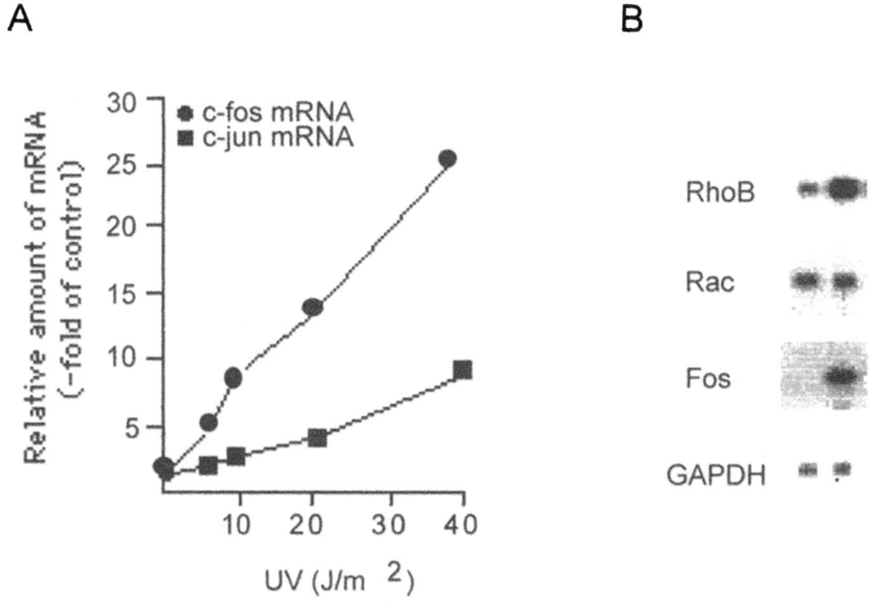

Figure 1. UV-irradiation results in rapid induction of genes encoding c-Fos, c-Jun and RhoB.
A: CHO cells were irradiated with 5-40 J/m^2 of UV-C light. 60 min after treatment RNA was isolated and subjected to Northern blot analysis using *c-fos* and *c-jun* specific hybridization probes, respectively. The filter was rehybridized with GAPDH probe. Autoradiograms were densitometrically analzed and the relative amount of c-fos and c-jun mRNA was calculated in relation to GAPDH mRNA
B: 45 min after UV-irradiation (40 J/m^2) of logarithmically growing NIH3T3 cells, RNA was isolated and subjected to Northern blot analysis using ^{32}P-labelled hybridization probes as indicated. Shown is the autoradiography

involved in cellular defense. Besides UV-C and IR, the simple methylating agents MMS and MNNG have been shown to be strong inducers of *fos* and *jun* [5]. Although the primary targets for UV-C, ionizing radiation and alkylating agent induced signaling are very likely not the same, signaling converges and finally may activate the expression of a similar pattern of genes. An important question is, whether the immediate-early inducible gene products Fos and Jun forming the heterodimeric transcription factor AP-1 are essential components in the induction of DNA repair genes, some of which are listed in Table 1. This issue will be addressed later.

TABLE 1. Inducibility of various genes after treatment with different types of DNA damaging agents. Mammalian cells were treated or not with UV-C (30 J/m^2), ionizing radiation (2-10 Gy) or the alkylating agent MNNG (10-50 μM). 30 min (*c-fos, c-jun*, rhoB) or 4-24 h (MGMT, MPG, APE, DNA ligase I) after treatment, RNA was isolated and subjected to Northern blot analysis using ^{32}P-labeled hybridization probes for the genes indicated. Induction of DNA ligase I by UV-C was reported previously [6].

	Treatment		
Gene	UVC	X-rays [1]	MNNG
c-fos	+	+	+
c-jun	+	+	+
RhoB	+	O	+
MGMT	+	+	+
MPG	Nd	-	-
APE	Nd	O	-
Ligase I	+	-	-

+ Strong induction, o Weak induction; - No induction; Nd not determined; [1] X-rays or radiomimetic agents such as H_2O_2

3. c-Fos Plays an Important Role in Cellular Defense

The fact that c-Fos is induced not only by growth factors and tumor promoters, but also by DNA damaging agents rised the hypothesis that Fos/AP-1 induction plays a role in cellular defense against the harmful effects of mutagens/carcinogens. In fact, inhibition of PKC by tyrosine kinase inhibitors (e.g. tyrphostin) rendered cells more sensitive to the killing effect of UV-C [7] which supported this view. However, treatment with an inhibitor may provoke various side effects because the inhibitor function may not be related solely to the downstream target considered. We have addressed this issue by utilizing cell lines not able to express c-Fos due to mutational inactivation of the c-*fos* gene (derived from c-*fos* knockout mice). These studies revealed a decisive role of c-Fos/AP-1 in cellular defense by showing that c-Fos deficient cells are significantly more sensitive to UV-C than c-Fos expressing cells [8]. As shown in Fig. 2, growth of c-*fos*-/- cells is clearly more inhibited than that of wild-type cells upon UV-C irradiation. Also,

UV-C had a much stronger effect on reproductive cell death in c-*fos* -/- than c-*fos* +/+ cells. Interestingly, cells derived from heterozygous mice expressing c-Fos to intermediary level (c-*fos* +/-) showed an intermediary killing response, indicating a dose effect of Fos expression (Fig. 2). Increased cell killing was also observed upon treatment with UV-mimeting agents (i.e. agents inducing bulky lesions) such as 4-nitroquinoline oxide (NQO) and benzo(a)pyrene diol epoxide (BPDE) as well as the methylating agents N-methyl-N'-nitro-N-nitrosoguanidine and methyl methanesulfonate. Furthermore, the cells deficient in c-Fos expression were hypersensitive regarding the induction of chromosomal aberrations upon treatment with these agents. Interestingly, this hypersensitivity did not pertain to ionizing radiation, against which c-fos -/- cells reacted in nearly the same way as the Fos proficient control [9].

To check the possibility, whether or not lack of c-Fos expression in c-*fos* -/- cells was compensated for by an increase in the expression of c-Jun, Northern blot analyses were performed. As shown in Fig. 3, both c-Fos deficient and c-Fos proficient cells responded to UV irradiation with the induction of c-*jun* mRNA. The level of induction (which is considered in relation to the housekeeping gene GAPDH) was similar in both cell types. Therefore it is concluded that deficiency in c-Fos leads to lower level of AP-1 induction. Also, Fos/Jun heterodimers can not be formed in these cells which, in the

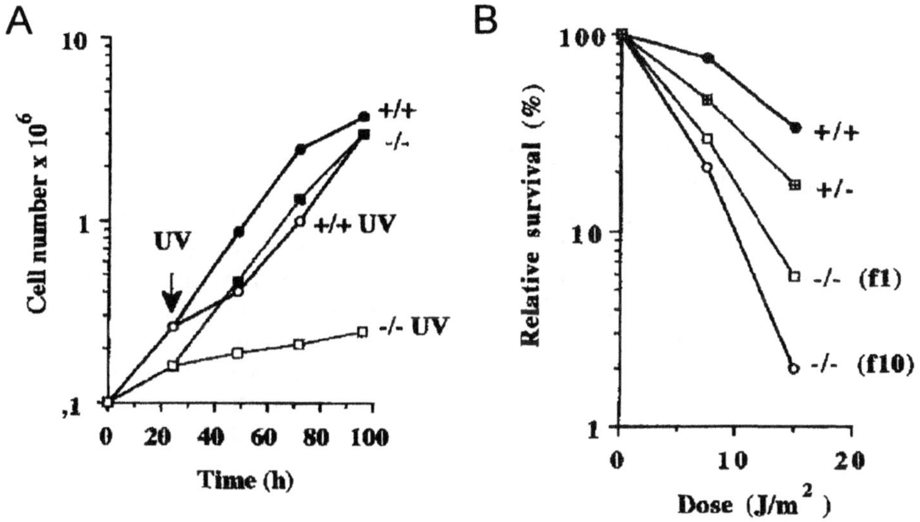

Figure 2: Effect of c-Fos deficiency on cellular growth and sensitivity to UV-C light.
A: 24 h after seeding, mouse fibroblast cell lines proficent (+/+) and deficient (-/-) for c-Fos were UV-irradiated or not. The time point of treatment is indicated by an arrow. Afterwards, cell growth was analyzed by counting cell number in intervals of 24 h.
B: c-Fos proficient (+/+) and deficient (-/-) cells, as well as cells heterozygous for *c-fos* (+/-) were seeded at low density (500 cells per dish) and irradiated 8 h later (15 J/m^2). After an incubation period of 8 d cells were fixed, stained and colonies were counted. The number of colonies in untreated control was set to 100%.

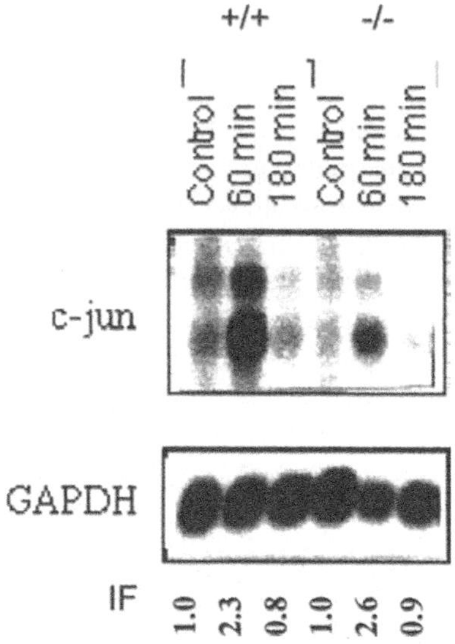

Figure 3: UV-stimulated expression of *c-jun* is similar in c-Fos deficient (-/-) and proficient (+/+) cells. Cells lacking (-/-) or harbouring (+/+) c-Fos were UV-irradiated (30 J/m^2). 60 min later, RNA was isolated and subjected to Northern blot analysis using a *c-jun* specific^{32}P-labeled hybridization probe. For quantitation of *c-jun* induction, filter was rehybridized with GAPDH probe. Fold increase (IF) in *c-jun* mRNA was determined after densitometrical analysis of the autoradiogram by relating the amount of *c-jun* mRNA to that of GAPDH mRNA.

wild-type, may target individual promoters thus transactivating particular genes involved in defense. Very likely, impairement of DNA repair gene activity is not the underlying basis of c-Fos mediated protection, because we did not find reduced repair activity in c-Fos deficient cells that could account for the hypersensitive phenotype. However, DNA replication upon DNA damage induction was impaired indicating that c-Fos/AP-1 triggers the synthesis of a protein involved in S-phase checkpoint control [8, 9]. Importantly, the hypersensitivity of c-Fos deficient cells was found to pertain to all agents whose clastogenic effect is S-phase dependent. Because IR induces directly DNA double-strand breaks giving rise to aberrations, the clastogenic effect of IR is largely independent on S-phase progression upon radiation which might explain why c-*fos* -/- cells were not significantly hypersensitive to γ-rays.

4. Is RhoB Involved in UV Defense?

To address the question of whether *rhoB* induction by UV-C exerts a protective effect, NIH3T3 cells were stably transfected with a pCDNA-*rhoB* expression construct. Single clones stably overexpressing RhoB were isolated and analyzed as to their cell killing response upon the induction of DNA damage. As shown in Fig. 4, cells overexpressing RhoB were clearly more resistant to UV-C light than control cells not overexpressing

RhoB, indicating that *rhoB* induction is part of the cellular defense program directed against radiation induced cytotoxicity.

5. Induction of DNA Repair Genes in Mammalian Cells by UV Light and IR

A very important question is whether DNA repair genes can be activated by radiation-induced genotoxic stress. The genotoxic, mutagenic and carcinogenic effects of UV-C and IR is related to the induction of DNA damage, and therefore increased repair of critical lesions induced by UV-C and IR would protect cells against the harmful effects of radiation. A couple of genes involved in repair of UV light-induced DNA damage has been cloned [10], but evidence has not been provided uptil now that either one of them is inducible by UV-C resulting in increase of overall or gene-specific DNA repair.

To address the question of repair gene activation by IR, we have analyzed the mRNA expression of a variety of DNA repair genes upon the induction of DNA damage by radiation. As shown in Fig. 5, X-rays (a dose of 2 Gy) induce very effectively an increase in MGMT mRNA. Also apurinic endonuclease (APE) mRNA is induced, although at lower level than observed for MGMT (Fig. 5A). There was no induction, at least under the conditions of radiation doses and recovery times used [11], of DNA polymerase ß (Polß), DNA ligase I and N-methylpurine-DNA glycosylase (MPG) (Fig. 5B).

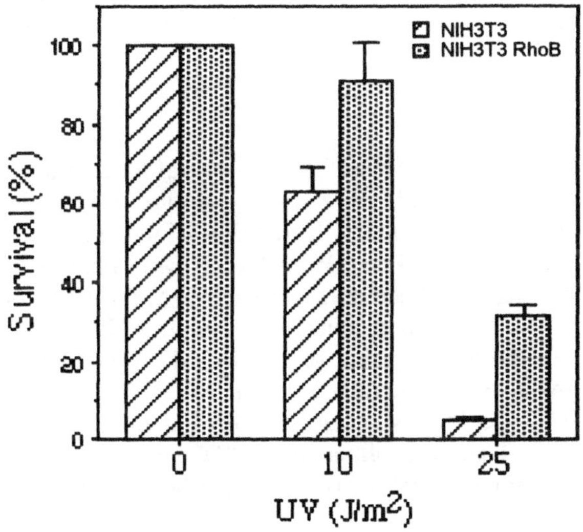

Figure 4. RhoB overexpression renders cells more resistant to UV-C light.
NIH3T3 cells were stably transfected with *rhoB* expression vector. Single cell clones were isolated and RhoB overexpression was analyzed by Western blotting using RhoB specific antibody. Transgenic and wildtype cells were analyzed as to their sensitivity to UV-C induced DNA damage measuring reproductive cell death in a colony formation assay system. The number of colonies in untreated control was set to 100%

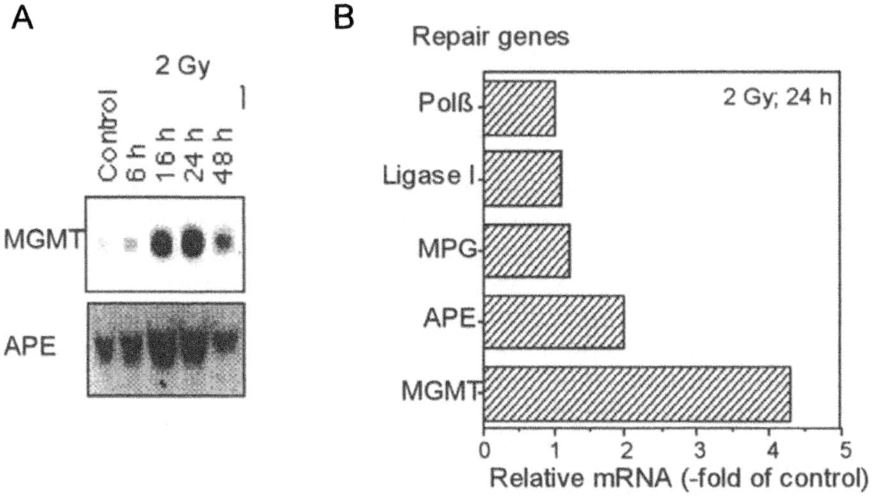

Figure 5. Inducible expression of DNA repair genes upon treatment with X-rays.
A: Logarithmically growing H4 cells were irradiated with 2 Gy. Up to 48 h after treatment RNA was isolated. 20 µg of total RNA were subjected to Northern Blot analysis using MGMT and APE specific hybridization probes.
B: Cells were irradiated with 2 Gy and total RNA was isolated 24 h after treatment. 20 µg of RNA were subjected to Northern Blot analysis using the hybridization probes indicated. For quantitation, filter was rehybridized with GAPDH probe. After densitometrical analysis of the autoradiograms, relative amount of RNA was determined by relating the mRNA of the corresponding repair gene to the level of GAPDH mRNA. Relative mRNA in untreated control was set to 1.0.

Induction of the repair gene MGMT occurs several hours after treatment with UV-C, X-rays or alkylating agents [11, 12] and can therefore be considered to be a late response upon induction of DNA damage. It would be of much interest to see whether the induction of this gene is triggered by AP-1 which is early induced by the same agents. The promoter of this repair gene harbours two AP-1 sites for which binding of AP-1 has been demonstrated [13]. Insertion of mutations disrupting either one or both AP-1 sites abolished Fos/Jun mediated activation of the MGMT promoter [13] indicating that Fos/Jun is indeed involved in the regulation of MGMT induction. It is important to note, however, that MGMT specifically repairs O^6-alkylguanine, which is not produced in DNA by UV-C and IR. Therefore, although MGMT is induced by radiation, it does not protect against the genotoxic effects of UV light and IR [12 and unpublished data].

6. Induction of MGMT by IR and MNNG Provokes Resistance to Alkylating Agents

In many studies on gene induction by genotoxic agents exceptional high mutagen doses, which are in itself highly cytotoxic, have been used in order to be able to detect a response. Therefore, often the question remained whether or not the effects observed were biologically significant in that induced gene expression is protective. If cells cultivated as mass culture were treated with a dose of γ-rays sufficient for activating MGMT gene expression, more than 50% of cells survived. If these cells were subsequently challenged with MNNG with a dose sufficiently high to induce gene mutations, the challenge dose-induced mutations were clearly reduced as compared to the mutation frequency observed in cells which received the challenge dose only (Fig. 6). Obviously, MGMT which was elevated in its level by pretreatment of cells with either IR or an alkylating agent, protected from mutation induction by an O^6-methylguanine generating compound. This effect is reminiscent to the adaptive response of *E. coli* [10]. Although in mammalian cells a broader spectrum of agents is obviously able to induce MGMT, as compared to bacteria, the mechanism of protection is the same; it is based on increased repair of O^6-methylguanine which is a principal premutagenic and pretoxic lesion.

7. Induction of Apurinic Endonuclease and the Adaptive Response to IR

A frequently reported phenomenon that has not yet been solved as to the underlying cause is the adaptive response of mammalian cells to the clastogenic effect of IR [14].

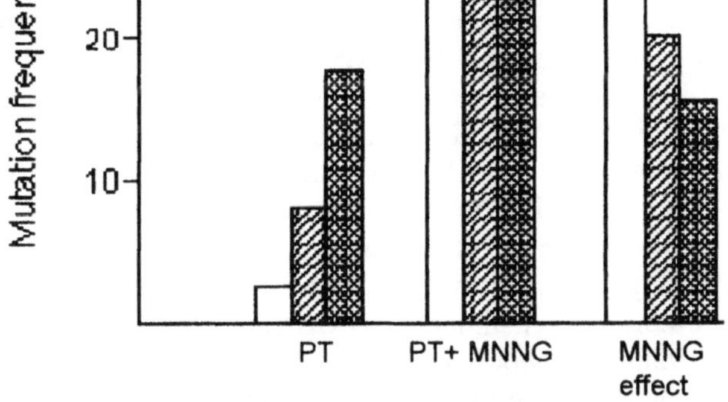

Figure 6. MGMT induction protects cells from the mutagenic effect of MNNG. H4 cells were left untreated (control) or were pretreated with X-rays (2 Gy) or MNNG (5 μM). 48 h after pretreatment, cells were exposed to a high dose of MNNG (15 μM). Shown is the mutation frequency on the HGPRT locus after pretreatment alone (PT), under conditions of pretreatment and MNNG challeng (PT + MNNG) and the corrected effect of the challenge dose induced mutations (MNNG effect). Data shown are mean values from 2-5 independent experiments.

As shown above, the repair enzyme APE is enhanced on mRNA expression level in rat liver cells (H4) upon their exposure to g-rays. This has also been found in CHO cells [22] and V79 fibroblasts [23] cells upon treatment with agents generating oxidative species such as hydrogen peroxide and sodium hypochlorite. Induction of APE was observed on the level of APE-promoter, mRNA, protein and enzyme activity. It resulted in lowering chromosomal aberration frequency induced by a challenge dose of the same agent [22]. Thus, induction of APE, which is required for repair of oxidative bases such as 8-oxy-guanine, appears to be involved in an adaptive response against oxidative DNA damaging agents. It remains to be determined, however, whether the adaptive response alluded to above, which has been observed in human lymphocytes exposed to low doses of IR and which is specifically directed against the clastogenic effect of IR, is caused by APE induction.

8. Signal Transduction: Concluding Remarks

Besides analyzing the physiological role of genotoxic stress-induced gene functions, the majority of studies has been focused on elucidating the signal pathways involved. A current view is shown in Fig. 7. It is hypothesized that UV radiation as well as the radiomimetic agent hydrogen peroxide causes activation of growth factor receptors, in particular the EGF-receptor [15] resulting in the activation of Ras and finally the activation of mitogen-activated protein kinases (MAPK) such as ERK1/2, JNK and p38. These protein kinases are known to phosphorylate various transcription factors, in particular c-Jun, c-Fos, Elk-1, ATF-2 and GADD153. There is evidence that phosphorylation of these proteins stimulates their transactivating activity [16], thus resulting in transcriptional activation of "immediate-early" genes such as c-*fos* and c-*jun* . Furthermore, atypical protein kinase C isoforms, which are targets of phosphatidylinositol-3 kinase, are also known to activate the Raf/ERK signaling cascade [17]. The extent of activation of either ERK or JNK-p38 MAP kinases is thought to determine whether the damaged cell undergoes necrotic or apoptotic cell death or whether it will survive [18]. Based on the observation that i) alkylating agents do not induce signaling via ERK [19] and ii) the activation of JNKs by UV-C but not by MMS is blocked by the EGF receptor binding drug suramin [19], it is speculated that the primary targets of UV-C and chemical DNA damaging agents are different. Whether ionizing radiation acts by activating essentially the same signaling pathway as UV-C is unclear. It should be noted that at least part of the effect of UV-C is mediated by intracellular radical formation, which makes it likely that X- or γ-rays whose genotoxic effects are largely due to radical formation, act in the same way. Interestingly, treatment of cells with both UV-C and X-rays results in the generation of cer amides which stimulate JNK/SAPKs activity and are discussed to be required for stress-induced apoptosis [20]. Induction of Fos/Jun/AP-1 appears to be involved in protection against cell killing and is required for maintaining genomic stability upon UV irradiation [8, 9]. In contrast to c-*fos* and c-*jun*, the UV-stimulated expression of *rhoB* appears to be regulated independent of the MAP kinase pathway. Furthermore, promoter analysis of *rhoB* indicates negative autoregulation of *rhoB* expression by RhoB protein itself [21]. Whether and to what extent the transient overexpression of RhoB is causally involved in cellular defense against cell killing and chromoso-

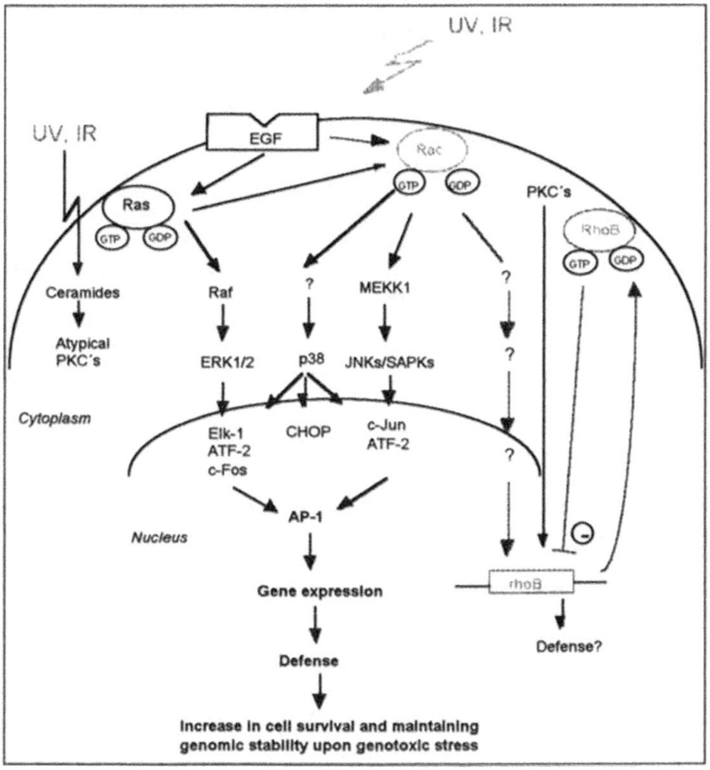

Figure 7: Signal pathways involved in the regulation of the expression of immediate-early genes

mal changes induced by UV-light, ionizing radiation and chemical mutagens by modulating one of the pathways triggering apoptosis and cell cycle progression upon the induction of DNA damage is currently under study.

Acknowledgements. We are grateful to U. Eichhorn for technical assistance. Various c-Fos deficient cell lines were kindly provided by E. Wagner, Vienna. Work was supported by DFG-KA 724/4-3, SFB 519, B4 and Stiftung Rheinland-Pfalz to B.K.and U.S. Public Health Service grants R01 ES08457, ES07572 and P01 AG10514 to S.M.

9. References

1. Herrlich, P., Ponta H., Rahmsdorf H.J. (1992) DNA Damage-Induced Gene Expression: Signal Transduction and Relation to Growth Factor Signaling. *Rev. Physiol. Biochem. Pharmacol.*, **119**, 187-212.
2. Rahmsdorf, H.J. (1996) Jun: transcription factor and oncoprotein. *J. Mol. Med.*, **74**, 725-745.
3. Angel, P., Karin M. (1991) The role of Jun, Fos and the AP-1 complex in cell proliferation and transformation. *Biochimica et Biophysica Acta*, **1072**, 129-157.

4. Fritz, G., Kaina B., Aktories K. (1995) The Ras-related small GTP-binding protein RhoB is immediate-early inducible by DNA-damaging treatments. *J. Biol. Chem.*, **270**, 25172-25177.
5. Dosch, J., Kaina B. (1996) Induction of c-fos, c-jun, junB and junD mRNA and AP-1 by alkylating mutagnes in cells dificient and proficient for the DNA repair protein O6-methylguanine DNA methyltransferase (MGMT) and tis relationship to cell death, mutation induction and chromosomal instability. *Oncogene*, **13**, 1927-1935.
6. Montecucco, A., Savini E., Biamonti G., Stefanini M., Focher F., Ciarrocchi G. (1995) Late induction of human DNA ligase I after UV-C irradiation. *Nucl. Acid Res.*, **23**, 962-966.
7. Devary, Y., Gottlieb R.A., Smeal T., Karin M. (1992) The mammalian ultraviolet response is triggered by activation of src tyrosine kinases. *Cell*, **71**, 1081-1091.
8. Haas, S., Kaina B. (1995) c-Fos is involved in the cellular defence against the genotoxic effect of UV radiation. *Carcinogenesis*, **16**, 985-991.
9. Kaina, B., Haas S., Kappes H. (1997) A general role for c-Fos in cellular protection against DNA-damaging carcinogens and cytostatic drugs. *Cancer Res.*, **57**, 2721-2731.
10. Friedberg, E.C., Wallner G.C. and Siede W. (1995) *DNA Repair and Mutagenesis*. ASM Press, Washington.
11. Grombacher, T., Kaina B. (1996) Isolation and analysis of inducibility of the rat N-methylpurine-DNA glycosylase promoter. *DNA Cell Biol.*, **15**, 581-588.
12. Fritz, G., Tano K., Mitra S., Kaina B. (1991) Inducibility of the DNA repair gene encoding O^6-methylguanine-DNA methyltransferase in mammalian cells by DNA-damaging treatments. *Mol. Cell. Biol.*, **11**, 4660-4668.
13. Boldogh, I., Ramana C.V., Chen Z., Biswas T., Hazra T., Grösch S., Grombacher T., Mitra S., Kaina B. (1998) Regulation of human O6-methylguanine-DNA methyltransferase via protein kinase C-mediated signalling. *Cancer Res.* **58**, 3950-3956
14. Shadley, J.D., Wolff S. (1987) Very low doses of X-rays can cause human lymphocytes to become less susceptible to ionizing radiation. *Mutagenesis*, **2**, 95-96.
15. Knebel, A., Rahmsdorf H.J., Ullrich A., Herrlich P. (1996) Dephosphorylation of receptor tyrosine kinases as target of regulation by radiation, oxidants or alkylating agents. *EMBO J.*, **15**, 5314-5325.
16. Canman, C.E., Kastan M.B. (1996) Three paths to stress relief. *Nature*, **384**, 213-214.
17. Akimoto, K., Takahashi R., Moriya S., Nishioka N., Takayanagi J., Kimura K., Fukui Y., Osada S., Mizuno K., Hirai S., Kazlauskas A., Ohno S. (1996) EGF or PDGF receptors activate atypical PKClambda through phosphatidylinositol 3-kinase. *EMBO J.*, **15**, 788-798.
18. Xia, Z., Dickens M., Raingeaud J., Davis R.J., Greenberg M.E. (1995) Opposing effects of ERK and JNK-p38 MAP kinases on apoptosis. *Science*, **270**, 1326-1331.
19. Wilhelm, D., Bender K., Knebel A., Angel P. (1997) The level of intracellular glutathione is a key regulatior for the induction of stress-activated signal transduction pathways including Jun N-terminal protein kinases and p38 kinase by alkylating agents. *Mol. Cell. Biol.*, **17**, 4792-4800.
20. Verheij, M., Bose, R., Lin, X.H., Yao, B., Jarvis, W.D., Grant, S., Birrer, M.J., Szabo, E., Zon, L.I., Kyriakis, J.M., Haimovitz-Friedman, A., Fuks, Z., Kolesnick, R.N. (1996) Requirement for ceramide-initiated SAPK/JNK signaling in stress-induced apoptosis. *Nature*, **380**, 75-79.
21. Fritz, G., Kaina B. (1997) rhoB encoding a UV-inducible Ras-related small GTP-binding protein is regulatio by GTPases of the Rho family an independent of JNK, ERK and p38 MAP kinase. *J. Biol. Chem.*, **272**, 30637-30644.
22. Grösch, S., Fritz, G., Kaina, B. (1998) Apurinic endonuclease (Ref-1) is induced in mammalian cells by oxidative stress and involved in clastogenic adaptation. *Cancer Res.* **58**, 4410-4416
23. Ramana, C.V., Boldogh, I., Izumi, T., Mitra, S. (1998) Activation of apurinic/apyrimidinic endonuclease in human cells and reactive oxygen species and its correlation with their adaptive response to genotoxicity of free radicals. *Proc. Natl. Sci. USA* **95**, 5061-5066

DIFFERENCES IN THE INDUCTION OF APOPTOSIS AFTER TREATMENT WITH γ-IRRADIATION AND C2-CERAMIDE

E. BÁRTOVÁ [1)] AND A. ŠPANOVÁ [2)]

[1)] *Institute of Biophysics, Academy of Sciences of the Czech Republic, Brno, Czech Republic*

[2)] *Department of Microbiology, Faculty of Sciences, Masaryk University, Brno, Czech Republic*

Abstract

Cell death in organisms can be caused by two distinct mechanism: apoptosis and necrosis. These two processes were studied in human leukemic cell lines HL-60 and U-937 after exposure to γ-radiation (6 Gy) and C2-ceramide (μM). Apoptosis was induced by C2-ceramide treatment in both cell lines, whereas after γ-irradiation it was detected only in U-937 cells, wheras HL-60 cells died by necrosis. The different behaviour of U-937 and HL-60 cells is apparently caused by differences in expression of some protooncogenes in the two cell lines. Two different apoptotic pathways may exist: one, observed only after the exposure to C2-ceramide, is ceramide-induced apoptosis and the other, observed after γ-radiation treatment, is p53-dependent.

1. Introduction

Apoptosis is a genetically determined cell suicide programme, which plays an essential role in the control of development and hemopoiesis. Apoptosis is a physiological process of cell death, while necrosis is a pathophysiological one. Apoptosis and necrosis can be detected by many methods, e.g. by morphological analysis, DNA fragmentation test, TUNEL test or with the aid of flow cytometry.

Apoptosis can be distinguished morphologically from necrosis by the presence of apoptotic bodies (Fig.1), which are swallowed by cells with phagocytic properties. Apoptosis is the natural death of the cells: it does not cause any inflammatory reactions in the organism. On the other hand, necrosis is a pathophysiological phenomenon, which results in cell lysis [3, 6] (Fig.1), destruction of cell organelles and inflammation responses. From the molecular-biological point of view apoptosis is a genetically controlled process, for whose induction or inhibition several genes are responsible, whereas genetic control of necrosis was not demonstrated [2, 7]. Several apoptotic pathways exist and their knowldge could be of great importance for their modulation, which might be exploited in tumour therapy. The aim of the present study was to investigate the activation of apoptosis in human leukemic cell lines HL-60 and U-937. The results of these experiments will be used in theoretical studies of apoptotic pathways.

C. Baumstark-Khan et al. (eds.),
Fundamentals for the Assessment of Risks from Environmental Radiation, 301-304.

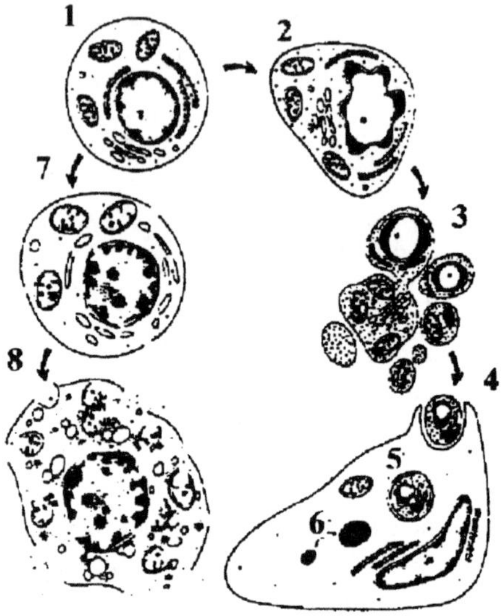

Figure 1. Sequence of ultrastructural changes in cells undergoing apoptosis (2-6) and necrosis (7-8). According to [7].

2. Material and Methods

2.1. CELL CULTIVATION

Human leukemic promyelocytic cell line HL-60 and human leukemic promonocytic cells were cultivated ***in vitro***. The cells suspended in RPMI-1640 medium complemented with 10 % of foetal calf serum were incubated in atmosphere containing 5 % CO_2 and 95 % of air atmosphere at 37 °C. Passages 15-45 were used in the experiments.

2.2. CELL TREATMENT

2.2.1. γ-Irradiation

The cells were irradiated with γ-rays (^{60}Co), dose 6 Gy. Apoptosis or necrosis were detected 24 hours after the exposure to radiation.

2.2.2. C2-Ceramide Treatment

C2-ceramide (N-acetyl sphingosine) was used in a concentration of 5 μM.

2.3. METHODS OF APOPTOSIS DETECTION

2.3.1. Morphological Analysis

The cells were stained with May-Grünwald/Giemsa-Romanowski solution and apoptotic cells were detected microscopically.

2.3.2. DNA Fragmentation Test

DNA was isolated according to Bártová et al. [1]. DNA fragmentation was detected by gel electrophoresis on 1.8 % agarose gel after staining with ethidium bromide (1 μM/ml) and visualized under UV light.

2.3.3. Flow Cytometric Analysis

The cells were fixed in 70% ethanol and stained with propidium iodide. Fluorescence intensity was detected with a flow cytometer COULTER EPICS using argon ion laser for excitation at 488 nm. Cell cycle phases were determined and in the case of apoptosis sub G_1 peak was detected here.

3. Results and Discussion

Using the fragmentation test, apoptosis was detected in both human leukemic cell lines HL-60 and U-937 after treatment with C2-ceramide at concentration of 5 μM (Fig. 2, lanes: B1, B2 (HL-60); C (U-937)). After γ-irradiation (dose 6 Gy), apoptosis was

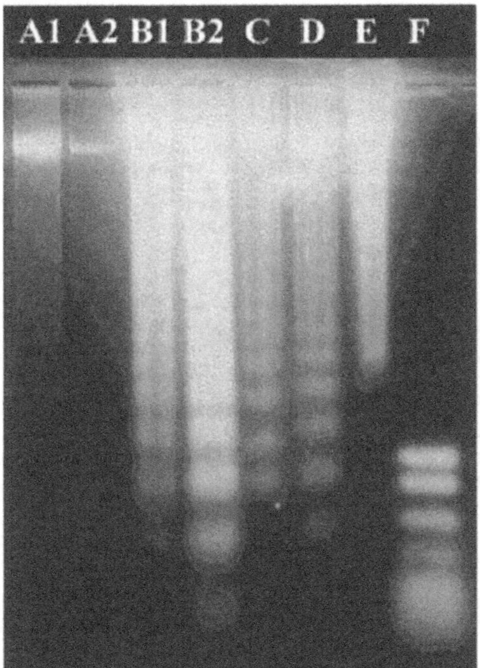

Figure 2. Results of the DNA fragmentation test in HL-60 and U-937 cells after γ-irradiation and C2-ceramide treatment.
A1: (control HL-60); A2: (control U-937); B1, B2: (HL-60/C2-ceramide); C: (U-937/C2-ceramide); D:(U-937/γ-irradiation); E: (HL-60/γ-irradiation); F: (DNA marker pBR/Msp I).

detected only in U-937 cells (Fig. 2, lane D), but not in HL-60 cells (Fig. 2, lane E), which died by necrosis. Nontreated cells were used as control (Fig. 2, lane A1 (HL-60); A2 (U-937)). DNA marker pBR 322/Msp I is shown in lane F of Fig. 2.

It is known from the literature that several apoptotic pathways exist. In the experiments described above we have encountered two apoptotic pathways: ceramide-dependent [4, 8, 9] observed after C2-ceramide treatment and a p53-dependent one [5] observed after γ-irradiation. We interpret our experiments on the basis of the fact that HL-60 cells are p53-deficient [10] and therefore apoptosis could not be detected after the action of γ-radiation. On the other hand, in γ-irradiated U-937 cells apoptosis occurred, because in these cells a residual expression of p53 exists.

In conclusion we would like to stress that apoptosis represents a very important process in cell and tissue renewal. It is a genetically controlled process that seems to be important in cancer therapy. Knowledge of apoptotic pathways should contribute to a better understanding of the process of programmed cell death and lead probably to more successful therapeutic approaches.

Acknowledgements. We thank Dr. J. Holá (Institute of Biophysics of the Academy of Sciences of the Czech Republic, Brno) and Š. Kuchtíèková (Institute of Pathological Physiology, Medical Faculty of the Masaryk University, Brno) for technical assistance. This work was supported by the Grant Agency of Czech Republic (No. 202/96/1718, 202/97/0874).

5. References

1. Bártová E., Španová A., Janáková L., Bobková M., Rittich B. (1997) Apoptotic damage of DNA in human leukemic HL-60 cells treated with C2-ceramide was detected after G1 blockade of the cell cycle. *Phys. Res.* **46,** 155-160.
2. Bissonnette R.P., Echeverri F., Mahboubi A., Green D.R.(1992), Apoptotic death induced by c-myc is inhibited by bcl-2. *Nature* **359**, 552-555.
3. Collins R.J., Harmon B.V., Gobé C.C., Kerr J.F.R. (1992) Internucleosomal DNA cleavage should not be the sole criterion for identifying apoptosis. *Int. J. Rad. Biol.* **61**, 451-453.
4. Hannun Y.A., Obei L.M. (1995) Ceramide: An intracellular signal for apoptosis. *TIBS* **20,** 73-77.
5. Kastan M.B., Kuerbitz S.J. (1993) Control of G1 arrest after DNA damage. *Environ. Health Persp.* **10**, 55-58.
6. Kerr J.F.R., Wyllie A.H., Currie A.R. (1972) Apoptosis: A basic biological phenomenon with wide ranging implication in tissue kinetics. *Br. J. Cancer* **26**, 239-257.
7. Schwartz L.M., Osborne B.A. (1995) Cell Death, Vol. **46,** Academic Press USA.
8. Obei L.M.,Linardic C.M., Karolak L.A., Hannun Y.A. (1993), Programmed cell death induced by ceramide. *Science* **259**, 1769-1771.
9. Raines M. A., Kolesnick R. N., Golde D.W. (1993) Sphingomyelinase and ceramide activate mitogen-activated protein kinase in myeloid HL-60 cells. *J. Biol. Chem.* **268,** 14572-14575.
10. Wolf D., Rotter V. (1985) Major deletion in the gene encoding the p53 tumor antigen causes lack of p53 expression in HL-60 cells. *Proc. Natl. Acad. Sci. USA* **82**, 790-794.

PENTOXIFYLLINE INHIBITS THE IRRADIATION INDUCED G_2/M BLOCK AND ALTERS DNA SYNTHESIS IN P53 MUTANT AND REPAIR DEFICIENT CELLS

L. BÖHM, T. THERON AND A. BINDER

Department of Radiation Oncology, Faculty of Medicine, University of Stellenbosch
P.O. Box 19063, Tygerberg, 7505, South Africa

1. Introduction

The G1/S and G2/M transitions have been recognised as important checkpoints in the response of cells to sublethal damage. It is commonly thought that delays in the G1 or G2 progression which typically occurs 10-18 hrs postirradion are protective mechanisms which prevent propagation of defective DNA. However, this view may be simplistic and is at variance with the fact that rejoining of DNA strand breaks is complete within 3-4 hrs postirradiation.

An alternative response to DNA damage is that cells do not return into the cycle, but die by apoptosis [1, 2]. A key molecule involved in the regulation of this damage response is the tumour suppressor gene p53 [1, 2]. Induction of this gene and rise of p53 protein after DNA damage initiates the G1/S delay and this route operates in p53 WT cells but not when p53 is disrupted or mutated [3]. Regulation of the G2/M transition appears to be less dependent upon p53 as it occurs in p53 WT and p53 mutant cell lines. In view of the importance of p53 status for checkpoint control and apoptosis [4], much attention has been focussed on drugs which abrogate cell cycle blocks and influence events controlled by p53. Abrogation of cell cycle blocks renders cells vulnerable to subsequent intervention with DNA damaging agent as shown for breast cancer MCT-7 cells in which p53 WT was disrupted by transfection with the p53 degrading E6 gene from human papilloma virus [5]. These cells did not respond to the induction of p53 and p21 by irradiation and were defective to arrest in G1 but not in G2. Subsequent inhibition of the G2/M block with Pentoxifylline (PENT) rendered the cells highly susceptible to cis-Platin [5].

The possibility of inhibiting damage induced cell cycle blocks with xanthine rugs and selecting new genotoxic agents for therapy [6]. We have exposed a number of human tumour cell lines to irradiation, determined the nature and maximum expression of cell cycle blocks and examined abrogation of the block with PENT, using BrdU incorporation for determination of S-phase activity. The results have been compared to p53 status to further evaluate the conditions required for targeted cell deregulation and susceptibility to subsequent damage by irradiation or cytotoxic drugs.

C. Baumstark-Khan et al. (eds.),
Fundamentals for the Assessment of Risks from Environmental Radiation, 305-310.

2. Material and Methods

Cell lines were maintained under standard conditions and were irradiated with 7 Gy ^{60}Co γ -irradiation, incubated for 40 hours and sampled at 2-hourly intervals. The cells were then trypsinized, centrifuged and fixed in 70% ethanol for at least 30 minutes at -20°C, stained with propidium iodide and analyzed by flow cytometry using LYSIS II software. BrdU-labelling, DNA denaturation, detection of the incorporated BrdU by immunofluorescence staining and analysis by double parameter flow cytometry was described elsewhere [7]. Percentage of active S-phase was calculated from the gated BrdU postive part of the dot plot of red fluorescence against green fluorescence.

3. Results

Growth inhibition studies on 3 human tumour cell lines indicated that 2mM PENT is distinctly subtoxic for tumour cells (not shown). Addition of PENT at maximum expression of the G2/M block shows that the drug inhibits the accumulation of cells in G2.

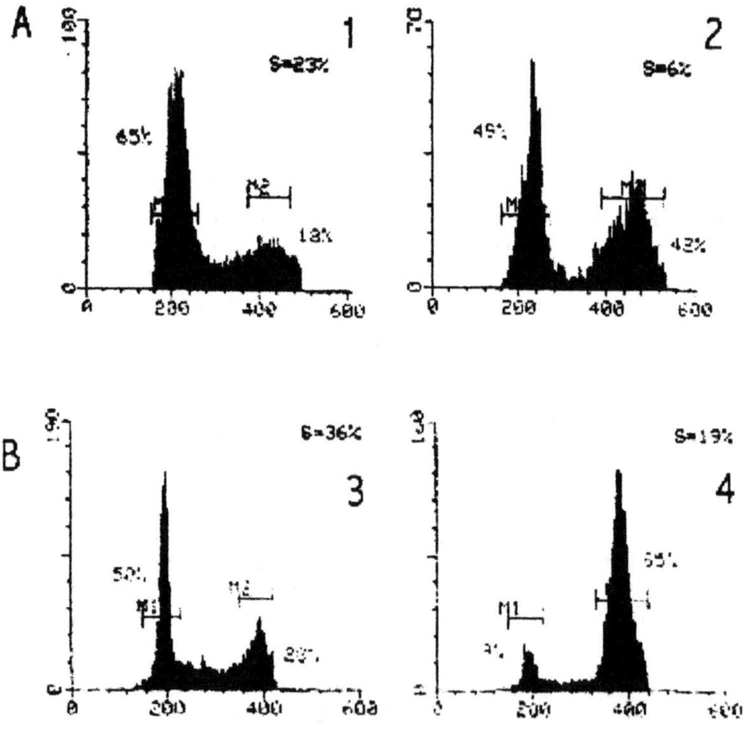

Figure 1: DNA histograms of A549 p53 WT cells (A) and MeWo p53 mutant cells (B). Panels 1, 3 show control cells and 2, 4 show irradiated cells after maximum expression of cell cycle blocks. No intervention by drugs.

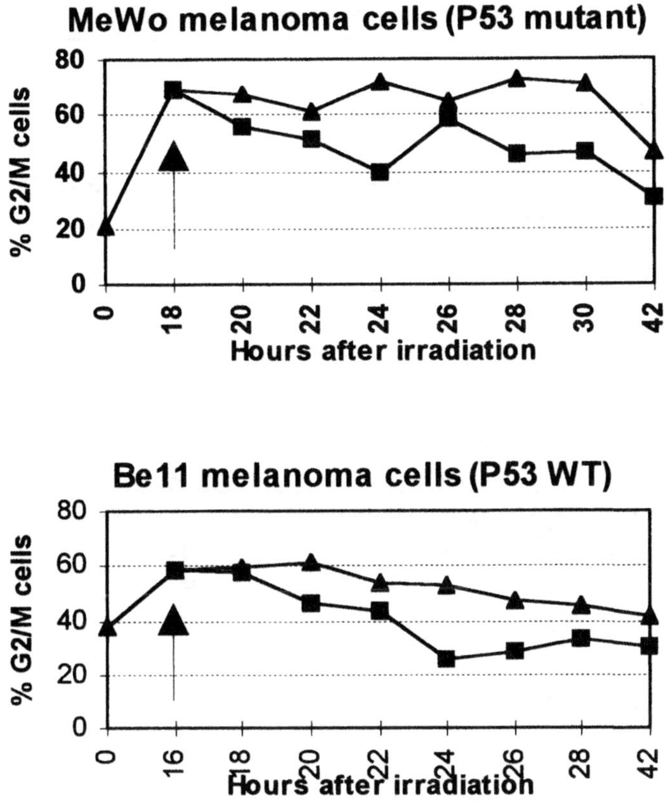

Figure 2: Influence of Pentoxifylline on irradiation induced cell cycle blocks in p53 WT and p53 mutant cells. Drug was added at maximum expression of cell cycle block (arrow). □ - Control Δ - +2 mM PENT

This would lead to early progression into M phase (Figure 2). P53 WT and p53 mutant cell lines differ in that the p53 WT cells show a G1 and a G2 block (Figure 1A, 1 and 2) whereas p53 mutant cell lines only express the G2 block (Figure 1B, 3 and 4). An influence of PENT on the G1 block may exist but was not examined here. Incorporation of BrdU into p53 WT cells after maximum expression of the G1 block showed that the number of active S-phases declines from 35% to 18% (Figure 3). This is in contrast to G2 blocked p53 mutant cells where the active S-phase cells increased from 20% to 45% (Figure 3). When PENT was added at maximum expression of the cell cycle block active S-phases for both category of cells showed a marginal reduction while the overall response pattern remained the same (Figure 3). In p53 WT cells resumption of DNA synthesis was found to synchronise more or less with the relaxation of the cell cycle blocks. In MeWo p53 mutant cells DNA-synthesis continues after irradiation damage

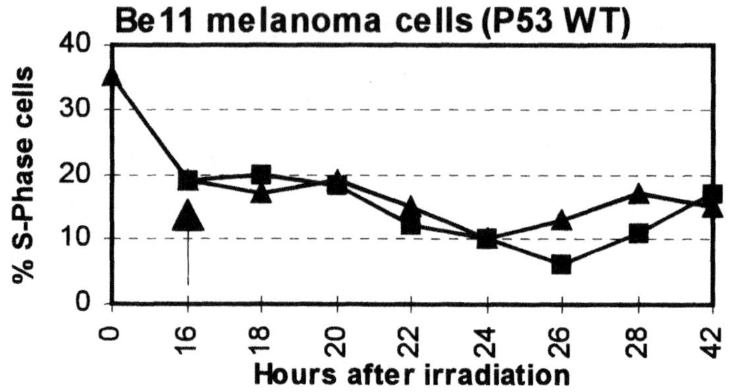

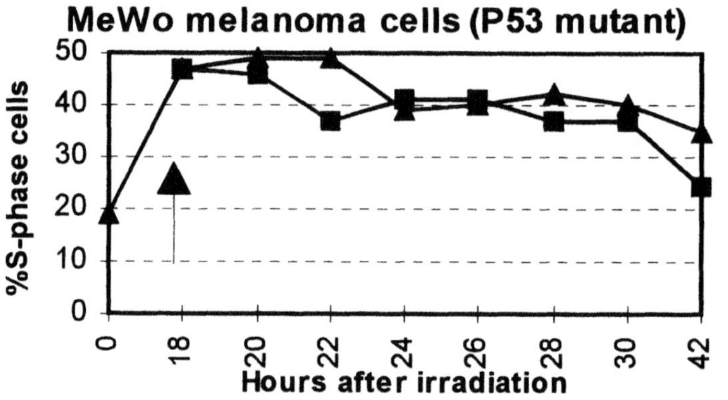

Figure 3: Active S-phases in p53 WT and p53 mutant cells recovering from irradiation damage measured by BrdU incorporation. Drug was added at maximum expression of the block (arrow). ▲ - Control ■ - + 2 mM PENT

(fig. 3) and this is in agreement with published data on the p53 mutant cell lines DLD-1, HT29 and SW80 and HCT 116 [6]. Irradiated p53 mutant cells also showed DNA at high channel numbers (polyploidy) consistent with the continuation of DNA synthesis previously noted in p21(-/-) cells [6]. This process was marginally reduced by PENT (not shown). Apoptosis was not assessed.

4. Discussion

We show that PENT inhibits G2 blocks in p53 WT cells and in p53 mutant cells. Similar results have been reported for caffeine [8, 9]. The G1 block seen in p53 WT cells gives rise to a relatively small increase of the G2 levels (Figure 1A) as compared to p53 mutants where the G2 block is large (Figure 1B). We found the G1 arrest difficult to monitor and studies using mitotic inhibitors have shown that this arrest does not occur in the first post-irradiation G1 phase [10]. Mutant and defective p53 cells show no G1 arrest [11, 12, 18] but a pronounced G2 arrest [5, 8]. Caffeine blocks the induction of p53 [13] and influences the activity of the CDC2 kinase [16] which is regulated by phosphorylation [14]. Suppression of CDC2 observed in cells where the G2 block was inhibited by PENT [16] would be in accord with the fact that PENT is a phosphatase inhibitor [15] and increases the levels of cyclic AMP [17]. The overriding impression of these experiments is that PENT effectively restores a key stimulus which promotes the G2/M transition. We have shown elsewhere that the drug upregulates cyclin B1 levels which are suppressed in the G2 block [19]. Since the G2/M block is a late response to irradiation which occurs approximately ½-1 cell cycle time after genotoxic damage, abrogation of the G2 block by PENT must differ from the documented effect of PENT to inhibit repair which is observed when the drug is present during irradiation [20]. Drug induced manipulation of cell cycle checkpoints is a promising approach for the selection of new chemotherapeutic agents and opens the possibility of selectively influencing the regulation between cell cycle arrest and apoptosis [21].

5. References

1. Jacks, T. and Weinberg, R.A. (1996) Cell-cycle control and its watchman, *Nature* **381**, 643-644.
2. O'Connor, P.M. (1997) Mammalian G1 and G2 phase chekpoints, *Cancer Surveys* **29**, 1-66, ICRF.
3. Greenblatt, M.S., Bennett, W.P., Hollstein, M. and Harris, C.C. (1994) Mutations in the p53 Tumor Suppressor Gene: Clues to Cancer Etiology and Molecular Pathogenesis, *Cancer Research* **54**, 4855-4878.
4. Polyak, K., Xia, Y., Zweier, J.L., Kinzler, K.W. and Vogelstein, B. (1997) A model for p53 induced apoptosis, *Nature* **389**, 300-305.
5. Fan, S., Smith, M.L., Rivet II, D.J. et al. (1995) Disruption of p53 Function Sensitizes Breast Cancer MCF-7 Cells to Cisplatin and Pentoxifylline, *Cancer Research* **55**, 1649-1654.
6. Waldman, T., Lengauer, C., Kinzler, K.W. and Vogelstein, B. (1996) Uncoupling of S phase and mitosis induced by anticancer agents in cells lacking p21, *Nature* **381**, 713-716.
7. Theron, T and Böhm, L. (1997) Pentoxifylline inhibits the radiation induced G2/M block and alters DNA synthesis in p53 mutant and repair deficient cells (unpublished).
8. Powel, S.N., De Frank, J.S., Connell, P. et al. (1995) Differential Sensitivity of $p53^{(-)}$ and $p53^{(+)}$ Cells to Caffeine-induced Radiosensitization and Override of G2 delay, *Cancer Research* **55**, 1643-1648.
9. Russel, K.J., Wiens L.W., Demers, G.W., Galloway, D.A., Plon, S.E. and Groudine, M. (1995) Abrogation of the G_2 Chekpoint Results in Differential Radiosensitization of G_1 Checkpoint-deficient and G_1 Checkpoint-competent Cells, *Cancer Research* **55**, 1639-1642.
10. Nagasawa, H., Li, C-Y., Maki, C.G., Imrich, A.C. and Little, J.B. (1995) Relationship between Radiation-induced G_1 Phase Arrest and p53 Function in Human tumor Cell, *Cancer Research* **55**, 1842-1846.
11. Kessis, T.D., Slebos, R.J., Nelson, W.G. et al. (1993) Human papillomavirus 16 E6 expression disrupts the p53-mediated cellular response to DNA damage, *Proceedings National Academy of Science USA* **90**, 3988-3992.
12. Kuerbitz, S.J., Plunkett, B.S., Walsh, W.V. and Kastan, M.B. (1992) Wild-type p53 is a cell cycle checkpoint determinant following irradiation, *Proceedings National Academy of Science USA* **89**, 7491-7495.
13. Kastan, M.B., Onyekwere, O., Sidransky, D., Vogelstein, B. and Craig, R.W. (1991) Participation of p53 Protein in the Cellular Response to DNA Damage, *Cancer Research* **51**, 6304-6311.

14. Morgan, D.O. (1995) Principles of CDK regulation, *Nature* **374**, 131-134.
15. Luke, D.R., Sarnoski, T.P., Bell, N. et al. (1993) Influence of pentoxifylline on renal function in HIV-seropositive patients, *Renal Failure* **15**, 181-188.
16. O'Connor, P.M., Ferris, D.K., Pagano, M., Draetta, G., Pines, J. and Hunter, T. (1993) G2 delay induced by mitogen mustard in human cells affects cyclin A/Cdk-2 and cyclin B1/Cdc-2 kinase complexes differently, *J. Biological Chemistry* **268**, 8298-8308.
17. Zabel, P., Schade, F.U. and Schlaak, M. (1993) Inhibition of Endogenous TNF Formation by Pentoxifylline, *Immunobiology* **187**, 447-463.
18. Zölzer, F., Hillebrandt, S. and Streffer, C. (1995) Radiation induced G_1-block and p53 status in six human cell lines, *Radiotherapy and Oncology* **37**, 20-28.
19. Theron, T. and Böhm, L. (1998) Pentoxifylline upregulates cyclin B1 in G2 blocked cells. *Cell Differentiation* (submitted).
20. Vernimmen, F., Verheye-Dua, F., du Toit, H. and Böhm, L. (1994) Effect of Pentoxifylline on Irradiation Damage and Tumor Growth, *Strahlentherapie und Onkologie* **10**, 595-601.
21. Waldman, T, Zhang, Y., Dillehay, L., Yu, J., Kinzler, K., Vogelstein, B. and Williams, J. (1997) Cell cycle arrest versus cell death in cancer therapy, *Nature Medicine* **3**, 1034-1036.
22. Merritt, A.J., Potten, C.S., Kemp, C.J. et al. (1994) The role of p53 in spontaneous and radiation-induced apoptosis in the gastrointestinal tract of normal and p53-deficient mice, *Cancer Research* **54**, 614-617.

GREEN FLUORESCENT PROTEIN (GFP) EXPRESSION IN MAMMALIAN CELLS AFTER UV-IRRADIATION

M. PALM, C. BAUMSTARK-KHAN AND G. HORNECK

Deutsches Zentrum für Luft- und Raumfahrt, Institut für Luft- und Raumfahrtmedizin, Abteilung Strahlenbiologie
D-51170 Köln, Germany

Abstract

Green fluorescent protein (GFP) is a reporter system for monitoring in-vivo gene expression superior to other assays as GFP fluorescence does not require specific substrates or cofactors. CHO cells were co-transfected with the pCX-GFP (β-actin/β-globin gene promoter) and a neomycin resistence transmitting plasmid. Resulting stable cell clones were selected and maintained using G418. CX-GFP codes for a red-shifted GFP-variant (488 nm excitation, 509 nm emission) enabeling visualization of GFP expression by spectroscopy, fluorescence microscopy or FACS-analysis (488 nm excitation, FITC filters).

GFP expressing cells were subjected to different fixation/staining protocols. Formaldehyde-fixed (FA) cells display GFP fluorescence comparable to living cells. Co-staining of DNA is possible in FA-fixed cells using Hoechst 33258 or DAPI. Propidium iodide (PI) staining of DNA is difficult, due to the fact that GFP-fluorescence in ethanol-fixed samples is not stable and in FA-fixed samples there is some overlapping of emission spectra from GFP and PI.

DNA repair proficient and deficient CHO strains were used for survival tests after UVC-irradiation. Cells carrying the GFP-construct show no difference in cellular survival compared to cells not carrying the construct.

After certain post-irradiation growth periods cells were harvested and the numbers of GFP expressing cells and fluorescence were determined by FACS analysis. Generally, GFP fluorescence in irradiated cells is not seen when cell membranes are damaged (leak-out of the soluble GFP). Irradiated cells without membrane damage express GFP continuously (increasing GFP contents).

1. Introduction

The introduction of reporter genes into cultured mammalian cells represents an important means of investigating cellular radiosensitivity. In studies of this kind, the gene or promoter element of interest is usually inserted into a vector, where the expression of

C. Baumstark-Khan et al. (eds.),
Fundamentals for the Assessment of Risks from Environmental Radiation, 311-316.

the reporter gene is either driven by a suitable selected heterologous promoter and other elements or is a consequence of cellular response to radiation.

Green fluorescent protein (GFP), originally isolated from the jellyfish Aequorea victoria, is such a useful reporter for monitoring gene expression in vivo [1]. Expression of GFP is superior to other reporter assays as GFP fluorescence only needs exposure to UV or blue light, unlike other bioluminiscent reporter system which require incubation of the cells with specific substrates or cofactors to emit light. GFP fluorescence can thus be monitored in living cells.

UV irradiation is known to induce certain classes of genes [2, 3]. The modulation of gene expression is thought to be mediated to some extent by promoter specific interactions and regulatory elements like transcription factors [4, 5]. We, therefore, are interested in the question, whether or not GFP expression is a suitable reporter for radiation-induced gene activation. Of special importance in this respect is the stability of the GFP molecule itself and of vector-driven GFP-expression after irradiation.

2. Materials and Methods

The CHO cell lines AA8 (ATCC CRL-1859, repair proficient) and UV5 (ATCC CRL-1865, defective in nucleotide excision repair gene ERCC2) grow under standard conditions (DMEM-medium, 5% FCS, 37°C, 5%CO_2, 95% air).]. The AA8 and UV5 cell lines were transfected using the pCX-GFP vector, where GFP expression is under controll of a strong hybrid promoter composed of CMV enhancer and β-actin/β-globin gene promoter [6] as described by Baumstark-Khan et al. [7]. GFP expressing cells were analyzed after different fixation procedures (4% paraformaldehyde in PBS; 70% ice-cold ethanol) and counterstaining with DNA dyes (propidium iodide).

UVC-irradiation was performed using a germicidal lamp (NN8/15, Heraeus), UV dosimetry was performed using a calibrated UVX-radiometer (UVP Inc.). For exposure with UVC, the cells were washed twice with phosphate buffered saline (PBS). Immediately after UV exposure, cells were incubated under standard conditions (DMEM-medium, 5% FCS, 37°C, 5%CO_2, 95% air).

Cellular UV-sensitivity of cultured CHO cells which were plated 16-24 hours before UV-exposure was tested using the colony forming ability test. Colonies grown after 14-18 days were stained with crystal violet (1 mg/ml in 4% paraformaldehyde solution) and scored. The AA8-pGFP and UV5-pGFP cell lines were harvested and analyzed on a FACScan flow cytometer (Becton-Dickinson), equipped for fluorescein isothiocyanate and propidium iodide detection at an excitation wavelength of 488 nm.

3. Results and Discussion

3.1. VISUALIZATION OF GFP-EXPRESSION

Like several other GFP's CX-GFP has a strong red-shifted excitation peak at 488 nm and an emission maximum at 509 nm wavelength (Figure 1). GFP fluorescence is stable and can be monitored non-invasively in living cells (Figure 2.). In paraformaldehyde-

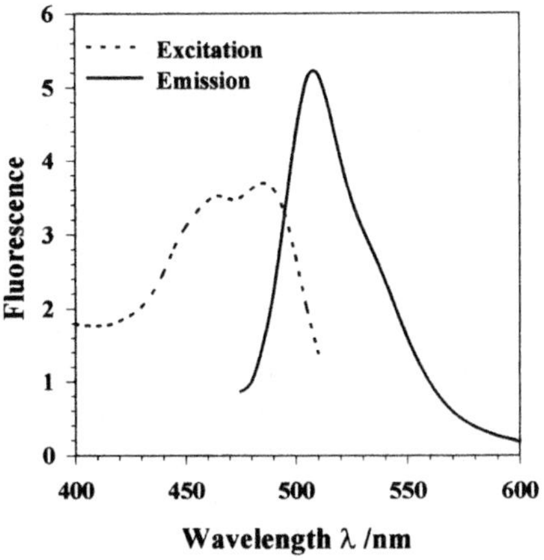

Figure 1. Excitation and emission spectra of GFP expressing cells (CHO AA8). The excitation data were obtained with emission at 540 nm and the emission date were obtained with excitation at 450 nm.

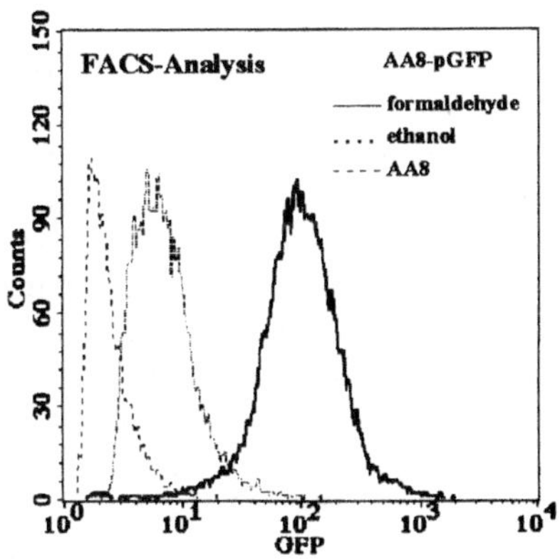

Figure 2. FACS-analysis of GFP-expression after fixation of AA8-pGFP cells using 5% paraformaldehyde or 70% ethanol. For comparison AA8 cells, not expressing GFP are included in the analysis

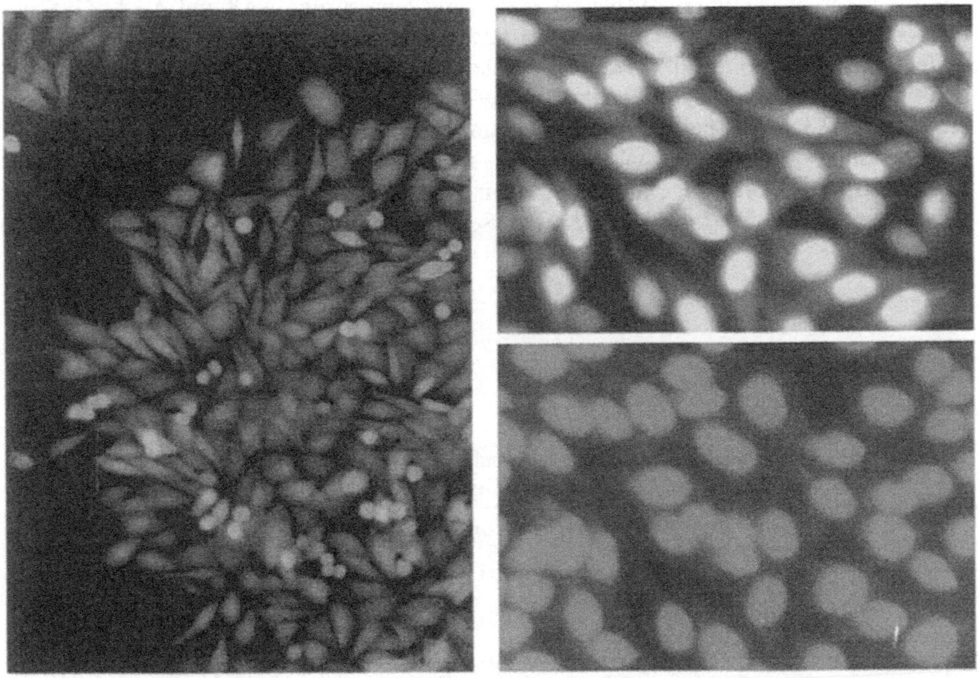

Figure 3. Colony of living cells of the stably transfected CHO AA8-pGFP cell clone IV. Photographs were taken with a 20x objective (left). Monolayer culture of cells fixated in 4% paraformaldehyde counterstained with propidium iodide (PI, right) after digestion with RNase for 30 min (32x objectiv).

fixed cells GFP fluorescence could be maintained, whereas applying fixation protocols using ethanol or methanol (Figure 3.), the GPF fluorescence is exhausted, probably due to the fact that the small soluble GFP molecules come out from cells which cellular membranes have been delipidized. The red-shifted spectra of the CX-GFP variant renders the protein superior to wild-type GFP for fluorescence activated cell sorting. Maximal emission is achieved using the excitation wavelength of 488 nm, the excitation line of argon lasers, routinely used in many FACS machines.

For double-labeling of GFP-expressing cells the DNA dyes of choice would either be DAPI or Hoechst 33258, both known as DNA-specific probes. The advantage of excitation (355 nm) and emission (455 nm) peaks far apart from the corresponding GFP peaks (488 nm, 509 nm) is extremely valuable in fluorescence microscopy. For FACS analysis, this advantage turns to a disadvantage, as FACS machines are not routinely equipped with a second laser for excitation. The DNA dye, routinely applied for FACS analysis is propidium iodide (PI), which requires complete digestion of RNA for reliable results. RNA digestion and PI fluorescence are usually applied for methanol-fixated or ethanol-fixated preparations, nevertheless paraformaldehyd-fixation followed by RNase digestion and counterstaining with PI gave satisfying results (Figure 2.).

3.2. CELLULAR SURVIVAL AFTER UV-IRRADIATION

Survival curves (Figure 3., Table 1.) of the wild-type strains AA8 and AA8-pGFP obtained after UVC-exposure are of curvi-linear shape with D_0 in the same range. Dose effect curves of the repair deficient mutants UV5 and UV5-pGFP are purely exponential. Cells carrying the GFP construct show no difference in cellular survival as measured by the CFA-test, neither in wild-type nor in repair proficient clones. Colonies, both of the stably transfected wild-type and mutant clones, were inspected for GFP-expression by fluorescence microscopic surveillance. In any case, surviving cells produced green colonies (results not shown). Sectoring within single colonies could not be observed.

TABLE. 1. Radiosensitivity of in vitro cultivated CHO cell clones with and without GFP-expression vector pCX-GFP.

Cell line	D_0	n	D_q	R^2
AA8	3.50±0.098	7.32±1.15	6.98±0.72	0.874
AA8-pGFP	3.79±0.086	5.95±0.68	6.77±0.57	0.954
UV5	0.96±0.018	1.10±0.18	0.09±0.16	0.952
UV5-pGFP	0.87±0.020	1.53±0.13	0.34±0.08	0.974

3.3. GFP EXPRESSION AFTER UVC-EXPOSURE

For using GFP as a marker for cell viability cells were harvested after certain post-irradiation growth periods and the amount of GFP expressing cells and fluorescence intensities were determined by FACS analysis. In non-irradiated populations, nearly all cells were found to be positive for GFP-expression. In short term culture (1-4 hours) the same results can be obtained for unirradiated cells and in cells grown after UV-exposure. For long term cultures (>4 hours), in UVC-exposed populations, two sub-populations of cells could be identified: (i) GFP-negative (GFP-content 10^0 to $3 \cdot 10^1$) and (ii) GFP-positive (GFP-content $>3 \cdot 10^1$) cells. With increasing fluences and increasing incubation periods after UVC-exposure, the amount of GFP-negative cells increases as the amount of GFP-positive cells decreases. The distributions for GFP-positiv cells are shifted to higher GFP-concentrations with increasing fluences. 72 hours after UVC-exposure (Figure 4.), about 40% of cells are GFP-negative for AA8-pGFP exposed to 25 Jm^{-2}, and for UV5-pGFP exposed to 7.5 Jm^{-2}. It is supposed, that, when cellular membranes are impaired, as it is the case in apoptotic cells, GFP fluorescence in irradiated cells could not be detected due to a leak-out of the soluble GFP molecules. Irradiated cells which exhibit no membrane damage express GFP continuously leading to a dose dependent increase in GFP fluorescence. This effect of irradiation on gene expression driven by constitutive promoters can presumably be explained by progressive protein synthesis of irradiated cells which fail to devide. In experiments inhibiting DNA synthesis by use of hydroxy urea, an inhibitor of ribonucleotide reductase, a similar effect can be achieved (results not shown).

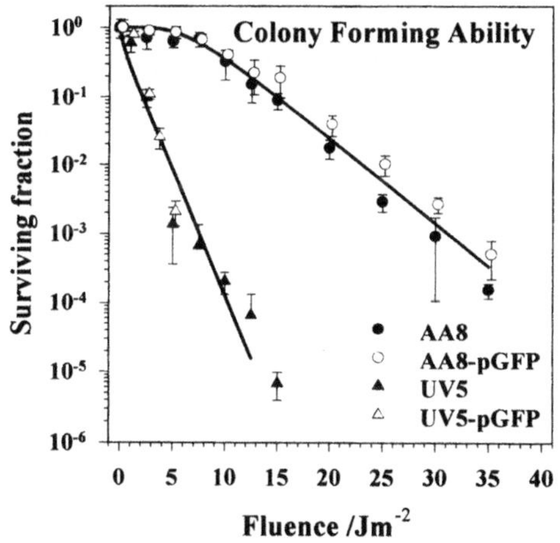

Figure 4. Survival curves of repair-proficient (AA8) and repair deficient (UV5) CHO cell strains with and without GFP-expression vector pCX-GFP.

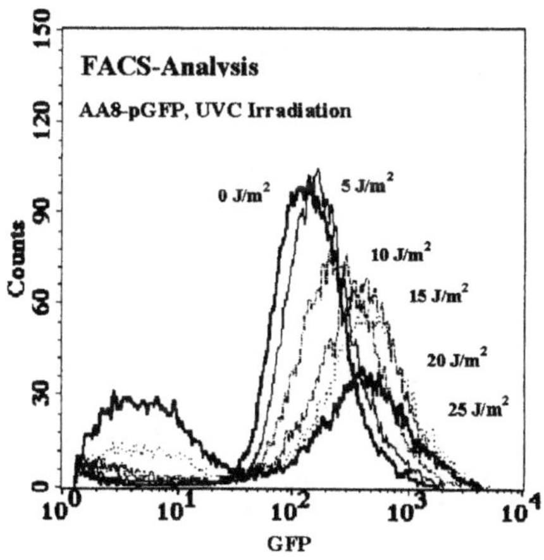

Figure 5. GFP-expression after UVC (254 nm) irradiation: the fraction of cells not expressing GFP increases slightly with dose (left), in GFP expressing cells the per cell GFP content increases (right)

Experiments describing the effect of ionizing radiation on the expression of the CAT gene under the control of different promoters [8] found for transient transfections vectors to be radiation-independent or -dependent, irrespectively of whether the promoters are constitutive or inducible. When stable transformats in which the transfected DNA has been stably integrated into the genome, locus specificeffects might be responsible for complex reactions from one cell clone to another. Other authors also found increased gene expression after X-irradiation or UV-exposure of stable transfectants [9, 10]. For certain viral promoters, the observed up-regulation after irradiation might be a common response with transfected vectors and it needs to be considered wehe radiation-induced responses are studied.

Acknowledgement. The experiments described here are part of the PhD thesis of M.P.

4. References

1. Chalfie M., Tu Y., Euskirchen G., Ward W.W., Prasher D.C. (1994), Green fluorescent protein as a marker for gene expression. *Science* **263**, 802-805
2. Filatov D., Bjorklund S., Johansson E., Thelander L. (1996), Induction of the mouse ribonucleotide reductase R1 and R2 genes in response to DNA damage by UV light. *J. Biol. Chem.* **271**, 23698-23704
3. Narayan S., He F., Wilson S.H. (1996). Activation of the human DNA polymerase beta promoter by a DNA-alkylating agent through induced phosphorylation of cAMP response element-binding protein-1. *J. Biol. Chem.* **271**, 18508-18513
4. Luethy J.D., Holbrook N.J. (1992). Activation of the gadd153 promoter by genotixic agents: a rapid and specific response to DNA damage. *Cancer Res.* **52**, 5-10
5. Tornaletti S., Pfeifer G.P. (1995). UV light as a footprinting agent: modulation of UV-induced DNA damage by transcription factors bound at the promoters of three human genes. J. Mol. Biol. **249**, 714-728
6. Akagi Y., Isaka Y., Akagi A., Ikawa M., Takenaka M., Moriyama T., Yamauchi A., Horio M., Ueda N., Okabe M., Imai E. (1997) Transcriptional activation of a hybrid promoter composed of cytomegalovirus enhancer and beta-actin/beta-globin gene in glomerular epithelial cells in vivo. *Kidney Int.* **51**, 1265-1269, 1997
7. Baumstark-Khan C., Palm M., J. Wehner J., Okabe M., Ikawa M., Horneck G. (1997), Green fluorescent protein (GFP) as a marker for cell viability after UV-irradiation. In W. Rettig (Editor): *International Conference on Methods and Applications of Fluorescence Spectroscopy* **5**, Berlin
8. Cheng X., Iliakis G. (1995). Effect of ionizing radiation on the expression of chloramphenicol acetyltransferase gene under the control of commonly used constitutive or inducible promoters. *Int. J. Radiat. Biol.* **67**, 261-267
9. Lin C.S., Goldthwait D.A., Samols D. (1990). Induction of transcription from long terminal repeat of moloney murine sarcoma provirus by UV-irradiation, X-irradiation and phorbol ester. *Proc. Natl. Acad. Sci. USA* **87**, 36-40
10. Ronai Z.A., Okin E., Weinstein I.B. (1988). Ultraviolet light induces the expression of oncogenes in rat fibroblast and human keratinocyte cells. *Oncogene* **2**, 201-204

THE IMPORTANCE OF SPONTANEOUS AND RADIATION INDUCED APOPTOSIS FOR STEM CELLS IN THE GASTROINTESTINAL TRACT

C. S. POTTEN

Department of Epithelial Biology, Paterson Institute for Cancer Research
Christie Hospital (NHS) Trust
Wilmslow Road, Manchester, M20 4BX, U.K.

Abstract

Apoptosis occurs at a low level in healthy unirradiated mice and also in humans. This spontaneous apoptosis is associated with cells in the stem cell region of the crypts of the small intestine. It is *p53*-independent and is assumed to be associated with the stem cell homeostatic processes. It is rare in the large intestine, probably because of the action of *bcl-2*. Radiation increases the levels of apoptosis at the stem cell location; peak levels are observed within 3-6 hours and the 6 cells in each crypt that appear to be susceptible to this mode of death are exquisitely radiosensitive. Other cells are reproductively sterilised and these have a greater resistance to radiation and are interpreted to be the clonogenic stem cell compartment. The apoptosis susceptible cells are interpreted to represent the ultimate stem cells which generate the clonogenic compartment by division. The radiation-induced apoptosis is an efficient protective mechanism, possibly accounting for the low cancer incidence figures in the small intestine. This protective mechanism is acting less efficiently in the large bowel, again because of the action of *bcl-2*.

1. Introduction

In 1906 Bergonié and Tribondeau [1] proposed what has become known as the basic law of radiobiology. This states in simple terms that the radiosensitivity of a cell is directly proportional to its rate of proliferation and its division potential and inversely proportional to its degree of differentiation. This statement, although made almost one century ago, still holds true as a generalisation.

Part of the gastrointestinal tract, and one or two other tissues such as the growing hair follicles in skin, are amongst the most rapidly proliferating tissues of the body and as such would be expected to be particularly radiosensitive. Radiation sensitivity of individual cells can be assessed by looking for evidence of cells that die and these are usually recognised by their morphological changes in appearance when viewed through the microscope. The most important process here is a the mode of cell death called apoptosis. An alternative approach to looking for cells that die is to study the loss of the

C. Baumstark-Khan et al. (eds.),
Fundamentals for the Assessment of Risks from Environmental Radiation, 317-326.

ability of cells to survive following exposure to a cytotoxic agent such as radiation and this is usually achieved by looking for the ability of cells to continue to divide and form colonies or clones of regenerating cells and this is the basis of the so called clonogenic assays. Clonal regeneration studies have been performed extensively in the haemopoietic tissue in the bone marrow and in the small intestine [2]. These clonal regeneration studies represent an *in vivo* test of the functional capacity of clonogenic stem cells and the use of these approaches has proved invaluable in dissecting the complexity of cell lineages in tissues such bone marrow, testes, intestine and epidermis. However, the approaches generally require the delivery of high doses of radiation. They demonstrate that the clonogenic cells have a radiosensitivity defined by a D_0 value on survival curves of about 1Gy and essentially by definition, look at the most resistant cells in the tissue, those that survive the high doses of radiation.

In the small intestine of the mouse, clonal regeneration studies performed 3-4 days after exposure to a range of radiation doses and analysed with a light microscope using transverse sections of the intestine have generated survival curves for the crypt-like colonies [3, 4]. These survival curves can be interpreted to provide information on the stem cell hierarchy in the crypts including information on the number of clonogenic stem cells [3, 4].

In 1977 we reported [7] that at the other end of the spectrum of radiosensitivity, there were cells in the crypts that were exquisitely radiosensitive and this could be detected by observing and counting the number of apoptotic cells. In theory, a precise inverse relationship would be expected between the number of cells that can be seen to die and the number that can be demonstrated to survive. However, the relationship between surviving clonogenic cells and dying apoptotic cells remains complex and poorly understood. Here, I wish to concentrate on the cells that are killed by exposure to low doses of radiation and to consider the consequences and cell biological importance of this apoptosis for the crypt homeostatic mechanisms and the susceptibility of stem cells to carcinogenic transformation.

Over the course of the last few years, we have been considering the questions of how apoptosis might be defined, recognised and quantitated and also what its biological significance is ie. which cells in the tissue are involved, does the tissue recognise that cells have been killed, and if so, how does the tissue respond and whether or not all regions or tissues respond in the same way? I shall be addressing these questions specifically in relation to the small and large intestine with a view to trying to understand the puzzling cancer incidence figures that are recorded for the gastrointestinal tract. In the Western world cancer of the colon is common and its prognosis is poor. The surprising feature is that in the more rapidly proliferating and much more extensive small intestine, cancer is very rare. There have been many attempts to provide explanations for the high cancer risk seen for colonic stem cells but rarely has the question been reversed to address what protects the small intestine so effectively from cancer induction.

2. The Biological Model System: The Intestine

The gastrointestinal mucosa is a simple columnar epithelium folded into complex 3-dimensional structures. In both the small and large bowel, proliferation is restricted to small flask-shaped structures called crypts. The cells produced in the crypts move into the differentiated functional zone of the mucosa which in the small intestine, consists of finger-like projections or villi and in the large intestine a relatively flat intercrypt table. Cells are lost at the end of their limited life span as senescent functional cells from the tip of the villus and the intercrypt table, by a process of extrusion into the lumen of the intestine. Their cellular constituents are then efficiently recycled. There is some evidence that, at the end of their short life span, these cells initiate some of the sequences involved in apoptosis. However, I shall not be concerned with this particular apoptotic process, but rather restrict my attention to cells within the proliferative units, the crypts. Here, extensive studies have indicated that proliferation is organised within the framework of a series of cell lineages and that each crypt has a limited small number of lineage ancestor stem cells [8]. These are undifferentiated proliferating cells that maintain their own numbers while generating progeny that enter an amplifying dividing transit population within which a range of differentiation options may be activated. The stem cells have a large division potential and are the most potent regenerators of the epithelium following injury [9]. They are also the long lived residents of the tissue and therefore the most likely candidates for carcinogenic target cells. The stem cells have no identifying morphological or staining characteristics but can be studied in this tissue by virtue of the fact that they are located at specific positions in the crypts. In the small intestine, the stem cell location within the flask-shaped structure is within an annulus of about 16 cells, approximately 4 cells positions from the base of the crypt, sitting upon differentiated Paneth cells. In the large intestine, the stem cell location is at the base of the crypt [8, 9].

3. Apoptosis

Since our initial publication in 1977, we have used a wide range of technical approaches from electron microscopy to specific staining including *in situ* end labelling approaches to identify and characterise apoptosis [10]. In the electron microscope, cells induced into apoptosis by small doses of radiation exhibit the classical morphological changes described by Kerr and colleagues in 1972 [11]. In the light microscope, apoptosis can be easily recognised in good quality haematoxylin and eosin stained sections by virtue of the chromatin condensation, the chromophilic cytoplasm and the fragmentation process. The levels of cell death in the crypts can be estimated by counting the number of cell positions at which apoptotic fragments are observed in longitudinal sections through the centre of the flask-shaped crypt. We have shown that the recording of the number of apoptotic events in this way provides a good approximation to the total number of cell death events that occur in the entire crypt [12, 13]. By recording the number of apoptotic events on a cell positional basis in crypt sections, frequency plots for the apoptotic yield at each cell position can be obtained and hence information obtained about the relative frequency of apoptosis at the stem cell positions.

3.1. SPONTANEOUS APOPTOSIS

If sections are cut of intestine from healthy adult mice or humans, it is noted that a low level of apoptosis can be detected in the small intestine. We have demonstrated that this *spontaneous* or *naturally occurring apoptosis* is not the consequence of random rare damage events in cells since it occurs in animals in which the *p53* "guardian of the genome" gene has been deleted [14]. If these *p53* null animals are exposed to radiation or some other DNA damaging agent, the damage-induced apoptosis is completely absent indicating that the initiation of this type of apoptosis is completely dependent on the *p53* gene. The *p53* null animals develop normally and apoptosis is known to be of great importance in a variety of developmental situations. So developmental apoptosis, like the spontaneous apoptosis, is not dependent on the *p53* gene and involves the removal of healthy undamaged cells. The spontaneous apoptosis that is seen in normal mice in the small intestine shows its highest frequency at the stem cell positions (4 cell positions from the base of the crypt), and so is associated with the stem cell population. In the large intestine the spontaneous apoptosis is extremely rare (10 times lower than in the small intestine) and does not have any specificity for the stem cell position. Following exposure to radiation, the p53 protein, as detected by antibody staining, is elevated in cells within the proliferative compartment of the crypt with some particularly intense p53 positive cells in the stem cell position, but not in adjacent apoptotic cells. p53 is believed to control cell cycle check points determining whether cells arrest in the cycle and undergo repair or commit suicide via apoptosis. We have interpreted the strongly p53 positive cells at the stem cell position to be the clonogenic compartment of the stem cell hierarchy. These clonogenic stem cells undergo repair, have a greater radio-resistance and as consequence survive to regenerate the crypt. Our interpretation of the low levels of spontaneous apoptosis in healthy mice and men is that this represents the stem cell homeostatic processes operating to ensure a stable stem cell population size. Mathematical modelling studies have suggested that the stem cells are likely to occasionally divide symmetrically generating stem cells excess to requirements [15]. These have to be removed, either by differentiation or apoptosis.

3.2. RADIATION-INDUCED APOPTOSIS

If an animal is exposed to low doses of radiation, the incidence of apoptosis in the crypt rises within a few hours. This is illustrated in fig.1. The maximum yield of apoptotic events occurs within 3-6 hours of irradiation. Depending on the dose, the yield of apoptosis then falls towards the control spontaneous levels which may be reached within 24 hours of low doses. Cells recognise that they have incurred DNA damage and respond rapidly by initiating the sequence of events that leads to the morphological changes that we recognise as apoptosis. These events involve a range of processes (damage recognition, damage response, metabolic and structural changes). This rapid sequence of events involves a variety of early response genes and apoptosis-related genes. *p53* is clearly one of the key players. If crypt sections are analysed and the yield of radiation-induced apoptosis occurring between 3-6 hours after radiation are analysed on a cell positional basis, it is evident that the greatest likelihood of observing apoptosis after radiation is again at position 4 in the small intestinal crypt (fig.2). In the large intestine, at least in the mid-colonic region, the radiation-induced apoptosis is lower in incidence and exhibits a stem cell specificity which can be seen by the highest yield occurring at cell positions 1-2 [7, 13, 14, 16].

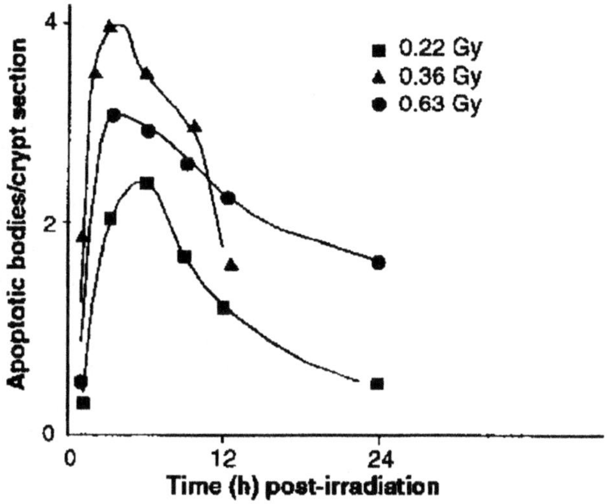

Figure1. Variation in the apoptotic yield in longitudinal small intestinal crypt sections with time after exposure of mice to Cs^{137} G rays (3.8 Gy/min, 4 mice per point, 50 crypts scored per mouse). Yields of apoptosis peak between 3 and 6h after irradiation.

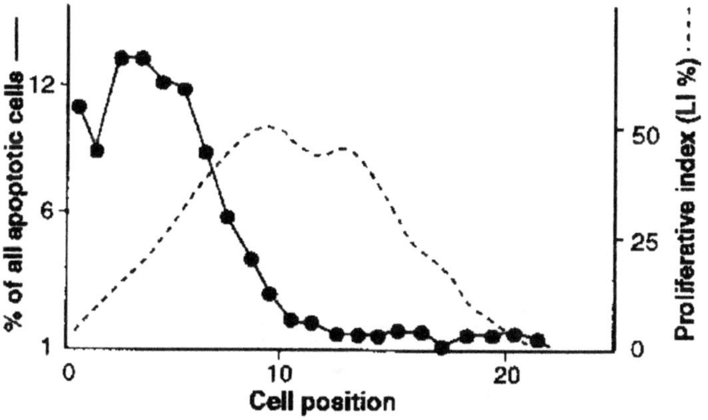

Figure 2. The percentage of all apoptotic bodies small intestinal crypts scored in 200 crypts from 4 mice exposed to 0.63 Gy and analysed about 4-5h later (left hand scale), together with the frequency plot for S phase cells [proliferative index measured as a tritiated thymidine labelling index (LI)]. Taken with modifications from ref.19. Peak yields of apoptosis are seen over cell positions 3-5 from the crypt base.

The most surprising feature about the apoptosis that is observed at early times after exposure to low doses of radiation is the dose-response relationship shown in fig.3. The two surprising features are firstly, the exquisite sensitivity of some cells within the stem cell zone:- doses as small as 0.01Gy can be detected, 0.05Gy significantly raises the levels. Secondly, the dose-response curve appears to saturate at about 0.5-1Gy at a level equivalent to between 2-4 cells per crypt section which is equivalent to 3-6 cells in the entire crypt or 2-4 cells at the stem cell position. The radiosensitivity exhibited here is roughly equivalent to a single ionising event occurring anywhere in the entire DNA (nucleus) ie. one hit and the cell dies. These results can be interpreted in terms of a survival curve for the susceptible population which lacks an initial shoulder suggesting a lack of repair capacity and is characterised by a sensitivity as described by a D_0 value of about 0.25Gy for low LET G rays or 0.06Gy for 14.7 MeV neutrons (fig.4) placing these intestinal stem cells amongst the most sensitive mammalian cells. The lack of any repair ability is further indicated by the lack of any dose rate or LET dependence (fig.5). These cells are so sensitive to radiation damage that the incorporation of tracer amounts of tritiated thymidine used to study their cell kinetics provides sufficient internal beta irradiation to trigger apoptotic cell death.

In the large intestine the radiation-induced apoptosis exhibits a specificity for the stem cell compartment but it occurs at a lower frequency at low doses (fig.5). However, the dose-response curve continues to increase up to doses of about 6Gy after which it may plateau or saturate.

3.3. APOPTOSIS RELATED GENES

Whether or not a cell dies via apoptosis appears to depend on the balance between conflicting survival or cell death genes. It has become clear that there is a large family of

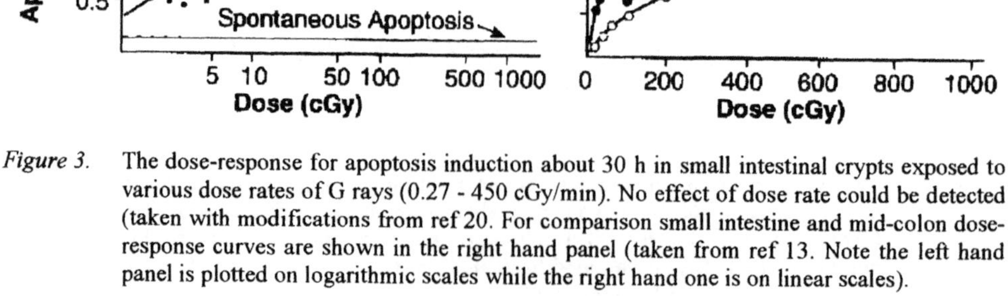

Figure 3. The dose-response for apoptosis induction about 30 h in small intestinal crypts exposed to various dose rates of G rays (0.27 - 450 cGy/min). No effect of dose rate could be detected (taken with modifications from ref 20. For comparison small intestine and mid-colon dose-response curves are shown in the right hand panel (taken from ref 13. Note the left hand panel is plotted on logarithmic scales while the right hand one is on linear scales).

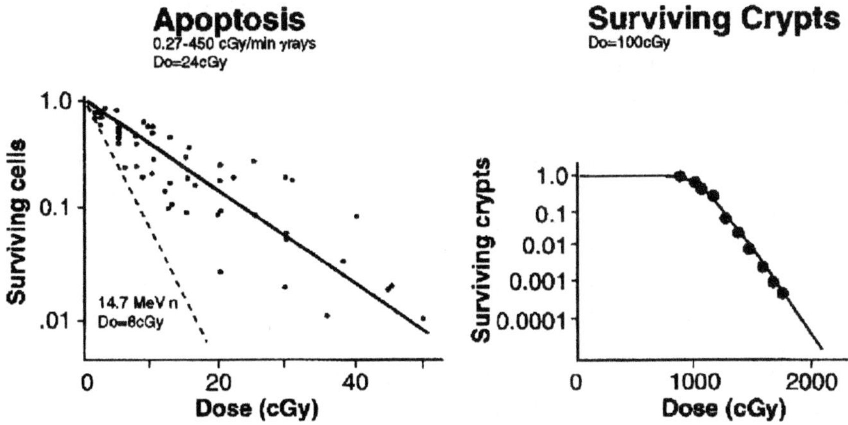

Figure 4. The data in the left hand panel of figure 3 can be replotted as a survival curve for the apoptosis susceptible cells. For G rays this survival curve has a D_0 of about 0.24 Gy while for 14.7 MeV neutrons the D_0 is 0.06 Gy. Taken with modification from ref 20. For comparison a crypt survivial curve which measures clonogenic stem cell survivial is shown on the right. Here the D_0 is 1.0 Gy.

related genes with some considerable redundancy and hence compensatory or overlapping potential. One of the key survival genes is *bcl-2* which interacts with a cell death gene *bax*. These interact by forming hetero- and homo- dimers. *Bad* is another cell death gene which is upregulated in the crypt very early following irradiation altering the *bcl-2/bax* balance in such a way to favour cell death. We have been determining the expression levels of a range of the survival and death genes in the crypts of the small and large bowel. It is becoming clear that the determination of survival or death is complex, involving many genes and factors including growth factors and extra-cellular matrix proteins.

Bcl-2 does not appear to be expressed in the small intestine while in the large intestine it is expressed in a rather variable fashion at the base of the crypts in the mouse and in a much more reproducible fashion in the base of the crypts in humans. The expression of *bcl-2* within the stem cell zone of the large intestinal crypt may explain the very low levels of spontaneous apoptosis and the lower levels of radiation-induced apoptosis observed in this region of the gut. If the *bcl-2* gene is deleted, the spontaneous and radiation-induced levels of apoptosis in the small intestine, where *bcl-2* was not expressed, remain unchanged in the *bcl-2* knockouts, while in the large intestine, both the spontaneous and radiation-induced apoptosis levels are increased [17].

3.4. APOPTOSIS AND CANCER RISK

It is believed that cancers arise within a permanent resident cell population in tissues, the stem cells. In the large intestine, these cells are at the base of the crypt and constitutively express the survival gene *bcl-2*. It might be expected therefore, that cancers in large intestine would consist of cells that continue to express *bcl-2* and this is indeed what is observed in the early stages of carcinoma development when *bcl-2* is expressed, but declines in the later stages [18]. The expression of *bcl-2* in cancers could be the explanation for the high chemo- and radio-resistance of these tumours.

In the small intestine, spontaneous apoptosis ensures that the stem cell population remains stable and the exquisite sensitivity of the stem cells following DNA damage ensures that damage, and the cells carrying the damage, are rapidly and efficiently removed from the tissue. This represents a very potent protective mechanism and may be the explanation for why cancer is so extremely rare in this rapidly proliferating tissue. In the large bowel, the action of *bcl-2* may be preventing the stem cell homeostatic process from acting efficiently and stem cell numbers may drift upward with the passage of time increasing the number of carcinogenic target cells at risk. The DNA protection mechanism may also be acting less efficiently, again due to the action of *bcl-2*. Thus, stem cells that have incurred DNA damage may repair the damage, mis-repair it or not recognise it in the first place and in this way increase the probability that DNA damage may persist and carcinogenic risk [16].

The reasons why these two regions of the gastrointestinal tract respond so differently to damage remains speculative, but may be related to the two different environments these two sites provide for stem cells. In the small intestine bacteria are rare and genotoxic agents are likely to be scarce. The proliferation here is very rapid and it would be important to delete the low levels of damage rapidly and efficiently replacing any dying cells by an occasional additional division in a neighbouring stem cell ie. by a process of local regenerative activity. In contrast, in the large intestine the levels of genotoxic agents may be considerably greater as a consequence of the higher bacterial content. If the same processes were operating here there would be a constant triggering of many stem cell deaths which would result in a continual process of regenerative proliferation. This constant stimulation of stem cells might be associated in its own right with carcinogenic risk, so in this region it might be beneficial to attempt to repair the damage rather than induce apoptosis and continually stimulate the stem cells.

3.5. STEM CELL ORGANISATION

Such studies together with many other radiation and cell kinetic experiments provide not only a possible explanation for the cancer incidence figures, but provide an insight into the cell biology of the proliferative hierarchies and into the complexities of the stem cell population and its regulation. As a consequence of such studies, we have formulated a model of the stem cell population involving an hierarchical organisation with 3 tiers. The cells that are highly susceptible to apoptosis represent the ultimate lineage ancestor stem cells ie. those responsible for day to day cell replacement. If all of these cells are killed by, for example a dose of radiation, there are a small number of other clonogenic stem cells which have a much greater radio-resistance (repair capacity) and therefore, these can regenerate the ultimate stem cell population and the entire crypt. These cells represent the immediate daughters of the ultimate stem cell compartment. The third tier consists of a further population of clonogenic stem cells which are recruited into action if all other tiers are destroyed (a situation that does not occur in nature). These third tier stem cells can regenerate the two earlier tiers, the entire crypt and ultimately the entire mucosa. The numbers of stem cells within each tier we believe to be less than 6 for each of the first 2 tiers and about 24 for the third tier, making a total of 36 actual and potential stem cells per crypt. Besides this, the crypts contains about 120 other rapidly proliferating cells that appear to have no stem cell functional capacity.

Acknowledgements. The work summarised in this paper was begun in the mid seventies and has inevitably involved a large number of collaborators and assistants and the support of the Cancer Research Campaign. Some of the work has also been supported by the Association for International Cancer Research. For all those who have helped over the years I am very grateful. I should also like to thank Linda Pomfret for her excellent help in preparing this and many other manuscripts.

References

1. Bergonié, J. et Tribondeau L. (1906) Interprétation de quelques résultats de la radiothérapie et essai de fixation d'une technique rationelle. *Comptes Rendu Acod. Sci. Paris.* **143**:983-985.
2. Potten, C.S. and Hendry, J.H. (1985) Cell Clones. *Churchill Livingstone*. pp.235.
3. Withers, H.R. and Elkind, M.M. (1970) Microcolony survival assay for cells of mouse intestinal mucosa exposed to radiation. *Int. J. Radiat. Biol.* **17**:261-267.
4. Potten, C.S. and Hendry, J.H. (1985) The microcolony assay in mouse small intestine. *In: Cell clones; manual of mammalian cell techniques. Churchill Livingstone Edinburgh.* Eds. Potten, C.S. and Hendry, J.H. pp.97-102.
5. Roberts, S.A. and Potten, C.S. (1994) The clonogen content of intestinal crypts: its deduction using a microcolony assay on whole mount preparations and its dependence on radiation dose. *Int. J. Radiat. Biol.* **65**:477-481.
6. Cai, W., Roberts, J. and Potten, C.S. (1997) The number of clonogenic cells in three region of murine large intestinal crypts. *Int. J. Rad. Biol.* **71**:573-579.
7. Potten, C.S. (1977) Extreme sensitivity of some intestinal crypt cells to X and γ irradiation. *Nature.* **269**:518-521.
8. Potten, C.S. (1995) Structure, function and proliferative organisation of mammalian gut. *In: Radiation and Gut.* Ed. Potten, C.S. and Hendry, J.H. *Elsevier, Amsterdam.* pp.1-31.
9. Potten, C.S. and Loeffler, M. (1990) Stem cells: attributes, cycles, spirals, uncertainties and pitfalls: lessons for and from the crypt. *Development.* **110**:1001-1019.
10. Merritt, A.J., Jones, L.S. and Potten, C.S. (1996) Apoptosis in murine intestinal crypts. *In: Techniques in Apoptosis.* Ed. Cotter, T. and Martin, S. pp.269-299.
11. Kerr, J.F.R., Wyllie, A.H. and Currie A.R. (1972) Apoptosis: a basic biological phenomenon with wide-ranging implications in tissue kinetics. *Br. J. Cancer.* **26**:239-257.
12. Potten, C.S. (1996) What is an apoptotic index measuring? A Commentary. *Brit. J. Cancer.* **74**:1743-1748.
13. Potten, C.S. and Grant, H. Unpublished Data. The relationship between ionising radiation-induced apoptosis and stem cells in the small and large intestine.
14. Merritt, A.J., Potten, C.S., Hickman, J.A., Kemp, C., Ballmain, A., Hall, P. and Lane, D. (1994) The role of *p53* paper in spontaneous and radiation-induced intestinal cell apoptosis in normal and *p53* deficient mice. *Cancer Res.* **54**:614-617.
15. Loeffler, M. Bratke, T., Paulus, U., Li, Y.Q. and Potten, C.S. (1997) Clonality and life cycles of intestinal crypts explained by a state development stochastic model of epithelial stem cell organisation. *J. Theoret. Biol.* **186**:41-54.
16. Potten, C.S. (1992) The significance of spontaneous and induced apoptosis in the gastrointestinal tract of mice. *Cancer & Metastasis Reviews.* **11**:179-195.
17. Merritt, A.J., Potten, C.S., Watson, A.J.M., Loh, D.O.Y. and Hickman J.A. (1995) Differential expression of *bcl-2* in intestinal epithelia: correlation with attenuation of apoptosis in colonic crypts and the incidence of colonic neoplasia. *J. Cell. Sci.* **108**:2261-2271.

18. Watson, A.J.M., Merritt, A.J., Jones, L.S., Askew, J.N., Anderson, E., Becciolini, A., Balzi, M., Potten, C.S. and Hickman, J.A. (1996) Evidence of reciprocity of *bcl-2* and *p53* expression in human colorectal adenomas and carcinomas. *Brit. J. Cancer.* **73**:889-895.
19. Potten, C.S. (1995) Effects of radiation on murine gastrointestinal cell proliferation. *In: Radiation and Gut*. Eds. Potten, C.S. and Hendry, J.H. *Elsevier, Amsterdam.* 61-84.
20. Hendry, J.H., Potten, C.S., Chadwick, C. and Bianchi M. (1982) Cell death (apoptosis) in the mouse small intestine after low doses: effects of dose-rate, 14.7 MeV neutrons and 600 MeV (maximum energy) neutrons. *Int. J. Radiat. Biol.* **42**:611-620.

DISTINCT PATTERN OF INDUCTION OF GENES RELATED TO CELL SURVIVAL AND APOPTOSIS AFTER LOW DOSE RADIATION

S.Z. LIU, Y. MU AND F. XIE

MH Radiobiology Research Unit, Norman Bethune University of Medical Sciences, 8 Xinmin Street, Changchun 130021, CHINA

1. Introduction

The stimulatory effect of low dose radiation (LDR) on immunity has been well documented in recent years [1-3]. The mechanisms of this effect have been studied at cellular, molecular and systemic levels. The consensus of opinion is that activation of the T cells occupies a central position in the LDR-induced immunologic stimulation [1-4]. It was once proposed that LDR might up-regulate immunity via the deletion or damage of the radiosensitive suppressor T cells (Ts) with an increase of the helper (Th) to Ts ratio [1] or that LDR might promote apoptosis of the radiosensitive precursor T cells leading to increased proliferation of the more mature T cells [4]. However, these hypotheses have not gained experimental support since there is no increase of the Th/Ts ratio in the lymphoid organs or peripheral blood [3,5] and thymocyte apoptosis is not accentuated but even decreased after LDR [6]. Actually there has been evidence indicating promotion of thymocyte renewal, maturation and differentiation after whole-body irradiation (WBI) with low dose X- and γ-rays [3]. It is, therefore, necessary to elucidate the molecular basis of such effects.

There have been many reports on the induction of genes by different doses of radiation [8-12]. Most of the reports are concerned with in vitro studies on cancer or normal cell lines in culture. The radiation doses used in many of these studies are above 0.1 or 0.2 Gy. Although data from these studies are important in dissecting the possible mechanisms of the effects of LDR, it is sometimes difficult to extrapolate the results to intact organisms exposed to environmental radiation. Therefore, it is appealing to design experiments in animals exposed to WBI to disclose the dose-effect relationship and/or the time course of the changes in the induction and expression of genes related to survival, maturation, differentiation and activation of the cells in the immune system. In the following sections of this paper some of the data obtained in our laboratory along this line will be presented.

2. The Early Genes

The induction of the immediate early genes (also called cell cycle- dependent or CCD genes) c-fos, c-jun and c-myc were examined after exposure of mice to WBI with X-rays. The results are shown in figure 1.

C. Baumstark-Khan et al. (eds.),
Fundamentals for the Assessment of Risks from Environmental Radiation, 327-337.

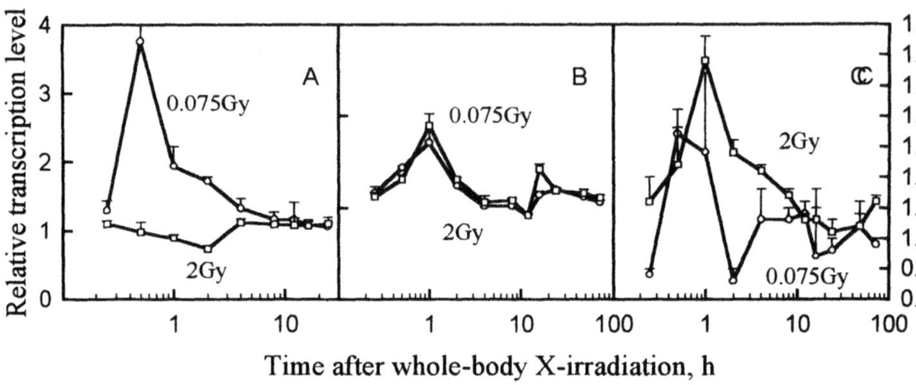

Figure 1. Induction of c-fos, c-jun and c-myc genes after WBI with 0.075 and 2 Gy X-rays
The transcription level of the genes was measured by slot blot hybridization with mGAP as reference; the average of the relative transcription level from 2-3 experiments was shown; the time intervals were in log scale in order to show the early changes clearly; A) c-fos, B) c-jun, C) c-myc.

The difference in the pattern of the induction and expression of these early genes after WBI with low versus high doses can be seen especially in the case of c-fos and c-myc. For the former WBI with 0.075 Gy caused an increased induction within 30' followed by its up-regulated protein expression 12-24h after irradiation (not shown). These changes reconcile very well with our previous morphological studies using in situ hybridization and immunohistochemistry [13]. For c-myc an opposite effect was observed, i.e., the level of both mRNA transcription and protein expression was stimulated by the higher dose of radiation. These changes in c-fos and c-myc genes would promote cell survival after LDR in the presence of an up-regulation of Bcl-2. It has been suggested that increased induction of c-myc may promote or inhibit cell survival depending on the state of other genes, especially the genes of the bcl-2 family [14]. In the next section the changes of bcl-2 gene transcription and Bcl-2 protein expression will be presented. The difference in c-jun transcription between the thymocytes from the low and high dose-irradiated animals was not marked. This may speak for the insignificant change in the activity of AP-1 in the lymphoid organs after LDR [13]. It was reported that irradiation of pig skin with γ-rays caused a marked increase of c-fos gene induction at doses of 0.05 to 0.5 Gy while doses of 2 Gy and above had no such effect, and induction of c-jun, junD and junB was not observed after LDR [15]. The authors thus suggested that c-fos induction in the pig skin might be a "marker" of LDR effect.

3. Genes Regulating Cell Survival

There are many genes regulating cell survival and apoptosis among which bcl-2, ICE, Rb, p53, gadd45 and FasL are some of the most important ones. The effect of WBI with different doses of X-rays on the induction/expression of these genes will be described in this section,

3.1. BCL-2 AND ICE

The bcl-2 gene is called the survival gene which is the mammalian homologue of the ced9 gene in the nematode C. elegance, and it has been found that the bcl-2 family consists of a series of genes some of which has similar actions and others demonstrate opposite effects. For example, The BAX protein has a killing effect on the cells when it forms homodimers, but in the presence of excessive amount of Bcl-2, the latter forms heterodimers with BAX abrogating its killing action. Therefore, the ratio of Bcl-2 to BAX is an important factor regulating cell survival. The ICE (interleukin-1β converting enzyme) gene, which is the mammalian homologue of ced3 in C. elegance, induces cell death. Now it is known that quite a number of proteases possess similar effects with ICE. WBI of mice with 0.075 Gy caused an up-regulation of bcl-2 and that with 2 Gy led to a down-regulation of its transcription (fig.2A). The transcription level of ICE after these two doses just showed an opposite relation (fig. 2B). The Bcl-2/BAX ratio in the thymocytes increased nearly 3 fold after 0.075 Gy and slightly decreased after 2 Gy (fig.2C). These experimental data give support to our previous findings that WBI with doses within 0.2 Gy led to a decrease of apoptosis to below the basal level while WBI with doses above 0.2 Gy caused a dose-dependent increase of thymocyte apoptosis [6]. This type of dose-effect relationship was observed in thymocytes when the apoptosis rate was quantified with fluorescence spectophotometry of DNA fragmentation. The same phenomenon was also observed in both the thymocytes and splenocytes when apoptosis was assessed with flow cytometry (FCM) measuring the number of apoptotic bodies (Liu and Ye, unpublished data).

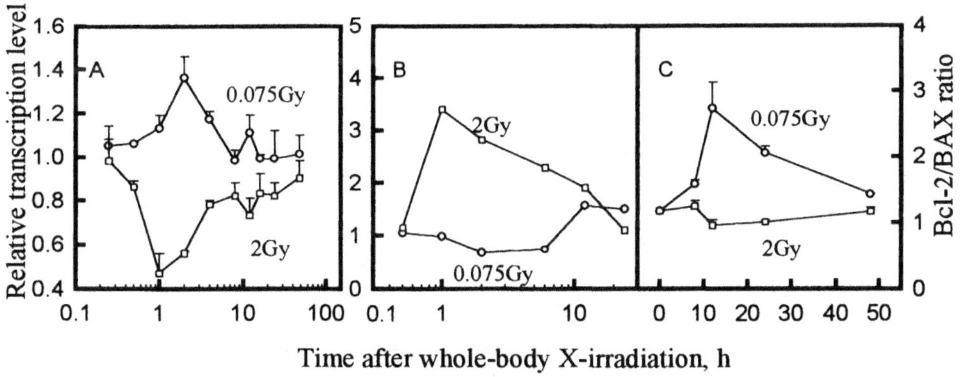

Figure 2. Changes in bcl-2, ICE and Bcl-2/BAX ratio after WBI with different doses of X-rays A) bcl-2 mRNA level in thymocytes was assessed with methods as described in fig. 1; B) ICE was measured using RT-PCR with β2-microglobulin as internal reference; C) Bcl-2 and BAX proteins were assessed using FCM with indirect immunofluorescnce.

3.2. RB AND P53

Rb and p53 are well known tumor suppressor genes which regulate cell survival. Many studies have shown that DNA damaging agents, including ionizing radiation, could induce the induction of these genes the protein products of which would affect the cell cycle progression and cell survival [16, 17]. Our previous data demonstrated that 24h after WBI with 2 Gy there was a significant increase in p53 protein in the thymocytes while 0.075 Gy led to its down-regulation [13]. Here it is shown in figure 3 that the transcription level of p53 gene in both the thymocytes (fig. 3A) and splenocytes (fig.3B) is up-regulated after WBI of mice with 2 Gy and WBI with 0.075 Gy had an opposite effect.

The level of expression of Rb protein was measured in thymocytes with FCM 8, 12, 24 and 48h after WBI and it was found that 0.075 Gy caused decrease in Rb expression during this time interval with the lowest level at 8h (73.5±6.11% of control, $p<0.001$) and 2 Gy led to its increase with the highest level at 48h (119.1±10.04%, $p<0.001$). The changes in p53 and Rb after WBI with X-rays also lend support at the molecular level to the increased survival of thymic and splenic lymphocytes after LDR.

3.3. GADD45 AND FASL

The gadd (growth arrest and DNA damage inducible) gene family has been studied extensively in recent years. Among its members the gadd45 gene has been found to be closely related to p53 [18-20]. It is known that p53 protein binds as a transcription factor to the third intron of the gadd45 gene and acts as an enhancer regulating the expression of gadd45 at the transcriptional level. The GADD45 protein probably prevents the cell to enter into the S phase through binding PCNA (proliferating cell nuclear antigen), thus decreasing the activity of cyclin D1 which forms a complex with PCNA to exhibit its action on cell cycle progression. As described in the last section, the transcription

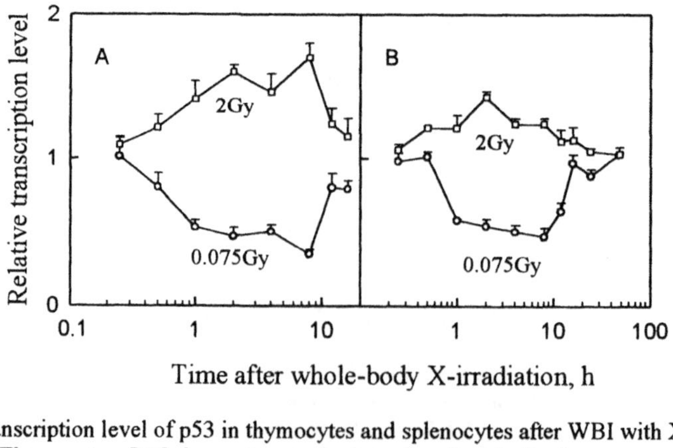

Figure 3. Transcription level of p53 in thymocytes and splenocytes after WBI with X-rays A) Thymocytes; B) Splenocytes; methods as described in fig. 1.

level of p53 changed in the opposite direction after low versus high dose radiation, so it could be predicted that GADD45 would have similar changes. This is actually so as seen in figure 4A.

Fas/APO1(CD95) is a type I transmembrane glycoprotein belonging to the NGF/TNF (nerve growth factor/tumor necrosis factor) receptor family widely distributed in various tissues including the lymphoid cells. FasL (Fas ligand) is a member of the TNF family expressing on activated T cells. A death signal is transmitted to the cells when FasL binds to Fas. As seen in figure 4B, FasL expression was up-regulated after WBI with 2 Gy X-rays while the low dose demonstrated an opposite effect.

Data presented in the previous sections could explain in part the J-shaped curve of thymocytes in response to WBI with different doses [6]. To fully elucidate the molecular mechanism of the cellular changes the signal transduction pathways should be examined.

4. Signal Transduction Pathways

It has been well documented that activation of the T lymphocytes involves the transmembrane signal transduction via the TCR/CD3 complex coupled to phospholipase C (PLC)-phosphatidylinositol biphosphate (PIP2) system [21,22]. Perturbation of the TCR/CD3 complex with antigenic or mitogenic stimulation would lead to activation of PLC which causes hydrolysis of PIP2 to form inositol trisphosphate (IP3) and diacylglycerol (DG) resulting in mobilization of intracellular calcium and activation of protein kinase C (PKC). In our previous work it was demonstrated that WBI of mice with low dose X-rays caused an increase in $[Ca2^{+}]_i$ as measured with FCM using Fluo-3AM as a probe and up-regulation of total PKC activity in T lymphocytes together with increased expression of TCR/CD3 complex on the thymocytes [23]. However, recent studies have shown that T cell activation involves multiple signal pathways among which the G

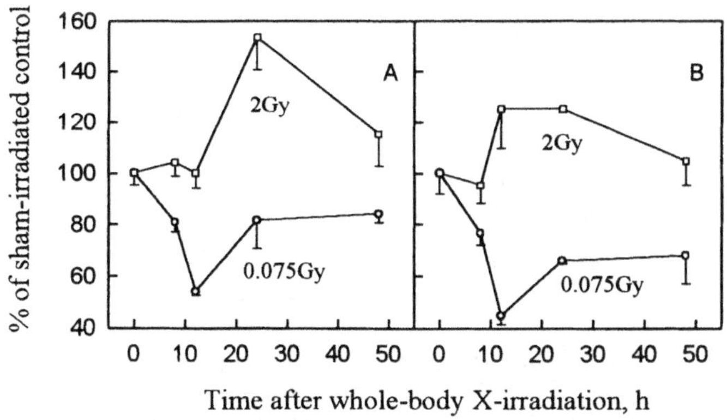

Figure 4. Changes in expression of GADD45 and FasL on thymocytes after WBI A) GADD 45; B) FasL; FCM analysis as described in fig. 2

protein-AC/GC (adenylate cyclase/guanylate cyclase) system is an important one [25,26]. In the present paper $[Ca^{2+}]_i$ is measured with Fura-2AM using fluorescence spectrophotometry and expression of PKC isoforms is measured using immunofluorescence with FCM to re-evaluate our previous results. At the same time changes in intracellular cAMP and cGMP content and protein kinase A (PKA) activity are observed in order to explore whether the G protein-AC/GC signal pathways are also involved together with the PLC-PIP2-IP3 pathway in the stimulation of T cell activity by LDR, and the dose-effect relationship is demonstrated.

4.1. DOSE-EFFECT RELATIONSHIP OF $[CA^{2+}]_I$ IN THYMOCYTES

Following WBI of mice with doses from 0.05 to 2Gy X-rays there was an increase in $[Ca^{2+}]i$ in the low dose range (0.075 and 0.1Gy, $p< 0.05$) which is more marked at the dose of 0.075Gy (Fig. 5, lower curve). This increase in $[Ca^{2+}]i$ was markedly accentuated in the presence of Con A (at a final concentration of 5 μg/ml). As shown in the upper curve of Fig. 5, there was a marked increase of $[Ca^{2+}]i$ after doses of 0.05, 0.075, 0.1 and 0.2Gy ($p<0.05$ to 0.01), which was most marked at the dose of 0.075Gy, reaching 228.6% of the sham-irradiated control. There was slight, but insignificant decrease of $[Ca^{2+}]i$ after 1 and 2Gy.

4.2. CHANGES IN EXPRESSION OF PKC ISOFORMS IN THYMOCYTES

As reported previously, the total PKC activity in the splenic T lymphocytes as measured by a PKC kit (Sigma BRL) was up-regulated after WBI with 0.075Gy, peaking at 12h following irradiation [23]. Here the expression of PKC isoforms α, β1 and β2 was estimated by immunofluorescence with FCM using specific antibodies. It was demonstrated that 0.075Gy caused increase in expression of the three isoforms of PKC, all peaking at

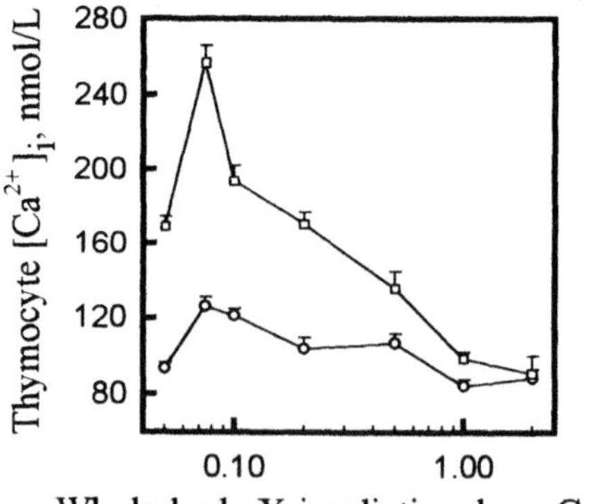

Figure 5. Changes in intracellular Ca ion in thymocytes after WBI with X-rays
Upper curve: with Con A (control=112.28±4.74); lower curve: without Con A (control = 86.25±2.88) ; measured with fluorescence assay

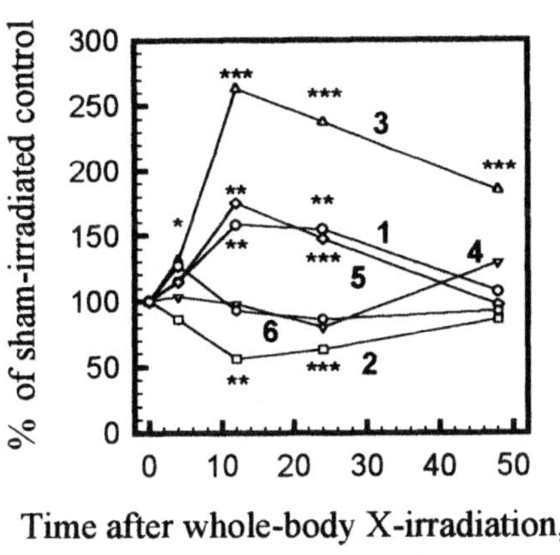

Figure 6. Changes in PKC isoforms in thymocytes after WBI with X-rays
1,2) PKC-α; 3,4) PKC-β1; 5,6) PKC-β2; 1,3,5) 0.075 Gy; 2.4.6) 4 Gy; *p<0.05, **p<0.01, ***p<0.001; methods as described in fig. 2.

12h after WBI, with most marked changes in PKC-β1 which showed an early onset, a higher peak and a longer duration of up-regulation (curve 3 in Fig. 6). The α and β2 isoforms only showed moderate increase in expression 12h and 24h after WBI with this low dose (curves 1 and 5 in fig. 6). Following WBI with 4Gy only the α-isoform showed significant decrease in expression in the thymocytes (curve 2 in fig. 6) while the changes in the other two isoforms were insignificant (curves 4 and 6 in fig. 6).

4.3. CHANGES IN CAMP, CGMP AND PKA IN THYMOCYTES AFTER WBI

WBI with low doses (0.05, 0.075 and 0.1Gy) caused significant decrease and WBI with high doses (1, 2 and 4Gy) caused marked increase in cAMP content in the thymocytes, thus leading to lowering of cAMP/cGMP ratio with low doses and its rise with high doses (fig. 7A). This gives rise to a J-shaped curve just as what was observed in the dose-effect curve of thymocyte apoptosis after WBI [6]. As seen in fig. 7B, the changes in thymocyte PKA (protein kinase A) activity after WBI with different doses showed a similar curve to that of cAMP/cGMP ratio.

It has been repeatedly demonstrated that WBI with 0.075 Gy X-rays would elicit activation of the T lymphocytes in mice within 12 to 24h expressed as increased thymocyte proliferation and up-regulated reaction of splenic lymphocytes to Con A [3]. To test the role of mobilization of calcium and involvement of the cAMP/cGMP system in the activation of T cells by LDR, experiments were designed with calcium antagonist TMB-8 and G protein stimulant CT (cholera toxin) added to cultures of thymocytes sampled 12h After WBI with 0.075 Gy to see their influence on the increased spontaneous incorporation of tritiated thymidine into the thymocytes after LDR. It is interesting to see that both TMB-8 and CT suppressed the up-regulation of DNA replication in the thymocytes induced by LDR (to be published separately). Therefore, it is evident that LDR-induced thymocyte activation involves both the PLC-PIP2-IP3 and G protein-

AC/GC pathways. Actually other signal pathways may also come into play after WBI, e.g., the possible changes in glucocorticoid receptor (GR) reaction with corresponding response elements in DNA is one of the candidates since a lowering of serum ACTH and corticosterone has been found after LDR with a putative decrease in binding sites of GR with dexamethasone in lymphocytes [3].

It is known that apoptosis of thymocytes is a constantly occurring process in the body regulating homeostasis and there can be seen a basal level of apoptosis rate in the thymus under normal conditions as assessed by either FCM or DNA fragmentation [6]. In a previous report it was demonstrated that WBI with low doses of X-rays (0.05-0.1 Gy) reduced the rate of thymocyte apoptosis to below its basal level [6]. This was also proven on morphological grounds using electron microscopy [7]. When mitogen-stimulated mouse splenic lymphocytes were adapted by a LDR (0.01 Gy), the extent of DNA fragmentation caused by a succeeding high dose was reduced [27]. The mechanism of the reducing effect of LDR on apoptosis of lymphoid cells has not been elucidated. Since thymocyte survival and apoptosis are governed by the interaction of different molecules of multiple signal pathways involving a number of genes, we intended to explore the changes in induction of relevant genes and signal pathways following different doses of radiation. The present paper is only concerned with the changes in some of the molecules which are important in two signal pathways, namely, the PLC-PIP2-IP3 system and the G protein-AC/GC system as well as some related genes, especially the proto-oncogens and tumor suppressor genes related to cell survival and death.

The influence of radiation on the inositol 1,4,5 trisphosphate receptor (IP3R) type 1 in human Epstein Barr Virus-transfected lymphoblastoid cells was studied and an increase in expression of cDNA corresponding to IP3R and its mRNA was found after exposure to 3Gy γ-rays [28]. There has been no other reports, to our knowledge, of the effect of LDR within 0.2Gy on signal molecules in thymocytes. Data in the present paper proved and extended our previous experimental results indicating the involvement of increased $[Ca^{2+}]_i$ and PKC activity in the stimulatory effect of LDR on thymocyte

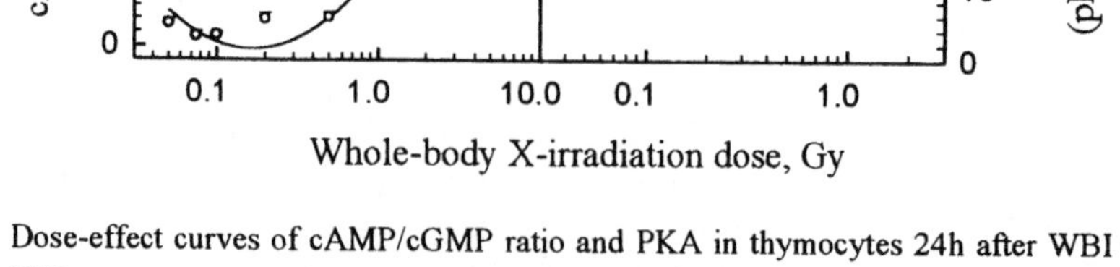

Figure 7. Dose-effect curves of cAMP/cGMP ratio and PKA in thymocytes 24h after WBI with X-rays
A) cAMP/cGMP ratio (control=2.19±0.27); B) PKA activity (control=18.63±0.24)

proliferation [23]. In the present study it was further established that the β1-isoform of PKC might contribute more than the other two isoforms examined in the facilitating effect of LDR on the IP3-PKC pathway and that the mobilization of intracellular calcium ions after WBI with low dose X-rays could be reinforced in the presence of mitogenic stimulation. This latter finding coincides well with our previous observation that the addition of Con A in vitro further reduced the lowered apoptpsis rate of thymocytes from low dose-irradiated mice in a dose dependent manner [6].

It has been documented that the AC-cAMP-PKA signal system plays an important role in determining the responses of lymphocytes to perturbation of CD3 molecules with monoclonal antibodies [29]. Here in the present study we investigated the changes in this signal system after WBI with different doses of X-rays. The results indicate that cAMP content of the thymocytes decreased and cGMP of the thymocytes increased after LDR while doses above 0.5Gy caused the opposite effect, giving rise to a J-shaped dose-response curve for the cAMP/cGMP ratio. The activity of PKA which acts as a donwstream molecule of cAMP was found to be down-regulated after LDR and stimulated after higher doses. It is known that PKA may exert a negative feedback effect on the PLC-PIP2-IP3 pathway, thus its down-regulation would facilitate the activity of the latter system resulting in further activation of the proliferation of the thymocytes. The role played by down-regulation of the G_s-AC-cAMP system in LDR-induced thymocyte proliferation is evidenced by the decreased amplitude of LDR-induced thymocyte proliferation with CT stimulation.

The dose-response relationship of the chief signal molecules in the two systems demonstrates the different effects of low versus high doses of irradiation on these molecules as shown in figures 5 and 7. It is interesting to note that the most prominent effect occurred at the dose of 0.075Gy which has been repeatedly shown to be the dose eliciting most marked stimulating effect on immune reactions [3,13,23].

5. Implications

Data presented in this paper chiefly concerns the molecular changes related to survival and differentiation of cells in the immune organs, especially the thymus, following WBI with different doses of X-rays. The thymocytes are well known to be highly radiosensitive. It is interesting to find that the dose-effect relationship of the apoptosis of this highly radiosensitive cell type does not follow a linear pattern but rather gives a J-shaped curve [6]. And this phenomenon at the cellular level has gained molecular basis from the examination of the induction and/or expression of the relevant genes as well as the changes in the signal pathways. The findings may have important implications in the explanation of the adaptive and defense responses induced by LDR. In risk assessment of low level radiation it has been assumed that cancer risk is best estimated by a linear, no-threshold model which is supposed to be a prudent way to protect the public. Such assumptions based on calculations chiefly from physical considerations may not be valid without detailed analysis of biological data obtained in recent years. It is now known that LDR could reduce the mutagenic effect of high dose radiation (HDR) [reviewed in 30]. It was reported that LDR decreased the frequency of neoplastic transformation to below the spontaneous rate in some clones of C3H 10T1/2 cells [31]. Pre-exposure with LDR of mice lowered the incidence of thymic lymphoma induced by

HDR [32,33]. These and other similar data should be considered in the analysis of risk assessment of low level radiation.

In the intact organism the damaged target cell is under very complex influence of the internal environment of the body. The consensus of opinion for carcinogenesis is that the multistage process of its progression takes a long time and not all the damaged cells are destined to become neoplastic. The pre-neoplastic cells may be eliminated by the existence of cell surveillance mechanisms among which immune responses take an active part. It is known that LDR up-regulates both the natural and acquired immunity against neoplasm [3]. The potentiation of the immune responses following LDR may last 1-2 months before being elicited by the challenge (unpublished data). Of course further studies are needed along this line to fully elucidate the nature of the adaptive and defense responses induced by LDR.

In conclusion, correlation of studies on the effects of LDR at systemic, cellular and molecular levels in animal models will promote our understanding of the fundamentals for the assessment of risk from environmental radiation.

Acknowledgment: This work is supported by a grant from NSFC.

References

1. Anderson, R.E., Lefkovits, T and Troup, G,M. (1980) Radiation-induced augmentation of immune response, *Contemporary topics of Immunology* **11**, 245-274.
2. Makinodan, T. and James, J (1990) T cell potentiation by low dose ionizing radiation: possible mechanisms, *Health Physics* **59**, 29-34.
3. Liu, S.Z. (1995) Current status of research on radiation hormesis in the immune system after low level radiation, *Journal of Radiation Research and Radiation Processing* **13**, 129-138.
4. Kondo, S. (1988) Altruistic cell suicide in relation to radiation hormesis, *International J. Radiation Biology* **53**: 95-102
5. Liu, S.Z., Qi, J and Zhang, Y.C. (1992) Low dose radiation does not cause ratio changes of regulatory T subsets in mouse spleen, *J Norman. Bethune University of Medical Sciences* **18:** 209-213.
6. Liu, S.Z., Zhang, Y.C., Mu, Y., et al (1996) Thymocyte apoptosis in response to low dose radiation, *Mutation Research*, **358**, 185-191
7. Shaposhnikova, V.V. and Korystove, Y.N. (1995) Thymocyte proliferation and Apoptosis induced by ionizing radiation, *Scanning Microscopy* **9**, 1203-1206.
8. Fornace, A.T. (1992) Mammalian genes induced by radiation: activation of genes associated with growth control, *Annual Review of Genetics* **26**, 507-526.
9. Woloshak, G.E., Chang-Liu, C-M., James, P.S., et al (1990) Modulation of gene expression in Syrian hamster embryo cells following ionizing radiation, *Cancer Research* **50**, 339-344.
10. Boothman, D.A., Majmudar, G. and Johnson, T. (1994) Immediate X-ray inducible responses from mammalian cells, *Radiation Research* **138**, S44-S46.
11. Woloshak, G.E., Chang-Liu, C-M., Panozzo, J, et al (1994) Low doses of neutrons induce changes in gene expression, *Radiation Research* **138**, S56-S59.
12. Ward, J.F. (1994) DNA damage as the cause of ionizing radiation-induced gene activation, *Radiation Research* **138**, S85-S88.
13. Liu, S.Z. (1997) Cellular and molecular basis of the stimulatory effect of low dose radiation on immunity, in Eds: Wei, L, Sugahara, T. and Tao, Z.,"High Levels of Natural Radiation: Radiation dose and Health Effects". Elsevier Science B.V., Tokyo, 342-353.
14. Chen, C.H., Zhang, J and Ling, C.C. (1994) Transfected c-myc and c-Ha-ras modulate radiation-induced apoptosis in rat cells, *Radiation Research* **139**, 307-315.

15. Martin, M., Crechet, F, Rampunet, B., et al (1995) Activation of c-fos by low dose radiation: a mechanism of the adaptive response in skin cells, *Radiation Research* **141**, 118-119.
16. Smith, M.L., Chen, I.T., Zhan, Q., et al (1994) Interaction of the p53 regulated protein Gaad45 with proliferation cell nuclear antigen, *Science* **226**, 1376-1379.
17. Goodrich, D.W. and Lee, W.H. (1992) Abrogation by c-myc of G1 phase arrest by Rb protein but not by p53, *Nature* **360**, 177-179.
18. Kastan, M.B., Zhan, Q., El-Deiry, S., et al (1995) A mammalian cell cycle check point utilizing p53 and Gadd45 is defective in Ataxia telangiectasia, *Cell* **71**, 587-597.
19. Yoshida, T., Okazaki, T, Hughes, P.E., et al (1996) Cloning of rat GADD45 gene and induction analysis following ionizing radiation in vivo, *FEBS Letters* **380**, 87-92.
20. Zhan, Q., Bae, I., Kastan, M.B., et al (1994) The p53-dependent γ-rays response of GADD45, *Cancer Research* **54**, 2755-2760.
21. Berridge, M.J. and Irvine, R.F. (1984) Inositol trisphosphate, a novel second messenger in cellular signal transduction, *Nature* **312**, 315-321.
22. Berridge, M.J. (1993) Inositol trisphosphate and calcium signaling, *Nature* **361**, 315-325.
23. Liu, S.Z., Su, X., Zhang, Y.C. , et al (1994) Signal transduction in T lymphocytes after low dose radiation. *International J. Occupational Medicine and Toxicology*, **3**, 107-117.
24. Tamir, A., Granot, Y. And Isakov, N. (1996) Inhibition of T lymphocyte activity by cAMP is associated with down-regulation of two parallel mitogen-activated protein kinase pathways, the extracellular signal-related kinase and c-jun N-terminal kinase, *Journal of Immunology*, **157**, 1514-1522.
25. Teschendorf, C., Trenn, G., Hoffkes, H.G., et al (1996) Differential effect of the activation of protein kinase A on the protein synthesis and secretion in the T helper 2 cell line D10.G4.1, *Scandinavian J. Immunology*, **44**, 150-156.
26. Lin, J., Gettys, T.W., Qin, L., et al (1995) Increased cAMP and cAMP-dependent protein kinase activity mediate anti-CD2 induced suppression of anti-CD3-driven interleukin-2 production and CD25 expression, *Pathobiology* **63**, 175-187.
27. Hyun, S.J., Yoon, M.Y.. Kim, T.H., et al (1997) Enhancement of mitogen-stimulated proliferation of low dose radiation-adapted mouse splenocytes, *Anticancer Research* **17**, 225-229
28. Yan, J., Khanna, K.K. and Lavin, M.F. (1995) Induction of inositol 1,4,5 trisphosphate receptor genes by ionizing radiation. *International J. Radiation Biology*, **69**, 539-546.
29. Li, M., Yang, J., Hu, Y.N., et al (1995) The influence of intracellular cAMP on T cell proliferation, *J. Tongji Medical University* **15**, 87-89.
30. Rigaud, O. and Moustacchi, E. (1996) Radioadaptation for gene mutation and the possible molecular mechanisms of the adaptive response, *Mutation Research* **358**, 127-134.
31. Azzam, E.I., de Toledo, S.M., Raaphorst, G.P., et al (1996) Low-dose ionizing radiation decreases the frequency of neoplastic transformation to a level below the spontaneous rate in C3H 10T1/2 cells, *Radiation Research* **146**, 369-173.
32. Ju, G.Z., Liu, S.Z., Li, X.Y., et al (1995) Effect of high versus low dose radiation on the immune system, Eds: Hagen, U., Harder, D., Jung, H. and Streffer, C. Congress Proceedings of 10th ICRR, Wurzburg, Germany, Aug. 27-Sept. 1, 709-714
33. Bhattacharjee, D. (1996) Role of radioadapation on radiation-induced thymic lymphoma, *Mutation Research* **358**, 231-235.

CELLULAR RESPONSES.

Recommendations for Future Research Needed

H. RINK [1], T. C. YANG [2] AND
L. BÖHM [3], R. GOVORUN [4], D. HÄDER [5], G. HORNECK [6],
B. KAINA [7], S. KOZUBEK [8], S. Z. LIU [9], C. POTTEN[10], L. PTITSYN [11]

[1] *Experimental Radiology and Radiation Biology, University of Bonn, D-53105 Bonn, Germany*
[2] *NASA Johnston Space Center, Radiation Biophysics, 2101 NASA Road, Houston, Texas, USA*
[3] *University of Stellenbosch, P.O.Box 19063, Tygerberg 7505, South Africa*
[4] *Joint Institute for Nuclear Physics, 141980 Dubna, Russia*
[5] *Botanical Institute, Friedrich-Alexander University, D-91058 Erlangen, Germany*
[6] *DLR, Institute of Aerospace Medicine, Radiobiology Section, D-51170 Köln, Germany*
[7] *Division of Applied Toxicology, Medical Faculty, University of Mainz, D-55131 Mainz, Germany*
[8] *Institute of Biophysics, Czech Academy of Sciences, 61265 Brno, Czech Republic*
[9] *MH Radiobiology Research Unit, Norman Bethune University of Medical Sciences, Changchun, 130021 China*
[10] *Department of Epithelial Biology, Paterson Institute, Christie Hospital Trust, Manchester M20 4BX, UK;*
[11] *State Research Centre of RF: GNII Genetica, 113545, Moscow, Russia.*

Brief Statement of Knowledge

This section deals with cellular mutations and transformations, adaptic and early gene responses, signal transduction pathways, cell cycle control and apoptosis, biological dosimeters and biomonitoring as well as aspects of the ecosystem and theoretical models. The role of chromosome aberrations and cancer genes in neoplastic cell transformation induced by radiation is still unclear. Most new results have been obtained by using techniques and methods of molecular biology. The group discussion showed that a huge amount of knowledge, which could fill several text books, is available on the topics mentioned above. On the other hand there are essential gaps in our knowledge.

C. Baumstark-Khan et al. (eds.),
Fundamentals for the Assessment of Risks from Environmental Radiation, 339-344.

The group concentrated more on these gaps than on the state of knowledge, which is mostly covered by the papers of this meeting.

The topic comprises various cell systems having different radiosensitivities. The variation of sensitivity depends on the organisms, the relevant organs, the cell system parameters such as cell cycle stage, state of differentiation and the differences in the mechanisms of apoptosis. We also see differences in cells taken from different individuals even with the same method. Of course, the differences between *in vitro* and *in vivo* systems have to be considered.

Under the topic of risk assessement, only low doses are considered. In this dose range it might be necessary to identify the cells which respond to irradiation whereas most of the cells in the population do not. The general radiation parameters, such as dose or fluence, dose rate, dose fractionation, radiation quality (LET as well as the wavelength of UV-light) and time-dependent effects, should always be taken into consideration. As far as ionizing radiations and nonionizing radiations are considered, there are no reasons to split up the topic into two subsections although a lot of differences exist especially in the field of indirect UV-light damage, which involves the action of oxygen radicals, as well as in the wavelength-dependent biological effects. As to the synergistic effects between UV- and ionizing radiation, there is a large gap in quantitative data as well as the information on mechanism(s).

Risk is mostly seen in mutagenesis and carcinogenesis, especially in skin and lung cancer and leukemia, but also in cataractogenesis, premature aging and impairment or disbalance of the immune system. For these reasons, not only mutation frequency but also the nature of mutation lesions and cell transformation events play an important role. Therefore, the mechanism(s) leading to cell mutation and transformation and to genomic instability need to be investigated. Additionally interactions between radiation and other environmental carcinogens or mutagens as well as synergistic effects between UV- and ionizing radiations are of special interest.

Cancer and other diseases start from cellular malfunctions and there is a general need to investigate the mechanism(s) of cellular responses to different radiations. The signals transduction and the factors, which have an influence on the processing of the radiation damage determining the cellular endpoints, as well as cell progression through the cell cycle, including regulation and the radiation specific delays at distinct transition points of the cell cycle, are of interest. The early gene response and the adaptive response comprising inducible genes for radiation protection and DNA repair, and the important balance between proliferating and resting, including surviving cells on the one hand and apoptosis and necrosis on the other hand, have to be considered and investigated. An understanding of the molecular mechanisms of all these events is clearly necessary, but the corresponding knowledge is very limited.

The early gene response (within minutes after irradiation) implicates an activation of pre-existing factors, such as EGF-receptor after UV-exposure, and a signal transduction via Ras, Raf, Map-kinases, p38 and Jun-kinase to pre-existing transcriptional factors, such as c-fos and c-jun which in turn become activated by phosphorylation. The next step is the activation of transcription factor AP-1, which is involved in the regulation of the early gene response and mediates late response (collagenase, gelatinase, stromelysin). An increased synthesis of collagenase promotes collagen degradation, which is required for tissue regeneration and may also be involved in cutaneous aging after excessive exposure to UV. Another interesting candidate of early gene response is the UV-C inducible rho-B, which seems to be independent from jun and fos. The biological meaning of the early gene response is rather unclear although distinct links point to cell cycle check point control and to DNA damage repair. Importantly, cells deficient in c-fos or c-jun are hypersensitive to UV-C, indicating the involvement of the correspondent proteins in DNA damage. An important question is whether DNA repair genes and more generally DNA damage defense functions, such as radical scavengers and DNA damage bypass mechanisms (including SOS functions), are induced by UV- and/or ionizing radiations. Until now only 4 genes (MGMT, DNA polymerase-beta, beta apurinic endonuclease, DNA ligase1) are known to be inducible by genotoxic agents. But none of these respond to UV- or ionizing radiation, nor do they seem to be specifically involved in radiation defense. In concert with other functions, the p53 gene controls genomic stability. Together with p21 (mdm-2, gadd45), it becomes activated and stabilized by DNA strandbreaks via phosphorylation and is also included in the cell cylce control via cyclins which is important for DNA damage repair.

The tumour suppressor protein p53 obviously plays an important role in the processes of DNA damage defense, maintenance of genomic stability, cell cycle control and apoptosis, which also involves c-myc and the bcl-2 gene family (30 memebers). Although apoptosis represents no marker for low dose radiation exposure, the factors involved (gene activation, repair efficiency, damage detection threshold, residual damage, misrepair) may be important for the determination of long term risk

Unsolved Problems

The number of genetic loci for mutation assays is very limited. Additional systems are required. The dose-response relationships obtained with rodent cell systems have to be confirmed with human cell systems other than human lymphocytes. Because radiation response can also be modified by cell-to-cell communication, co-culture systems (fibroblast / epithelial cells) should be examined.

In general low-dose rate exposure is less effective than high-dose rate exposure. However, some mutation data suggest an inverse effect with low-LET radiations. The existence of such an inverse dose rate effect should be examined for oncogenic cell transformation and carcinogenesis in animals as well as for UV-radiations. Data for UV-B dose responses are limited and not available for cell transformation. Additionally the role of cancer genes in neoplastic cell transformation remains to be answered. Under the aspect of risk assessement, especially for space radiations, the effects of single traversal of high-LET charged particles have to be determined in detail. In this connection the existing biophysical models for radiogenic mutation, oncogenic transformation, and cell survival have to be tested experimentally.

At cellular level the primary signal, which induces events such as the early gene response, as well as the genes involved in the defense systems such as DNA repair, cell cycle checkpoint control and apoptosis/survival functions (especially after ionizing radiation), have to be identified. DNA repair is highly regulated. This occurs at three levels : i) gene activation, ii) post-transcriptional activities, and iii) activity of factor proteins (accessory proteins). The regulation of the complex repair network needs more knowledge about the number and functions of these proteins. Cloning of repair genes (stress genes, checkpoint control genes) is essential in understanding the molecular base of cellular defense.

The genomic instability, which can be induced many generations after exposure to ionizing radiation, also needs more investigation and explanation. Because apoptosis differs among various cell types in terms of yield, time course, genes, endonuclease activcity, morphology, it affords the investigation of the underlying mechanisms.

There is a further need for quick and cheap, specific and sensitive biological dosimeter systems which can be applied for different purposes. The big advantage of such systems is given by its complementary to physical dosimetry and by the fact that radiation effects, accumulated over longer time periods, can be recorded. Such dosimeters should be able to weight the different components of radiation to their biological efficacy.

Already in use are systems which rely on chromosomal aberrations in lymphocytes or on minisatellite mutations in germ line cells. Other possible systems detect biomarkers, antibodies or typical metabolites (lipids, peroxides, proteins, enzymes, cholesterol, melatonin) in the blood and/or in the urine. These systems should be improved, linked to each other and to physical dosimetry. Changes in hair follicles and lens opacifications or lens fluorescence should also be tested in order to establish their sensitivities and usefulnes as biological dosimeters.

A special case is the highly developed DLR film like UV-dosimeter, which can be used as an individual dosimeter for the monitoring of solar UV-lights. Other extrinsic systems using bacteria, viruses or simple plant systems, e.g., Ames test, umuC test, SOS-chromotest and SOS-lux test, belonging to the class of biosensors used mostly for the detection of substances with genotoxic effects, are also useful for the detection of radiation effects (DNA damage, impairment of photosynthesis, reduced acticities, cell killing) and should be developped and improved. Other *in situ* biomarkers should be searched for in order to construct dosimetric systems with well characterized mechanisms.

It may be necessary to monitor the UV-exposure of distinct persons at risk or of groups of persons at risk. In this case biological dosimeters should be at hand.

The depletion of stratospheric ozone with an increase of UV-B radiation may have a dramatic influence on the whole biosphere and the ecosystem. Therefore continuous and global, longterm monitoring of UV-light spectra and intensities as well as the biomonitoring of the aquatic biomass are necessary and of high importance. This work is partly performed already by different satellites and should be continued and improved. Decreases in biomass production may affect both aquatic ecosystems as well as crop plants leading to decreased food production. The greenhouse effect may be augmented, because a decrease in aquatic biomass will increase the amount of CO_2, which is normally removed from the atmosphere by this system.

Because of the different sensitivity and adaptability of species to UV-radiation, changes in species composition could occur as a result of increasing UV-B radiation. The interference of UV-light with the molecular processes of photosynthesis, the UV-induction of repair and of other defective mechanisms, and biological counter-measures which might enhance UV-B resistance (in crop plants), therefore, should also be investigated in more detail.

Last but not least theoretical aspects have been discusssed. The oncogenic and mutagenic effect of a single traversal of high-LET charged particles is yet to be understood and should be studied carefully. In this field any model which contributes to a better understanding of biological effects of high LET-radiations, is desirable. The dependence of mutation induction on LET and specific energy can be very complex. In a short range of LET, the mutation frequency can change by more than one order of magnitude. This suggests that the calculation of the equivalent dose may be not appropriate if only LET-spectra are used.

Justification of Further Research

A number of the unsolved problems as mentioned above gives strong justification for further research. More than 50 genes are known to be inducible by genotoxic stress. Only some of them have been functionally identified and shown to play a role in DNA damage removal, bypass of lesions and checkpoint control. Another group of inducible genes and proteins determines the viability of damaged cells (survival and apoptotic functions). They may be highly important in determining the mutagenic and carcinogenic risk of UV- and ionizing radiation. By large, inducible responses may be considered to be protective, influencing the yield of mutations in a cell population.

Recommended Approaches

PART I: MUTATION AND TRANSFORMATION

- Develop human cell model systems for quantitative assay of radiation induced cell transformation, additional assays to study mutation in various loci on different chromosomes, and transgenic cell systems for evaluating the role of tumour suppressor genes.
- Apply techniques to determine gene alterations which represent „signatures“ of different radiation qualitites.
- Determine mutagenic and oncogenic effects of a single heavy ion traversal through a mammalian cell.
- Determine possbile mechanisms in the inverse dose rate effect for low LET-radiation.
- Determine the mechanisms of genomic instability in radiation induced cell transformation.
- Test biophysical models which can be used to assess UV-, low- and high-LET radiation risk at low doses and low dose rates.
- Determine wavelength dependency for UV-B in cell transformation and cell mutation.
- Investigate mechanisms of synergistic effects between UV- and ionizing radiation.

PART II: EARLY GENE RESPONSE AND CELL CYLCE CONTROL

- Identify primary targets which trigger the inducible cell responses.
- Compare different qualities of UV- and ionizing radiations in order to find threshold doses for inducible cell responses in different cell types and individuals.
- Identify the signal transduction pathway and the downstream target genes.
- Analyze and compare the processes of apoptosis after UV- and ionizing radiation.
- Elucidate the interaction of particular inducible functions and the genetic endpoints (cell killing, mutation, chromosomal aberrations, and oncogenic transformation) .
- Study the regulation of DNA repair at gene and protein level.
- Study cell cycle checkpoint control upon UV- and ionizing radiation.
- Create transgenic and knockout cell lines and individuals (mice) exhibiting alterations in UV- and ionizing radiation induced signaling and genetic responses.
- Modulate the gene expression of repair, checkpoint control and apoptosis by means of antisense and recombinant techniques in order to elucidate their involvement in cellular defense strategies.

PART III: BIOLOGICAL MONITORING

- Efforts should be made to develop and apply *in situ* biomarkers for monitoring of individual radiation exposure.
- Extrinsic dosimeters or indicators should be included in long-term monitoring of the enviromental UV-radiation.
- Biosensors should be developped which respond to different components of environmental radiation.

UV CARCINOGENESIS: FROM EXPERIMENT TO RISK ASSESSMENT.

From Hairless Mice to Humans

F.R. DE GRUIJL

Dermatology; University Hospital Utrecht
PO Box 85500; NL-3508 GA Utrecht

1. Abstract

UV radiation that reaches us from the sun is very photochemically active, and damages organic macromolecules such as DNA. UV radiation is the most abundant and prominent exogenous carcinogenic factor in our natural environment to which the human skin is necessarily well adapted. Nevertheless, skin cancer may occur in the long run (i.e. at old age), and it is by far the most frequent form of cancer in USA (over 1 million cases a year). By combining epidemiological, experimental animal and cellular data a fundamental understanding of UV carcinogenesis is steadily growing: mutation spectra from human skin cancer provide the most direct evidence that UV radiation is the culprit.

Murine models can be used to establish the wavelength dependence, and the dose-time relationship for skin carcinoma induction by chronic UV radiation. Epidemiological data on skin cancer incidences can be used to adjust the murine model to humans, and then apply the result in risk assessments for altered UV exposures (e.g. from an atmospheric ozone depletion). The murine models, especially with transgenic mice, can further help us in identifying and validating important bio-indicators of risk which can refine our assessments, perhaps even to the extent that an individual's risk can be estimated.

2. Introduction

Ultraviolet (UV) radiation from the sun with wavelengths below 290 nm does not reach us mainly because it is effectively absorbed by ozone in the stratosphere. This short-wave UV radiation is very cytotoxic, and blocking it is likely to be an important prerequisite for the successful evolution of organisms on the primordial earth surface. The part of the UV-B band (wavelengths 280-315 nm) that passes through the atmosphere is still enough to give us sunburn. Hence, the natural background level of UV radiation may not always be considered 'low level': it can be quite sufficient to induce a toxic effect.

C. Baumstark-Khan et al. (eds.),
Fundamentals for the Assessment of Risks from Environmental Radiation, 345-357.

3. Human Skin Cancers and Sun Exposure

Skin cancer, with well over 1,000,000 cases a year, is the most frequent type of cancer in the USA [1], and the incidences are considerably higher in the Southern, more sunny, States than in the Northern ones. Three main types of skin cancer can be distinguished: basal cell carcinoma (BCC), squamous cell carcinoma (SCC) and cutaneous malignant melanoma (CMM). All of these skin tumors originate from the epidermis, i.e. the outermost viable cell layer with a typical thickness of 70 μm (note that the most carcinogenic solar UV radiation hardly penetrates any deeper than this, see sections 4.11 and 6.1). BCC are most common in white Caucasians (typically about 90 cases per 10^5 people per year in NW Europe, py), followed by SCC (about 15 cases per 10^5 py), and CMM (about 10 per 10^5 py) [2]. CMM stem from the pigment cells (melanocytes) and are the most aggressive type: they can rapidly metastasize. SCC and BCC stem from the keratinocytes, and these tumors are less aggressive, but they do grow invasively, and if SCC is neglected it can metastasize, which is quite a rare event with BCC. Mortality is about 20-25% of the CMM cases, whereas it is 1-3% for SCC and less than 1% for BCC. Early and easy detection of these tumors on the body surface clearly works in favor of the therapeutic success, especially for SCC and BCC. Although CMM is a minority in the skin cancer incidence, it strongly dominates the mortality from skin cancer.

The contention that SCC and BCC are related to extensive sun exposure dates back to the 19-th century, when it was noted that these tumors occurred commonly on people with outdoor professions (people in vineyards, and fishermen). A possible link between CMM and sun exposure has always been more the subject of debate. The common feature of people who run the highest risk of skin cancer is their sun (UV) sensitivity, i.e., they burn easily and never tan ('skin type I'). SCC appears to have the most straightforward relationship to sun exposure: these tumors occur on the regularly exposed skin (head, neck, back of the hands) [3] and the risk goes up with the life-long accumulated exposure [4, 5]. For a long time the etiology of BCC was believed to be largely similar to that of SCC, but more recent epidemiological studies have pointed out distinct differences. Although BCC on patients are also mainly located on the most exposed areas (80-90% on the head [3]) and the risk has been reported to go up with the total sun exposure [5], BCC -- in contrast to SCC -- hardly occurs on the backs of the hands but relatively more frequently on the trunk [3, 6], and the sun exposure in the decades prior to removal of the tumor are not related to the risk [4]. Apparently, BCC shows a predilection for certain skin sites (sebaceous skin?), and sun exposure is probably related to an early event in the genesis of the tumor. Recent Australian data from a population survey [6] show that the number of recalled sunburn episodes was positively related to the risk, especially sunburns that occurred in childhood. BCC in some respects appears more similar to CMM. Except for the Hutchinson Melanotic Freckel Melanoma (a minority of CMM, about 10%, with an etiology more like SCC), CMM is known not to be related to the life-long accumulated sun exposure, but more to intermittent over-exposure and to high levels of ambient UV radiation during childhood [7]. The latter result on childhood exposure is obtained from studies on migration from a temperate to a subtropical climate, whereas the relation to intermittent over-exposures hinges purely on recall, which may be biased in (ex-)patients, and will be confounded by an individual's sensitivity, for which statistical corrections need to be made.

Epidemiology has provided us with valuable data on associations between skin cancer and sun exposure, but it cannot by itself establish with any certainty which factors play a causal role in the pathogenesis: e.g. which part of the solar spectrum is really causing these tumors? Clearly, to this end the epidemiological data need to be supplemented with relevant experimental data on UV-induced skin carcinogenesis.

4. UV Genotoxicity and Carcinogenesis

The central paradigm in cancer research is that cancer arises from a clonal expansion of a cell with defects in a combination of genes that are crucial in the regulation of the cell cycle and differentiation. These defects arise as a result of DNA damage caused by endogenous and exogenous agents. The initiation of such a clone and its outgrowth can be stimulated by epigenetic ('promoting') effects, e.g. an altered microenvironment of the cell may provide a persistent growth signal, or a protection against the elimination of an aberrant cell. In this section we will focus on the 'classical' genotoxic route from UV-induced DNA damage to tumor formation, but there are indications that UV radiation may also contribute to carcinogenesis by epigenetic alterations (e.g. see section 5).

4.1 DNA DAMAGE

UV-B radiation is very potent in starting up all kinds of photochemical reactions: from every day life we are well acquainted with 'photochemical smog' and the photodegradation of plastics and paints. The conjugated bonds (alternating double and single bonds) in organic molecules are sites that effectively absorb the UV radiation between 200 and 250 nm; in long chain repeats or in ring structures the absorption band of these bonds shift to longer wavelengths. In DNA all the nucleic acids are aromatic and contribute to the absorption. The maximum absorption by DNA lies around 260 nm and drops off by 2 orders of magnitude toward 300 nm. Nevertheless, this absorption in the UV-B band is still high enough to cause cytotoxicity in naked cells exposed to sunlight. Even the substantial absorption of overlying layers of dead and differentiated cells [8], does not fully protect the germinative basal layer of cells (about 70 μm deep) in the human epidermis from accumulating significant amounts of DNA damage during a UV-B exposure just enough to cause a mild sunburn [9].

Although all bases in DNA contribute to the absorption, the sites with two or more neighboring pyrimidine bases in a strand are preferentially targeted by UV-B radiation to form dimers: either a cyclobutane pyrimidine dimer, CPD, or a (6-4)pyrimidine-pyrimidone photoproduct, 6-4PP. The quantum yield (average number of lesions per absorbed photon) is dependent on the nucleotides flanking the target site. Although the absorption by DNA becomes vanishingly small with increasing wavelengths in the UV-A band (315-400 nm), the long-wave UV-A radiation may still damage the DNA indirectly by reactive oxygen species, ROS, generated through the absorption by other molecules than DNA.

4.2 CELLULAR RESPONSES

UV-induced cellular damage evokes cellular responses to remove the damage, and to restore normal cell functions. DNA damage in particular needs to be repaired because it can block transcription and lead to errors in DNA replication: transcription blockage is a direct threat to cell survival and a non-lethal mutation in a daughter cell may be a step toward malignant transformation.

A complex enzyme process, called nucleotide excision repair, NER, is responsible for the removal of 'bulky adducts' and the UV-induced CPD and 6-4PP from the genomic DNA in mammals. In human cells the adduct is cut out in a oligonucleotide of about 30 bases length, and the gap in the DNA strand is filled in using the complementary strand as a template. Most of the dimers are removed 24 hours after a UV exposure, but the rate of repair appears to vary considerably from one individual to the next [9]. The 6-4PP are removed much faster than CPD [10], and the latter lesions are removed faster from a transcribed strand, TS, than from a non-transcribed strand, NTS. This is a manifestation of so-called transcription coupled repair, TCR. This contrast between removal of CPD from TS and NTS is larger in mice, where TCR of CPD is very proficient but the (global) repair (GR) of CPD is deficient [11]; the sensitivity of mice to the UV induction of skin tumors (section 6.2) is likely to be for a large part attributable to this deficiency in GR. In humans a defect in NER, especially GR, (in patients with the rare disease Xeroderma pigmentosum) is associated with a dramatically increased risk of SCC, BCC and CMM on sun-exposed skin areas, and these patients often succumb to their skin cancers before reaching adulthood [12]. Under daily UV exposure mice with a complete deficiency in NER (XP-A knockout mice) got their skin carcinomas about 4 times faster than their wild type littermates [13]. A deficiency in TCR in humans (patients with Cockayne Syndrome) are not known to be associated with an increased risk of skin cancer, but these patients usually die in early youth from other causes (e.g. infections). In a model with CS-B knockout mice it turns out that a defect in TCR also enhances the susceptibility to UV carcinogenesis [14] (but apparently less dramatically than with XP-A knockout mice).

After a UV exposure the cell cycle is slowed down: the DNA synthesis is transiently suppressed [15]. This allows for more time for DNA repair, before replication is continued. An important protein in regulating cell responses after a genotoxic insult is the p53 gene product; it appears to mediate an arrest in the late G1 phase, presumably to keep cells from entering the S phase in which they would replicating their (damaged) DNA [16]. Additionally, the p53 protein was found to be involved in forcing an overly damaged cell into a preprogrammed suicide (apoptosis): this would reduce the risk of malignant cell transformation [17]. The immuno histochemically detectable nuclear p53 expression peaks about 24 hours after a UV-B exposure (between 12 and 24 hours after a 365-nm exposure) [15, 18]. A defect in p53 is likely to jeopardize the genomic integrity.

4.3 MUTATIONS AND CELL TRANSFORMATION

Like many internal tumors, SCC and BCC often bear a mutation in the p53 gene; in about 90% of SCC [17] and 50% of BCC [19] in the USA. Unlike internal tumors, nearly all of these mutations occur at sites with neighboring pyrimidines (UV-B targets), and most of them (> 70%) are transitions of cytosine, C, to thymine, T, sometimes even CC to TT tandem transitions. The same mutation spectrum as was earlier found after UV irradiation of cells in culture, and our group found it in a majority (50-70%) of experimentally UV-induced SCC in hairless mice [20]. Interestingly, the mutations in the murine tumors were predominantly associated with di-pyrimidine sites on the NTS, a bias that is not found in the human tumors and corresponds to the poor repair of CPD in the NTS of murine cells. Thus, the solar UV-B radiation appears to have left its signature in the genome of cells from human SCC and BCC. These mutations appear to occur early in genesis of SCC, as they are found in the precursor lesions, actinic keratosis [17]. Our group found microscopic foci of epidermal cells with mutant p53 in daily UV-exposed murine skin long before tumors occurred [18], such foci were also found in non-neoplastic human skin [21].

In most of the UV-induced carcinomas only one allele appears to be mutated, while the other, wildtype, allele appears to be still present. The p53 protein in its functional form, a homotetramer, is a transcription factor (e.g. for p21/WAF1, mdm-2 and bax); one mutated allele may render most of the tetramers dysfunctional. The effect of a mutated allele may be quite different from the effect of a loss of one of the alleles; the latter appears to diminish the induction of apoptosis after UV exposure [17]. In transgenic mouse a mutated p53 gene appears to increase the tumor multiplicity (yield, see section 6.2) by a factor of 2 with only a marginally faster induction of the tumors [22] (the latter is to be expected with a mere two fold increase in the hazard, see section 6.2, eq. 1). Our group recently found that the complete loss of one functional p53 allele (in heterozygous p53 knockout mice) only speeds up tumor induction at high daily UV doses, but that this effect is lost with low daily UV doses [23] (when apoptosis probably becomes less important).

As a combination of genes needs to become dysfunctional for a malignant transformation of a cell, other genes than p53 only are likely to play a role in UV carcinogenesis. Moreover, in some SCC, in a sizeable fraction of BCC and in most CMM no aberrations were found in the p53 gene; in CMM it is sometimes seen in metastatic stages. Constitutive overexpression of bcl-2 is quite common in BCC and CMM [25], and is known to counteract apoptosis. Bcl-2 could substitute for mutant-p53 [26]. Recently, the human ptch gene has been identified, and allelic loss of this gene was found to be associated to the Gorlin syndrome which cause patients to contract multiple BCC [27]. Ptch was also found to be mutated in 'spontaneous' BCC [28]. In cell lines of CMM the β-catenine gene was found to be mutated [29]. These mutations in ptch and β-catenine could be related to UV radiation, but the 'UV-signature' was not as clear-cut as in the earlier results on p53 in SCC and BCC.

In mouse experiments our group found that mutations in the p53 gene is less prominently present in skin carcinomas induced by UV-A radiation [23] and in skin carcinomas induced by UV-B radiation in NER-deficient XP-A knockout mice [23]. Hence, the type of UV radiation and the repair characteristics will influence the carcinogenic pathway, i.e., the genes affected in order to arrive at the malignant cell transformation.

5. UV Immunosuppression

Experimentally UV-induced skin tumors are commonly found to be highly antigenic, i.e., such tumors are rejected upon transplantation into unirradiated, syngeneic mice [30]. After a series of UV exposures, however, the mice will accept these implants long before any primary UV-induced tumors can be observed. These tumor implants are also accepted when the immune system of the mice is generally suppressed. A striking analogy is found in the observation that the skin cancer risk is dramatically increased in people who are on medication to suppress the immunological rejection of a kidney transplant [31].

UV irradiation can suppress cellular immunity both in mice and in humans, it can even induce an antigen-specific tolerance (i.e. the immune system becomes selectively 'blind' for one particular antigen). People who have been treated for SCC or BCC are particularly prone to develop such a tolerance [32], indicating that this trait is a risk factor for skin carcinoma.

6. The Three Physical Dimensions to UV-Induced Tumor Formation

Tumor formation is a multi-step dynamic process which cannot be properly understood if one of the important variables is only studied at one fixed value: e.g. ascertaining the tumor burden at a single time point in groups with different exposures will be totally inadequate as it does not provide information on one of the most important variables, 'time'. In UV carcinogenesis the wavelength of the irradiation (spectrum), the dose (exposure regimen) and time (tumor latency) are the most important physical variables.

6.1. WAVELENGTH

As discussed earlier (section 4.1), short-wave UV radiation is directly absorbed by DNA and causes the formation of dimers at di-pyrimidine sites. Above 300 nm the efficiency of induction of these DNA adducts drops off steeply, and correspondingly, the mutation rate per J/m^2 drops off. Our group and the Photobiology Group of the former Skin and Cancer Hospital in Philadelphia have performed a series of skin carcinogenesis experiments on hairless SKH-1 mice with various broadband UV sources. From these data we have mathematically extracted the wavelength dependency of UV carcinogenesis, dubbed the SCUP-m action spectrum (Skin Cancer Utrecht-Philadelphia-murine) [33]. The epidermis absorbs strongly below 300 nm [8], which limits the penetration of the UV radiation, and counteracts the increasing absorption by DNA and the corresponding increase in DNA damage toward 260 nm (section 4.1). One would, therefore, expect the SCUP-m action spectrum to reach a maximum somewhere below 300 nm, but considerably higher than 260 nm (maximum in absorption spectrum of DNA): the maximum was indeed found at 293 nm. By correcting for differences in transmission of murine and human epidermisses we estimated the carcinogenic action spectrum for humans, dubbed the SCUP-h action spectrum (h stands for human) [34]; this correc-

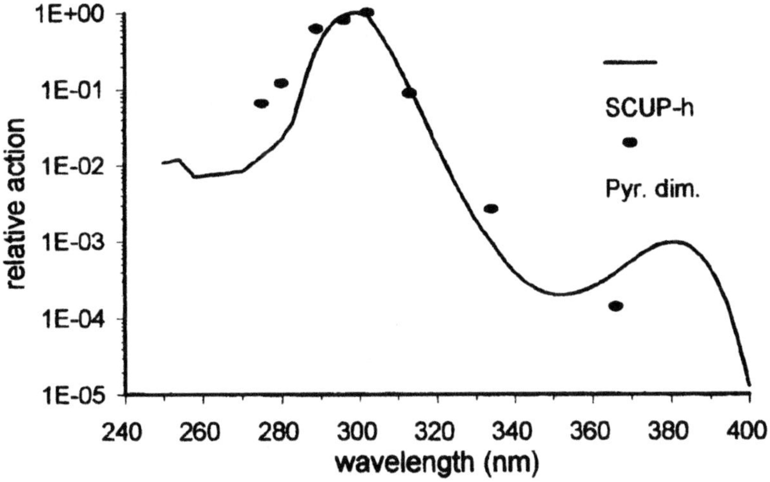

Figure 1: Comparison of the SCUP-h action spectrum [34] and the action spectrum for the induction of pyrimidine dimers in human skin [35].

tion shifted the maximum to 299 nm. The SCUP-h action spectrum is depicted in Fig. 1 together with the measured action spectrum for the induction of pyrimidine dimers in the human skin by Freeman et al. [35]. The two action spectra resemble each other remarkably well, especially in the UV-B, which confirms that the UV-induced genotoxicity is indeed the major driving force behind UV carcinogenesis.

6.2. TIME VERSUS DOSE

Tumor formation can be envisaged as a process with multiple rate limiting steps (mutations in oncogenes and tumor suppressor genes), some of which are UV driven and some of which are not (the latter may be caused by endogenous metabolic processes which generate ROS). The likelihood of the occurrence of a UV-dependent step will increase with the accumulated UV dose, whereas the likelihood of a UV-independent step will simple increase with the lapse of time. Clearly, the UV-independent steps will render the whole process of carcinogenesis less depend on the UV dose: in mouse experiments the tumor induction time will not be shortened by a factor of 2 if the daily dose is increased by a factor of 2, i.e. there will be no direct reciprocity between daily dose and tumor induction time. This reciprocity can also be modified by adaptive processes, e.g. epidermal hyperplasia which diminishes the penetration to the germinitive basal cells. This lack of reciprocity is indeed found in experiments of SCC induction in hairless mice by daily UV exposure. The average number of tumors per mouse, the yield Y (or tumor multiplicity) can be written as

$$Y = (H/H_0)^{p1} (t/t_0)^{p} \tag{1}$$

where H is the daily UV exposure (in $J/m^2/d$), t is time (in days), and H_o, t_o, p1 and p are constants (when $H=H_o$ and $t=t_0$ then Y=1, i.e. an average of 1 tumor per mouse; note that if $H=2H_o$ then $t=t_0/2^{p1/p}$ for Y=1). In a straighforward interpretation, the power of time, p, is likely to be proportional (not necessarily equal to) the total number of rate limiting steps that occur in the course of time (including those steps that depend on the daily UV exposure). The power of the daily UV exposure, p1, is likely to be proportional to the number of UV driven steps; which implies that $p1 \leq p$ [36].

If all mice in a group are comparable (in sensitivity and treatment) then in terms of the probability theory Y is a cumulative hazard function, more particularly, a cumulative Weibull hazard function with H and t as variables. The probability, P, for a mouse of contracting a first tumor then becomes

$$P = 1 - \exp(-Y) \tag{2}$$

which, in absence of any intervening deaths, equals the prevalence of tumor bearing animals in a (large) group. Our group found that these relationships hold rather well in the experiments with albino hairless mice, where $p = 7.2 \pm 0.8$ and $p1 = 4.3 \pm 0.5$ for early tumors of 1 mm in diameter [36, 37]. In experiments with pigmented hairless mice $p1 = 2.1 \pm 0.2$ appears to hold (I computed that $p1/p = 0.3$ from data in [38]); this lower dependency on the daily UV exposure is probably due to a better adaptation by pigmentation of the animals (at low daily exposures the tumor induction becomes comparable to that in their albino counterparts, [38] and personal communication R.J.W. Berg).

When the daily exposure of the albino mice is discontinued after a couple of weeks, long before the appearance of tumors, then Y increases with t to the power 2.8 ± 0.2, which power equals p-p1 ($=2.9 \pm 0.3$), the power related to the UV-independent steps [36].

In looking for intermediate biological endpoints, our group found that the percentage of inbred albino hairless mice that accepted a syngeneic implant of a UV-induced tumor could also be fitted with a Weibull distribution with $p = 2.8 \pm 0.7 \approx p1$, which implies reciprocity between the daily exposure and the time till conversion of the mouse to a state in which the tumor is accepted [39]. At high daily exposures this state precedes the occurrence of the induction of primary skin tumors, but extrapolation indicates that this is not the case at low daily exposures. We are currently studying the induction kinetics of microscopic foci of cells with mutant p53 (section 4.3) as another intermediate biological endpoint in the process of UV carcinogenesis.

In human populations the age-specific incidences of cancers are measured, which equal the increase in Y per unit time (e.g. per year) if the fraction of patients with multiple tumors is small. In Fig. 2 the probability (crudely calculated according to equation 2) of SCC in Dutch males ($p = 6.6 \pm 0.4$, 8.9 ± 0.7 for females) versus log(time) is compared to that of albino hairless mice under a level of daily exposure comparable to the estimated average among Dutch males: the two curves run parallelly but the tumors in the mice occur about 250 times faster. This appears to indicate that the tumor kinetics are very similar, but that the developmental rate of SCC (the mutation rate) is much higher in the mice. By combining measurements of ambient UV loads and epidemiological data from USA it can be inferred that $p1 = 2.6 \pm 0.8$ for human SCC [34, 40]; a

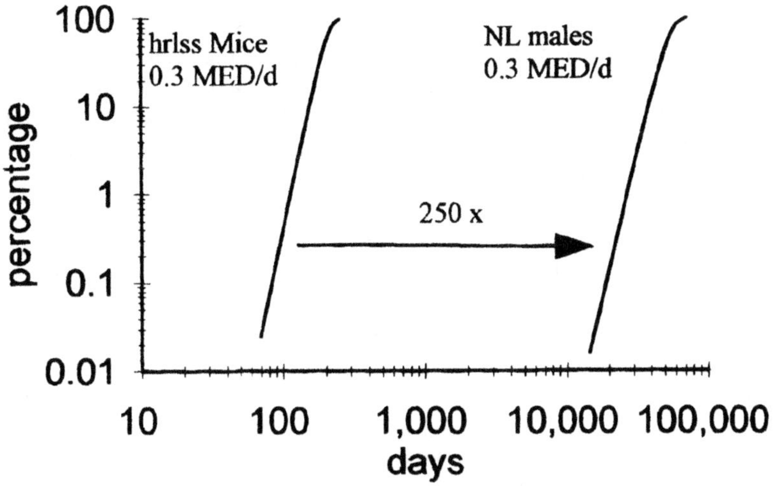

Figure 2: Comparison of the probability of SCC among Dutch males and hairless mice under similar average levels of chronic UV exposure.

value of p1 that is very similar to the one for pigmented hairless mice. For BCC p ranges between 4.5 and 5.5 and p1 = 1.4 ± 0.5 [40].

7. Risk Assessment

The basis for a good risk assessment is obviously a reliable quantitative model of tumor induction. In experiments it is fairly straightforward to extract such a model for mice, but this cannot be done for humans. The epidemiological data need a sound interpretation in order to arrive at a reliable model for humans. From the above it appears that the mouse model is excellently suited to derive a model of UV carcinogenic risk for humans: There appears to be a similarity between mouse and man in UV-induced oncogenic events and tumor kinetics; most of all for SCC. BCC are not easily induced by UV radiation in mice, but the nature of the p53 mutations indicates that UV-B radiation plays an important role (section 4.3). For CMM, however, very little is known on the relevant UV-induced oncogenic changes, and experiments in fish even indicate that UV-A radiation in sunlight may dominate over UV-B radiation in the initiation of these tumors [41]. The uncertainty about the action spectrum for CMM undermines any attempt to make risk assessments for CMM.

7.1. ALTERED INCIDENCES AND RISK

The easiest approach to quantify the effect of an increased UV exposure, such as caused by a depletion of the ozone layer, is to compare two stationary situations: one with a life-long low UV exposure and the other a life-long high UV exposure. Assuming that the White Caucasian populations in various locations in the USA are sufficiently comparable (in behavior and genetic composition) and in a stationary condition (in relation to UV exposure and skin tumor incidences), one can use the relationship between ambient UV loads and incidences for the assessment: for each percent of ozone depletion SCC would increase by 3%, BCC by 1.7%, and CMM by 0.5-1%, if the SCUP-h action spectrum is also applicable to CMM [42-44]. The complication here is that the incidences of BCC and CMM in particular are not stationary in the USA [43].

Transitional changes in incidences due to changes in UV exposure are much harder to estimate. This requires much more detailed knowledge, e.g. on whether UV acts early, late or continuously in the process of skin carcinogenesis. From the above (sections 3, 4.3 and 6.2) it appears that UV acts throughout the genesis of SCC, but mainly in an early stage of BCC and CMM development. For scenario studies on dynamic changes the tumor yield, Y, in equation 1 may be factorized in a contribution from events driven by the total UV exposure, written as $(TH/H_o.t_o)^{p1}$, and a contribution from events purely dependent on time, taken as $(t/t_o)^{p-p1}$, where TH (= H.t) is the total UV exposure. For SCC these two factors are simply multiplied to calculate the yield, but for BCC and CMM the two factors need to be convoluted in time because the UV-dependent events are preceding the purely time dependent events (assuming that a small fraction of UV-initiated foci becomes tumors). Thus, scenario studies of changes in ozone can be made [45]: without restrictions the skin cancer incidence could quadruple by the end of the next century, and even initial restrictions under the Montreal Protocol would still yield a two-fold increase. With the Copenhagen amendments the incidence is projected to increase by a maximum of 10% around the year 2060. Clearly, the risk model needs to be refined further, but these calculations, despite their wide error margins, emphasize the disastrous potential of an uncontrolled increase of the ambient UV-B load.

The impact presented above needs to be viewed on a population scale (section 3): in NW Europe the risk of contracting a BCC before the age of 75 is about 1 in 20, and about 1 in 100 for SCC and CMM (the risk of skin cancer is about 2 in 3 in Australia!). Although the personal risk in a temperate climate is small, the impact on public health (and care) is very substantial, and worthwhile to curb by a simple moderation of our sun exposure.

7.2. BIO-INDICATORS OF RISK; SEEKING REFINEMENT

In the previous section we considered the incidence in the total population and the risk of an 'average' person in that population. However, it is well known that the risk can vary dramatically between groups of people: i.e., 'sun sensitive' or 'freckled' people run a higher risk. Instead of ill-defined subjective judgements it would be better to make objective measurements on a person or a group of people in order to calculate reliably well differentiated risks. Clearly, a person's DNA repair proficiency (various sub-types of NER), or UV-immuno-suppressibility, or frequency of p53-mutant patches could be important indicators of risk. One could, for example, consider the probability of contracting a tumor before the age of 75 without anything particular known at age a, or the conditional probability when it is known that a person at age a has more than 2 foci of mutant p53 cells per cm^2 and/or has a DNA repair capacity below a certain level. Quantitative information like this would be worthwhile for the refinement of risk models and for proper identification of high-risk individuals; the latter could, for instance, be very useful in deciding whether or not an individual should have photo therapy or an other therapy.

8. Conclusions

Epidemiology has revealed various ways in which different types of skin cancers are related to sun exposure. In combination with data from (animal) experiments it is possible to construct a simple probability model of tumor induction as function of UV exposure and time. Unraveling the molecular events leading up to tumor formation provides a fundamental understanding of UV carcinogenesis. This in turn may refine the mathematical risk models that we have available at the moment: e.g., measuring intermediate biological endpoints in the process of UV carcinogenesis may, with proper models, be exploited as predictors of certain levels of risk for particular groups of people, or even individuals.

9. References

1. Miller, D.L. and Weinstock, M.A. (1994) Nonmelanoma skin cancer in the Untited States: incidence, *J. Amer. Acad. Dermatol.* **30**, 774-778.
2. Health Council (1994) UV radiation from sunlight. Report no. 1994/05E, pp.83-118, The Hague, the Netherlands.
3. Scotto, J. and Fears, T.R. (1981) Incidence of Nonmelanoma Skin Cancer in the United States. Publ. no. NIH 82-2433, U.S. Dept. of Health and Human Services, Washington, DC.
4. Vitasa, B.C., Taylor, H.R., Strickland, P.J., Rosenthal, F.S., West, S., Abbey, H., Ng, S.K., Munoz, B. and Emmett, E.A. (1990) Association of nonmelanoma skin cancer and actinic keratosis with cumulative solar ultraviolet exposure in Maryland watermen. *Cancer* **65**, 2811-2817.
5. Vitaliano, P.P. and Urbach, F. (1980) The relative importance of risk factors in nonmelanoma carcinoma. *Arch. Dermatol.* **116**, 454-456.
6. Kricker, A., Armstrong, B.K., English, D.R. and Heenan, P.J. (1995) Does intermittent sun exposure cause basal cell carcinoma? A case-control study in Western Australia. *Int. J. Cancer* **60**, 489-494.
7. Holman, C.D.J. and Armstrong, B.K. (1984) Cutaneous malignant melanoma and indicators of total accumulated exposure to the sun: an analysis separating histogenic types. *J. Natl. Cancer Inst.* **73**, 75-82.
8. Bruls, W.A.G., Van Weelden, H. and Van der Leun, J.C. (1984) Transmission of UV radiation through human epidermal layers as a factor influencing the minimal erythema dose. *Photochem. Photobiol.* **39**, 63-87.
9. Berg, R.J.W., De Bueger, S.C., Guikers, K., Van Weelden, H., Van Vloten, W.A., Van der Leun, J.C. and De Gruijl, F.R. (1995) Induction and disappearance of thymine dimers in human skin exposed to UV-B radiation: flow cytometric measurements in replicating and non-replicating epidermal cells, *Photochem. Photobiol.* **62**, 970-975.
10. Mitchell, D.L., Haipe, C.A. and Clarkson, J.M. (1985) (6-4) photoproducts are removed from the DNA of UV-irradiated mammalian cells more efficiently than cyclobutane dimers, *Mutation Res.* **143**, 109-112.
11. Ruven, H.G.T., Seelen, C.M.J., Lohman, P.H.M., Van Kranen, H.J., Van Zeeland, A.A. and Mullenders, L.H.F. (1994) Strand-specific removal of cyclobutane pyrimidine dimers from p53 gene in the epidermis of UV-B-irradiated hairless mice, *Oncogene* **9**, 3427-3434.

12. Kraemer, K.H., Lee, M.M. and Scotto, J. (1987) Xeroderma pigmentosum; cutaneous, ocular, and neurologic abnormalities in 830 published cases. *Arch Dermatol.* **123**, 241-250.
13. Berg, R.J.W., De Vries, A., Van Steeg, H. and De Gruijl, F.R. (1997) Relative Susceptibilities of XPA Knockout Mice and Their Heterozygous and Wild-Type Littermates to UV-B-induced Skin Cancer, *Cancer Res.* **57**, 581-584.
14. Van der Horst, G.T.J., Van Steeg, H., Berg, R.J.W., Van Gool, A.J., De Wit, J., Weeda, G., Morreau, H., Beems, R.B., Van Kreijl, C.F., De Gruijl, F.R., Bootsma, D. and Hoeijmakers, J.H.J. (1997) Defective Transcription-Coupled Repair in Cockayne Syndrome B Mice Is Associated with Skin Cancer Predisposition, *Cell* **89**, 425-435.
15. De Laat, J.M.T., Kroon, E.D. and De Gruijl, F.R. (1997) Cell Cycle Effects and Concommittant p53 Expression in Hairless Murine Skin after Longwave UV-A (365 nm) Irradiation: A Comparison with UV-B Irradiation, *Photochem. Photobiol.* **65**, 730-735.
16. El-Deiry, W.S., Harper, J.W., O'Connor, P.M., Velculescu, V.E., Canman C.E., Jackman, J., Pieterpol, J.A., Burrell, M., Hill, D.E., Wang, Y., Wiman, K.G., Mercer, W.E., Kastan, M.B., Kohn, K.W., Elledge, S.J., Kinzler, K.W. and Vogelstein, B. (1994) WAF1/CIP1 is induced in p53-mediated G1-arrest and apoptosis, *Cancer Res.* **54**, 1169-1174.
17. Ziegler, A., Jonason, A.S., Leffel, D.J., Simon, J.A., Sharma, H.W., Kimmelman, J., Remington, L., Jacks, T. and Brash, D.E. (1994) Sunburn and p53 in the onset of skin cancer, *Nature* **372**, 773-776.
18. Berg, R.J.W., Van Kranen, H.J., Rebel, H.G., De Vries, A., Van Vloten, W.A., Van Kreijl, C.F., Van der Leun, J.C. and De Gruijl, F.R. (1996) Early p53 alterations in mouse skin carcinogenesis by UV-B radiation: Immunohistochemical detection of mutant p53 protein in clusters of preneoplastic epidermal cells, *Proc. Natl. Acad. Sci. USA* **93**, 274-278.
19. Ziegler, A.D., Leffel, D.J., Kunala, S., Sharma, W., Gailani, M.R., Simon, J.A., Halperin, A.J., Shapiro, P.E., Bale, A.E. and Brash, D.E. (1993) Mutation hotspots due to sunlight in the p53 gene of nonmelanoma skin cancers. *Proc. Natl. Acad. Sci. USA* **90**, 4216-4220.
20. Dumaz, N., Van Kranen, H.J., De Vries, A., Berg, R.J.W., Wester, P.W., Van Kreijl, C.F., Sarasin, A., Daya-Grosjean, L. and De Gruijl, F.R. (1997) The role of UV-B light in skin carcinogenesis through the analysis of p53 mutations in squamous cell carcinomas of hairless mice. *Carcinogenesis* **18**, 897-904.
21. Ren, Z.-P., Hedrum, A., Pontèn, F., Nister, M., Ahmadian, A., Lundeberg, J., Uhlén, M. and Pontèn, J. (1996) Human epidermal cancer and accompanying precursors have identical p53 mutations different from p53 mutations in areas of clonally expanding non-neoplastic keratinocytes. *Oncogene* **13**, 765-773.
22. Li, G., Ho, V.C., Berean, K. and Tron, V.A. (1995) Ultraviolet Radiation Induction of Squamous Cell Carcinomas in p53 Transgenic Mice, *Cancer Res.* **55**, 2070-2074.
23. De Gruijl FR (1997) Photomutagenicity and photocarcinogenicity: experimental aspects. *Mutation Res.* **377** (Suppl.), 185.
24. Van Kranen, H.J., De Laat, J.M.T., Van de Ven, J., Wester, P.W., De Vries, A., Berg, R.J.W., Van Kreijl, C.F. and De Gruijl, F.R. (1997) Low Incidence of p53 Mutations in UV-A (365-nm)-induced Skin Tumors in Hairless Mice, *Cancer Res.* **57**, 1238-1240.
25. Morales-Ducret, C.R., Van der Rijn, M., Le Brun, D.P., Smoller, B.R. (1995) Bcl-2 expression in privacy malignancies of the skin. *Arch. Dermatol.* **131**, 909-912.
26. Wikonkal, N.M., Berg, R.J.W., Van Haselen, C.W., Horkay, I., Remenyik, E., Begany, A., Hunyadi, J., Van Vloten, W.A., De Gruijl, F.R. (1997) Bcl-2 versus p53 expression and apoptotic rate in human non-melanoma skin cancers. *Arch. Dermatol.* **133**, 599-602.
27. Hahn, H., Wicking, C., Zaphiropoules, P.G., Gailini, M.R., Shanley, S., Chidambaram, A., Vorechovsky, I., Holmberg, E., Unden, A.B., Gillies, S., Nagus, K., Smyth, I., Pressman, C., Leffell, D.J., Gerrard, B., Goldstein, A.M., Dean, M., Toftgard, R., Chenerix-Trench, G., Wainwright, B. and Bale, A.E. (1996) Mutations of the human homologue of Drosophila patched in the Nevoid Basal Cell Carcinoma Syndrome. *Cell* **85**, 841-851.
28. Gailini, M.R., Stahle-Bäckdahl, M., Leffell, D.J., Glynn, M., Zaphiropoulos, P.G., Pressman, C., Undén, A.B., Dean, M., Brash, D.E., Bale, A.E. and Toftgard, R. (1996) The role of the human homologue of Drosophila patched in sporadic basal cell carcinomas. *Nature Gen.* **14**, 78-81.

29. Rubinfield, B., Robbins, P., El-Gamil, M., Albert, I., Porfiri, E. and Polakis, P. (1997) Stabilazation of β-catenin by genetic defects in melanoma cell lines. *Science* **275**, 1790-2.
30. Kripke, M.L. (1974) Antigenicity of murine skin tumors induced by ultraviolet light, *J. Natl. Cancer Inst.* **53**, 1333-1336.
31. Hardie, I.R., Strong, R.W., Hartley, L.C.J., Woodruff, P.W. H. and Clunie, G.J.A. (1980) Skin cancer in Caucasian renal allograft recipients living in a subtropical climate. *Surgery* **87**, 177-180.
32. Yoshikawa, T., Rae, V., Bruin-Slot, W., Van der Berg, J.W., Taylor, J.R. and Streilein, J.W. (1990) Susceptibility to effects of UV-B radiation of contact hypersensitivity as a risk factor for skin cancer in humans. *J. Invest. Dermatol.* **95**, 530-536.
33. De Gruijl, F.R., Sterenborg, H.J.C.M., Forbes, P.D., Davies, R.E, Cole, C., Kelfkens, G., Van Weelden, H., Slaper, H. and Van der Leun, J.C. (1993) Wavelength dependence of skin cancer induction by ultraviolet irradiation of albino hairless mice. *Cancer Res.* **53**, 53-60.
34. De Gruijl, F.R. and Van der Leun, J.C. (1994) Estimate of the wavelength dependency of ultraviolet carcinogenesis in humans and its relevance to the risk assessment of a stratospheric ozone depletion. *Health Phys.* **67**, 314-325.
35. Freeman, S.E, Hachan, H., Gange, R.W., Maytum, D.J., Sutherland, J.C. and Sutherland, B.M. (1989) Wavelength Dependence of pyrimidine dimer formation in DNA of human skin irradiated in situ with ultraviolet light, *Proc. Natl. Acad. Sci. USA* **86**, 5605-5609.
36. De Gruijl, F.R. and Van der Leun, J.C. (1991) Development of skin tumors in hairless mice after discontinuation of ultraviolet irradiation, *Cancer Res.* **51**, 979-984.
37. De Gruijl, F.R., Van der Meer, J.B. and Van der Leun, J.C. (1983) Dose-time dependency of tumor formation by chronic UV exposure, *Photochem. Photobiol.* **37**, 53-62.
38. Davies, R.E. and Forbes, P.D. (1988) Retinoids and photocarcinogenesis: a review, *J. Toxicol. Cut. Ocul. Toxicol.* **7**, 241-253.
39. Sontag, Y., Garssen, J., Van Loveren, H., Van der Leun, J.C., Van Vloten, W.A. and De Gruijl, F.R. (1997) Time and dose dependence of acceptance of UV-induced syngeneic tumor implants in chronically UV-exposed hairless mice, *Photochem. Photobiol.* **65**, 342-346.
40. De Gruijl, F.R. and Forbes, P.D. (1995) UV-induced skin cancer in a hairless mouse model, *BioEssays* **17**, 651-660.
41. Setlow, R,B,, Grist, E,, Thompson, K. and Woodhead, A.P. (1993) Wavelengths effective in induction of malignant melanoma. *Proc. Natl. Acad. Sci. USA* **90**, 6666-6670.
42. Longstreth, J.D., De Gruijl, F.R., Kripke, M.L., Takizawa, Y. and Van der Leun, J.C. (1995) Effects of increased solar ultraviolet radiation on human health, *Ambio* **24**, 153-165.
43. De Gruijl, F.R. and Van der Leun, J.C. (1993) Influence of ozone depletion on the incidence of skin cancer. In Environmental UV Photobiology, A.R. Young, L.O. Björn, J. Moan and W. Nultsch, Eds, pp. 89-112. Plenum Press, New York.
44. Scotto, J., and Fears, T.R. (1987) The association of solar ultraviolet and skin melanoma incidence among Caucasians in the Unites States. *Cancer Invest.* **5**, 275-283.
45. Slaper, H., Velders, G.J.M., Daniel, J.S., De Gruijl, F.R. and Van der Leun, J.C. (1996) Estimates of ozone depletion and skin cancer incidence to examine the Vienna Convention achievements. *Nature* **384**, 256-258.

FREE RADICALS AND DERMAL DAMAGE IN PHOTOAGING AND PHOTOCARCINOGENESIS

K. SCHARFFETTER-KOCHANEK, P. BRENNEISEN, J. WENK, R. BLAUDSCHUN, M. SCHAUEN, M. WLASCHEK

Department of Dermatology, University of Cologne
Joseph-Stelzmann Str. 9, D-50924 Cologne Germany

Abstract

The increase in UV irradiation on earth due to the stratospheric ozone depletion represents a major environmental threat to the skin increasing its risk of photooxidative damage by reactive oxygen species (ROS). ROS have been implicated in several photo dermatological disorders including photoaging. The clinical manifestations of cutaneous photoaging are due to quantitative and qualitative alterations of the dermal connective tissue. Therefore, we were interested in a better understanding of the involvement of ROS in the up-regulation of matrix metalloproteinases (MMP) responsible for connective tissue degradation in photoaging, tumor invasion and metastasis. For this purpose fibroblast monolayer cultures have been subjected to various UV modalities. In parallel fibroblasts have been challenged by distinct ROS generating systems. Using scavengers, enhancers, and inhibitors for the activity of ROS detoxifying enzymes, using stably transfected fibroblast cell lines overexpressing ROS detoxifying enzymes, and using iron chelators blocking the Fenton reaction, distinct ROS have been increased peri- or intracellularly. A time and dose dependent increase in mRNA and protein levels for collagenase was observed after exposure of fibroblasts to UV-B, UV-A, $^1O_2$, H_2O_2, $O_2^{\cdot-}$, or $^{\cdot}OH$ generating systems. After UV-B irradiation inhibition of the Fenton reaction using iron chelators abrogated the MMP induction substantially. Scavengers for the hydroxyl radical reduced the MMP induction after UV-B, thus pointing to the importance of the Fenton reaction and the hydroxyl radical in the UV-B response. By modulating the lifetime of $^1O_2$ by enhancers and quenchers, singlet oxygen could be identified as a crucial mediator of the UV-A response. In addition, the inhibition of glutathion peroxidase, catalase and the Fenton reaction enhanced MMP mRNA induction whereas inhibition of superoxide dismutase diminished the MMP mRNA increase. Furthermore, exposure to UV-A of a stably transfected fibroblast cell line overexpressing the manganese superoxide dismutase resulted in a substantial increase in MMP mRNA. These results indicate that following UV-A irradiation, beside $^1O_2$, H_2O_2 is mainly responsible for the induction of MMP synthesis. In conclusion, we have identified distinct ROS involved in matrix-degrading enzyme induction resulting in tissue degradation follow-

C. Baumstark-Khan et al. (eds.),
Fundamentals for the Assessment of Risks from Environmental Radiation, 359-365.

ing UV irradiation. This will allow a rational risk assessment for the dermal damage of different UV modalities in photoaging and photocarcinogenesis.

1. Introduction

Ultraviolet irradiation (UV) is known to elicit a variety of pathological effects on the human organism as photoaging and photocarcinogenesis of the skin. Photoaging of the skin secondary to chronic UV exposure is mainly due to damage of the dermal extracellular matrix resulting in impaired mechanical and biophysical properties of the skin. Clinical symptoms include wrinkle formation, loss of recoil capacity, increased fragility and impaired wound healing of the skin [1-4]. For photocarcinogenesis, a strong association exists between exposure to sunlight and human non-melanoma skin cancer [5-8]. In the last years an increasing incidence of aggressive, highly invasive non-melanoma skin cancer has been reported. This increasing incidence of skin malignancies and photoaging most likely is due to the increase in environmental UV-load and in sunbathing and tanning behavior due to cosmetic reasons [9]. This is particularly relevant in terms of UV-irradiation predicted to increase on earth's surface upon ozone depletion [10-12]. In addition, chronic UV-A irradiation alone or in combination with UV-B has been reported to increase the incidence of non-melanoma skin cancer in humans. Because UV-irradiation is known to generate distinct reactive oxygen species the risk of photooxidative damage of the skin has increased substantially [13-16]. Photoaging and photocarcinogenesis share common features, however, reveal unique molecular characteristics which finally determine the fate of the cell and its host.

The effects of different reactive oxygen species (ROS) have been implicated in the pathogenesis of photoaging and photocarcinogenesis. The term reactive oxygen species not only collectively includes oxygen-centered radicals such as the superoxide anion ($O_2^{\cdot-}$) and the hydroxyl radical ($^{\cdot}OH$) but also some non-radical species, such as hydrogen peroxide (H_2O_2) and singlet oxygen ($^1O_2$) all being produced upon UV irradiation and inflammation. A complex antioxidant defense system has evolved in the skin and protects against ROS [17]. However, following UV irradiation or chronic inflammation ROS are generated substantially compromising the enzymatic and non enzymatic antioxidant defense system of the skin, thus tilting the balance towards a pro-oxidant state [18]. As well as causing permanent genetic changes involving protooncogenes and tumor suppressor genes, reactive oxygen species activate cytoplasmic signal transduction pathways related to growth, differentiation, senescence and tissue degradation determining the malignant invasive and metastatic phenotype in non-melanoma skin cancer as well as the photodamaged phenotype in photoaged tissue [13, 14, 19]. In fact, upregulation of matrix-degrading metalloproteinases responsible for connective tissue degradation in photoaging, tumor invasion and metastasis is regulated by different reactive oxygen species generated by UV irradiation or produced by tumor and inflammatory cells. The identification of distinct ROS in the signal transduction pathways leading to the induction of matrix-metalloproteinases will allow a rational risk assessment for the dermal damage in photoaging and photocarcinogenesis.

2. Role of Matrix-Metalloproteinases in Connective Tissue Degradation in Photoaging and Photocarcinogenesis

Matrix-metalloproteinases, responsible for the breakdown of interstitial collagens and other connective tissue components, form a family of until now 19 enzymes with substrate specifities for different extracellular matrix compounds, secretion of a latent precursor form and activation by mercury or proteases. While MMP-1 cleaves collagen type I, MMP-2 is able to degrade basement membrane compounds including collagen Type IV and type VII [20, 21]. As to their proteolytic activity, UV-induced MMPs may contribute to the dissolution of the basement membrane and dermal structural proteins leading to tumor invasion and metastasis at distant anatomical sites as has been shown for a variety of tumors including skin cancer [22, 23]. The activity of MMPs is inhibited by a special class of inhibitors of metalloproteinases (TIMPs). Similar to MMPs, TIMPs are synthesized by ordinarily resident fibroblasts [24]. In photodamaged skin quantitative and qualitative alterations of dermal extracellular matrix proteins such as elastin, glycosaminoglycans and interstitial collagens are involved. In particular, collagen type I which belongs to a family of closely related but genetically distinct proteins [25] providing the dermis with tensile strength and stability, has been found to be reduced in actinically or photoaged skin [1]. Because the collagen amount in skin apart from collagen synthesis is controlled by degradation it has been postulated that the UV-induction of matrix-metalloproteinases is responsible for the connective tissue damage in photoaged skin. Previous work has shown that fibroblast derived matrix-metalloproteinases were dose-dependently induced in vitro and in vivo by UV-B and UV-A irradiation [26-33]. By contrast, the synthesis of the tissue inhibitor of metalloproteinases (TIMP) remains almost unchanged following UV irradiation. The resulting imbalance between matrix-degrading metalloproteinases and their inhibitors may contribute to the connective tissue degradation related to photoaging. It has been demonstrated, that skin tumors are able to synthesize matrix-metalloproteinases or to induce the synthesis of MMPs in peritumoral fibroblasts closely associated to the invasion front of the tumor. In contrast to basal cell carcinomas, squamous cell carcinomas synthesize the 92-kDa type IV collagenase. Differential expression of metalloproteinases and their physiological inhibitors, TIMPs, might be responsible for the distinct biological behavior of these skin tumors. The signaling steps resulting in the above mentioned changes in gene expression in photoaging and photocarcinogenesis are not fully understood. Due to their short half lives reactive oxygen species represent at least ideal signaling molecules and in this capacity are involved in several physiological and pathological states [34]. ROS are able to modulate the expression at different genes through specific mechanisms [19]. Since UV irradiation generates reactive oxygen species in the skin a special role of ROS in photoaging and photocarcinogenesis of the skin has been postulated. There is now circumstantial evidence suggesting UV-generated ROS to play a major role in UV induction and progression of skin pathological conditions. In addition, UV-irradiation results in the infiltration of inflammatory cells into the connective tissue of the skin thus increasing the ROS load by several orders of magnitude. While it is well accepted that low levels of reactive oxygen species are continuously produced in vivo and are involved in signal transduction pathways, cell activation, differentiation and growth control, there is accumulating indirect evidence for the damaging effect of higher concentrations of reactive oxygen species generated in vivo following UV-A and UV-B-irradiation of the skin [13, 35-37]. Beside direct absorption of UV-B-photons by DNA and subsequent structural changes, generation of reactive oxygen species following irradiation with UV-A and UV-B has been reported requiring the absorption of photons by endogenous photosensitizer molecules.

3. Cellular Signaling in the ROS Mediated Induction of Matrix-Metalloproteinases

To better understand the involvement of ROS in the up-regulation of matrix-metalloproteinases (MMP) responsible for connective tissue degradation in photoaging, tumor invasion and metastasis, we were interested in the molecular mechanisms regulating the signal transduction pathways. For this purpose human dermal fibroblasts have been subjected to various UV modalities or in parallel challenged by distinct ROS generating systems. Exposure of fibroblasts to ROS generating systems represents a model which reflects the situation that tumor cells or inflammatory cells may release high concentrations of ROS and thus stimulate synthesis of collagenase in the peritumoral fibroblasts or in fibroblasts adjacent to inflammatory cells. Peri- and intracellular concentrations of ROS were modulated by using scavengers and enhancers for the stability of different ROS and using inhibitors for the activity of ROS detoxifying enzymes. Using stably transfected fibroblast cell lines overexpressing ROS detoxifying enzymes, and using iron chelators blocking the Fenton reaction, distinct ROS pathways were modulated. A time and dose dependent increase in mRNA and protein levels for collagenase was observed after exposure of fibroblasts to UV-B, UV-A, $^1O_2$, H_2O_2, $O_2^{\cdot-}$, or $^{\cdot}OH$ generating systems [26, 32, 38]. In contrast, mRNA levels of TIMP-1, the tissue inhibitor of metalloproteinases 1, remained unchanged. In addition to direct UV-B effects on DNA the generation of reactive oxygen species following UV-B irradiatiation modulates the oxidant state of the cell resulting in changes in the signal transduction pathways [39]. Inhibition of the Fenton reaction by iron chelators substantially abrogated the MMP induction after UV-B-irradiation. Additionally, scavengers for the hydroxyl radical reduced the MMP induction after UV-B, thus pointing to the importance of the Fenton reaction and the hydroxyl radical in the UV-B response. By modulation of the lifetime of singlet oxygen by enhancers and quenchers, singlet oxygen was identified as a crucial mediator of the UV-A response. Interestingly, generating singlet oxygen outside the cells by irradiation of rose bengal-coated resin particles resulted in the induction of interstitial collagenase suggesting that extracellularly generated singlet oxygen may be responsible for this response [40, 41]. In addition, the inhibition of glutathion peroxidase, catalase and the Fenton reaction enhanced MMP mRNA induction whereas inhibition of superoxide dismutase diminished the MMP mRNA increase. Furthermore, exposure to UV-A of a stably transfected fibroblast cell line overexpressing the manganese superoxide dismutase resulted in a substantial increase in MMP mRNA. These results indicate that following UV-A irradiation, beside $^1O_2$, H_2O_2 is mainly responsible for the induction of MMP synthesis. There is now circumstantial evidence that the UV-A and UV-C induced up-regulation of matrix-metalloproteinases is mediated by signaling peptides comprising growth factors, cytokines and interleukines. Using antisense strategies, function blocking antibodies and bioassays, we have provided evidence for an UV-A-induced cytokine network of interleukin-1alpha, interleukin-1beta and interleukin-6 in human dermal fibroblasts which via interrelated autocrine loops induced interstitial collagenase [42, 43]. Similarly, Kraemer et al. [44] showed that UV-C irradiation of HeLa cells induced the synthesis and release of IL-1alpha and basic fibroblast growth factor (bFGF) which together stimulate the synthesis of collagenase. Obviously, collagenase is mediated by different cytokines depending on the UV-spectra and possibly on the cell type. This interleukin signaling network is initiated by distinct reactive oxygen species [41]. Recently, we showed that the iron driven Fenton reaction by generation of ˙OH and lipid peroxides dramatically enhances the UV-B dependent JNK2 activity and c-jun transcription as well [46]. By contrast, no induction of the mitogen-activated protein kinase (MAPK), ERK-1 and ERK-2, was reported following UV-treatment [27], though in UV-independent settings ERKs are involved in the induction of c-fos expression and AP-1 activity [45]. In summary we have identified distinct reactive oxygen species involved in matrix-degrading enzyme induction resulting in tissue degradation following UV irradiation allowing enhanced metastasis and invasion of tumor cells. The changes in the gene expression of distinct matrix-metalloproteinases in photocarcinogenesis and photoaging most likely are causally linked to the generation of ROS, depending not only on the UV quality but also the interaction of histogenetically different cells in UV-damaged tissue and in tumor cells.

4. Conclusion

Reactive oxygen species play a major role in the UV-dependent changes of matrix-metalloproteinase gene expression. The expected increase in the UV-irradiation on earth due to the depletion of the stratospheric ozone layer therefore will contribute to the ROS load of the skin and will have consequences for matrix-metalloproteinase expression in different pathological states. The analysis of the underlying molecular and cellular mechanisms will allow a rational risk assessment for the dermal damage of different UV modalities in photoaging and photocarcinogenesis.

5. References

1. Gilchrest, B.A. (1995) Photodamage. Blackwell Science, Cambridge MA
2. Kligman, L.H. (1992) UV-A induced biochemical changes in hairless mouse skin collagen: A contrast to UV-B effects. In: Urbach, F. (ed.) Biological Responses to Ultraviolet A Radiation. Valdemar Publishing, Overland Park, pp 209-216.
3. Oikarinen, A. (1990) The aging of skin: chronoaging versus photoaging. *Photodermatol. Photoimmunol. Photomed.* 7, 3-4.
4. Scharffetter-Kochanek, K., Wlaschek, M., Bolsen, K., Herrmann, G., Lehmann, P., Goerz, G., Mauch, C., and Plewig, G. (1992) Mechanisms of cutaneous photoaging. In: Plewig, G., Marks, R. (eds.). The environmental threat of the skin. Martin Dunitz Publishers, London, pp 72-82.
5. Brash, D.E. (1997) Sunlight and the onset of skin cancer. *Trends Genetics* **13**, 410-414.
6. Scotto, J., and Fears, T.R (1987) The association of solar ultraviolet and skin melanoma incidence among caucasians in the United States. *Cancer Invest.* **5**, 275-283.
7. Gallagher, R.P., Hill, G.B., Bajdik, C.D, Fincham, S., Goldman, A.J., McLean, D.I., and Threlfall, W.J. (1995) Sunlight exposure, pigmentary factors, and risk of nonmelanocytic skin cancer. I. Basal cell carcinoma. *Arch. Dermatol.* **131**, 157-163.

8. Gallagher, R.P., Hill, G.B., Bajdik, C.D., Fincham, S., Goldman, A.J., McLean, D.I., and Threlfall, W.J. (1995) Sunlight exposure, pigmentary factors, and risk of nonmelanocytic skin cancer. II. Squamous cell carcinoma. *Arch. Dermatol.* **131**, 164-169.
9. Spencer, J.M., and Amonette, R.A. (1995) Indoor tanning: Risks, benefits and future trends. *J. Am. Acad. Dermatol.* **33**, 288-298.
10. Coldiron, B.M. (1992) Thinning of the ozone layer: facts and consequences. *J. Am. Acad. Dermatol.* **27**, 783-785.
11. Henriksen, T., Dahlback, A., Larsen, S.H.H., and Moan, J. (1990) Ultraviolet-radiation and skin cancer. Effect of an ozone layer depletion. *Photochem. Photobiol.* **51**, 579-582.
12. Slaper, H., Velders, G.J.M., Daniel, J.S., de Gruijl, F.R., and van der Leun, J.C. (1996) Estimates of ozone depletion and skin cancer incidence to examine the Vienna Convention achievements. *Nature* **384**, 256-258.
13. Black, H. (1987) Potential involvement of free radical reactions in ultraviolet light-mediated cutaneous damage. *Photochem. Photobiol.* **46**, 213-221.
14. Cerutti, P.A. (1994) Oxy-radicals and cancer. *Lancet* **344**, 862-863.
15. Scharffetter-Kochanek, K. (1997) Photodamage of the skin; prevention and therapie. In: Antioxidants in Disease - Mechanisms and Therapeutic Strategies. Sies, H. (ed), Series: Advances in Pharmacology. Academic Press, San Diego, pp 639-655.
16. Scharffetter-Kochanek, K., Wenk, J., Brenneisen, P., Blaudschun, R., and Wlaschek, M. (1997) Molecular Role of Reactive Oxygen Species in Photoaging and Tumor Progression. In: Altmeyer, P., Hoffmann, K., Stöcker, M. (eds.) Skin Cancer and UV Radiation. Springer Verlag, Berlin - Heidelberg New York, pp 115-127.
17. Darr, D., and Fridovich, I. (1994) Free radicals in cutaneous biology. *J. Invest. Dermatol.* **102**, 671-675.
18. Sies, H. (1991) Oxidative stress: oxidants and antioxidants. Academic Press, New York.
19. Remacle, J., Raes, M., Toussaint, O., Renard, P., and Rao, G. (1995) Low levels of reactive oxygen species as modulators of cell function. *Mutat. Res.* **316**, 103-122.
20. Birkedal-Hansen, H., Moore, W.G, Bodden, M.K., Windsor, L.J., Birkedal-Hansen, B., De Carlo, A., and Engler, J.A. (1993) Matrix metalloproteinases: A Review. *Crit. Rev. Oral Biol. Med.* **4**, 197-250.
21. Stetler-Stevenson, W.G., Hewitt R., and Corcoran, M. (1996) Matrix metalloproteinases and tumor invasion: from correlation and causality to the clinic. *Sem. Cancer Biol.* **7**, 147-154.
22. Boyd, D. (1996) Invasion and metastasis. *Cancer Metas. Res.* **15**, 77-89.
23. 23. Majmudar, G., Nelson, B.R., Jensen, T.C., Voorhees, J.J., and Johnson, T.M. (1994) Increased expression of stromelysin-3 in basal cell carcinomas. *Mol. Carcinogen.* **9**, 17-23.
24. Woessner, J.F.Jr. (1991) Matrix metalloproteinases and their inhibitor in connective tissue remodeling. *FASEB J.* **5**, 2145-2154.
25. Rest, van der M., and Garrone, R. (1991) Collagen family of proteins. *FASEB J.* **5**, 2814-2823.
26. Brenneisen, P., Briviba, K., Wenk, J., Wlaschek, M., and Scharffetter-Kochanek, K. (1997) Hydrogen peroxide (H_2O_2) increases the steady-state mRNA levels of collagenase/MMP-1 in human dermal fibroblasts. *Free Radic. Biol. Med.* **22**, 515-524.
27. Fisher, G.J., Talwary, H.S., Wang Z.Q., McPhillips, F., Kang, S., and Voorhees, J. (1996) UV-B activates stress - activated protein kinases, Fos/Jun driven dermis degrading proteinases in human skin in vivo. *J. Invest. Dermatol.*, **107**, 449.
28. Herrmann, G., Wlaschek, M., Lange, T.S., Prenzel, K., Goerz, G., and Scharffetter-Kochanek, K. (1993) UV-A irradiation stimulates the synthesis of various matrix-metalloproteinases (MMPs) in cultured human fibroblasts. *Exp. Dermatol.* **2**, 92-97.
29. Herrmann, G., Wlaschek, M., Bolsen, K., Prenzel, K., Goerz, G., and Scharffetter-Kochanek, K. (1996) Pathogenic implication of matrix-metalloproteinases (MMPs) and their counteracting inhibitor TIMP-1 in the cutaneous photodamage of human porphyria cutanea tarda (PCT). *J. Invest. Dermatol.* **107**, 398-403.
30. Koivukangas, V., Kallioinen, M., Autio-Harmainen, H., and Oikarinen, A. (1994) UV irradiation induces the expression of gelatinases in human skin in vivo. *Acta Derm. Venereol.* **74**, 279-282.

31. Petersen, M.J., Hansen, C., and Craig, S. (1992) Ultraviolet A irradiation stimulates collagenase production in cultured human fibroblasts. *J. Invest. Dermatol.* **99**, 440-444.
32. Scharffetter, K., Wlaschek, M., Hogg, A., Bolsen, K., Schothorst, A., Goerz, G., Krieg, T., and Plewig, G. (1991) UV-A irradiation induces collagenase in human dermal fibroblasts in vitro and in vivo. *Arch. Dermatol. Res.* **283**, 506-511.
33. Stein, B., Rahmsdorf, H.J., Steffen, A., Litfin, M., and Herrlich, P. (1989) UV-induced DNA damage is an intermediate step in UV-induced expression of human immunodeficiency virus type I, collagenase, c-fos, and metallothionein. *Mol. Cell. Biol.* **9**, 5169-5181.
34. Nonakada, Y., Iwagaki, H., Kimura, T., Fuchimoto, S., and Orita, K. (1993) Effect of reactive oxygen intermediates on the in vitro invasive capacity of tumor cells and liver metastasis in mice. *Int. J. Cancer* **54**, 983-986.
35. Taira, J., Mimura, K., Yoneya, T., Hagi, A., Murakami, A., and Makino, K. (1992) Hydroxyl radical formation by UV-irradiated epidermal cells. *J. Biochem.* **111**, 693-695.
36. Dalle-Carbonare, M., and Pathak, M.A. (1992) Skin photosensitizing agents and the role of reactive oxygen species in photoaging. *J. Photochem. Photobiol. B* **14**, 105-124.
37. Jurkiewicz, B.A., Bissett, D.L., and Buettner, G.R. (1995) Effect of topically applied tocopherol on ultraviolet radiation-mediated free radical damage in skin. *J. Invest. Dermatol.* **104**, 484-488.
38. Scharffetter-Kochanek, K., Wlaschek, M., Briviba, K., and Sies, H. (1993) Singlet oxygen induces collagenase expression in human skin fibroblasts. *FEBS Lett.* **331**, 304-306.
39. Halliwell, B., and Aruoma, O.I. (1991) DNA damage by oxygen-derived species. Its mechanism and measurement in mammalian systems. *FEBS Lett.* **281**, 9-19.
40. Wlaschek, M., Briviba, K., Stricklin, G.P., Sies, H., and Scharffetter-Kochanek, K. (1995) Singlet oxygen may mediate the ultraviolet A induced synthesis of interstitial collagenase. *J. Invest. Dermatol.* **104**, 194-198.
41. Wlaschek, M., Wenk, J., Brenneisen, P., Briviba, K., Schwarz, A., Sies, H., and Scharffetter-Kochanek, K. (1997) Singlet oxygen is an early intermediate in cytokine-dependent ultraviolet-A induction of interstitial collagenase in human dermal fibroblasts in vitro. *FEBS Lett.* **413**, 239-242
42. Wlaschek, M., Bolsen, K., Hermann, G., Schwarz, A., Wilmroth, F., Heinrich, P.C., Goerz, G., and Scharffetter-Kochanek, K. (1993) UV-A-induced autocrine stimulation of fibroblast-derived-collagenase by IL-6: A possible mechanism in dermal photodamage? *J. Invest. Dermatol.* **101**, 164-168.
43. Wlaschek, M., Heinen, G., Poswig, A., Schwarz, A., Krieg, T., and Scharffetter-Kochanek, K. (1994) UV-A-induced stimulation of fibroblast-derived collagenase/MMP-1 by interrelated loops of interleukin-1 and interleukin-6. *Photochem. Photobiol.* **59**, 550-556.
44. Kraemer, M., Sachsenmaier, C., Herrlich, P., and Rahmsdorf, H.J. (1993) UV irradiation-induced interleukin-1 and basic fibroblast growth factor synthesis and release mediate part of the UV response. *J. Biol. Chem.* **268**, 6734-6741.
45. Minden, A., Lin, A., Smeal, T., Derijard, B., Cobb, M., Davis, R., and Karin, M. (1994) C-jun N-terminal phosphorylation correlates with activation of the JNK subgroup but not the ERK subgroup of mitogen-activated protein kinases. *Mol. Cell Biol.* **14**, 6683-6688.
46. Brenneisen, P., Wenk, J., Lkozu, O., Wlaschek, M., Briviba, K., Krieg, T., Sies, H., Scharffetter-Kochanek, K. (1998) Central role of ferrous/ferric iron in the ultraviolet B irradiation-mediated signaling pathway leading to increased interstitial collagenase (matrix-degrading metalloprotease (MMP-1) and stromelysin-1 (MMP-3)) m-RNA levels in cultured human dermal fibroblasts. *J. Biol. Chem.* **273**, 5279-5289

THE DLR-BIOFILM AS PERSONAL UV DOSIMETER

P. RETTBERG, R. SIEF AND G. HORNECK

Deutsches Zentrum für Luft- und Raumfahrt, Institut für Luft- und Raumfahrtmedizin, Abteilung Strahlenbiologie
D-51170 Köln, Germany

Abstract

In general the biological efficiency of UV radiation increases with decreasing wavelengths. This has to be taken into account by calculation if solar UV radiation is measured with physical methods. UV dosimetry with biological UV dosimeters utilize biological objects as UV sensitive targets which *per se* weight the incident UV radiation according to its biological effectiveness thereby integrating over the UV wavelengths and the exposure time. The wavelength sensitivity of a biological UV dosimeter is described by the so-called action spectrum. The biological UV dosimeter 'biofilm' uses dry spores of the bacterium *Bacillus subtilis* immobilized on the surface of a plastic sheet as UV sensor. It weights the UV radiation according to its DNA damaging potential. Measurement areas on a biofilm in an exposure housing adapted to the measurement purpose are exposed to the UV source which can be the sun but also artificial UV lamps. Unexposed areas on each biofilm are used afterwards for calibration with 254 nm UV-C. The inhibition of biological activity by the biologically effective UV radiation is determined as relative optical density after development and staining of the biomass formed inside the biofilm. The biologically effective dose of each measurement area is determined with the help of the calibration curve. Now the biofilm is also adapted to its application as personal UV dosimeter. For the personal UV dosimetry different groups of persons were selected according to their UV exposures during their work or leisure activities. In measurement periods at different seasons volunteers carried a personal biofilm UV dosimeter during the whole day and made a protocol of their activities. First results of these measurement campaigns are shown.

1. Introduction

Solar UV radiation has beneficial as well as deleterious effects on humans. All these effects e.g. vitamin D synthesis in the human skin, sunburn, premature skin aging, immune suppression, induction of skin cancer, show a strong wavelength dependency. This has to be taken into account by calculation if solar UV radiation is measured with physical methods at a specific location [1, 2]. From these data, however, it is difficult to

C. Baumstark-Khan et al. (eds.),
Fundamentals for the Assessment of Risks from Environmental Radiation, 367-370.

extrapolate the individual biologically weighted UV doses persons are exposed to during their normal life. Up to now most informations are obtained by estimations from retrospective questionnaires, measurements with small electronic dosimeters with wavelength-sensitivities not exactly characterized or polysulphone film badges.

The biologically weighting UV dosimeter 'biofilm' was originally developed for long term UV measurements to investigate the trends in the UV radiation climate as a result of changes in the stratospheric ozone concentration [3-5]. The UV sensitive target are dried spores of the bacterium *Bacillus subtilis*, which weight the UV radiation according to its DNA damaging potential. This biofilm action spectrum is very similar to the action spectrum of the human skin with respect to erythema formation. Now the biofilm is also adapted to its application as personal UV dosimeter.

2. The personal biofilm UV dosimeter

The following requirements have to be fulfilled by a personal UV film dosimeter:

- large dynamic range due to large individual variations in sun exposure
- avoidance of shadowing of the measurement areas by the dosimeter housing itself while the person is moving relative to the UV source
- protection of the UV sensor against physical and chemical damages during use
- light-protected calibration areas and dark control areas
- comfortable wearing
- easy handling

The biofilm technique was used for the construction of a personal UV dosimeter which fulfills these requirements [6]. Fig. 1 shows a drawing of the dosimeter housing made of plastics. Polyamide gazes which serve as neutral density filters are integrated in the frame for the separation of the individual measurement areas.

This housing contains the biofilm as UV sensor in a water-tight biofilm stack. As cut-off filter over 3 of the 6 measurement areas on the biofilm a foil (Mylar®) is used which cuts off UV-B so that the parallel measurement of UV-A+B and UV-A is possible with the same biofilm. One half of a biofilm serves as calibration and dark control area.

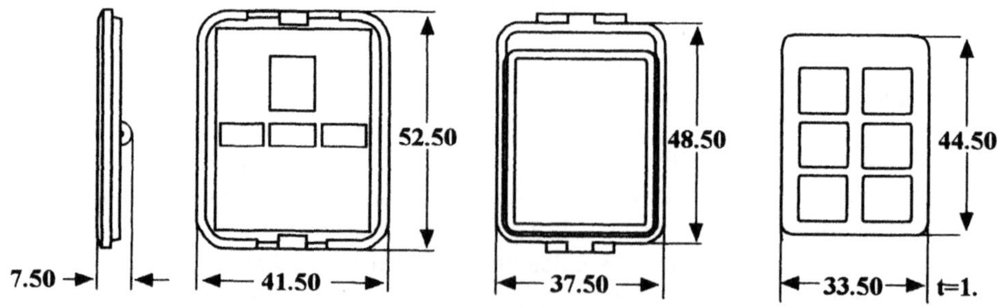

Figure 1. Drawing of a personal biofilm UV dosimeter housing (dimensions are given in mm)

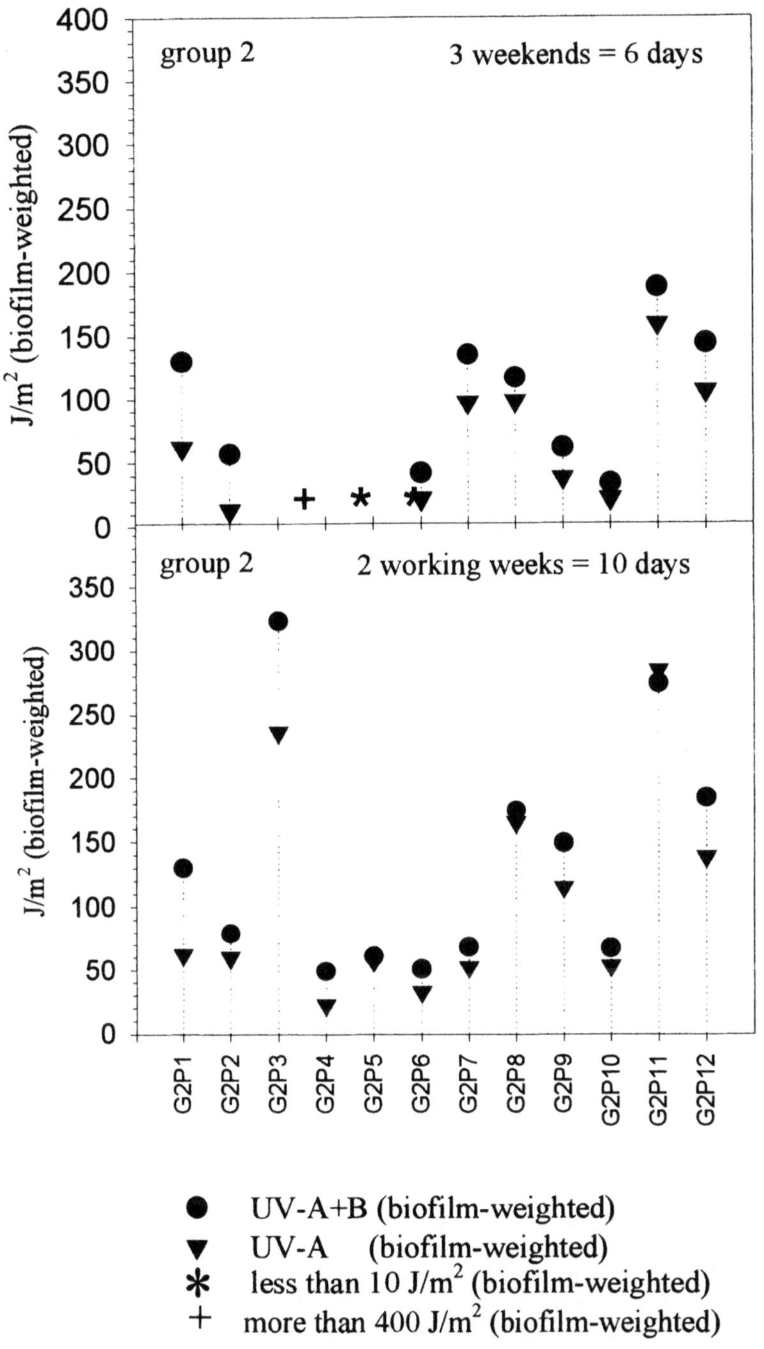

Figure 2. Individual UV doses of school children in May/June 1997

It is folded backwards and additionally protected against UV exposure by light-tight cardboard layers. The whole stack is hermetically sealed in an UV transparent plastic foil.

3. Application of the personal biofilm UV dosimeter

These biofilm dosimeters were used to determine the UV exposure in different groups of the German population, e.g. school children, indoor workers. Measurement periods of two weeks including three weekends were chosen in different seasons, thereby excluding holiday periods. One biofilm dosimeter was worn during the working days, another one during the weekends to obtain informations about the magnitude and distribution of UV exposure during normal work and during leisure activities. In addition, a protocol of outdoor periods should be written with the accuracy of half an hour. Stationary UV measurements with biofilms, but also with RB-meters and a spectro radiometer were performed in parallel. Figure 2 shows exemplary the results for a group of school children in May/June 1997. In general the UV exposure in each group was very inhomogeneous due to the strong dependence on behavior, ranging from more than 400 Jm^{-2}(biofilm-weighted) to less than 10 Jm^{-2}(biofilm-weighted) for the sum of the three weekends.

4. Conclusions

Up to now no data of the average UV exposure of different groups in the German population are available. With these experiments the DLR-biofilm has proven to be suitable for the measurement of the individual UV dose of persons and to be applicable for the determination of the missing baseline data which will allow the estimation of possible risks of UV solar radiation.

Acknowledgements: This work was supported by BMBF, 07UVB54A and EC EV5V-CT-93-0342.

5. References

1. Horneck, G.(1995) Quantification of the biological effectiveness of environmental UV radiation, *J. Photochem. Photobiol. B: Biol.* **31**, 43-49.
2. Horneck, G. (1997) Biological UV dosimetry. In: D.-P. Häder (ed.). The Effects of Ozone Depletion on Aquatic Ecosystems, Academic, Landes, Austin, pp. 119-142.
3. Quintern, L.E., Horneck, G., Eschweiler, U., Bücker, H. (1992) A biofilm used as ultraviolet dosimeter, *Photochem. Photobiol.* **55**, 389-395.
4. Quintern, L. E., Puskeppeleit M., Rainer P., Weber S., El Naggar S., Eschweiler U., Horneck G. 1994. Continuous dosimetry of the biologically harmful UV-radiation in Antarctica with the biofilm technique. *J. Photochem. Photobiol. B: Biol.* **22**, 59-66.
5. Horneck, G., Rettberg, P., Rabbow, E., Strauch, W., Seckmeyer, G., Facius, R., Reitz, G., Strauch, K., J.-U. Schott (1996) Biological UV dosimetry of solar radiation for different simulated ozone column thicknesses. *J. Photochem. Photobiol. B: Biol.:* **32**, 189-196.
6. Rettberg, P., Horneck, G. 1998. Messung der UV-Stralenbelastung mit dem DLR-Biofilm. *Derm* **4**, 180-184.

HUMAN HEALTH CONSEQUENCES IN SOUTHEASTERN OF BELORUSSIA UPTO 11 YEARS AFTER THE CHERNOBYL ACCIDENT.

The Situation in the District of Krasnopolje

J. FÜLLER [1], M. W. TROJANOW [2], S. SCHWEITZER [1]

[1] *Department of Radiation Oncology, University Hospital, Friedrich-Schiller-University Jena*
Bachstr. 18, D-07740 Jena, Germany
[2] *District Hospital Krasnopolje*
Region Mogiljow, BY-213560 Krasnopolje, Belorussia

Abstract

Within the framework of humanitary the amount of radiation injury and its effect on the population in the highly contaminated district of Krasnopolje, Belorussia, has been continuously analyzed for the past 4 years. The existence of radiation damage to the people (i.e. increasing incidence of carcinoma, hypothyroidism, immuno deficiency syndromes and psychogenic disorders) has been proven since 1990. Additional systematic analysis of a wider population are of extreme importance in order to obtain valid epidemiologic data.

1. Introduction

On the occasion of the sad 10th anniversary of the Chernobyl accident, a great number of scientific as well as popular publications tried to give a preliminary survey of the situation in the affected regions. These publications show considerable discrepancies with regard to biological effects. This might be due to regional differences, problems with obtaining and delivering data, especially considering the present political, economical and social situation, and to the application of differing statistical methods. Bearing this in mind the present situation in the highly contaminated district of Krasnopolje (region of Mogiljow) will be analyzed.

2. Methods

Within the framework of regular support campaigns of the Jena-based association "Aid for the children of Chernobyl" and the University of Jena, we have received substantial help with the analysis of the radiometric and radiobiological situation in the district of

C. Baumstark-Khan et al. (eds.),
Fundamentals for the Assessment of Risks from Environmental Radiation, 371-378.

Krasnopolje, located 100 km from Gomel. This retrospective analysis of data is based on the annual epidemiologic reports of the district physician with a cut off date of December, 31, 1996.

3. Results

3.1. DEMOGRAPHIC SITUATION

On December 31st, 1996, 14,414 individuals (included are 3,092 children under the age of 15 years) were living in this rural district, which comprises the district town as well as 101 villages. The district covers an area of 177 km^2. Knowing the number of inhabitants in 1985, which was 22,600, there has been a relocation of 8,200 persons from 72 villages to less contaminated regions of Belorussia. Relocation though did not start until 1988. Comparing the years 1994 to 1996 and 1985 to 1986, the annual birth rate has gone down, whereas a moderate increase in the general mortality can be seen in the same time period (fig. 1). Compared with data from the less contaminated district Mogiljow and the whole of Belorussia it becomes evident that the death rate in the district of Krasnopolje is somewhat higher than in the other areas, however mortality has been higher compared to these districts even before the catastrophe (fig. 2).

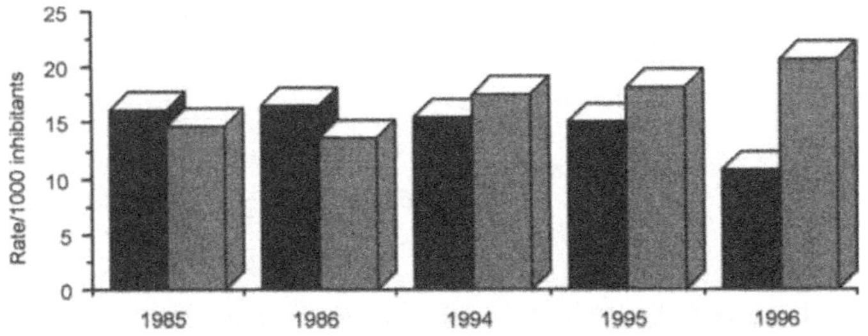

Figure 1. Yearly birth-rate and mortality in the district of Krasnopolje

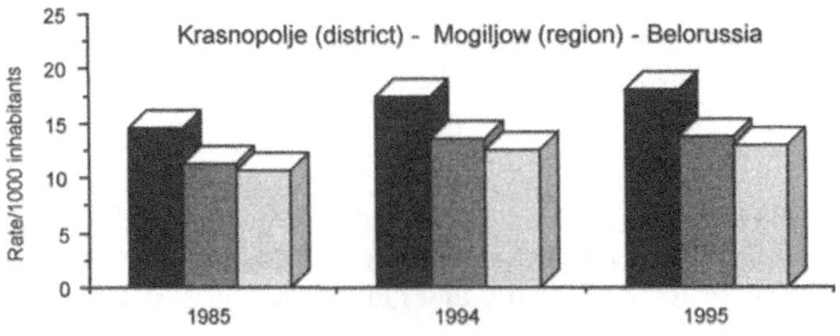

Figure 2. Comparison of general mortality, region dependence

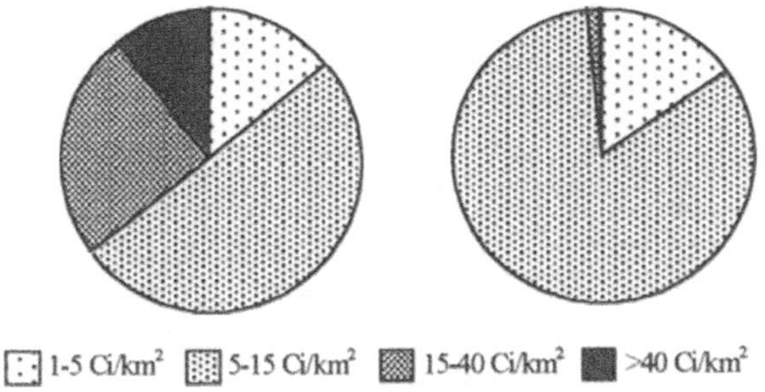

Figure 3. Areas of the contaminated zones (left) and population distribution (right) in the district of Krasnopolje

3.2. CONTAMINATION AND ACTUAL DOSE RATE

As seen on figure 3 (left), half of the territory shows a contamination with ^{137}Cs of 5-15 Ci/km^2 (185-555 kBq/m^2). For one quarter of the total area the rate of contamination measures between 15 and 40 Ci/km^2 (meaning significantly reduced habitable conditions). 11 % of the district shows dose levels of over 40 Ci/km^2 (>1480 kBq/m^2) and will not allow neither resettlement nor agricultural use in the future. Only 14 % of the territory, comprising parts of the district town (7,000 inhabitants), measure dose rates of 1-5 Ci/km^2 (37-185 kBq/m^2). Altogether 83 % of all inhabitants live in areas with a contamination of 5-15 Ci/km^2 (fig. 3 right).

Control measurements of the actual gamma-dose rates on April, 16, 1996 could be obtained in form of random samples only (n=144). In areas inhabited, the dose rate fluctuates between 0.2 and 1.0 µSv/h (n=83). In evacuated areas ("forbidden zone", n=61) the dose rate are, as expected, higher with a maximum of 8.4 µSv/h (at ground level). Comparing corresponding measurement points in 1994 and 1995 no change can be seen. The dose rate in the woodland (about 30 % of the whole territory) and on the meadowland were somewhat higher than those on neighboring fields. Sealed ground (i.e. asphalt or concrete) resulted in dose rates 10 times less than those measured on adjacent land.

3.3. FOOD CONTAMINATION

Regular dosimetric measurements (ca. 3,500 per year) of provisions available through government controlled trade are done in the small laboratory-facilities of the hospital. These provisions are from less contaminated areas at the most. Due to measurements done in 1993 and 1995 (Table 1) there are almost no exceedings of dose limits [2]. Privately produced food, coming from higher contaminated areas as well as black market provisions are hardly controlled and can therefore be considered a serious problem. It is well known that especially mushrooms, berries and game are highly contaminated with a wide range of measured values. Officially, the consumption of these provisions is prohibited in the district of Krasnopolje. Considering the poor food supply in these areas it is not surprising that this prohibition is not generally obeyed.

TABLE 1. Contamination of selected foodstuffs in Bq/kg

foodstuffs	Medium	Minimum	Maximum
Bread	20	20	20
Potatoes	20	5	30
Cabbage	20	8	24
Milk	36	20	464
Apples	55	20	188
Honey	120	15	444
Beef	44	20	186
Pork	57	20	498
Game	2965	20	16132
Berries	2557	20	11118
Mushrooms	3975	20	24372
Dry mushrooms	83222	6348	406058

3. 4. INCIDENCE OF REPRESENTATIVE DISEASES

There is no complete data about tumor and other disease incidence in the years 1987-1991. This is why the following analysis is generally based on comparisons between 1985 and 1986 on the one and 1992 respective 1994 to 1996 on the other hand [2, 10].

3.4.1. Malignant diseases

As seen in fig. 4, a strong increase in the incidence of malignant tumors can first be seen in 1993, 7 years after the catastrophe (with an incidence being 1.8 times as much as in 1986, falling to 1.4 in 1996). It is striking that there is quite a low incidence of newly diagnosed cancer in 1991 and 1992.

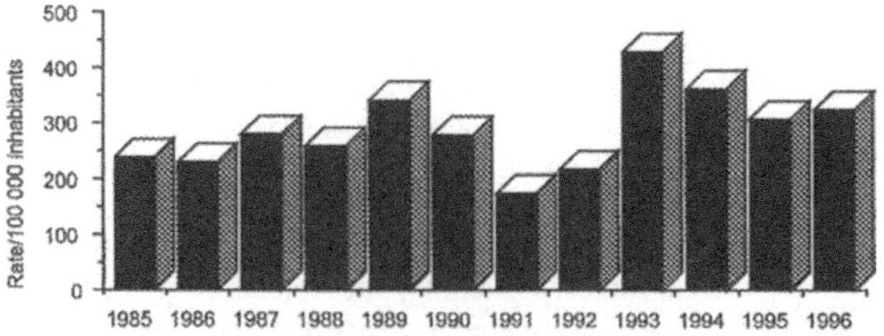

Figure 4. Yearly cancer incidence in the district of Krasnopolje

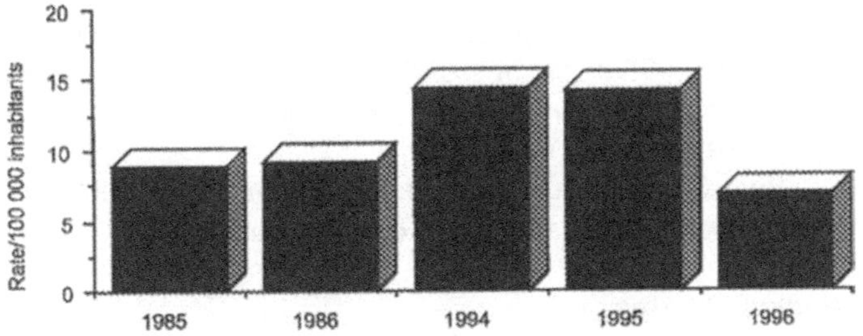

Figure 5. Yearly incidence of leukemia in the district of Krasnopolje

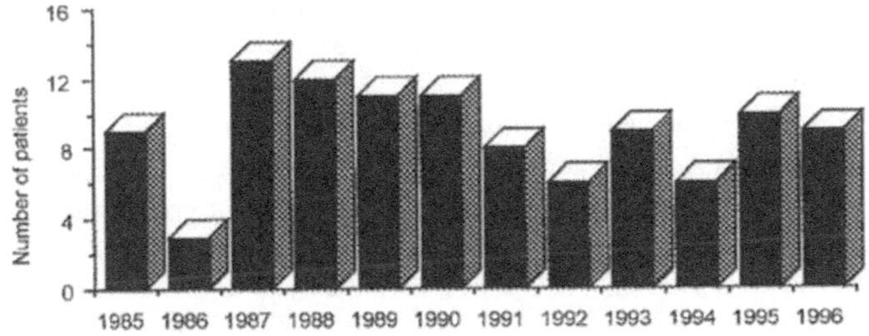

Figure 6. Yearly incidence of lung cancer in the district of Krasnopolje

The exact onset of newly diagnosed representative malignant tumors is shown in table 2 and in fig. 5 and 6. Generally a fluctuating incidence of newly diagnosed cancer

TABLE 2. Incidence of malignant cancer (absolute) in the district of Krasnopolje

Years	85/86	87/88	89/90	91/92	93/94	95/96
Gastrointestinal cancers	36	42	39	33	42	23
Lung cancers	12	25	22	14	15	19
Carcinoma of urogenitary tract	9	5	10	4	14	14
Breast cancer	12	8	3	6	2	7
Head and neck cancer	10	12	15	6	6	11
Thyroid gland cancer	1	0	1	1	3	0
Malignant lymphomas	7	9	6	1	5	3
Skin cancer	13	13	18	11	8	11
Other tumors	4	3	10	5	13	3

can be seen, with a rising tendency in gastro-intestinal, genito-urinary and bronchial carcinomas when comparing 1987 to 1990 with 1985 to 1986. With leukemia, there is an increase in 1994 and 1995 of factor 1.6, showing falling tendency again in 1996 (0.8; all compared to 1985/86). Unfortunately, we could not get hold of any data concerning leukemia incidences between 1987 and 1993.

Four out of five thyroid carcinomas (both papillary and follicular) seen between 1989 and 1994 were diagnosed in childhood or adolescence.

3.4.2. Non malignant diseases

A significant increase in the incidence of endocrine disorders can be seen in fig. 7. Especially compared to 1985/86, endocrine disorders in infants occur significantly more often. A steep rise of newly diagnosed illnesses is seen in 1995 and 1996. Hypothyroidism comprises more than 90% of all endocrine diseases. The others are hyperthyroidism, including autonomic adenoma, diabetes mellitus, adrenal and pituitary gland disorders, malfunction of the ovaries or testicles as well as growth disturbances due to endocrine malfunction.

Rising tendency is also seen with childhood anemia, gastric or bowels disorders (i.e. gastric and duodenal ulcer, chronic gastritis) as well as inflammatory airway diseases (chronic bronchitis), autonomic neuropathy, arterial hypertension and myocardial infarction (fig. 8-11). With tuberculosis, only a slight, not significant, rise in incidence is observed.

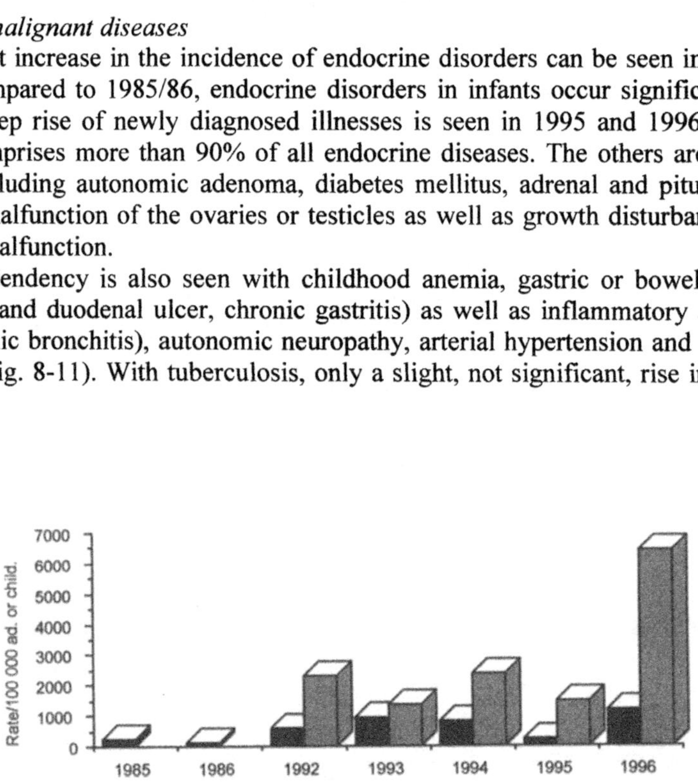

Figure 7. Yearly incidence of endocrine disorders in adults and children in the district of Krasnopolje

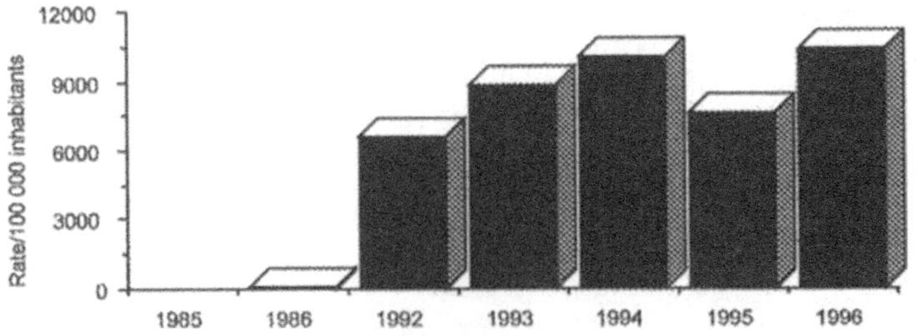

Figure 8. Incidence of anemia in children in the district of Krasnopolje

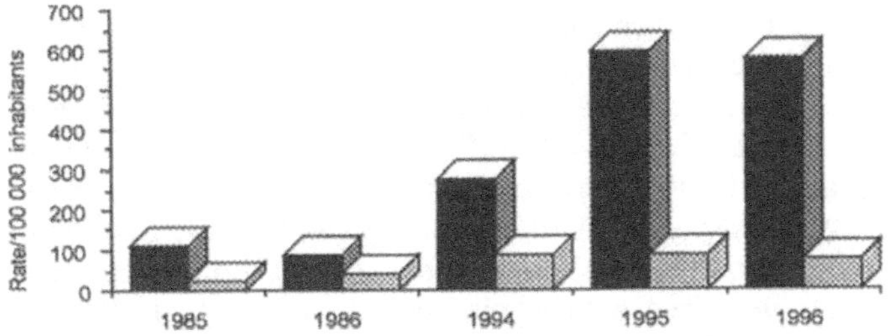

Figure 9. Yearly incidence of hypertension and heart attacks in the district of Krasnopolje

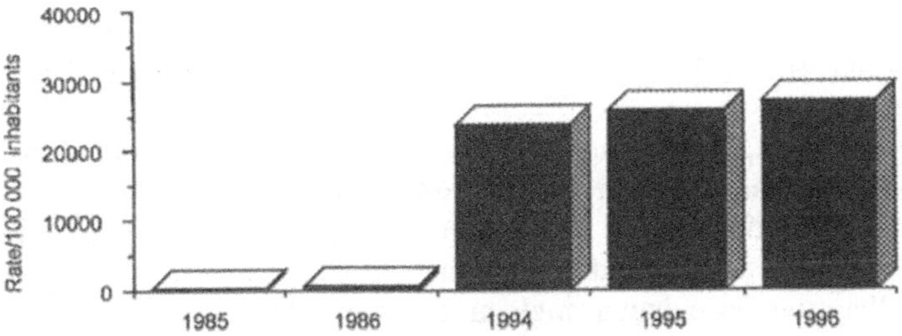

Figure 10. Yearly incidence of lung diseases in the district of Krasnopolje

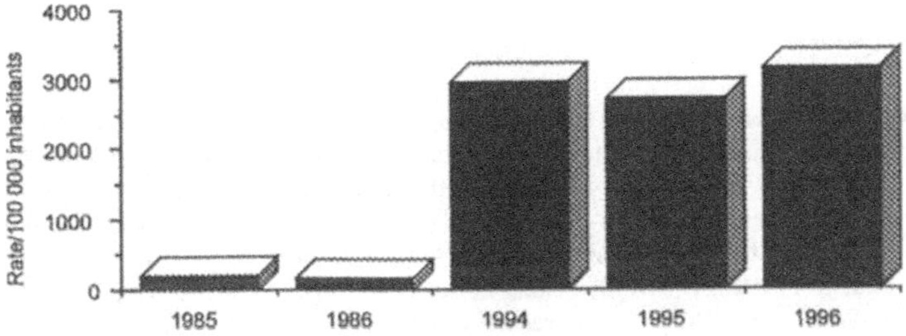

Figure 11. Yearly incidence of gastrointestinal diseases in the district of Krasnopolje

4. Discussion

The district of Krasnopolje, due to its geographical position and rate of contamination, can in fact be assumed as being representative for the highly contaminated southeastern area of Belorussia. In spite of the obvious negative effects on people's health comparing 1985 to 1986 and 1992 to 1996, definite statements with regard to significance should be carefully considered due to the small number of cases. This is especially important when discussing the rate of malignant tumors. Things become even more difficult when we try to look at the people evacuated into less contaminated areas. An additional increase in tumor incidence might be expected, but will hardly be provable. Additional difficulties in evaluating the data derive from the changes within the public health system. Due to help from other countries, provisional and logistic circumstances have changed for better since the early 1990's.

Calculations of the mean yearly dose rates adding external and internal contamination lead to measured values of >40 mSv for areas evacuated (1986) and <1 mSv in areas contaminated with not more than 5 Ci/km^2 (1994). The increasing incidence in malignant diseases is especially striking considering the dose rates and the short latent period to onset, especially since recent publications [9] do not postulate a significant increase in tumor incidence with whole body doses less than 200 mSv (children 50 mSv). The increase in childhood thyroid carcinoma and hypothyroidism gives evidence for the high local dose rates within the thyroid gland [6, 7]. Especially thyroid carcinoma in children is otherwise a rare finding. The increase in childhood anemia, disorders of the stomach and the intestines, chronic bronchitis and hypertension has to be considered as a complex combination of chronic radiation exposure and poor general living conditions (i.e. lack of vitamins and proteins, alcoholism, economic fears, lack of perspective).

Chromosomal analysis cannot be performed within the district of Krasnopolje, therefore data evaluating genetic effects are not available. Altogether, an increase in chromosomal aberrations of up to 1.8 times is expected for all contaminated areas of Belorussia. It has been suggested though, that chronic low dose exposure might increase the rate of alteration more than a single high dose exposure [1].

Further systematic analysis in the contaminated regions, associated with further support, is of great importance in order to obtain valid epidemiologic data and to gain a better knowledge of the radiobiological effects of chronic radiation exposure.

5. References

1. Dubrova YE, Nesterov VN, Krouchinsky NG, Ostapenko VA, Neumann R, Neil DL, Jeffreys AJ. (1996): Human minisatellite mutation rate after the Chernobyl accident. *Nature* **380**, 683-686.
2. Füller, J. and M.W. Trojanow (1996): Southeastern of Belorussia 10 years after Chernobyl - Radiobiological situation in the district of Krasnopolje, 28. Congress of German Association of Radiation Protection, October, 23-25, 1996, Hannover, Germany.
3. Heinemann, K. (1996): Umweltmessung in der GUS, *Strahlenschutzpraxis* **2**, 8-10.
4. Kasakow, W.S. et al. (1994): Basic results of research received in the medical section of the state programme on the elimination of the Chernobyl disaster after-effects. II. International Congress "The World after Chernobyl", April, 18-22, 1994, Minsk, Belorussia.
5. Koepp, R., T. Koepp-Schewyrina (1996): Tschernobyl-Katastrophe und Langzeitfolgen, B.G. Teubner Verlagsgesellschaft Stuttgart - Leipzig - Zürich, pp 100-108.
6. OECD (1994): Chernobyl -10 years on radiological and health impact, Paris, 60-64.
7. Reiners, C. (1994): Schilddrüsenkrebs bei Kindern in der Umgebung von Tschernobyl, *Spektrum der Wissenschaft* **12**, 117-120.
8. Sauermann, P. (1996): Die gegenwärtige Strahlenbelastung der Bevölkerung des Bezirkes Gomel, *Strahlenschutzpraxis* **2**, 10-13.
9. Streffer, C. (1991): Strahlenrisiko im niedrigen Dosisbereich - Wie ist der Wissensstand ?, *Deutsches Ärzteblatt (Ausgabe D)* **88**, 2066-2074.
10. Trojanow, M.W. (1994, 1995, 1996): Epidemiological reports of the district of Krasnopolje, Belorussia.

RADIATION INDUCED CANCER IN ANIMALS

J.J. BROERSE

IRI/TNO Centre for Radiological Protection and Dosimetry, Delft, and Department of Clinical Oncology, Leiden University Medical Centre Leiden, The Netherlands

1. Introduction

Under normal circumstances radiation levels in the environment due to industrial applications do not exceed those from natural radiation. Detrimental consequences of background radiation in man i.e., exposure to an effective dose of 2 mSv have never been proven. The reactor accident at Chernobyl was a catastrophe with large social and economical consequences. Statements that more than 100,000 persons would already have died due to this accident, however, have never been validated. The highest levels of radioactive contamination mainly iodine and caesium, occurred in Belarus. Up to 1996 more than 300 excess cases of thyroid cancer were observed in children [1]. The minimum latency period between exposure and diagnosis of thyroid cancer was four years. Since most cancers could be cured, the number of fatal cases was restricted to three [2].

Apprehension of cancer risks accrued from cancer incidence data of the Japanese population exposed to relatively large doses in excess of 0.5 Gy. This information has to be extrapolated on a world-wide scale to risks in the low dose region. The present contribution summarizes some uncertainties in human epidemiology and attempts to elaborate on the added value of animal results. The subsequent sections relate to risk factors at high doses, dose-effect relationships for low- and high-LET radiation, radiation weighting factors in connection with relative biological effectiveness, dependence on the temporal delivery of the dose and tissue weighting factors in relation to differences in susceptibility for tumour induction. This review is based on a selected number of experiments and is therefore necessarily not exhaustive.

A common approach in the analysis of tumour induction studies in animals is to score the fraction of animals that develop a tumour in course of time. A bare statement of tumour incidence is, however, of limited or no value. The incidence results should be analyzed by actuarial methods, correcting for loss of animals from the experiment due to intercurrent disease or other reasons. The carcinogenic effect can subsequently be described by hazard functions. Such descriptions have been performed by applying non-parametric models e.g. the proportional hazards model [3] or analytical models e.g. the Weibull distribution [4].

C. Baumstark-Khan et al. (eds.),
Fundamentals for the Assessment of Risks from Environmental Radiation, 379-388.

2. Risks at High Doses

Since 1972 a number of scientific bodies have made major risk evaluations resulting in estimates associated with 1 Gy of acute low-LET uniform total body irradiation. These bodies include the Committee on the Biological Effects of Ionizing Radiation of the US National Academy of Sciences (BEIR), the International Commission on Radiological Protection (ICRP), the Nuclear Regulatory Commission of the United States (NUREG) and the United Nations Scientific Committee on the Effects of Atomic Radiation (UNSCEAR). Values on excess lifetime mortality based on the Japanese A-bomb survivors are listed in Table 1. The risk factors up to 1987 were based on the tentative dosimetry of 1965, TD65. The later values were derived with the dosimetry system of 1986, DS86. The ICRP [5] concludes that estimates based on the additive and multiplicative model have come closer together with time. The risk estimates according to the 1977 recommendations [6] have now been increased by about 3-4 times. The 1988 estimate of life time fatal cancers using the preferred multiplicative risk projection model was $8\text{-}11 \times 10^{-2}$ per Sv, at high doses for the exposed populations (more than 90.000 survivors) at Hiroshima and Nagasaki of whom more than half in the epidemiological study were still alive. The number of excess solid cancers in the additional 11 year follow-up period increased from 135 to 260 [5]. In its 1993 report UNSCEAR [7] mentions a number of about 350 excess cancer deaths.

Medical studies including patients treated for ankylosing spondylitis and for cancer of the uterine cervix revealed lower risk factors [8, 9]. Alternative methods of modelling and dose assessment include the age attained model [10] and possible changes in the dose contribution from neutrons in Hiroshima [11]. These findings could be translated into lower risk factors in the high dose range.

Non-human primates show an appreciate number of similarities with man in their response to ionizing radiation and it is therefore of value to communicate results on tumour induction in monkeys. Acute effects in rhesus monkeys were investigated after

TABLE 1. Excess life-time mortality from all cancers attributable to 1 Gy acute low-LET total body irradiation of the general population of Japan [5].

Source of estimate	Probability of death (10^{-2} per Gy) Additive risk projection model	Multiplicative risk projection model
BEIR I, 1972	1.2	6.2
UNSCEAR, 1977	2.5	--
BEIR III, 1980	0.8 - 2.5	2.3 - 5.0
NUREG, 1985	2.9	5.2
UNSCEAR, 1988	4.0 - 5.0	7.0 - 11.0
BEIR V, 1990	--	8.85

For full references, see ICRP publication 60 [5].

exposure to relative high doses of X rays (average dose 6.7 Gy) or fission neutrons (average dose 3.4 Gy). The protective effect of autologous bone marrow transplantation was demonstrated [12]. The stochastic risks of total body irradiation were investigated by keeping the long term surviving monkeys under continuous observation. Rhesus monkeys of a comparable age distribution served as a control group [13]. Information on the number of animals involved and the types of malignancies are summarized in Table 2. In many cases, multiple tumours were observed in the irradiated animals, whereas in the control group, 2 out of 21 animals developed single malignancies over a total observation period of 482 monkey years.

The histogenic origin of the tumours was diverse, with a preponderance of renal cortical tumours in the irradiated group. Furthermore, 2 out of the 9 neutron-irradiated monkeys died with glial tumours after relatively short post-irradiation intervals of 4 and 6 years. The tumour induction results have been analyzed in terms of the number of animals developing tumours per group as a function of the total observation period for the entire group and the average absorbed dose. In this way, risk factors for tumour induction for X rays of (10/243 - 2/482) 6.7-1 = 61×10^{-4} $year^{-1}$ Gy^{-1} and (7/92 - 2/482) 3.4-1 = 216×10^{-4} $year^{-1}$ Gy^{-1} for fission neutrons can be derived. For a risk period of 15 years the life time risk factor for X-rays would be 9×10^{-2} Gy^{-1} which is rather close to the value for man at high doses. Another study on long term mortality in rhesus monkeys [14] after irradiation with protons and X rays at widely different dose levels has also revealed some information on the distribution of malignant tumours among various organs. In the group of 33 and 7 animals irradiated with protons or X rays, respectively, the site-specific cancers included 12 brain tumours, six adenocarcinoma of the small intestine and three renal carcinomas.

TABLE 2. Malignant tumours induced in rhesus monkeys after high-dose irradiation with X-rays and fission neutrons.

Type of irradiation	Animals with tumours/total number at risk	Latency periods (years)	Total observation period (years)	Type of malignancy
X rays 5.5-8.0 Gy	10/20	7-16	243	6x cortical carcinoma, kidney 2x osteosarcoma 2x adenocarcinoma, ileum or colon 1x multiple myeloma 1x malignant glomus tumour 1x carcinoma thyroid 1x malignant Schwannoma
fission neutrons 2.3-4.4 Gy	7/9	4-18	92	3x cortical carcinoma, kidney 3x malignant glomus tumour 2x osteosarcoma 2x brain tumours 1x carcinoma, liver 1x carcinoma, thyroid 1x synovial sarcoma

3. Dose-Effect Relationships

Extensive reviews of dose-response relationships for radiation carcinogenesis in animals have been provided by UNSCEAR [7, 8]. The induction of skin cancer in rats and mice by local low-LET irradiation requires high doses and apparent or real thresholds cannot be excluded, nor can they be established. For the induction of bone sarcoma after incorporation of Ra-226 and Sr-90 in dogs and mice, a practical threshold is indicated. When all appropriate corrections have been applied, the dose-effect relationships between tumour incidence, I(D) and absorbed dose D can, in their most general way, be described as:

$$I(D) = \left(I_D + a_1 D + a_2 D^2\right) \cdot \exp\left(-b_1 D - b_2 D^2\right) \qquad (1)$$

where I_0 is the spontaneous incidence, a_1 and a_2 are coefficients for the linear and dose-squared terms of cancer initiation and the exponential term is the probability of survival of transformed cells. For tumour induction in most tissues, the influence of the cell kill factor will not be very important in the range of doses up to 1 to 2 Gy of low-LET radiation (or their high-LET equivalent). For a full apprehension of the dose-response curves a relatively large dose range should be studied. Ullrich and Preston [15] reported a linear dose-response curve for induction of myeloid leukaemia in mice. By extending the doses up to 4 Gy, however, Mole [16] derived a biphasic curve with an initial quadratic dose dependence (see Fig. 1).

For three other experimental endpoints, notably lung adenocarcinoma in mice [18], life shortening in mice [19] and mammary carcinoma in rats [20] approximately pure quadratic dose-response relationships are reported for low-LET radiation. On the basis of microdosimetric considerations the apparent absence of a linear term in the dose-response equation for tumour induction after X irradiation is difficult to explain. Apparently, repair processes at the (sub)cellular level and systemic factors can modify the

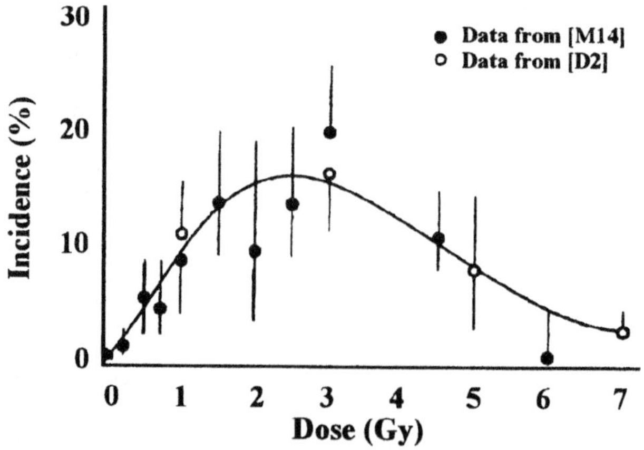

Figure 1: Incidence of myeloid leukaemia in male CBA/H mice following brief exposures to X rays. Closed symbols [16], open symbols [17].

primary process of cancer initiation. Characteristics of the initial part of dose-response curves for carcinogenesis in various tissues after exposure to radiation of different quality are summarized elsewhere [21]. It can be concluded that for low- and high-LET radiation, respectively, the tumour incidence curves generally show a linear-quadratic and a linear dependence on dose in the range of low absorbed doses.

4. Radiation Weighting Factors and Relative Biological Effectiveness

It is a well known fact that equal doses of radiation of different quality do not necessarily lead to the same level of biological damage. Radiobiological studies have shown that the relative biological effectiveness (RBE) is dependent on factors such as the nature of the detrimental effect, the absorbed dose, the dose rate and physiological conditions. For radiation protection applications emphasis should be placed on the risk for late stochastic effects after low-dose irradiation. For this condition the ICRP [6] introduced the radiation weighting factor w_R which was slightly amended in the recent recommendation [5]. For photons and electrons $w_R = 1$, for α-particles $w_R = 20$ and for neutrons w_R varies between 5 and 20, with the highest value in the energy range of 0.1 - 2.0 MeV.

For specific endpoints in animals and in man RBE values in excess of 100 have been reported [22]. It is important to verify the type of reference radiation. When γ rays would be used as the reference radiation, instead of X rays, appreciably higher RBE values would be derived, especially at low doses, since the RBE of γ radiation can be appreciably smaller than unity. It should further be realized that higher RBE values of neutrons do not necessarily imply an extremely high risk factor per unit dose for this radiation modality but that they reflect more the low efficacy for tumour induction by low-LET radiation. This can be exemplified by the results of studies on mammary carcinogenesis after exposure to Cs-137 gamma radiation [13]. The effect of fractionated irradiation with 120 fractions of 2.5 and 10 mGy (at intervals of 12 hours) was compared with that of a single acute exposure with doses of 0.3 and 1.2 Gy. A mathematical analysis of the relative excess hazard reveals quadratic dose-response curves without a significant linear component for the induction of carcinomas. On the basis of the dose-response curves for acute gamma irradiation in comparison to the linear dependency for 0.5 MeV neutrons (see Fig. 2) the RBE would approach infinity in the region of very low doses (below 0.3 Gy γ rays). The recommended w_R values [5] can be considered as a safe approach for radiation protection. It was recently suggested [23] that the w_R of 5 for protons with energies in excess of 2 MeV and of 20 for alpha particles from radon progeny are both too high by at least a factor of two.

5. Influence of the Time Factor

When the radiation is administered in a number of fractions or at a reduced dose rate, the biological response is different from that obtained after single acute doses. For cell survival, the modifying processes include repair of sublethal damage, repopulation, and redistribution over the cell cycle. For tumour induction, the nature of modifying mechanisms is not yet well understood, and it is rather difficult to interpret all experimental results obtained after fractionated or protracted exposures.

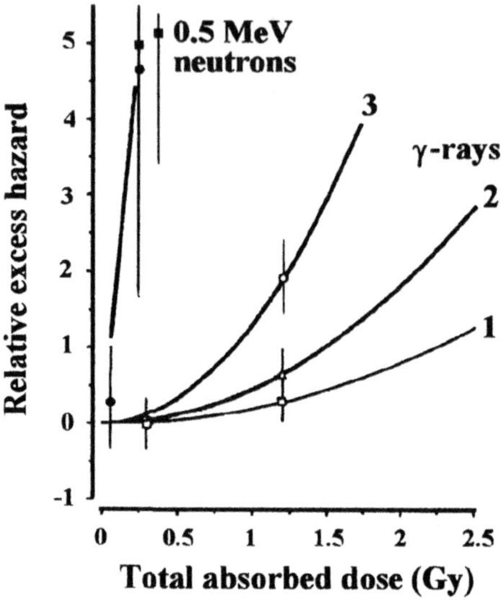

Figure 2: Relative excess hazard as a function of the total absorbed dose for the induction of mammary carcinomas in WAG/Rij rats after irradiation with 0.5-MeV neutrons and Cs-137 gamma rays. Curve 1, fractionated irradiation: curve 2, single-dose irradiation at 17 weeks of age; curve 3, single-dose irradiation at 8 weeks of age.

In general, for low-LET radiation the effectiveness is reduced with respect to the acute exposure when fractionation or protraction is applied. One of the few unexplained exceptions is the work of Maisin et al. [19] which indicated that exposure to ^{137}Cs γ radiation delivered in 8 fractions (interval 3 hr) and 10 fractions (interval 24 hr) would be more effective per dose unit in causing leukaemias and all carcinomas and sarcomas in mice than a single-session irradiation.

For high-LET radiation, the effects of fractionation and protraction are different for the tumour types, presumably because of the different mechanisms of tumorigenesis that are involved. For the induction of ovarian tumours in mice Ullrich [24] observed that fission neutron irradiation was less effective when delivered at low dose rates in comparison with high dose rates. However, the mammary carcinogenic effect of neutrons was enhanced at low dose rates, a finding similar to that of Vogel and Dickson [25]. Studies on the induction of mammary carcinoma in WAG/Rij rats [4] after single and fractionated irradiations with X rays and 0.5-MeV neutrons indicate that for equal total absorbed dose the tumours appeared at approximately the same age. It should be noted that the experimental studies on mammary carcinogenesis are generally based on whole-body irradiations of the animals. The induction of mammary cancer is highly susceptible to modification by hormonal factors, and it might well be that specific endocrinological conditions necessary for mammary tumour induction are influenced to a lesser extent by fractionated or protracted exposures than by a single dose.

The induction of lung cancer after single or protracted irradiation with α particles was investigated independently in two species. Lundgren et al. [26] studied the effect of inhalation of ^{239}Pu oxide in mice after single or repeated exposure. For similar cumulative doses from α particles, an approximately four times greater incidence of pulmonary tumours was observed for protracted exposure than for the single inhalation. The effect of dose rate on the induction of lung cancer in Syrian hamsters was studied by Little et al. [27] after intratracheal instillation of ^{210}Po. Protraction of the α irradiation over 120 days was slightly more carcinogenic at lower total lung doses but slightly less carcinogenic at higher doses, in comparison to an exposure limited to a 10-day period. Animal studies have not yet provided an unambiguous answer about the influence of fractionation or protraction of the dose administration on the carcinogenic effects of high-LET radiation and the epidemiological data from human exposures do not provide any reliable answer either. In view of its importance for radiation protection this aspect will certainly need further investigation in a variety of animal species.

For low-LET radiation, however, there is general consensus that the number of cancers induced per unit dose at low doses and low-dose rates, is reduced from that obtained at high doses and high-dose rates by a factor in the range of 2-10 [28]. The ICRP [5] called this reduction factor the dose and dose-rate effectiveness factor (DDREF). For the studies on induction of mammary carcinoma in rats shown in Figure 2 a DDREF of 4.5 can be derived. For high-LET radiation UNSCEAR [7] recommends a DDREF of 1.

6. Tissue Weighting Factors

Tissue weighting factors, w_T, were introduced by the ICRP in 1977 [6] to indicate the proportion of the stochastic risk resulting from specific tissues (T) to the total risk, when the whole body would be uniformly irradiated. In the 1990 recommendations [5], the list of w_T values was extended and in some cases appreciably modified, based on the analysis of the cancer mortality data in the Japanese atomic bomb survivors, employing the multiplicative risk model. Focussing on radiation carcinogenesis, in the older recommendations the highest tissue weighting factors are quoted for breast (0.15), red bone marrow (0.12.) and lung (0.12). According to the 1990 recommendations, red bone marrow, lung, colon and stomach are ranked among tissues with high w_T values (0.12), followed by a second category, bladder, breast, liver, oesophagus and thyroid with w_T values of 0.05. The other remaining tissues have w_T values of 0.01 to 0.005. It will be worthwhile to investigate whether differences in susceptibility for tumour induction could be substantiated by animal studies. Although attempts are made [29] to coordinate the collection and interpretation of radiation carcinogenesis data in animals, only a limited number of studies allow a comparison of relative radiogenic risks in various organs.

In a study on various radiogenic neoplasms in mouse strains, Storer et al. [30] showed that the relative risk model was superior over the absolute model for five tumour types: lung, breast, liver, ovary and adrenal. Both models appeared to fit myeloid leukaemia and neither provided good fits for thymic lymphoma. As shown in Table 3, the relative risk of death from various neoplasms at a dose of 1 Gy gamma radiation does not vary more than a factor of 3.

TABLE 3. Fatal probability coefficients (FPC) or relative risks (RR with confidence intervals) for radiogenic tumours in rodents per Gy X or gamma radiation

Neoplasm	Storer et al. [30]	Covelli et al. [31]		Bartstra et al. [32]
	RR	RR	FPC	RR
Lung carcinoma	1.77 (1.48 - 2.17)	1.6	0.95	
Breast carcinoma	2.00 (1.62 - 2.60)			1.78 (1.50 - 2.06)
Myeloid leukaemia	3.18 (1.72 - 10.25)		1.74	
Thymic lymphoma	3.20 (2.63 - 3.95)			
Liver tumours	3.60 (2.60 - 5.10)	1.2	1.59	
Adrenal carcinoma	4.30 (2.32 12.00)			
Gastro-intestinal tract			0.15	
Kidney tumours			0.15	

Covelli et al. [31] derived from the studies at Casaccia information on the fatal probability coefficient (FPC) and the relative risks in $BC3F_1$ mice. For neoplasms in the gastrointestinal tract and the kidneys, the values for the FPC are rather low, however, these values are largely strain dependent. Finally, Table 3 contains information on the relative risks for induction of mammary carcinoma in rats [32]. It would be useful to complement the list of relative risk factors for radiogenic tumours in rodents.

7. Conclusions

- Studies in long-term surviving monkeys after relatively high doses of X rays have revealed a risk factor of 9×10^{-2} Gy^{-1} for tumour induction. A relatively high number of tumours in the kidney and brain were observed. These tissues are classified by the ICRP [6] with low tissue weighting factors.
- Dose-effect relations can be described by biphasic curves with maxima at doses of 4 to 8 Gy depending on the endpoint.
- High RBE values for tumour induction after neutron irradiation do not necessarily imply an extremely high risk factor for this radiation modality, but reflect more the low efficacy for tumour induction by low-LET radiation.
- Relative risks for radiogenic tumours in most rodent strains do not vary more than a factor of three for various neoplasms. The variation in tissue weighting factors for radiogenic tumours by a factor of 24 [5] and the correlation with various organs in experimental animals require a more sound scientific basis.

8. References

1. No authors listed. (1996), Chernobyl's legacy to science. Editorial. *Nature* **380**, 653.
2. IAEA (1996). Joint IAEA/EC/WH International Conference: One Decade after Chernobyl: Summing up the Consequences of the Accident.
3. Kellerer, A.M. and Chmelevsky, D. (1982). Analysis of tumor rates and incidences - A survey of concepts and methods. In J.J. Broerse and G.B. Gerber (eds.) *Neutron Carcinogenesis*. pp. 209-231. Commission of the European Communities, Luxembourg.
4. Broerse, J.J., Hennen, L.A. and Zwieten, M.J. van (1985). Radiation carcnogenesis in experimental animals and its implications for radiation protection. *Int. J. Radiat. Biol.* **48**, 167-187.
5. ICRP (1990). Recommendations of the International Commission on Radiological Protection. ICRP Publication 60, Annals of the ICRP 21, 1-3. Pergamon Press, Oxford.
6. ICRP (1977). Recommendations of the International Commission on Radiological Protection. ICRP Publication 26, Annals of the ICRP 1, 3. Pergamon Press, Oxford.
7. UNSCEAR (1988). *Sources, Effects and Risks of Ionizing Radiation,* United Nations Scientific Committee on the Effects of Atomic Radiation. United Nations, New York.
8. UNSCEAR (1993). Sources, *Effects and Risks of Ionizing Radiation,* United Nations Scientific Committee on the Effects of Atomic Radiation. United Nations, New York.
9. Broerse, J.J. and Dennis, J.A. (1990). Dosimetric aspects of exposure of the population to ionizing radiation. *Int. J. Radiat. Biol.* **57**, 633-645.
10. Kellerer, A.M. and Barclay, D. (1992). Age dependences in the modelling of radiation carcinogenesis. *Radiat. Prot. Dosim.* **41**, 273-281.
11. Straume, T. (1996). Risk implications of the neutron discrepancy in the Hiroshima DS86 dosimetry system. *Radiat. Prot. Dosim.* **67**, 9-12.
12. Broerse, J.J., Bekkum, D.W. van, Hollander, C.F. and Davids, J.A.G. (1978). Mortality of monkeys after exposure to fission neutrons and the effect of autologous bone marrow transplantation. *Int. J. Radiat. Biol.* **34**, 253-264.
13. Broerse, J.J., Bekkum, D.W. van, Zoetelief, J. and Zurcher, C. (1991). Relative biological effectiveness for neutron carcinogenesis in monkeys and rats. *Radiat. Res.* **128**, S128-S135.
14. Wood, D.H. (1991). Long-term mortality and cancer risk in irradiated rhesus monkeys. *Radiat. Res.* **126,** 132-140.
15. Ullrich, R.L. and Preston, R.J. (1987). Myeloid leukemia in male RFM mice following irradiation with fission spectrum neutrons or γ rays. *Radiat. Res.* **109,** 165-170.
16. Mole, R.H. (1984). Dose-response relationships. In J.D. Boice, J.F. Fraumeni. (eds.). *Radiation Carcinogenesis: epidemiology and biological significance*. New York, Raven Press, 403-420.
17. DiMajo, V., Coppola, M., and Rebessi S. et al. (1986). Dose-response relationships of radiation-induced Harderian gland tumours and myeloid leukaemia of the CBA/Cne mouse. J. *Natl. Cancer Inst.* **76,** 955-963.
18. Coggle, J.E. (1988). Lung tumor induction in mice after X-rays and neutrons. *Int. J. Radiat. Biol.* **53,** 585-598.
19. Maisin, J.R., Wambersie, A., Gerber, G.B., Mattelin, G., Lambiet-Collier, M., Coster, B. de, Gueulette, J. (1988). Life-shortening and disease incidence in C57Bl mice after single and fractionated gamma and high-energy neutron exposure. *Radiat. Res.* **113**, 300-317.
20. Bekkum, D.W. van, Broerse, J.J., Hennen, L.A. and Solleveld, H.A. (1986). The gene transfer-misrepair hypothesis of radiation carcinogenesis tested for induction of mammary tumours in rats. *Leukem. Res.* **10,** 761-766.
21. Broerse, J.J. (1989). Influence of physical factors on radiation carcinogenesis in experimental animals. In K.F. Baverstock and J.W. Stather (eds.) *Low Dose Radiation: Radiological Bases of Risk Assessment*. pp. 181-194. Taylor & Francis, London.
22. Lowry Dobson, R. and Straume, T. (1982). *Cancer Risks and neutron RBE's from Hiroshima and Nagasaki.* Report EUR 8084 en. Commission of the European Communities.

23. Curtis, S.B. (1995). Are radiation weighting factors relevant for protons and heavier charged particles? In U. Hagen, D. Harder, H. Jung, C. Streffer (eds.) *Radiation Research*, Congress Proceedings. pp. 161-164. Universitätsdruckerei H. Stürtz AG, Würzburg.
24. Ullrich, R.L. (1984). Tumor induction in Balb/c mice after fractionated or protracted exposures to fission-spectrum neutrons. *Radiat. Res.* **97**, 587-597.
25. Vogel, H.H. and Dickson, H.W. (1982). Mammary neoplasia in Sprague-Dawley rats following aute and protracted irradiation. In J.J. Broerse and G.B. Gerber (eds.) *Neutron Carcinogenesis.* pp. 135-154. Commission of the European Communities, Luxembourg.
26. Lundgren, D.L., Gillett, N.A., Hahn, F.F., Griffith, W.C. and McClellan, R.O. (1987). Effects of protraction of the α-dose to the lungs of mice by repeated inhalation exposure to aerosols of $^{239}PuO_2$. *Radiat. Res.* **111**, 201-224.
27. Little, J.B., Kennedy, A.R. and McGandy, R.N. (1985). Effects of dose rate on the induction of experimental lung cancer in hamsters by α radiation. *Radiat. Res.* **103**, 293-299.
28. NRPB (1995). Risk of Radiation-Induced Cancer at Low Dose and Low Dose Rates for Radiation Protection Purposes. Chilton, Didcot.
29. IRAA (1996). International Radiobiology Archives of Long-Term Animal Studies.
30. Storer, J.B., Mitchell, T.J. and Fry, R.J.M. (1988). Extrapolation of the relative risk of radiogenic neoplasms across mouse strains and to man. *Radiat. Res.* **114**, 331-353.
31. Covelli, V., Coppola, M., DiMajo, V., Rebessi, S. and Bassani, B. (1988). Tumor induction and life shortening in $BC3F_1$ female mice at low doses of fast neutrons and X rays. *Radiat. Res.* **113**, 362-374.
32. Bartstra, R.W., Bentvelzen, P.A.J., Zoetelief, J., Mulder, A.H., Broerse, J.J. and Bekkum, D.W. van. (1998) Induction of mammary tumours in rats after single dose gamma irradiation at different ages. *Radiat. Res.* **150**, 442-450

METABOLIC AND HORMONAL CHANGES IN IRRADIATED ANIMALS

I. AHLERS

Institute of Animal Physiology, P. J. Šafárik University
04167 Košice, Slovak Republic

Abstract

This is a short review on metabolic and hormonal changes in irradiated experimental animals. These changes are intensive and extensive in rats, less expressive in mice and dogs and of lesser degree in other animals. The research on metabolic changes associated with various forms of ionizing radiation action peaked within fifties and seventies, than decreased; now the use of nuclear magnetic resonance could partially restore the interest on metabolic changes, especially in experimental radiation oncology. The review will be focused on results of our group and on a special form of irradiation-the continuous application of gamma rays.

1. Lipids

Increased lipolysis and lipomobilization, hyperlipemia and hyperlipoproteinemia, increased formation of triacylglycerols, cholesterol and of lipid peroxides predominated within the tissue lipid changes in irradiated rats [2]. Increased levels of non-esterified fatty acids (NEFA) were found in the serum white adipose tissue and liver or male Wistar rats irradiated with lethal [5] or non-lethal dose of X-rays [8] and in rats, irradiated continuously with medium daily doses (25-30 days, daily 0.57 Gy) of gamma rays [4]. The release of NEFA from white adipose tissue pads was enhanced 2 and 48 h after single irradiation of rats with 14.4 Gy of X-rays [43] and in rats irradiated continuously in our fundamental model [40]. High NEFA levels occured in the serum of rats in early period after irradiation with 14.4 Gy of X-rays; the concentration of triacylglycerols (TG), cholesterol (CH) and of phospolipids (PL) were rather low at this time [5]. Later especially the levels of TG and CH increased. This mixed hyperlipemia could be attributed to changes in lipoprotein lipase activity decrease. The activity of this enzyme in whole-body 14.4 Gy X-ray irradiated rats was dependent on the nutritional state of animals and the changes in white adipose tissue correlated with serum TG levels [39]. Lipoprotein lipase activity was altered in the tissues of continuously irradiated rats, too [41]. The proportion of individual lipoprotein classes was changed in irradiated rats: low density and (lesser) very low density lipoproteins increased, the high density family decreased. Essentially, the same patterns were observed in rats irradiated with 14.4 Gy

C. Baumstark-Khan et al. (eds.),
Fundamentals for the Assessment of Risks from Environmental Radiation, 389-394.

of X-rays [1] as in rats continuously irradiated by daily dose of 9.6 mGy of gamma rays up to 100 days [38]. Feliste et al. [23] described a defect in the apolipoprotein composition in the blood of irradiated rabbits, but later no attempts were made to elucidate the role of altered protein part of lipoproteins in the origin of postradiation hyperlipemia.

Enhanced synthesis of cholesterol belongs to the fundamental biochemical signs of irradiated organism. Originally there was an increase of CH synthesis from acetate [25], the site of radiation action was before mevalonate synthesis with the stimulation of hydroxymethylglutarylCoA reductase. Increased concentration of CH was found in various tissues of lethally irradiated rats [5, 6, 7] with exception of adrenals, where CH is used for synthesis of cortical hormones. The turnover of phospholipids was stimulated by ionizing radiation in experimental animals. Otherside the radiation-induced damage of cellular membranes was accompanied by loss of PL and in this way it was possible to find a fall in PL concentration/content in radiosensitive tissues of experimental animals. A correlation between the decrease in cellularity and the fall in PL concentration was found in single whole-body [32] and in continuously irradiated rats [33]. Increased concentration/content of triacylglycerols was found in the serum and various tissues of irradiated rats, e.g. the concentration of liver TG correlated with the degree of radiation-induced fatty infiltration. The decrease in fatty acid oxidation is probably the main cause of TG increase in irradiated tissues [46]. The bone marrow represent here a special case: the increase in TG concentration could be attributed to the increase in adipocyte formation. We have shown, that the increase in TG concentration of bone marrow is one of the best biochemical markers of radiation damage in rats. It occured in lethally [6] as in non-lethally [8] X-irradiated rats. TG accumulation in bone marrow was found in a wide range of continuously irradiated animals with equal daily dose of 960 mGy [2] as in 9.6 mGy daily dose [1], with clear dose-range arrangement with higher daily doses used [2]. Continuous irradiation with decreasing daily doses induced an increase in bone marrow TG concentration, too [3, 10]. Bone marrow TG accumulation remained by analysis 6 mounths after the end of 30-day continuous irradiation of rats with daily dose 0.57 Gy of gamma rays [2] and the normalization was seen in this model only 12 month after the end of the exposure [42]. Extensive lipid changes could be detected in the thymus. The decrease in PL content correlated well with radiation induced thymus involution (decrease of mass). TG content in thymus decreased, markedly in continuously irradiated rats [9].

The increase in lipid peroxidation plays a serious role in the development of radiation injury. The concentration of lipid peroxides in the liver was rather diminished [5], in other tissues varied in the dependence on the time after exposure [7]. Marked alterations in tissue lipid peroxides were seen immediately after the end of continuous irradiation of rats by daily dose 0.57 Gy of gamma rays up to 25 days [4].

Generally, the lipogenesis from ^{14}C glucose or ^{14}C acetate was increased in white adipose tissue and in the liver of single whole-body X-irradiated rats [45] and in continuously irradiated rats [44].

2. Carbohydrates

Hyperglycemia and liver glycogen accumulation were the earliest biochemical-metabolic signs attributed to radiation action. They were found in rats subjected to lethal (14.4 Gy) dose of X-rays [12] and the increase in liver glycogen was more pronounced in rats irradiated after preceding overnight fasting. The rise in serum glucose and liver glycogen and their duration was dose-dependent within the range of 4.8 - 14.4 Gy of gamma rays applied to male Wistar rats [18]. Pair-fed rats were irradiated with 2.4 Gy of X-rays [14] : serum glucose concentration rose immediately after the exposure, liver glycogen accumulation lasted for 21 days after irradiation end. In continuously irradiated rats (50 days, daily dose 0.57 Gy of gamma rays) liver glycogen accumulation was detected even 6 month after the irradiation end [13]. Accelerated gluconeogenesis was the main source for blood glucose and liver glycogen increase. Increased incorporation of ^{14}C -alanine into blood glucose and liver glycogen were found in rats irradiated with 14.4 Gy of X-rays. Liver phosphoenolpyruvate carboxykinase activity increased and a negative correlation between glucose-6-phosphatase activity and liver glycogen concentration was observed [35]. During the continuous irradiation (0.57 Gy daily up to 30 days) an increased incorporation of ^{14}C-alanine and low activity of glucose-6-phosphatase were seen in rats [36]. Irradiation with fractions (2.4 Gy weekly) up to 14.4 Gy total dose of gamma rays led to increased incorporation of ^{14}C-glucose into the liver, heart and skeletal muscle glycogen but not into the brain glycogen [37].

Liver glycogen accumulation might be considered as one of the most relevant metabolic marker in irradiated animals.

3. Hormones

In whole-body lethally irradiated rats (14.4 Gy of X-rays) a decrease in serum insulin was seen early after the exposure (6-24 h), later-in the terminal period-blood glucose raised to a „diabetic“level and serum insulin increased, too [11]. In this period an increase in serum glucagon was observed and this observation could contribute to the unsolved problem of glucose utilization in irradiated rats [15]. Serum insulin was not altered in single low dose (2.4 Gy of X-rays) irradiation of male Wistar rats [14]. Generally, the effect of irradiation on serum insulin was more pronounced in older rats [20].

Adrenocortical axis is activated during the irradiation of experimental animals. A two-phased elevation of circulating corticosterone was described in irradiated rats : the first peak was related to the early signs of radiation stress and the second one was attributed to the gut injury in irradiated rats and was dose-dependent [24]. The first peak of corticosterone increase could be inhibited in lethally irradiated (14.4 Gy of X-rays) rats by the administration of dexamethasone [31]. Serum and adrenal corticosterone was elevated in the course of our fundamental model of continuous irradiation of rats [30].

Thyroid gland is involved in the origin, extent and reparation of the radiation injury. Lethal dose (14.4 Gy of X-rays) caused in the early period after irradiation (1-6 h) a decrease in the concentrations of thyroxine (T_4) and triiodthyronine (T_3). Later the parallel sharp decrease of both hormones was seen in irradiated and non-irradiated fasted rats; additionally an increase in the concentration of reverse triiodthyronine (rT_3) was noted [15]. There were differences in the reaction of thyroid hormones in the same model between young and older rats; the concentration of serum thyrotropin was elevated [20]. The dependence of the changes in thyroid hormone levels on the dose was confirmed in rats whole-body irradiated by single doses 4.8, 7.2, 9.6 and 14.4 Gy of gamma rays, respectively. Only the T_3 level decreased by the lowest dose used. In the medium doses used an initial increase in both basic hormones was followed by their fall at the end of observation. By the highest dose the decrease of T_4 and T_3 levels was described during the whole period of observation [19]. Using the fractions (0.96 or 2.4 Gy weekly) up to total doses of 4.8 - 9.6 - 14.4 Gy of gamma rays the T_4 level decreased with the increase of total dose, T_3 decreased only at the highest total dose used [17]. In rats irradiated continuously 120 days with the daily dose 0.19 Gy the levels of both basic thyroid hormones increased; in rats irradiated continuously by daily dose 0.57 Gy of gamma rays for 90 days only the T_3 level increased in first 30 days and the rT_3 level mirrored this pattern. In the highest daily dose (0.96 Gy up to 35 days) an uniform decrease in T_4 and T_3 level with concomittant increase in rT_3 level was noted [16].

The role of pineal gland in the pathogenesis of radiation injury was not defined. The reaction of pineal was two-phased in male Wistar rats whole-body irradiated by a single dose 9.6 Gy of gamma rays : the initial decrease was followed by an increase in the terminal period of observation [26]. Similarly did change the activity of serotonin-N-acetyltransferase rays, key enzyme of melatonin synthesis, in lethally irradiated rats [27]. Two-phased changes in pineal melatonin was noted in rats, irradiated with 14.4 Gy of gamma rays locally on the head [28]. The changes in serum melatonin, corticosterone and thyroid hormones were dependent on the time of day and on lighting regimen used in radiation experiments [21].

Further research is needed for discovering the changes of central and peripheral catecholamines in the radiation action. Extensive changes of catecholamines were found in the ovine hypophysis, hypothalamus and adrenals after 7 days of continuous irradiation up to a total dose of 6.7 Gy of gamma rays [34]. A marked decrease in pineal catecholamines was seen in rats, irradiated on the head with 14.4 Gy of gamma rays [29].

Other metabolic and hormonal changes in irradiated animals are cited in well-known textbooks on radiation biochemistry [47, 22, 48].

4. Conclusion

Metabolic and hormonal changes are integral parts in the development of radiation injury in experimental animals. The study of these parameters should be continued using new progressive methods, suitable for completing the picture of radiation injury.

5. References

1. Ahlers, I. (1984) The effect of ionizing radiation and of other space flight factors on tissue lipids in rats. DrSc Thesis, P.J.Šafárik University, Košice.
2. Ahlers, I. (1994) Changes in whole-body metabolic parameters associated with radiation. *Advances in Space Research* **14**, 531-539.
3. Ahlers, I., Ahlersová, E. (1990), Bone marrow and thymus lipids in rats irradiated continuously with gradually decreasing dose rates of gamma radiation (In Slovak). *Bratislavské Lekárske Listy* **91**, 357-362
4. Ahlers, I., Ahlersová, E., Sedláková, A., Praslicka, M. (1967) Influence of chronic irradiation on organ lipids of rat. *Studia Biophysica* **2**, 467-472.

5. Ahlers, I., Ahlersová, E., Sedláková, A., Praslicka, M. (1973) Tissue lipids in lethally X-irradiated rats. I.Changes in serum, liver, white and brown adipose tissue. *Folia biologica (Praha)* **19**, 124-129.
6. Ahlers, I., Ahlersová, E., Sedláková, A., Praslicka, M. (1973), Tissue lipids in lethally X-irradiated rats.II. Changes in bone marrow, thymus, spleen and testes. *Folia biologica (Praha)* **19**, 130-134.
7. Ahlers, I., Ahlersová, E., Sedláková, A., Praslicka, M. (1973), Tissue lipids in lethally X-irradiated rats. III.Changes in adrenals, kindeys, heart, skeletal muscle and lungs. *Folia biologica (Praha)* **19**, 135-139
8. Ahlers, I., Ahlersová, E., Sedláková, A., Praslicka, M. (1981), Metabolic changes after non-lethal X-irradiation of rtas. II. Serum and tissue lipids. *Folia biologica (Praha)* **27**, 413-421.
9. Ahlers, I., Ahlersová, E., Toropila, M., Datelinka, I. (1992), Thymus lipids in continuously irradiated rats. *Physiological Research* **41**, 417-421
10. Ahlersová, E., Ahlers, I. (1993), Studies of continuous irradiation with gradually decreasing daily doses on tissue lipids in rats. Folia Facultatis Medicae Universitatis Šafarikianae Cassoviensis L, 119-126
11. Ahlersová, E., Ahlers, I., Praslicka, M. (1979), Changes of serum insulin level in lethally irradiated rats. *Endocrinologia experimentalis* **13**, 111-114
12. Ahlersová, E., Ahlers, I., Paulíková, E., Praslicka, M. (1980), Tissue glycogen and blood glucose in irradiated rats.I. Effect of single lethal X-irradiation. *Folia biologica (Praha)* **26**, 415-422
13. Ahlersová, E., Ahlers, I., Praslicka, M. (1980), Tissue glycogen and blood glucose in irradiated rats.II.Effect of non-lethal doses of continuous gamma irradiation. *Folia biologica (Praha)* **26**, 423-431
14. Ahlersová, E., Ahlers, I., Slavkovská, E., Praslicka, M. (1981), Metabolic changes after non-lethal X-irradiation of rats.I.Carbohydrates, hormones. *Folia biologica (Praha)* **27**, 404-412
15. Ahlersová, E., Ahlers, I., Praslicka, M. (1985), Thyroid and pancreatic hormones in lethally irradiated rats (In Slovak). *Bratislavské Lekárske Listy* **83**, 433-439
16. Ahlersová, E., Ahlers, I., Toropila, M., Datelinka, I. (1986), Thyroid hormones in continuously irradiated rats. In: Institute of Radiobiology (ed.), Selected papers from Radiobiology. Medical Faculty, Hradec Králové (In Slovak), pp.131-139
17. Ahlersová, E., Molcanová, A., Ahlers, I. (1987), The effect of fractionated irradiation on some hormones and carbohydrates in laboratory rats (In Slovak). *Sborník lékarský* **89**, 200-206
18. Ahlersová, E., Ahlers, I., Molcanová, A. (1988), Serum glucose and liver glycogen in gamma irradiated rats. *Radiobiologia-Radiotherapia* **29**, 165-169
19. Ahlersová, E., Molcanová, A., Ahlers, I. (1988), Thyroid hormones in gamma irradiated rats. *Radiobiologia-Radiother* **29**, 507-512
20. Ahlersová, E., Ahlers, I., Šmajda, B. (1997), Age dependence of metabolic and hormonal changes in lethally irradiated rats. *Folia Medica Cassoviensis* **54**, 96-103
21. Ahlersová, E., Kassayová, M., Ahlers, I. (1997), Serum melatonin, corticosterone and thyroid hormones in irradiated rats: effect of exposure during various times of day and of light regimen. *Journal of Physiology and Pharmacology* **48**, 435-442
22. Altman, K.I., Gerber, G.B., Okada, S. (1970), Radiation biochemistry, Academic Press, New York, vol.I,II
23. Feliste, R., Dousset, N., Carton, M., Douste-Blazy, L. (1981) Changes in plasma apolipoproteins following whole-body irradiation in rabbits. *Radiation Research* **87**, 602-612.
24. Flemming, K., Geierhaas, B. (1970), Radiation effects and adrenal cortex. V. Investigation on the „second" corticosteroid increase in whole-body X-irradiated rats. *Atomkernenergie* **15**, 169-173
25. Gould, R.G., Bell, V.L., Lilly, R.H. (1959) Stimulation of cholesterol biosynthesis from acetate in rat liver and adrenals by whole body X-irradiation. *American Journal of Physiology* **196**, 1231-1237.
26. Kassayová, M., Ahlersová, E., Ahlers, I. (1996), The changes of key pineal indices in laboratory rat after whole-body irradiation with gamma rays. In: Radiation Biology 1996, M.Toropila, P.Fedorocko, B.Šmajda (eds.). Data Help, Košice (In Slovak), pp.51-54

27. Kassayová, M., Ahlersová, E., Ahlers, I., Milárová, R. (1993), Changes of pineal N-acetyltransferase activity in gamma irradiated rats. *Physiological Research* **42**, 167-169
28. Kassayová, M., Ahlersová, E., Ahlers, I., Milárová, R. (1993), The changes in pineal N-acetyltransferase activity, pineal and serum melatonin concentration in rats after irradiation of the head. *Physiological Research* **42**, 419-422
29. Kassayová, M., Ahlersová, E., Pástorová, B., Ahlers, I. (1995), The early response of pineal N-acetyltransferase activity, melatonin and catecholamine levels in rats irradiated with gamma rays. *Physiological Research* **44**, 315-320
30. Malátová, Z., Ahlers, I., Praslicka, M. (1978), Adrenal and plasma corticosterone changes in continuously irradiated rats. I.The dynamics in the course of irradiation. *Biológia (Bratislava)* **33**, 177-182
31. Malátová, Z., Ahlers, I., Praslicka, M. (1981), Effect of dexamethasone on pituitary-adrenocortical response to acute irradiation stress in rats. *Endocrinologia Experimentalis* **15**, 205-210
32. Molcanová, A., Ahlers, I. (1986), Infuence of single total body gamma radiation on cellularity and lipids in bone marrow of rats. *International Journal Radiation Biology* **49**, 723-724
33. Molcanová, A., Ahlers, I. (1989), The changes of cellularity and lipid concentration in bone marrow of continuously gamma-irradiated rats (In Russian). *Radiobiologiya* **29**, 271-274
34. Pástorová, B., Arendarcik, J., Molnárová, M. (1985), Catecholamine levels in ovine hypophysis, hypothalamus and adrenals after whole-body gamma irradiation (In Slovak). *Veterinary Medicine (Praha)* **30**, 485-492
35. Paulíková, E., Ahlers, I., Prasliccka, M. (1983), Gluconeogenesis in lethally X-irradiated rats. *Physiologia bohemoslovaca* **32**, 73-79
36. Paulíková, E., Ahlers, I., Praslicka, M. (1985), Changes in liver gluconeogenesis in the course of continuous irradiation of rats. *Acta Facultatis Medicae Universitatis Brunensis* **26**, 501-506
37. Paulíková, E., Sedláková, A. (1988), Glycogenesis and lipogenesis from ^{14}C-glucose in vivo in rats irradiated with fractionated doses of gamma rays. *Physiologia bohemoslovaca* **37**, 333-341
38. Racek, L., Ahlers, I., Hozová, E. (1993) Metabolic changes in rats irradiated continuously with low daily doses (9.6 mGy). Folia Facultatis Medicae Universitatis Šafarikianae Cassoviensis L, 127-134.
39. Sedláková, A., Ahlers, I., Praslicka, M. (1979) The effect of a single lethal X-irradiation exposure on the activity of lipoprotein lipase in the tissues of the rat. *Folia biologica (Praha)* **25**, 254-265.
40. Sedláková, A., Ahlers, I., Praslicka, M. (1980) Changes in lipolysis in rat adipose tissue during continuous irradiation. *Folia biologica (Praha)* **26**, 211-216.
41. Sedláková, A., Ahlers, I., Praslicka, M. (1980), Changes in lipoprotein lipase activity in the adipose tissue, heart and liver of continuously irradiated rats. *Physiologia bohemoslovaca* **29**, 535-542.
42. Sedláková, A., Ahlers, I., Praslicka, M. (1980), The changes of serum and tissue lipids after the end of continuous irradiation of rats with gamma rays (In Russian). *Radiobiologiya* **20**, 884-848
43. Sedláková, A., Malatová, Z., Ahlers, I., Praslicka, M. (1980) Lipid mobilization and lipolysis in adipose tissue of single X-irradiated rats. *Folia biologica (Praha)* **26**, 204-209.
44. Sedláková, A., Malátová, Z., Ahlers, I., Praslicka, M. (1981), Changes in rat liver and adipose tissue lipogenesis in the course of continuous gamma irradiation. *International Journal of Radiation Biology* **39**, 1-8
45. Sedláková, A., Paulíková, E., Ahlers, I. (1985), The role of adipose tissue in glucose utilization in irradiated rats. *Physiologia bohemoslovaca* **34**, 527-533
46. Snyder, F. (1965), Fatty acid oxidation in irradiated bone marrow cells. *Nature* **206**, 733
47. Streffer, C. (1969), Strahlen-Biochemie, Springer, Berlin
48. Walden, T.L., Farzaneh,N.K. (1990), Biochemistry of Ionizing Radiation, Raven Press, New York

ABSCOPAL EFFECTS ON THE SELECTED IMMUNOLOGICAL PARAMETERS IN RAT AFTER IRRADIATION OF THE HEAD

E. PAULÍKOVÁ, A. KOCIKOVÁ, D. VAJÁNYIOVÁ

Institute of Animal Physiology, Faculty of Science, P.J. Šafárik University
Moyzesova 11, 041 67 Košice, Slovak Republic

Abstract

Although the main part of the central nervous system is a radioresistant tissue, the activation of the neuro-endocrine axis and the subsequent stimulation of the radiosensitive recirculating lymphocytes may play a crucial role in the abscopal effects observed after irradiation of the head (IH).Therefore, we investigated the hematological parameters and phagocytic activity (PA) of leukocytes from peripheral blood (PB) and peritoneal exudate (PE) in rat after IH. Two experimental designs were used: 1. an early response was investigated at 30 and 120 minutes after IH with a dose of 14.35 Gy; 2. a late response was investigated at days 1,7, 14, 21, 35, 40 and 46 after IH with a dose of 10 Gy. Analyses at the early time points after the irradiation revealed a significant increase in the numbers of erythrocytes and neutrophils in PB and increased PA as well as the phagocytic index (PI) of PB leukocytes.The late effects were observed only at the day 40 and 46 after IH. We found significantly increased numbers of erythrocytes, increased absolute and differential numbers as well as the PI of PB leukocytes. Furthermore, we observed significantly decreased numbers of mast cells and an increased PI of the macrophages in PE obtained at the days 7 and 46 after exposure. Taken together, our results indicate that in both models, irradiation of the head induced significant quantitative and qualitative changes in PB and PE of rat.

1. Introduction

The radiation therapy of the patients with the head cancer is widely used. The abscopal effects after IH, however, are not completely understood and the animal model to study these effects is not available. It has been shown recently that the exposure of the head to the X-rays in a dose of 20 Gy does not induce the vascular degeneration in rat brain [8]. However, due to the irradiation, neural cells may be stimulated to produce substances, e.g. neurohormones, which may activate the peripheral leukocytes and thus, lead to the abscopal effects in the body. Therefore, we attempted to establish a model to study these effects. Here, we demonstrate that a single dose of gamma irradiation of the head leads to the changes of the selected immunological parameters.

C. Baumstark-Khan et al. (eds.),
Fundamentals for the Assessment of Risks from Environmental Radiation, 395-400.

2. Material and Methods

Male Wistar rats 8 weeks old, weighing 180-190 g were used. They were arranged into the three groups: intact animals, sham-irradiated and head irradiated animals. Each group consisted of 8 animals. Both sham- and head irradiated rats were immobilized in special boxes shielding the rest of the body. The head was exposed to the gamma rays from a ^{60}Co source with a single dose of 14.35 and 10 Gy, respectively (dose rate 0.39 Gy/minute). To investigate the early effects, animals were killed by decapitation at 30 and 120 min. after irradiation. To investigate the late effects, animals were killed at days 1, 7, 14, 21, 28, 35, 40 and 46 after exposure. Cell counts, morphology and the phagocytic activity were determined in the PB and PE of these animals. For this purpose, PB was collected into the Eppendorf tubes containing heparin (5 I.U./ml). These samples were used to: i) determine the numbers of erythrocytes and leukocytes (by means of the hemocytometer chamber); ii) determine differential counts of the leukocytes (blood smears); iii) evaluate the phagocytic activity of leukocytes (blood smears). Blood smears were dried out and stained with May-Grünwald and Giemsa-Romanowsky solutions . More than 200 cells per slide were counted. To determine the phagocytic activity, 1 ml of the PB was incubated with 10^7 yeast cells in the 5 ml tubes at 37°C (continuously shaked). After 1 h, blood smears were prepared and stained as described above. The percentage of phagocyting cells (phagocytic activity) and the numbers of phagocyted yeast per phagocyte (phagocytic index) were determined. More than 100 phagocyting cells per slide were counted. Peritoneal exudate cells (PEC) were obtained as described elsewhere [2]. Briefly, 5 ml of PBS was injected i.p. After a gentle massage of the peritoneum (3 min), fluid rich in PEC was harvested, cells were washed and resuspended in RPMI 1640 medium supplemented with 5% autologous serum. PE smears were stained and served for determination of differential counts and morphology of PEC. More than 500 cells per slide were counted. Phagocytic activity was determined as described for PB. All data are expressed as mean values ± SEM. Non-paired t-test was used to evaluate the significance of experimental versus control groups.

3. Results

Early-effects after irradiation of the head. Rats were irradiated with a dose of 14.35 Gy. Significant increase in the numbers of erythrocytes were found in the PB of sham- and head irradiated animals. Absolute numbers of PB leukocytes did not change neither after the immobilization, nor after the head irradiation. However, the numbers of lymphocytes decreased and the numbers of neutrophils increased significantly at 120 minutes after irradiation. In addition, the PI of the leukocytes from both sham- and head irradiated animals markedly increased as compared to the control group. Immobilization stress of the sham-irradiated animals was mirrored in the increase in the the numbers of NK cells. Summarized results of peripheral blood are in Figure.1.

Late effects after irradiation of the head. Rats were irradiated with a dose of 10 Gy. Except for the significant decrease in the erythrocytes at day 14 and their increase at day 40 after irradiation, we did not observe any changes in their numbers.

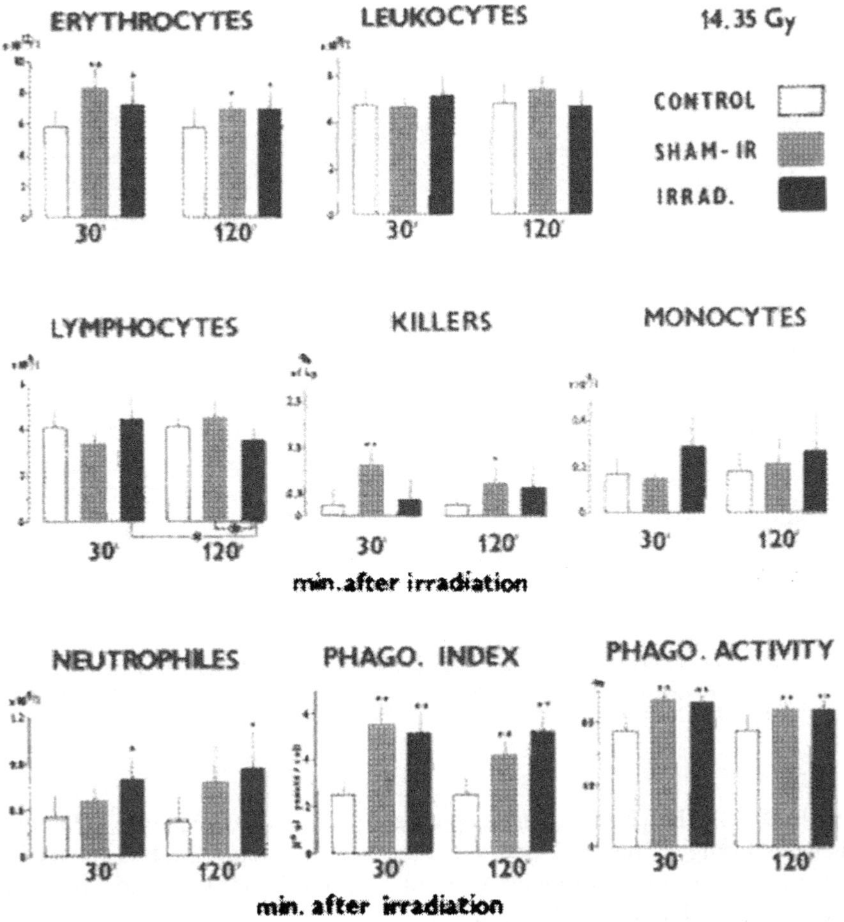

Figure 1. Alterations in peripheral blood parameters and their phagocytic function early after irradiation of the head (14.35 Gy). Results are expressed as mean values ± SEM, n = 8. The probability of significant differences when comparing groups was : * $p < 0.05$, ** $p < 0.01$.

The numbers of PB leukocytes were constantly decreased untill day 21 after exposure, most likely due to the decreased numbers of lymphocytes and neutrophils (Figure 2). Lymphopenia was probably the signal for the intensive proliferation of cells in thymus, since an increase in thymus weight was observed. During days 28-35, the numbers of PB leukocytes increased and reached the numbers observed in the control animals. This was accompanied by an extensive increase in the numbers of lymphocytes and neutrophils, which peaked at day 40 and 46 after irradiation. The percentage of the phagocyting cells in PB was constantly decreased since day 1 after irradiation. However, an increase in the PI was observed at days 40 and 46. Furthermore, we found an

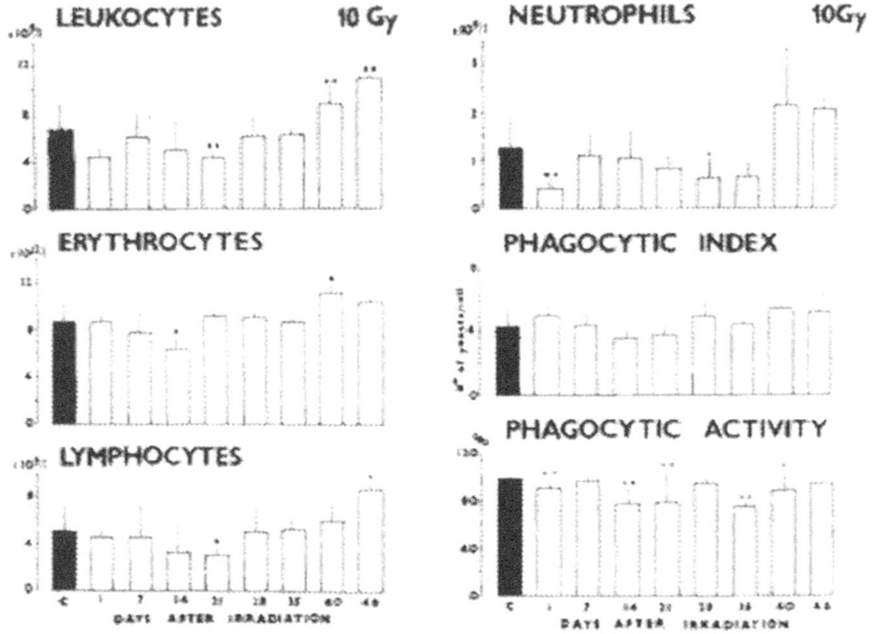

Figure 2. Influence of irradiation of the head (10 Gy) on peripheral blood parameters and their phagocytic function at various intervals after exposure. C-control animals (black columns). Other details as in *Figure 1*.

increase in PA of the peritoneal macrophages at day 1 after exposure. This declined significantly during the next 1-2 weeks and reached the control levels. However, the PI of the peritoneal macrophages increased already at day 7 after irradiation and remained constantly enhanced until day 46 (Figure 3).

4. Discusion

Under normal, healthy circumstances, the majority of leukocytes remains in the circulation. However, in the case of inflammation, leukocytes migrate through the blood vessels to the site of the inflammation. For this event, cell adhesion molecules are crucial. After irradiation the cytokine level changes [7] might induced an enhanced expression of cell adhesion molecules on leukocytes and their ligands. This could be one of the reasons why the leukocyte numbers decrease in the PB. An overproduction of tumor necrosis factor (TNF) by human peripheral blood mononuclear cells was found after their X-irradiation (9Gy) [5].

We have seen a decline in peritoneal mastocytes at day 1 and 46 after irradiation.It is an evidence for their degranulation. It was shown that the released mediators from mastocytes [1] activate macrophages with the following stimulation of TNF and interleukin 1 (IL-1) production [6].

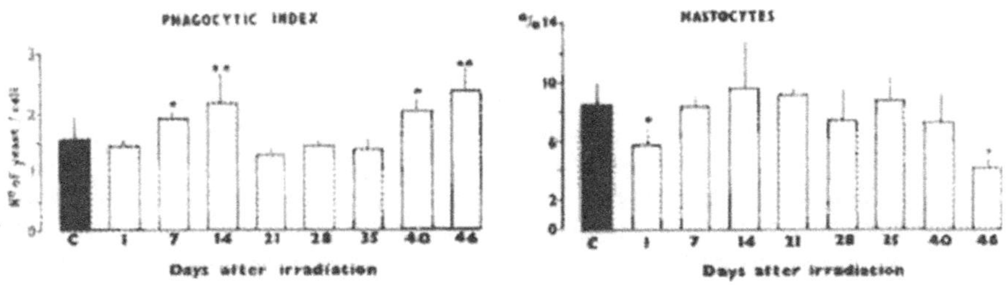

Figure 3. Influence of irradiation of the head (10 Gy) on peritoneal exudate cells phagocytosis (left side). Ordinate = Number of yeast cells per phagocyting cell. Right side - percentage of mastocytes from 500 PEC. Other details as in *Figure* 2

During the irradiation, animals are stressed by the immobilization (increase in body temperature, anoxia) and by the irradiation itself [3]. Indeed, increase in the amount of corticosterones was observed in the sera of animals after irradiation of the head [4]. After irradiation of the head (10 Gy), at day 21 postexposure, in spite of lymphopenia we have found killer cells (1% of lymphocytes). In our experiments, peripheral blood and peritoneal exudate cells from the head irradiated rats phagocytosed yeast cells more intesively than cells from intact controls at day 40 and 46 after exposure. It is reasonable to assume that the findings might be caused by late inflammation after single irradiation of the head.

However, other factors, for example heat shock proteins (HSP) may play an important role in our model. HSP are present in almost all cells in the body and their amount increases rapidly when cells become stressed. Whether changes of the immunohematological parameters we have observed in our model are of any clinical relevance, remains to be identified. If so, the abscopal effect in rats after irradiation of the head should not be neglected.

Acknowledgements. Project No.1-1236-94 was supported by the Grant Agency of Slovak Republic.

5. References

1. Czarnetzki, B.M., Grabbe, J., Kolde, G., Kruger-Krasagakes, S.,Welker, P., Zuberbier, T. (1995) Mast cells in the cytokine network: the what, where from and what for. *Experiemntal Dermatology* **4**, 221-226.
2. Edelson, P.J., Cohn, Z.A. (1976) Purification and Cultivation of Monocytes and Macrophages. In: B.R.Bloom, J.R.David (eds)., In Vitro Methods in Cell-Mediated and Tumor Immunity. Acad. Press New York, 333-340.
3. Kandasamy, S.B., Hunt, W.A., Mickley, G.A. (1988) Implication of prostaglandins and histamine H1 and H2 receptors in radiation induced temperature responses of rats, *Radiation Research* **114**, 42-53.
4. Kassayová , M., Ahlersová ,E., Pástorová , B., Ahlers, I. (1995) The early response of pineal N-acetyltransferase activity, melatonin and catecholamine levels in rats irradiated with gamma rays, *Physiol.Res.* **44**, 315-320.

5. Krivenko, S.I., Dryk, S.I.,Komarovskaja, M.E., Karkanitsa, L.V. (1992) Ionizing radiation increases TNF production by human peripheral blood mononuclear cells in vitro, *International Journal of Hematology* **55**, 127-130.
6. Lambert, L.E., Paulnock, D.M. (1987) Modulation of macrophage function by gamma-irradiation. Acquisition of the primed cell intermediate stage of the macrophage tumoricidal activation pathway, *Journal of Immunology* **139**, 2834-2841.
7. Neta, R., Oppenheim, J.J., Douches, S.D. (1988) Interdependence of the radioprotective effects of human recombinant interleukin alpha, tumor necrosis factor alpha, granulocyte-colony-stimulating factor, and murine recombinant granulocyte-colony-stimulating factor. *Journal of Immunology* **140**, 108-111
8. Reyners, H., Gianfelici de Reyners, E., Hopewell, J.W. (1993) Early vascular indicators of the radiation syndrome in the adult rat brain, 25th Annual Meeting of the European Society for Radiation Biology L 08:03

THE EFFECT OF SINGLE WHOLE-BODY IRRADIATION ON PINEAL FUNCTIONS

M. KASSAYOVÁ, E. AHLERSOVÁ, I. AHLERS

Institute of Animal Physiology, Faculty of Science, P.J.Šafárik University
Moyzesova 11, 041 67 Košice, Slovak Republic

Abstract

The influence of the single lethal dose of gamma rays on the functional parameters of the pineal gland was studied in experiments on Wistar rats. Male rats, adapted to an artificial light:dark regimen (LD) 12:12 h, were exposed to whole-body irradiation with gamma rays at a dose of 9.6 Gy. Irradiation, sham-irradiation and decapitation were carried out in darkness. The animals were analyzed 60 min, 1, 3 and 5 days after irradiation.

The ionizing radiation increased the activity of the rate-limiting enzyme of melatonin synthesis - serotonin N-acetyltransferase (NAT) - on day 3 post-irradiation, without any change in the remaining intervals investigated.The changes in pineal melatonin concentration occured in two phases: a decrease at 60 min was followed by an increase, significant on day 5 after exposure. The serum concentration of melatonin started to increase on day 3 post-irradiation and on day 5 was significantly higher in comparison with the control.

The results obtained in the present experiment and in our studies published previously suggest that there are two phases in the response of the rat pineal parameters to the single whole-body lethal irradiation: during the early phase up to 2 hours post-irradiation the concentration of pineal melatonin decreases, without any change of NAT. The reason may be the inbalance, induced by ionizing radiation, between N-acetylation and oxidation of serotonin. During the late phase after exposure the NAT activity increases, as well as the pineal and serum melatonin concentration, which is considered an adaptive response of the pineal gland to the radiation stress.

1. Introduction

The pineal gland and its principal hormone - melatonin (Mel) plays an important role in regulation of many physiological functions [16], particularly in timing of circadian and circaannual rhythms in mammals. Information about the effect of ionizing radiation on the pineal functions, which is scarce and incomplete [2,6], is of importance not only for basic research but especially in the clinical practice in patients treated by actinotherapy.

C. Baumstark-Khan et al. (eds.),
Fundamentals for the Assessment of Risks from Environmental Radiation, 401-405.

We aimed our studies at the effect of different doses of gamma rays on the key parameters of pineal function: pineal and serum Mel concentration and activity of the rate-limiting enzyme of Mel synthesis - serotonin N-acetyltransferase (NAT). Acute whole-body irradiation with a dose of 4.8 Gy of gamma rays did not change above mentioned pineal parameters (unpublished results), exposure to the lethal dose of 14.4 Gy caused the marked changes in NAT activity and Mel concentration depending on the period of the post-exposure examination [9,10]. Therefore, we performed the single whole-body irradiation of rats with the lower lethal dose of 9.6 Gy of gamma rays.

2. Material and Methods

Male Wistar rats weighing 100 g were adapted to an artificial light-dark (LD) regimen 12:12 h under standard vivarium conditions. Cool light of 150 lux intensity in each cage was automatically switched on at 0700 h. The rats had free access to water and food (ST pellets, Velaz Prague) until irradiation. The animals were single whole-body irradiated with the dose of 9.6 Gy of gamma rays from the ^{60}Co source (exposure rate 0.38 $Gy.min^{-1}$). After the irradiation and sham-irradiation the rats were pair-fed, i.e. the same amount of chow was given to the irradiated and control group, which was found in the preliminary experiment by the measurement of the day chow consumption in rats irradiated with the dose of 9.6 Gy. The group of intact rats was fed ad libitum till the analysis. Sixty min., 1, 3 and 5 days after irradiation the rats were killed by rapid decapitation between 2200 and 0130 h. Irradiation, sham-irradiation and decapitation were carried out in darkness under dim red light of less than 1 lux intensity. The pineal NAT activity was determined radioenzymatically according to Deguchi and Axelrod [5] as modified by Parfitt et al. [13]. The concentration of pineal and serum Mel was assessed radioimmunologically according to Charron et al. [4], the serum corticosterone concentration was determined fluorimetrically according to Guillemin et al. [8]. Each group consisted of 8 rats and results were evaluated by non-paired t-test.

3. Results

Activity of pineal NAT exhibited no marked changes at 60 min and on day 1 post-irradiation, on day 3 after exposure a significant increase was detected ($P<0.01$), followed by decline to the control values on day 5. The changes in pineal Mel concentration in irradiated animals occured in two phases: an initial decrease at 60 min ($P<0.05$) was replaced with a gradual increase, significant on day 5 post-exposure ($P<0.01$) in comparison with the control (Figure 1.).

The serum Mel concentration was on the increase from day 3 post-irradiation, and on day 5 was significantly higher ($P<0.01$) in comparison with the control. The serum corticosterone (CS) level increased significantly at 60 min ($P<0.01$) in irradiated rats; the second peak of values was recorded 3 days after exposure. After that, the concentration decreased down to the control values by day 5 (Figure 2.). Neither the absolute nor relative average weight of the pineal gland was affected by irradiation.

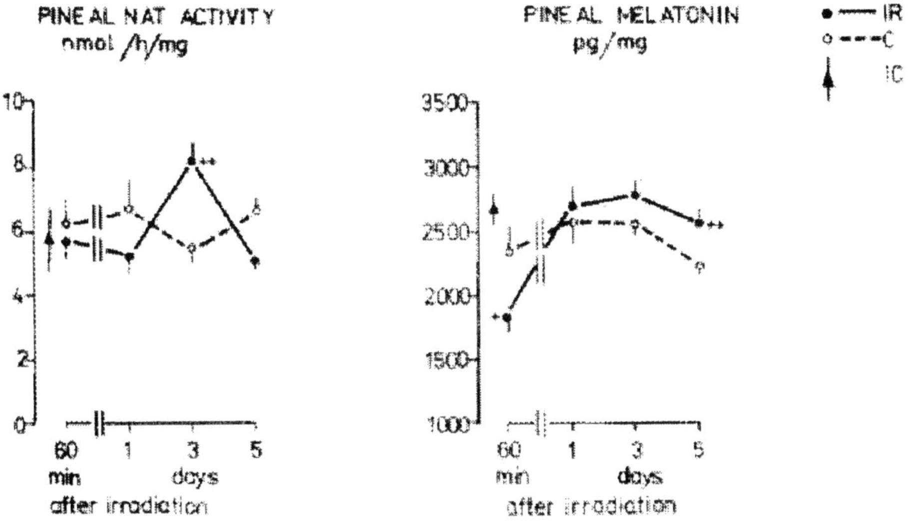

Figure 1. Pineal serotonin N-acetyltransferase activity (NAT) and pineal melatonin concentration in irradiated (IR) and sham-irradiated (C) rats after single whole-body irradiation with a dose of 9.6 Gy of gamma rays. Irradiated and sham-irradiated rats were analyzed in darkness from 60 min to 5 days after irradiation. Triangles represents values of intact rats (IC). Results are expressed as means ± S.E.M., n=8 (+ $P<0.05$ and ++ $P<0.01$, IR vs C).

4. Discussion

The single whole-body exposure to a dose of 9.6 Gy increased the activity of the rate-limiting enzyme in melatonin production - NAT- on day 3 post-irradiation, while no changes were observed in the other intervals investigated. The pineal gland responded to whole-body irradiation with a dose mentioned by the two-phase changes in the Mel concentration: immediately after exposure, at 60 min we recorded the decrease, during the next few days the increase of concentration, which was significant on day 5 after irradiation. Similar trend of changes we recorded in rats single whole-body irradiated with the higher lethal dose of 14.4 Gy of gamma rays [1,9,10]. Up to 2 h after exposure, pineal Mel concentration decreased, on days 3 and 4 we observed the signs of increased Mel synthesis, the rise of pineal NAT activity and serum Mel concentration. The changes in pineal Mel concentration that we observed in lethally irradiated rats, are in a good agreement to the changes in terminal enzyme activity of Mel synthesis, HIOMT, which decreased 1-24 h after exposure to a dose of 350 R of X-rays, and subsequently increased between days 4 and 19 [6].

Serum corticosterone (CS), as an indicator of radiation stress intensity, responded to acute whole-body irradiation with a dose of 9.6 Gy by the increase of concentration 60

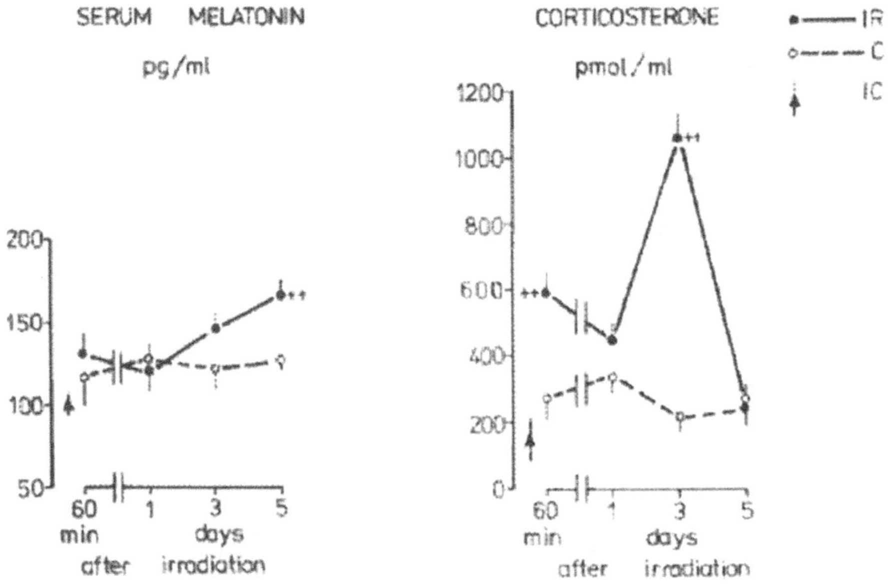

Figure 2. Serum melatonin and corticosterone concentration in irradiated and sham-irradiated rats after single whole-body irradiation with a dose of 9.6 Gy of gamma rays. Other details as in Figure 1.

min and 3 days after exposure. This finding is in agreement with the classical conception of Flemming et al. [7] about the two-phased response of CS to the radiation stressor. Glucocorticoids which are released to the circulation under the stress stimuli, supress the immune functions in humans and other animal species [3], whereas Mel increases them [11]. In our study we can see the increase of pineal and serum Mel concentration 5 days post-exposure, when CS level returned to the normal values.

Mechanisms of the effect of ionizing radiation on the pineal function have not been explained as yet. Under the physiological conditions, in the light part of day, when pineal melatonin synthesis is very low, in the pineal gland the oxidative deamination of serotonin, catalyzed by monoaminooxidase (MAO) dominate, with the high production of 5-hydroxyindole derivates of serotonin. In the dark part of day an alternative metabolic serotonin pathway - N-acetylation - is activated, with the high NAT activity and multiply increased Mel synthesis [12]. We assume that ionizing radiation induced the inbalance between N-acetylation and oxidation of serotonin. Whole-body irradiation increases the total activity of MAO in the hypothalamus of sheep [14] and in the brain of rabbits [15]. The increased MAO activity leads to the preference of serotonin metabolization to 5-hydroxyindole products, which may be a cause of the decline of Mel concentration immediately after irradiation, recorded in our studies. The increase of pineal Mel synthesis in the later period after whole-body exposure to gamma rays we are not able to explain yet. We may consider the adaptive response of the pineal gland to the radiation stressor which is presented by the increased production of the hormone with immunomodulating, antioxidative and antistressing properties.

Acknowledgement. The project No. 1/1236/94 was supported by the Grant Science Agency of the Slovak Republic.

5. References

1. Ahlersová, E., Kassayová, M., Ahlers, I. (1997), Serum melatonin, corticosterone and thyroid hormones in irradiated rats: Effect of exposure during various times of day and of light regimen. *Journal of Physiology and Pharmacology* **48**, 435-442.
2. Barfuss, D.W., Tait, G.R., Ellis, L.C. (1969), Melatonin synthesis in rat pineal gland inhibited by X- irradiation. *Proceedings of Society for Experimental Biology and Medicine* **131**, 35-36.
3. Black, P.H. (1994), Central nervous system - immune system interactions: psychoneuroendocrinology of stress and its immune consequences. *Antimicrobial Agents Chemotherapy* **38**, 1-6.
4. Charron, G., Souloumiac, I., Fournier, M.C., Canivenc, R. (1991), Pineal rhythm in N-acetyltransferase activity and melatonin in the male badger Meles meles L. under natural daylight: relationship with the photoperiod. *Journal of Pineal Research* **11**, 80-85.
5. Deguchi, T., Axelrod J. (1972), Sensitive assay for serotonin N-acetyltransferase activity in rat pineal. *Analytical Biochemistry* **50**, 174-179.
6. Ellis, Z., Barfuss, D.W., Van Kampen, K.R. (1970), Modification of rat testicular response to X-irradiation by various lighting regimens. I. The pineal gland and melatonin synthesis as a possible mediator of the response. *Life Sciences* **9**, 1011-1020.
7. Flemming, K., Geierhaas, B., Hemsing, W., Mannigel, U. (1968), Strahlenwirkungen und Nebennierenrinde. IV. Untersuchungen an Ratten über den ersten Corticosteroidanstieg nach Ganzkörper-Röntgenbestrahlung. *International Journal of Radiation Biology* **14**, 93-105.
8. Guillemin, R., Clayton, G.W., Smith, J.D., Lipscomb, H.S. (1958), Measurement of free corticosteroids in rat plasma. Physiological validation of a method. *Endocrinology* **63**, 349-355.
9. Kassayová, M., Ahlersová, E., Ahlers, I. (1993), Changes of pineal N- acetyltransferase activity in gamma-irradiated rats. *Physiological Research* **42**, 167-169.
10. Kassayová, M., Ahlersová, E., Pástorová, B., Ahlers, I. (1995), The early response of pineal N-acetyltransferase activity, melatonin and catecholamine levels in rats irradiated with gamma rays. *Physiological Research* **44**, 315-320.
11. Maestroni, G.J., Conti, A., Pierpaoli, W. (1986), Role of the pineal gland in immunity. Circadian synthesis and release of melatonin modulates the antibody response and antagonizes the immunosupressi-ve effect of corticosterone. *Journal of Neuroimmunology* **13**, 19-30.
12. Mefford, I.N., Chang, P., Klein, D.C., Namboodiri, M.A.A., Sugden, D., Barchas, J. (1983), Reciprocal day/night relationship between serotonin oxidation and N-acetylation products in the rat pineal gland. *Endocrinology* **113**, 1582-1586.
13. Parfitt, A., Weller, J.L., Klein, D.C. (1975), Blockade by ouabain or elevated potassium ion concentration of the adrenergic and adenosine cyclic 3,,5,-monophosphate-induced stimulation of pineal serotonin N-acetyltransferase activity. *Molecular Pharmacology* **11**, 241-255.
14. Pástorová, B., Arendarèík, J. (1988), The effect of protracted exposure to gamma radiation (6.7 Gy) on the activity of monoaminooxidase in the hypothalamus of ewes in the anoestral period (in Slovak). *Veterinary Medicine (Prague)* **33**, 735-740.
15. Pausescu, E., Popescu, M., Paun, M., Teodosiu, T. (1976), Dynamics of the changes in the tissue levels of cyclic AMP after 60Co gamma irradiation. *Strahlentherapie* **151**, 165-171.
16. Reiter, R.J. (1991), Melatonin: that ubiquitously acting pineal hormone. *News in Physiological Sciences* **6**, 223-227.

THE CHANGES OF PINEAL FUNCTION IN FRACTIONALLY IRRADIATED RATS

E. AHLERSOVÁ [1], B. PÁSTOROVÁ [2], M. KASSAYOVÁ [1], I. AHLERS [1], B. ŠMAJDA [1]

[1] *Institute of Animal Physiology, P.J. Šafárik University Moyzesova 11, 04167 Košice, Slovak Republic*
[2] *Department of Physiology, University of Veterinary Medicine Komenského 73, 04181 Košice, Slovak Republic*

Abstract

The effect of ionizing radiation on pineal parameters in mammals or human have not yet been explained sufficiently, we lack in particular the data in laboratory rats subjected to fractionated irradiation. Male Wistar rats adapted to an artificial lighting regimen 12 h light : 12 h darkness were fractionally whole-body irradiated with a dose of 2.4 Gy of gamma rays twice a week up to total doses of 4.8, 9.6 and 14.4 Gy. Irradiation and sham-irradiation were performed at late afternoon, the rats were sacrified between 24.00 and 01.00 h in darkness. The analyses of serotonin N-acetyltransferase (NAT) activity and melatonin (Mel) concentration in pineal gland and serum were done 6 h, 3 and 5 days after the exposure. Pineal monoamine oxidase (MAO) activity was determined only in rats irradiated up to dose of 14.4 Gy 1 and 3 days postexposure.

The activity of NAT in some intervals was significantly lower in rats irradiated with 4.8 and 9.6 Gy without changes of Mel concentration. Irradiation with a dose of 14.4 Gy decreased pineal NAT activity and the concentration of pineal and serum Mel 6 h resp. 3 days postexposure. The activity of MAO was increased significantly 3 days after irradiation.

The fractionated irradiation of the rats up to dose of 14.4 Gy caused a decrease in pineal Mel concentration what could be the consequence of preferred oxidative deamination of serotonin in comparison with the N-acetylation route of melatonin synthesis.

1. Introduction

The knowledge of the effects of ionizing radiation on pineal parameters is important not only in the basic research but especially in clinical practice, e.g. in actinotherapy. The older data about the effect of ionizing radiation on pineal parameters are sparse [4]. The response of the most important biochemical parameter in pineal gland, the melatonin (Mel) and the key enzyme of its synthesis - serotonin N-acetyltransferase (NAT) on

C. Baumstark-Khan et al. (eds.),
Fundamentals for the Assessment of Risks from Environmental Radiation, 407-410.

ionizing radiation was investigated in our laboratory. Fractionated whole-body irradiation was chosen as a model for a study of the chronic effect of ionizing radiation, closely related to clinical practice.

2. Material and methods

Male Wistar rats (Velaz, Prague) weighing 200 g were adapted to an artificial light/dark cycle of 12 : 12 h (light on at 0700) four weeks before irradiation. All rats received food and Water ad libitum. The rats were fractionally whole-body irradiated from a ^{60} Co source (exposure rate 0.29 Gy . min $^{-1}$). The doses of 2.4 Gy were applied twice a week up to total accumulated doses of 4.8, 9.6 and 14.4 Gy. Irradiation and sham-irradiation of the controls were performed between 3.30 and 6.30 p.m. After the exposure the rats had free access to food and water. Irradiated and control rats were quickly decapitated in dark, 6 h, 3 and 5 days after the exposure within 2400 - 0100 h using dim red light of less than 1 lux intensity. The following parameters were determined : pineal NAT activity (radioenzymatically according to Deguchi and Axelrod [3] in modification of Parfitt et al. [8]), the concentration of pineal and serum Mel (radioimmunologically according to Charron et al. [2]), pineal monoamine oxidase (MAO) activity (radioenzymatically according Wurtman and Axelrod [11]). The activity of MAO was analyzed only on the 1st and 3rd day in rats irradiated with total dose of 14.4 Gy.

3. Results

The concentrations of pineal and serum Mel were not changed after irradiation up to total doses of 4.8 and 9.6 Gy. After total dose of 14.4 Gy the concentration of Mel in pineal gland decreased significantly at 6 h and 3 days (Figure 1), the serum level of Mel decreased 6 h postexposure. In rats irradiated with the accumulated doses of 4.8 and 9.6 Gy a significant decrease in pineal NAT activity was observed 5 days (4.8 Gy) and 6 h postexposure, in rats irradiated up to dose of 14.4 Gy NAT activity decreased significantly at 6 h , a decrease at day 3 postexposure is of borderline significance (Figure 2).

The activity of pineal MAO exhibited no marked changes at 1st day but at 3rd day after irradiation up to total dose of 14.4 Gy about twofold increase was noted in comparison with the control group (Figure 3).

4. Discussion

Mel is synthetized from aminoacid tryptophan through serotonin which is converted to N-acetylserotonin. This reaction is catalyzed by the enzyme serotonin NAT. Transformation of N-acetylserotonin to melatonin is catalyzed by the enzyme hydroxyindole-O-methyl-transferase (HIOMT). This metabolic pathway is activated in the dark part of day [7]. Other possibility of serotonin transformation is its oxidative deamination to 5-hydroxyindole derivates catalyzed by MAO.

The single whole-body irradiation with a dose of 14.4 Gy of gamma rays in our previous experiments decreased the concentration of pineal Mel 30 to 120 min postexposure without changes in the activity of NAT. Later (3-4 days) after exposure the signs of increased synthesis of pineal Mel were observed:pineal NAT activity and concentration of Mel in the serum increased significantly [5 ,6, 1]. The changes of pineal and serum concentration of Mel in our experiments are in a good agreement with the changes of HIOMT pineal activity in X-irradiated rats [4].

The changes of all followed parameters were notet in the present experiment only at the highest accumulated dose (14.4 Gy). The decrease in NAT activity, in the concentration of Mel and the increase in MAO activity up to 3 days after irradiation indicate a drop in Mel synthesis in consequence of changes in serotonin metabolism after irradiation. Ionizing radiation increased the MAO activity in the brain of the rabbit [10] and of sheep [9]. The increased MAO activity resulted in preference of oxidative deamination of serotonin in comparison with N-acetylation pathway of Mel synthesis. This could be a reason for the observed fall in pineal Mel concentration in irradiated rats [6]. For the

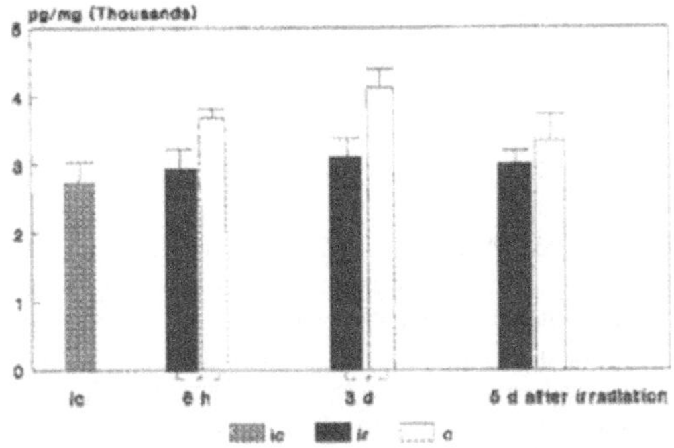

Figure 1. Pineal melatonin concentration in rats 6 h to 5 days after fractionated irradiation-IR with 2 x 2.4 Gy weekly up to total dose 14.4 Gy of gamma rays and after sham-irradiation-C. IC - values of intact rats. Values are given as the mean ± S.E.M., + ($P < 0.05$), ++ ($P < 0.01$).

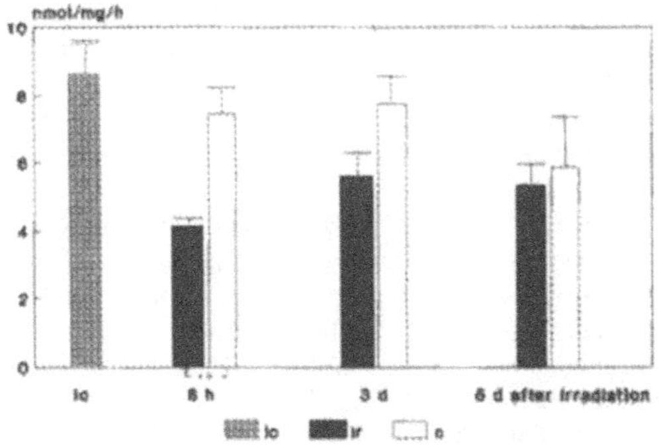

Figure 2. Pineal N-acetyltransferase (NAT) activity in fractionally irradiated and sham-irradiated rats. Other details as in figure 1.

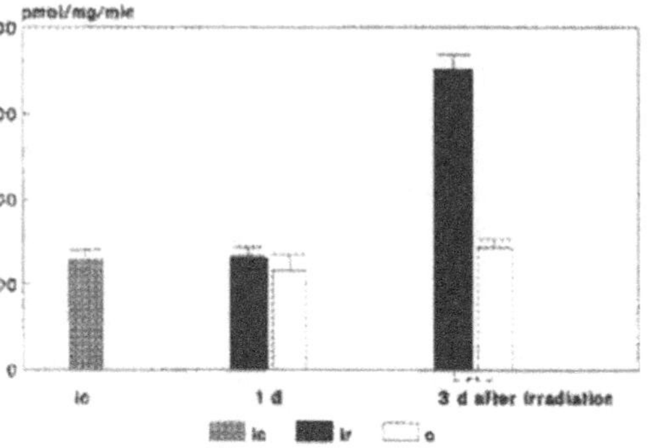

Figure 3. Pineal monoamine oxidase (MAO) activity in rats 1 and 3 days after fractionated irradiation. Other details as in figure 1.

final approvement of this variant the determination of 5-hydroxyindole derivates directly in the pineal gland of irradiated rats would be required.

5. References

1. Ahlersová, E., Kassayová, M., Ahlers, I. (1997), Serum melatonin, corticosterone and thyroid hormones in irradiated rats : effect of exposure during various times of day and of light regimen. *Journal of Physiology and Pharmacology* **48**, 435-442
2. Charron, G., Souloumiac, I., Fournier, M.C., Canivenc, R. (1991), Pineal rhythm in N-acetyltransferase activity and melattonin in the male badger Meles meles L. under natural daylight:relationship with the photoperiod. *Journal of Pineal Research* **11**, 80-85
3. Deguchi, T., Axelrod, J. (1972), Sensitive assay for serotonin N-acetyltransferase activity in rat pineal. *Analytical Biochemistry* **50**, 174-179
4. Ellis, L.C., Barfuss, D.W.Van Kampen, K.R. (1970), Modification of rat testicular response to X-irradiation by various lighting regimens:I.The pineal gland and melatonin synthesis as possible mediator of the response. *Life Sciences* **9**, 1011-1020
5. Kassayová, M., Ahlersová, E., Ahlers, I. (1993), Changes of pineal N-acetyltransferase activity in gamma irradiated rats. *Physiological Research* **42**, 167-169
6. Kassayová, M., Ahlersová, E., Pástorová, B., Ahlers, I. (1995), The early response of pineal N-acetyltransferase activity, melatonin and catecholamine levels in rats irradiated with gamma rays. *Physiological Research* **44**, 315-320
7. Mefford, I.N., Chang, P., Klein, D.C., Namboodiri, M.A., Sugden, D., Barchas, J. (1983), Reciprocal day/night relationship between serotonin oxidation and N-acetylation products in the rat pineal gland. *Endocrinology* **113**, 1582-1586
8. Parfit, A., Weller, J.L., Klein, D.C. (1975), Blockade by ouabain or elevated potassium ion concentration of the adrenergic and adenosine cyclic 3′,5′-monophosphate-induced stimulation of pineal serotonin N-acetyltransferase activity. *Molecular Pharmacology* **11**, 241-255
9. Pástorová, B., Arendarèík, J. (1988), The effect of protracted exposure to gamma radiation (6.7 Gy) on the activity of monoamine oxidase in the hypothalamus of ewes in the anestral period (In Slovak). *Veterinary Medicine (Praha)* **33**, 735-740
10. Pausescu, E., Popescu, M., Paun, M., Teodosiu, T. (1976), Dynamics of the changes in the tissue levels of cyclic AMP after ^{60}Co gamma irradiation. *Strahlentherapie* **151**, 165-171
11. Wurtman, R.J., Axelrod, J. (1963), A sensitive and specific assay for the estimation of monoamine oxidase. *Biochemistry and Pharmacology* **62**, 1439-1440

PHARMACOLOGICAL RADIATION PROTECTION

M. POSPÍŠIL

Institute of Biophysics, Academy of Sciences of the Czech Republic
Královopolská 135, 612 65 Brno, Czech Republic

Abstract

Historically, the development of radioprotective agents has been dominated by the investigation of effects of chemical compounds, particularly aminothiols and phosphorothioates. In the last 15 years, there has been an increasing interest in biological modifiers of radiation injury. While chemical protectors seem to influence the development of the initial radiation damage, biological protectors may modulate the repair, recovery and repopulation of critical cell elements. Mechanisms of action, efficacy and practical utilization of the chemical and biological radioprotectors are reviewed.

1. Introduction

The ultimate aim of many medical actions is to protect people against various adversely acting factors. Attempts to protect an organism against radiation damage include also modes of treatment which utilize the principles of pharmacological and biological intervention. The pharmacological approach to radiation protection accomplishes this task by the search for various chemicals, drugs and biological products capable of reducing the radiation damage. Taking into account the biological complexity, the radioprotective action can be realized on the chemical, macromolecular, cellular and organismal levels. Its nature may be either more chemical, modulating the basic radiochemical events and decreasing the extent of the initial damage, or more biological, acting via regulatory systems and enhancing the repair and recovery potential of cells and tissues.

2. Chemical Radioprotection

In the late forties Patt et al. [1] showed that the sulfhydryl-containing amino acid cysteine, injected several minutes before irradiation to mice, can reduce the lethality induced by ionizing radiation. Two years later Bacq et al. [2] demonstrated that cysteamine (beta-mercaptoethylamine) was five times as active as cysteine in equimolecular quantities. After these discoveries an explosion of studies occurred, which tested various classes of substances for their radioprotective ability (for reviews see [3-6]. In 1959 the US Army initiated its Antiradiation Drug Development Program. The most

C. Baumstark-Khan et al. (eds.),
Fundamentals for the Assessment of Risks from Environmental Radiation, 411-420.

significant development emerging from that program was the synthesis and testing of the phosphorothioate class of radioprotective compounds. It is this class of agents from which the current leading drug was developed. It is chemically S-2-(3-aminopropylamino)-ethyl phosphorothioic acid and has been given a Walter Reed code number WR-2721. It is also known under the name „ethiofos". WR-2721 is a prodrug, which is converted to a free thiol compound, ethanethiol (WR-1065), by membrane-bound alkaline phosphatase [7]. The mechanisms of radioprotection by thiols are not clearly understood. The most generally accepted theory is based on free radicals scavenging by thiols. Thiols, thiolate anions and disulfides react rapidly with $OH^{\cdot}$ and thus protect cells before this radical can react with DNA. Another possibility involves hydrogen ion donation. It is hypothesized that an H atom, $H^{\cdot}$, or an electron are transferred from the thiol to a DNA radical induced by radiation. The third possibility is that the presence of the protector may induce cellular hypoxia, which could be involved in the protection. Thiols may undergo autooxidation, resulting in the depletion of oxygen and a decrease in O_2-fixation reactions. In addition, in vivo experiments demonstrated that these radioprotectors can induce transient depression of respiration and hypotensive effects which may further decrease the oxygenation of tissues. Nonradical processes may be also involved. Brown [8] stated that a delay in the first division of cells after irradiation induced by thiols may allow greater time for more extensive repair or allow a greater fidelity of repair. Grdina and Nagy [9] exposed V79 cells to WR-1065 and observed a significant alteration of the cell cycle progression with cells accumulated in S-phase. Because cell division is an important step in the fixation of mutational and transformational events, agents that can prolong the time required for cell division might be expected to be effective as antimutagens or anticarcinogens.

The highest protective effectiveness of thiols is obtained 15-30 min after their injection. Protection factor as high as 2.7 against 30-day lethality in mice has been achieved [10]. Experimental evidence indicates that there exist variations in the degree of protection among different tissues [11, 12]. The highest protection was observed in bone marrow. Ethiofos is able to protect not only early responding tissues (bone marrow, gastrointestinal tract), but also the late responding ones (e.g., lung). Ethiofos does not cross the blood-brain barrier and for this reason does not protect the central nervous system. Experimental findings showing that ethiofos protects normal tissues more than tumors might be important in radiotherapy of tumors, especially in situations where damage to normal tissues is dose-limiting. The reason for the preferential radioprotection of normal tissues can be differences in uptake of the protector, differences in the oxygenation of tissues (WR-2721 offers relatively poor protection to hypoxic cells), etc. However, the application of these findings in strategies of tumor radiotherapy raises several questions. These include the effectiveness of radioprotection under conditions of fractionated irradiation, evidence of cumulated toxicity of the repeatedly administered protector, lower effectiveness of chemical radioprotection when applied before smaller radiation doses, etc.

It is interesting to note that sulfhydryl compounds could be useful even in tumor chemotherapy. Yuhas [13] showed that ethiofos reduced toxicity associated with alkylating agent therapy. Glover et al. [14] have reported on clinical trials, suggesting that ethiofos may potentiate the antitumor activity of cisplatin in metastatic melanoma.

Encouraging are data suggesting that sulfhydryl class of agents can modulate damage associated with nonlethal radiation events, i.e. mutagenesis, transformation and cancerogenesis. Maisin et al. [15] have shown that these agents markedly reduced the incidence of radiation-induced thymic lymphoma in mice. Milas et al. [16] have found a reduced frequency of tumor induction in irradiated mice protected with phosphorothioates. The free thiol WR-1065 induced antimutagenic and antitransformation activity in cells in vitro [17]. Antimutagenic effects were observed even when the protective agent was administered up to 3 h after irradiation. These observations illustrate the promising possibility to utilize the treatment with thiols even after accidental exposure to radiation. The problem is opened for further research.

A serious disadvantage of sulfhydryl compounds, including ethiofos, is their relatively high toxicity. Thiols can induce profound changes in cell metabolism, including effects on cell progression, DNA synthesis, and protein synthesis. The combined effects are often referred to as biochemical shock. Protective doses of thiols induce hypothermia and performance decrement in rodents. Studies of ethiofos effects in humans have shown limiting nausea and vomiting, or fall in blood pressure when i.v. doses have approached levels equipotent to those demonstrated to be radioprotective in mice. As stated by Maisin and Bacq [18], the goal of nontoxic chemical radioprotection appears difficult to be attained because protective action and toxicity seem to be intimately linked in all organisms, particularly in mammals.

3. Biological Radioprotection

The term biological radioprotection was coined to include the action of various agents of biological or synthetic origin mediating the radioprotective effects via physiological modulation of cellular and intercellular regulatory functions. Such agents were recognized already earlier, but renewed interest in this area appeared in the beginning of 80's and coincided with the increased evidence that toxicity of chemical protection might limit its usefulness. The first and longest-studied agents of this class were endotoxins, i.e. lipopolysaccharide complexes associated with the cell wall of gram-negative bacteria, known as *immunomodulators.* The discovery of their radioprotective action dates back for more than 40 years. As reviewed by Ainsworth [19], (a) the endotoxin protection phenomenon works in several animal species; (b) in studies on mice (LD50/30) the maximum protection effect is produced by injecting endotoxin 1 or 2 days before irradiation, the protection factor is about 1.1-1.3; (c) even an early postirradiation treatment with endotoxin is slightly effective; (d) multiple injections of endotoxin were either not effective or less effective than single doses (development of endotoxin tolerance due to its enhanced clearance?); (e) under conditions of endotoxin protection an accelerated recovery of hematopoiesis, without modification of the radiosensitivity of hematopoietic stem cells in terms of D_o was observed.

While the radioprotective (and also toxic) properties of endotoxin are associated with its lipid component [20], also polysaccharides from various biological sources, exhibiting immunomodulating properties, were successively tested for their radioprotective efficacy [21-24]. Among them, the greatest attention was given to glucan, a beta-1-3-polyglucose isolated from the cell wall of Saccharomyces cerevisiae. Particulate, as well as the less toxic soluble forms of glucan have shown similar effects like endotoxins: (a) in mice optimal protection was observed with glucan injected 20-24 h prior to irradiation, the protection factor being about 1.2 or even more; (b) glucan induces slight positive effects even after the early postirradiation administration; (c) in sublethally irradiated mice glucan accelerates the repopulation of both pluripotent and committed hematopoietic stem cells without modifying their radiosensitivity based on D_o [21, 22, 25-28]. In recent years several other immunomodulating agents, like, e.g., aqueous extract from chlorococcal algae [29]), bacterial extract broncho-vaxom [30], synthetic muramyl tripeptide [31] have been reported to be radioprotective in mice. The most fundamental problem is the mechanism of the radioprotective action of immunomodulators. Generally, these agents enhance host resistance to a variety of biological injuries through mechanisms involving macrophage activation, which results in cytokine production. Thus, it is logical that the current research is directed toward defining the role of the isolated cytokines and other intercellular regulatory factors in the radioprotective effect.

Cytokines are hormone-like proteins or glykoproteins synthetized and released primarily by the cells of the immune system, namely macrophages and lymphocytes. These factors represent biochemical signals, which affect essentially three parameters: the growth, differentiation, and effector functions of various cells. They act either in strictly localized or in a generalized manner, and play an essential role in the defense and recovery from various forms of injury. Neta with her group [32, 33] were first to demonstrate that recombinant murine and human interleukin-1 (IL-1) are radioprotective in mice. Maximal radioprotection in terms of survival was obtained when IL-1 was administered 20 h before irradiation; protection factor of 1.2-1.25, analogous to that achieved by immunomodulators, was obtained. Granulocyte-macrophage colony-stimulating factor, or interleukine-2 did not induce radioprotection under comparable conditions. However, another inflammatory cytokine, tumor-necrosis factor (TNF), induced radioprotection, even though doses of TNF 40-100-fold higher than those of IL-1 were required. Combined administration of IL-1 and TNF resulted in additive radioprotection. The use of neutralizing antibodies to cytokines indicated that cooperation of both cytokines is necessary [34]. Neta et al. [35] proposed an explanation of the protective action of IL-1. They found that this cytokine induced bone marrow hematopoietic cells to enter the cycle. Thus, the protective effect may be related to the large numbers of bone marrow cells in the late S phase which is considered to be radioresistant. Another explanation of the IL-1 action might be the induction of the antioxidant enzyme manganese-superoxide dismutase [36]. Furthermore, it has been shown that the stem cell factor is also important in the radioprotection and synergizes with IL-1 [37]. Interleukin-12, a potent stimulator of hematopoietic progenitor cells was found to act as a radioprotector against the lethal hematopoietic syndrome in mice, but also as a sensitizer of the intestinal tract to radiation effects [38]. Interestingly, Potten [39] demonstrated that intestinal clonogenic stem cells can be protected from radiation damage by interleukin-11. Recently, acidic fibroblast growth factor was found to act as a radioprotector increasing survival and hematopoietic recovery in mice [40]. While granulocyte colony-stimulating factor does not seem to be radioprotective in single physiological, i.e. granulopoiesis - enhancing doses, about 20-fold higher dose of this factor, administered 3 or 24 h before irradiation, has been protective in mice [41].

Attention was also paid to another important class of tissue mediators, namely *eicosanoids,* metabolites of arachidonic acid. Prostaglandins (PGE_2), products of the cyclooxygenase metabolic pathway, were shown to induce radioprotection in murine intestinal and hematopoietic tissues when given shortly before irradiation [42, 43]. Similarly, leukotriene C_4, formed via the lipoxygenase pathway, was found to be an effective radioprotective agent for hematopoietic stem cells in vivo [44]. The mechanisms of these actions are not well understood, but may include changes in cell metabolism, as well as hypoxia-mediated effects induced by high vasoactive doses of exogenous prostaglandins or leukotrienes. Another aspect was introduced by experiments using the inhibition of eicosanoid production by means of pharmacological agents known as non-steroidal anti-inflammatory drugs (NSAIDs). It was shown that not only the postirradiation treatment with indomethacin [45], but also the preirradiation administration of NSAIDs inhibiting prostaglandin production, i.e. indomethacin, diclofenac, flurbiprofen, or nordihydroquaiaretic acid, enhanced the recovery of hematopoietic functions in mice [46-51]. Esculetin, an inhibitor of leukotriene production, did not induce protective effects [48]. The mechanism of the radioprotective action of NSAIDs is probably linked with the proposed role of prostaglandins in the negative feedback regulation of myelopoiesis. The pharmacological inhibition of the prostaglandin action could thus lead to the preponderance of the positive control and an enhancement of hematopoietic recovery without modulating the intrinsic radiosensitivity of hematopoietic cells. The possibility of using NSAIDs in combination with radiotherapy of tumors was investigated in experiments of Furuta et al. [52], clinical aspects were reviewed by Michalowski [53]. However, the potential utilization of NSAIDs in protection of hematopoiesis seems to be limited by their undesirable side effects on the gastrointestinal tract. Newly developed inhibitors of prostaglandin production, devoid of these side effects, could find use in radioprotection.

Over the past few years, our group re-evaluated and newly conceived the older observations of Langendorff and Langendorff [54] on the radioprotective action of *adenine nucleotides.* The radioprotective effectiveness of these compounds is not dependent on the degree of their phosphorylation and the effects induced are comparable to those of *adenosine.* Thus, the decisive factor seems to be represented by adenosine, which is released extracellularly by ectonucleotidase activity. At present the physiological role of adenosine in experimental pharmacology and medicine is widely explored [55, 56]. It has been ascertained that extracellular adenosine may interact with specific receptors on the cell surface, modulate adenylate cyclase activity and thus modify the cAMP formation. This signaling pathway initiates various physiological actions of adenosine. Among them, at least two effects can be of importance in radioprotection, namely systemic hypoxia resulting from the cardiovascular action of adenosine (bradycardia-vasodilation-hypotension), and stimulation of growth in the cell renewal systems, including hematopoiesis. In order to induce an enhanced adenosine receptor signaling, a combination of two drugs was used in our experiments in mice: dipyridamole, a drug inhibiting the cellular uptake of adenosine and thus increasing its extracellular level, and adenosine monophosphate, which serves as an adenosine prodrug. Such a combination was found to exert the best radioprotective effects [57]. The role of the proposed mechanisms, i.e. hypoxia and stimulation of hematopoietic recovery, could be deduced from experiments using different preirradiation timing of the combined drug administration (30-60 min) and monitoring the hypoxic and hematostimulatory effects. Both mechanisms seemed to be equipotent in inducing radioprotection (protection factor about 1.1) [58] and to act independently [59]. Protective effects of the drug combination were observed not only in mice subjected to single radiation exposure, but also to fractionated radiation regimens [58, 60]. The protection was found to be effective also under conditions of low (1 Gy) radiation doses [61, 62]. Furthermore, it was found that adenosine effects resulting in stimulation of hematopoiesis are more effective under conditions of fractionated irradiation [63]. Concerning the hematopoiesis-stimulating effect of extracellular adenosine it was found that adenosine signaling synergizes with the stimulatory effects of granulocyte colony-stimulating factor, and can thus play a priming role in the action of specific growth signals [64]. The exploration of adenosine receptor signaling deserves attention with regard to its clinical applicability. It is worth noting that pharmacologists and pharmaceutical industry have been searching for new potent adenosine receptor agonists and that this area of research may open new perspectives also in the field of radioprotection.

As shown, the approaches of biological radioprotection proved their usefulness. Further progress in this field is dependent on the better understanding of regulatory mechanisms which participate in the physiological control of cell renewal systems. When compared with chemical protection, biological protection is less effective, but seems to be more suitable for clinical use. However, more studies are needed to define the practical possibilities of these modes of radioprotection. It should be noted that the action of biological response modifiers and biologicals on the mutagenic and carcinogenic effects of radiation has not yet been discussed in scientific forums.

4. Regimens Based on Combined Agents

The evidence for existence of different radioprotective mechanisms offered the possibility to combine differently acting agents to obtain an enhanced radioprotective efficacy. Furthermore, it could be hypothesized that by such an approach it might be possible to reduce the doses of individual components of the combination, to decrease thus their toxicity and to obtain a relatively good protection factor. Already earlier it has been found that endotoxin acts additively with an aminothiol protector [19]. It was shown that sodium selenite and vitamin E enhance the protective effectiveness of sulfhydryl compounds [65, 66]. Patchen et al. [67] combined ethiofos with early postirradiation administration of glucan and obtained and enhanced hematopoietic recovery and survival of mice. Similar effects were observed when combining peroral administration of cystamine with i.p. injection of glucan [68], and ethiofos with an immunomodulator broncho-vaxom [69]. Also, drugs inhibiting prostaglandin production synergized with thiol protection [49, 70-71]. Experiments of our group demonstrated also the possibility of combining various biological protectors. It was shown that the combined administration of glucan and diclofenac enhanced the hematopoietic reconstitution in singly [72] and repeatedly [73] irradiated mice. Similarly, indomethacin was found to enhance radioprotective effects of the immunomodulator broncho-vaxom [74] and of the synthetic muramyl tripeptide [75]. The use of combinations of various agents seems to be a valid and promising concept whose possibilities have not yet been exhausted.

The literature on pharmacological radiation protection is extensive and it is not possible to be fully covered in this review. Only the most discussed problems are reviewed here. More data can be found in monographs and reviews of Bacq [2], Foye [76], Sweeney [4], Kuna [77], Livesey and Reed [5], Grdina and Sigdestad [6] and Murray and McBride [78].

Acknowledgement. Work in the author's laboratory was supported in part by the Grant Agency of the Czech Republic (Grans No. 306/93/0051 and 306/96/1721).

5. References

1. Patt HM, Tyree EB, Straube RL, Smith DE (1949) Cysteine protection against X irradiation. *Science* **110**, 213-214
2. Bacq ZM, Herve A, Lecomte J, Fisher P, Blavier J, Dechamps G, Le Bihan H, Rayet P (1951) Protection contre le rayonnement X par la beta-mercaptoethylamine. *Arch Intern Physiol* **5**, 442-447
3. Bacq ZM (1965) Chemical Protection against Ionizing Radiation. Charles C Thomas, Springfield, Ill
4. Sweeney TR (1979) A Survey of Compounds from the Antiradiation Drug Development Program of the U.S. Army Medical Research and Development Command. Walter Reed Inst Res, Washington, DC
5. Livesey JC, Reed DJ (1987) Chemical protection against ionizing radiation, in Advances in Radiation Biology, vol 13, Academic Press, Inc, pp 285-353
6. Grdina DJ, Sigdestad CP (1989) Radiation protectors: The unexpected benefits. *Drug Metabol Rev* **20**, 13-42
7. Purdie JW (1979) A comparative study of the radioprotective effects of cysteamine, WR-2721 and WR-1065 in cultured human cells. *Radiat Res* **77**, 303-311
8. Brown PE (1967) Mechanism of action of aminothiol radioprotectors. Nature 213, 363-364
9. Grdina DJ, Nagy B (1986) The effect of S-[(aminopropyl)amino]ethanthiol (WR-1065) on radiation-induced DNA damage and repair and cell progression in V79 cells. Br J Cancer 54, 933-941
10. Yuhas JM, Storer JB (1969) Chemoprotection against three modes of radiation death in the mouse. *Int J Radiat Biol* **15**, 233-237
11. Milas L, Murray D, Brock WA, Meyn RE (1988) Radioprotectors in tumor radiotherapy: factors and settings determining therapeutic ratio. *Pharmac Ther* **39**, 179-187
12. Rasey JS, Spence AM, Badger CC, Krohn KA, Vera DM, Livesey JC (1988) Specific protection of different normal tissues. *Pharmac Ther* **39**, 33-43
13. Yuhas JM (1979) Differential protection of normal and malignant tissue against the cytotoxic effects of mechlorethamine. *Cancer Treat Rep* **63**, 971-976
14. Glover D, Glick JH, Weiler C, Fox K, Guerry DP (1987) WR-2721 and high-dose cisplatin: an active combination in the treatment of metastatic melanoma. *J Clin Oncol* **5**, 574-578
15. Maisin JR, Decleve A, Gerber GB, Mattelin G, Lambiet-Collier M (1978) Chemical protection against the long-term effects of a single whole-body exposure of mice to ionizing radiation. II. Causes of death. *Radiat Res* **74**, 415-435
16. Milas L, Hunter N, Stephens CL, Peters LJ (1984) Inhibition of radiation carcinogenesis by S-2-(3-aminopropylamino) ethyl phosphorothioic acid. *Cancer Res* **44**, 5567-5569
17. Grdina DJ, Nagy B, Sigdestad CP (1988) Radioprotectors in treatment therapy to reduce risk in secondary tumor induction. *Pharmac Ther* **39**, 21-25
18. Maisin JR, Bacq ZM (1975) Toxicity, International Encyclopedia of Pharmacology and Therapeutics: Sulfur-Containing Radioprotective Agents, ZM Bacq (ed), Pergamon, NY, pp 15-39
19. Ainsworth EJ (1988) From endotoxins to newer immunomodulators: Survival-promoting effects of microbial polysaccharide complexes in irradiated animals. *Pharmac Ther* **39**, 223-241

20. Langendorff H, Langendorff M, Weckesser J, Steinbach KH (1971) Die Wirkung einzelner Strukturkomponenten von bakteriellen Lipopolysacchariden auf die Strahlenempfindlichkeit der Maus. *Strahlentherapie* **141**, 457-463
21. Pospíšil M, Jarý J, Netíková J, Marek M (1982) Glucan-induced enhancement of hemopoietic recovery in gamma-irradiated mice. *Experientia* **38**, 1232-1233
22. Patchen ML, DiLuzio NR, Jacques P, MacVittie TJ (1984) Soluble polyglycans enhance recovery from cobalt-60-induced hemopoietic injury. *J Biol Response Modif* **3**, 627-633
23. Maisin JR, Topalova S, Kondi-Tamba A, Mattelin G (1988) Radioprotection by polysaccharides. *Pharmac Ther* **39**, 255-259
24. Chirigos MA, Patchen ML (1988) Survey of newer biological response modifiers for possible use in radioprotection. *Pharmac Ther 39*, 243-246
25. Patchen ML, D'Alesandro MM, Brook I, Blakely WF, MacVittie TJ (1987) Glucan: Mechanisms involved in its radioprotective effect. *J Leukoc Biol* **42**, 95-105
26. Patchen ML, D'Alesandro MM, Chirigos MA, Weiss JF (1988) Radioprotection by biological response modifiers alone and in combination with WR-2721. *Pharmac Ther* **39**, 247-254
27. Pospíšil M, Šandula J, Pipalová I, Hofer M, Viklická Š (1991) Hemopoiesis stimulating and radioprotective effects of carboxymethylglucan. *Physiol Res* **40**, 377-380
28. Hofer M, Pospíšil M, Pipalová I, Holá J, Šandula J (1995) Haemopoiesis-enhancing effects of repeatedly administered carboxymethylglucan in mice exposed to fractionated irradiation. *Fol Biol (Praha)* **41**, 249-256
29. Vacek A, Rotkovská D, Bartoníèková A (1990) Radioprotection of hemopoiesis conferred by aqueous extract from chlorococcal algae (Ivastimul) administered to mice before irradiation. *Exp Hematol* **18**, 234-237
30. Fedoroèko P, Brezáni P, Macková NO (1992) Radioprotection of mice by the bacterial extract Broncho-vaxom: haemopoietic stem cells and survival enhancement. *Int J Radiat Biol* **61**, 511-518
31. Fedoroèko P (1994) Liposomal muramyl tripeptide phosphatidylethanolamine (MTP-PE) promotes haemopoietic recovery in irradiated mice. *Int J Radiat Biol* **65**, 465-475
32. Neta R, Douches S, Oppenheim JJ (1986) Interleukin 1 is a radioprotector. *J Immunol* **136**, 2483-2485
33. Neta R (1988) Role of cytokines in radioprotection. *Pharmac Ther* **39**, 261-266
34. Neta R, Oppenheim JJ, Schreiber RD, Chizzonite R, Ledney GD, MacVittie TJ (1991) Role of cytokines (interleukin 1, tumor necrosis factor, and transforming growth factor β) in natural and lipopolysaccharide-enhanced radioresistance. *J Exp Med* **173**, 1177-1182
35. Neta R, Sztein MB, Oppenheim JJ, Gillis S, Douches SD (1987) The in vivo effects of interleukin-1. I. Bone marrow cells are induced to cycle after administration of interleukin-1. *J Immunol* **139**, 1861-1866
36. Eastgate J, Moreb J, Nick HS, Suzuki K, Taniguchi N, Zucali JR (1993) A role for manganese superoxide dismutase in radioprotection of hematopoietic stem cells by interleukin-1. *Blood* **81**, 639-646
37. Neta R, Oppenheim JJ, Wang J-M, Snapper CM, Moorman MA, Dubois CM (1994) Synergy of IL-1 and stem cell factor in radioprotection of mice is associated with IL-1 up-regulation of mRNA and protein expression for c-kit on bone marrow cells. *J Immunol* **153**, 1536-1543
38. Neta R, Stiefel SM, Finkelman F, Herrmann S, Ali N (1994b) IL-12 protects bone marrow from and sensitizes intestinal tract to ionizing radiation. *J Immunol* **153**, 4230-4237
39. Potten CS (1996) Protection of the small intestinal clonogenic stem cells from radiation-induced damage by pretreatment with interleukin 11 also increases murine survival time. *Stem Cells* **14**, 452-459
40. Ding I, Wu T, Matsubara H, Magae J, Shou M, Cook J, Okunieff P (1997) Acidic fibroblast growth factor (FGF_1) increases survival and haematopoietic recovery in total body irradiated C_3H/HeNCr mice. *Cytokine* **9**, 59-65
41. Uckun FM, Souza L, Waddick KG, Wick M, Song CW (1990) In vivo radioprotective effects of recombinant human granulocyte colony-stimulating factor in lethally irradiated mice. *Blood* **75**, 638-645

42. Hanson WR, Ainsworth EJ (1985) 16,16-dimethyl prostaglandin E_2 induces radioprotection in murine intestinal and hematopoietic stem cells. *Radiat Res* **103**, 196-203
43. Walden TL, Patchen M, Snyder SL (1987) 16,16-dimethyl prostaglandin E_2 increases survival in mice following irradiation. *Radiat Res* **109**, 440-448
44. Walden TL (1992) Leukotriene C_4-induced radioprotection: the role of hypoxia. *Radiat Res* **132**, 359-367
45. Pospíšil M, Netíková J, Kozubík A (1986) Enhancement of hemopoietic recovery by indomethacin after sublethal whole-body gamma irradiation. *Acta Radiol Oncol* **25**, 195-198
46. Kozubík A, Pospíšil M, Netíková J (1989) The stimulatory effect of single-dose pre-irradiation administration of indomethacin and diclofenac on haemopoietic recovery in the spleen of gamma-irradiated mice. *Stud biophys* **131**, 93-101
47. Kozubík A, Hofmanová J, Holá J, Netíková J (1993) The effect of nordihydroguaiaretic acid, an inhibitor of prostaglandin and leukotriene biosynthesis, on hematopoiesis of gamma-irradiated mice. *Exp Hematol* **21**, 138-142
48. Kozubík A, Hofmanová J, Pospíšil M, Netíková J, Holá J, Lojek A (1994) Effect of drugs inhibiting prostaglandin or leukotriene biosynthesis on postirradiation haematopoiesis in mouse. *Int J Radiat Biol* **65**, 369-377
49. Nishiguchi I, Furuta Y, Hunter N, Murray D, Milas L (1990) Radioprotection of hematopoietic tissues in mice by indomethacin. *Radiat Res* **122**, 188-192
50. Hofer M, Pospíšil M, Pipalová I (1996a) Radioprotective properties of flurbiprofen. *Fol Biol (Praha)* **42**, 267-269
51. Hofer M, Pospíšil M, Pipalová I, Holá J (1996b) Modulation of haemopoietic radiation response of mice by diclofenac in fractionated treatment. *Physiol Res* **45**, 213-220
52. Furuta Y, Hunter N, Barkley T, Hall E, Milas L (1988) Increase in radioresponse of murine tumors by treatment with indomethacin. *Cancer Res* **48**, 3008-3013
53. Michalowski AS (1994) On radiation damage to normal tissues and its treatment. II. Anti-inflammatory drugs. *Acta Oncol* **33**, 139-157
54. Langendorff H, Langendorff M (1972) Adenosin-Nukleotide und Strahlenempfindlichkeit. *Strahlentherapie* **144**, 451-456
55. Burnstock G (1993) Physiological and pathological roles of purines: An update. *Drug Develop Res* **28**, 195-206
56. Abbracchio MP (1996) P1 and P2 receptors in cell growth and differentiation. *Drug Develop Res* **39**, 393-406
57. Pospíšil M, Hofer M, Netíková J, Viklická Š, Pipalová I, Bartoníèková A (1992a) Effect of dipyridamole and adenosine monophosphate on cell proliferation in the hemopoietic tissue of normal and gamma-irradiated mice. *Experientia* **48**, 253-257
58. Pospíšil M, Hofer M, Netíková J, Pipalová I, Vacek A, Bartoníèková A, Volenec K (1993a) Elevation of extracellular adenosine induces radioprotective effects in mice. *Radiat Res* **134**, 323-330
59. Pospíšil M, Hofer M, Vacek A, Netíková J, Pipalová I, Viklická Š (1993b) Noradrenaline reduces cardiovascular effects of the combined dipyridamole and AMP administration but preserves protective effects of these drugs on hematopoiesis in mice. *Physiol Res* **42**, 333-340
60. Pospíšil M, Hofer M, Znojil V, Vácha J, Netíková J, Holá J (1995a) Radioprotection of mouse hemopoiesis by dipyridamole and adenosine monophosphate in fractionated treatment. *Radiat Res* **142**, 16-22
61. Hošek B, Boháèek J, Šikulová J, Pospíšil M, Vacek A (1992) Protection of early cellular damage in 1 Gy-irradiated mice by the elevation of extracellular adenosine. *Radiat Environ Biophys* **31**, 289-297
62. Boháèek J, Hošek B, Pospíšil M (1993) Postirradiation administration of adenosine monophosphate combined with dipyridamole reduces early cellular damage in mice. *Life Sci* **53**, 1317-1324
63. Hofer M, Pospíšil M, Netíková J, Znojil V, Vácha J, Holá J (1995a) Radioprotective efficacy of dipyridamole and AMP combination in fractionated radiation regimen, and its dependence on the time of administration of the drugs prior to irradiation. *Physiol Res* **44**, 93-98

64. Pospíšil M, Hofer M, Znojil V, Vácha J, Netíková J, Holá J (1995b) Synergistic effect of granulocyte colony-stimulating factor and drugs elevating extracellular adenosine on neutrophil production in mice. *Blood* **86**, 3692-3697
65. Weiss JF, Kumar KS, Walden TL, Neta R, Landauer MR, Clark EP (1990) Advances in radioprotection through the use of combined agent regimens. *Int J Radiat Biol* **57**, 709-722
66. Srinivasan V, Weiss JF (1992) Radioprotection by vitamin E: injectable vitamin E administered alone or with WR-3689 enhances survival of irradiated mice. *Int J Radiat Oncol Biol Phys* **23**, 841-845
67. Patchen ML, MacVittie TJ, Jackson WE (1989) Postirradiation glucan administration enhances the radioprotective effects of WR-2721. *Radiat Res* **117**, 59-69
68. Pospíšil M, Netíková J, Pipalová I, Jarý J (1991) Combined radioprotection by preirradiation peroral cystamine and postirradiation glucan administration. *Fol Biol (Praha)* **37**, 117-124
69. Fedoroèko, P, Brezáni P, Macková NO (1994) Radioprotective effects of WR-2721, Bronchovaxom[R] and their combinations: survival, myelopoietic restoration and induction of colony-stimulating activity in mice. *Int J Immunopharmac* **16**, 177-184
70. Kozubík A, Pospíšil M, Netíková J (1990) Enhancement of haemopoietic recovery in sublethally gamma-irradiated mice by the joint use of indomethacin and cystamine. *Fol Biol (Praha)* **36**, 291-300
71. Kozubík A, Pospíšil M, Netíková J (1991) Possibilities of the combined use of non-steroidal anti-inflammatory drugs and sulfhydryl compounds in radioprotection. *Strahlenther Onkol* **167**, 186-190
72. Pospíšil M, Hofer M, Pipalová I, Viklická Š, Netíková J, Šandula J (1992) Enhancement of hematopoietic recovery in gamma-irradiated mice by the joint use of diclofenac, an inhibitor of prostaglandin production, and glucan, a macrophage activator. *Exp Hematol* **20**, 891-895
73. Hofer M, Pospíšil M, Viklická Š, Vacek A, Pipalová I, Bartoníèková A (1993) Hematopoietic recovery in repeatedly irradiated mice can be enhanced by a repeatedly administered combination of diclofenac and glucan. *J Leukoc Biol* **53**, 185-189
74. Fedoroèko P, Macková NO (1996a) Radioprotective effects of combination broncho-vaxom, a macrophage activator, and indomethacin, an inhibitor of prostaglandin production: relationship to myelopoiesis. *Eur J Haematol* **56**, 54-61
75. Fedoroèko P, Macková NO (1996b) Combined modality radioprotection: enhancement of survival and hematopoietic recovery in gamma-irradiated mice by the joint use of liposomal muramyl tripeptide phosphatidylethanolamine (MTP-PE) and indomethacin. *Int J Immunopharmac* **18**, 329-337
76. Foye WO (1969) Radiation-protective agents in mammals. *J Pharm Sci* **58**, 283-300
77. Kuna P (1985) Chemical Radioprotection. Avicenum, Prague (in Czech)
78. Murray D, McBride WH (1996) Radioprotective agents, in Kirk-Othmer Encyclopedia of Chemical Technology, Fourth Edition, vol 20, pp 963-1006

CLINICALLY AVAILABLE DRUGS AS POTENTIAL CURATIVE MEANS FOR TREATMENT OF RADIATION-INDUCED MYELOSUPPRESSION

M. HOFER, M. POSPÍŠIL

Institute of Biophysics, Academy of Sciences of the Czech Republic
Královopolská 135, 612 65 Brno, Czech Republic

Abstract

The bone marrow syndrome represents the most probable manifestation of the acute radiation disease following medical application of ionizing radiation, as well as contingent nuclear accidents. Protection and treatment of marrow damage induced by radiation exposures in the range of sublethal and near lethal doses is, at present, a serious medical problem. Moreover, radiation-induced haematopoietic suppression may serve as a model for studying the effects of other bone marrow damaging factors and medical procedures, including cytostatic chemotherapy.

Among the compounds tested as potential stimulators of mammalian haematopoiesis damaged by ionizing radiation in the Institute of Biophysics, Brno, Czech Republic, also clinically available medicaments belonging to non-steroidal anti-inflammatory drugs (NSAIDs) or drugs used in cardiovascular medicine were used.

NSAIDs act on the principle of inhibition of prostaglandin production. Prostaglandins operate in negative feedback control of haematopoiesis, especially granulopoiesis. Removal of this feedback enables to enhance production of functional blood cells. Indomethacin, diclofenac, flurbiprofen, and nitroxybutylester of flurbiprofen have been successfully tested as haematopoietic stimulators in irradiated mice. Administration of flurbiprofen nitroxybutylester, a newly synthesized flurbiprofen derivative, appears to be especially promising from the point of view of decreased gastrointestinal toxicity of this compound.

Dipyridamole (DP) and adenosine monophosphate (AMP) used clinically for decreasing platelet aggregation (DP) and as vasodilators and cardioprotectors (DP, AMP) operate as enhancers of extracellular concentration of adenosine. Receptor-based extracellular action of adenosine has been found to stimulate haematopoiesis on the levels of stem and progenitor cell populations. Interesting results on synergistic action of DP + AMP and granulocyte colony-stimulating factor (G-CSF) on mouse granulopoiesis have been obtained as well. Haematopoiesis-enhancing effects of drugs elevating extracellular adenosine may be of clinical importance both from the point of view of medical benefit as well as from the standpoint of contingent financial savings obtained when using these unexpensive drugs.

C. Baumstark-Khan et al. (eds.),
Fundamentals for the Assessment of Risks from Environmental Radiation, 421-426.

1. Introduction

Myelosuppression accompanied by increased risk of infections, bleeding, and anaemia represents the most clinically significant syndrome of the acute radiation disease. Consequently, endeavours aimed at pharmacological reduction of haematopoietic radiation damage and acceleration of the regeneration of haematopoietic structures and functions following irradiation are of first-rate importance. Many types of pharmacological means have been tested up to now in animal studies, and often rather optimistic results have been obtained.

However, from the point of view of contingent practical utilization of haematopoiesis-stimulating agents in human clinical practice, attention should be focused primarily at preparations which are already used in medicine. Such drugs do not need to pass through long-term and expensive licensing procedures. Also the knowledge of their contingent undesirable side effects in man is of importance.

In the last several years, two groups of clinically available drugs have been tested at the Institute of Biophysics, Brno, Czech Republic, from the standpoint of their abilities to stimulate murine haematopoiesis suppressed by ionizing radiation, namely non-steroidal anti-inflammatory drugs (NSAIDs) and some drugs used in cardiovascular medicine for decreasing platelet aggregation (dipyridamole - DP) and as vasodilators and cardioprotectors (DP, adenosine monophosphate - AMP).

The mechanism supposed to be responsible for the haematopoiesis-enhancing effects of NSAIDs, which act on the principle of inhibiting prostaglandin production, is the removal of a negative feedback control of haematopoiesis mediated by prostaglandins [20, 5, 22]. We have tested this mechanism of NSAIDs action in situation of radiation-suppressed haematopoiesis.

Findings obtained in experiments performed *in vitro* with a number of cell types have shown that purine nucleosides and nucleotides are capable of modulating several cellular functions including cell proliferation and differentiation [33, 34]. Purinoceptors have been found to be expressed on myeloid cell lineages [2], and Gualtieri *et al.* [6] have shown that the addition of adenosine or adenosine monophosphate (AMP) to murine long-term bone marrow cultures resulted in increased granulopoiesis. Our studies have applied the above findings on experiments performed *in vivo* and tested a hypothesis about extracellular proliferation-stimulating action of adenosine.

This article briefly summarizes the results obtained and points at contingent possibilities of practical utilization of the pharmacological procedures tested.

2. Material and Methods

Experimental animals were mice, mostly males of the (CBAxC57BL/10)F_1 strain.
Irradiation was performed by a ^{60}Co gamma-ray source, dose rate was 0.3 to 0.4 Gy/min.

Drugs: Indomethacin sodium salt (Confortid, Copenhagen, Denmark), diclofenac sodium salt (Sigma, St. Louis, U.S.A.), flurbiprofen (Sigma), flurbiprofen nitroxybutylester (kindly provided by Dr. P. Del Soldato, Pharmaceutical Discovery Service, Milan, Italy), adenosine 5'-monophosphate sodium salt (AMP, Sigma, precursor of adenosine), dipyridamole (DP, Sigma), recombinant human granulocyte colony-stimulating factor (G-CSF, Neupogen, F. Hoffman-LaRoche, Basel, Switzerland).

Haematological methods: Blood cell counts and numbers of nucleated cells in haematopoietic organs were determined using a Coulter Counter. Differential counts of peripheral blood, bone marrow, and spleen were performed on smear preparations. Bone marrow progenitor cells committed to granulocyte-macrophage development (GM-CFC) were assayed by means of a semisolid plasma-clot technique. Pluripotent haematopoietic progenitor cells were determined by endogenous spleen colony technique.

Survival of animals irradiated with lethal radiation doses was observed until day 30 after irradiation.

3. Results

3.1. NSAIDS

Clinically routinely used NSAIDs (indomethacin, diclofenac, or flurbiprofen) have been shown to influence positively haematopoiesis in sublethally irradiated animals when given before irradiation with a single dose [16], after irradiation with a single dose [19, 15, 30, 31], before each fractional dose in the course of fractionated irradiation [9, 10], as well as when administered in a combination with sulfhydryl compounds [17, 18] and an immunomodulator (glucan) [25, 12]. Our results have been confirmed also by others [21 - application before irradiation; 35 - application after irradiation; 3 - combined administration of indomethacin with an immunomodulator (Broncho-Vaxom)].

However, a combination of NSAIDs administration with lethal radiation doses has been observed to cause a decreased survival of experimental animals [11, 10, 4]. This undesirable effect has been suggested to be raised by damaging action of NSAIDs on gastrointestinal tissues afflicted by intestinal radiation syndrome evoked by lethal irradiation. A positive finding in this direction has been obtained in our laboratory when a newly synthesized flurbiprofen derivative with a decreased gastrointestinal toxicity, flurbiprofen nitroxybutylester, was tested; this new compound has retained the positive influence of NSAIDs on haematopoiesis after sublethal irradiation, and its administration has not led to an increased mortality following lethal radiation doses [14].

3.2. DP + AMP (DRUGS ELEVATING EXTRACELLULAR CONCENTRATION OF ADENOSINE)

In 1989, the finding was published that DP enhances the radioprotective effectiveness of AMP [32]. Since DP inhibits the cellular uptake of adenosine, it followed from this result that the radioprotective effects of adenosine are extracellular, receptor-mediated. Elevated extracellular level of adenosine has been found to act radioprotectively when DP + AMP were given either singly before a single irradiation [24, 26, 23] or repeatedly before a single irradiation [7] or repeatedly before each fractional radiation dose in the course of fractionated radiation regimen [28, 8].

The combination of AMP and DP has also decreased early cellular damage after a low dose of 1 Gy ascertained by measuring the level of free thymidine in plasma and free polynucleotides in thymus and spleen when given either before [13] or after [1] irradiation.

Drugs elevating extracellular adenosine concentration have been found also to synergize in an enhancement of haematopoiesis with G-CSF both in normal [29] and irradiated [27] mice.

4. Discussion and Conclusions

From the point of view of an evaluation of the possibilities of contingent practical utilization of some of the pharmacological procedures tested in stimulating haematopoiesis damaged by ionizing radiation in man, the following brief considerations can be mentioned:

- Among the NSAIDs studied, the best drug seems to be the newly synthesized nitroxybutylester of flurbiprofen. This substance retains the haematopoiesis-stimulating effects of NSAIDs and, at the same time, it exerts much lower gastrointestinal side effects in comparison with other NSAIDs [36].
- As concerns the drugs elevating extracelluar adenosine (DP + AMP), as the most promising procedure appears to be the combined administration of these compounds with G-CSF. In this way it could be probably possible to obtain both clinically favourable effects and financial savings by decreasing the consumption of the expensive G-CSF.

Acknowledgements. This work was supported by the Grant Agency of the Czech Republic (Grants 306/93/0051 and 306/96/1721) and the Grant Agency of the Academy of Sciences of the Czech Republic (Grants 60402 and 604104).

5. References

1. Bohácek, J., Hošek, B., and Pospíšil, M. (1993) Postirradiation administration of adenosine monophosphate combined with dipyridamole reduces early cellular damage in mice. *Life Sciences* **53**, 1317-1324.
2. Cronstein, B.N., Rosenstein, E.D., Kramer, S.B., Weissmann, G., and Hirschhorn, R. (1985) Adenosine: A physiological modulator of superoxide anion generation by human neutrophils. Adenosine acts via an A_2 receptor on human neutrophils. *J. Immunol.* **135**, 1366-1371.
3. Fedorocko, P., Macková, O.N., Brezáni, P., and Kopka, M. (1994) Administration of the bacterial extract Broncho-Vaxom enhances radiation recovery and myelopoietic regeneration. *Immunopharmacol.* **28**, 163-170.
4. Floersheim, G.L. (1994) Allopurinol, indomethacin and riboflavin enhance radiation lethality in mice. *Radiat. Res.* **139**, 240-247.
5. Fontagné, J., Adolphe, M., Semichon, M., and Lechat, P. (1980) Effect of *in vivo* treatment with indomethacin on mouse granulocyte-macrophage colony-forming cells in culture (CFUc). Possible role of prostaglandins. *Exp. Hematol.* **8**, 1157-1164.
6. Gualtieri, R.J., Berne, R.M., McGrath, H.E., Huster, W.J., and Quesenberry, P.J. (1986) Effect of adenine nucleotides on granulopoiesis and lithium-induced granulocytosis in long-term bone marrow cultures. *Exp. Hematol.* **14**, 689-695.

7. Hofer, M., Pospíšil, M., Netíková, J., Znojil, V., and Vácha, J. (1997) Enhancement of the haemopoietic spleen colony formation by drugs elevating extracellular adenosine: Effects of repeated in vivo treatment. *Physiol. Res.*, **46**, 285-290.
8. Hofer, M., Pospíšil, M., Netíková, J., Znojil, V., Vácha, J., and Holá, J. (1995) Radioprotective efficacy of dipyridamole and AMP combination in fractionated radiation regimen, and its dependence on the time of administration of the drugs prior to irradiation. *Physiol. Res.* **44**, 93-98.
9. Hofer, M., Pospíšil, M., and Pipalová, I. (1996) Radioprotective properties of flurbiprofen. *Fol. Biol. (Praha)* **42**, 267-269.
10. 10. Hofer, M., Pospíšil, M., Pipalová, I., and Holá, J. (1996) Modulation of hemopoietic radiation response by diclofenac in fractionated treatment. *Physiol. Res.* **45**, 213-220.
11. Hofer, M., Pospíšil, M., Tkadlecek,L., Viklická, Š., Pipalová, I., and Holá, J. (1992) Low survival of mice following lethal gamma-irradiation after administration of inhibitors of prostaglandin synthesis. *Physiol. Res.* **41**, 157-161.
12. Hofer, M., Pospíšil, M., Viklická, Š., Vacek, A., Pipalová, I., and Bartoníncková, A. (1993) Hematopoietic recovery in repeatedly irradiated mice can be enhanced by a repeatedly administered combination of diclofenac and glucan. *J. Leukoc. Biol.* **53**, 185-189.
13. Hošek, B., Bohácek, J., Šikulová, J., Pospíšil, M., and Vacek, A. (1992) Protection of early cellular damage in 1Gy-irradiated mice by the elevation of extracellular adenosine. *Radiat. Environm. Biophys.* **31**, 289-297.
14. Juchelková, L., Hofer, M., Pospíšil, M., and Pipalová, I. (1998) Radioprotective effects of flurbiprofen and its nitroderivative. *Physiol. Res.* **47**, 73-80.
15. Kozubík, A., Hofmanová, J., Pospíšil, M., Netíková, J., Holá, J., and Lojek, A. (1994) Effects of drugs inhibiting prostaglandin or leukotriene biosynthesis on postirradiation haematopoiesis in mouse. *Int. J. Radiat. Biol.* **65**, 369-377.
16. Kozubík, A., Pospíšil, M., and Netíková, J. (1990) The stimulatory effect of single-dose pre-irradiation administration of indomethacin and diclofenac on hemopoietic recovery in the spleen of gamma-irradiated mice. *Stud. Biophys.* **131**, 93-101.
17. Kozubík, A., Pospíšil, M., and Netíková, J. (1990) Enhancement of haemopoietic recovery in sublethally-irradiated mice by the joint use of indomethacin and cystamine. *Fol. Biol. (Praha)* **36**, 291-300.
18. Kozubík, A.., Pospíšil, M., and Netíková, J. (1991) Possibilities of the combined use of non-steroidal anti-inflammatory drugs and sulfhydryl compounds in radioprotection. *Strahlenther. Onkol.* **167**, 186-190.
19. Kozubík, A., Pospíšil, M., Viklická, Š., Tkadlecek, L., and Netíková, J (1987) Effect of indomethacin on proliferative and differentiation activity in the haemopoietic system of lethally irradiated mice after syngeneic bone marrow transplantation. *Fol. Biol. (Praha)* **33**, 418-423.
20. Kurland, J., and Moore, M.A.S. (1977) Modulation of hemopoiesis by prostaglandins. *Exp. Hematol.* **5**, 357-373.
21. Nishiguchi, I., Furuta, Y., Hunter, N., Murray, D., and Milas, L. (1990) Radioprotection of hemopoietic tissues by indomethacin. *Radiat. Res.* **122**, 188-192.
22. Pelus, L.M. (1989) Modulation of myelopoiesis by prostaglandin E_2: demonstration of a novel mechanism of action *in vivo*. *Immunol. Res.* **8**, 176-184.
23. Pospíšil, M., Hofer, M., Netíková, J., Pipalová, I., Vacek, A., Bartoníncková, A., and Volenec, K. (1993) Elevation of extracellular adenosine induces radioprotective effects in mice. *Radiat. Res.* **134**, 323-330.
24. Pospíšil, M., Hofer, M., Netíková, J., Viklická, Š., Pipalová, I., and Bartoníèková, A. (1992) Effect of dipyridamole and adenosine monophosphate on cell proliferation in the hemopoietic tissue of normal and gamma-irradiated mice. *Experientia* **48**, 253-257.
25. Pospíšil, M., Hofer, M., Pipalová, I., Viklická, Š., Netíková, J., Šandula, J. (1992) Enhancement of hematopoietic recovery in gamma-irradiated mice by the joint use of diclofenac, an inhibitor of prostaglandin production, and glucan, a macrophage activator. *Exp. Hematol.* **20**, 891-895.
26. Pospíšil, M., Hofer, M., Vacek, A., Netíková, J., Pipalová, I., Viklická, Š. (1993) Noradrenaline reduces cardiovascular effects of the combined dipyridamole and AMP administration but preserves radioprotective effects of these drugs on hematopoiesis in mice. *Physiol. Res.* **42**, 333-340.

27. Pospíšil, M., Hofer, M., Znojil, V., Netíková, J., Vácha, J., Holá, J., Vacek, A. (1998) Granulocyte colony-stimulating factor and drugs elevating extracellular adenosine synergize to enhance haematopoietic reconstitution in irradiated mice. *Eur. J. Haematol.* **60**, 172-180
28. Pospíšil, M., Hofer, M., Znojil, V., Vácha, J., Netíková, J., Holá, J. (1995) Radioprotection of mouse hemopoiesis by dipyridamole and adenosine monophosphate in fractionated treatment. *Radiat. Res.* **142**, 16-22.
29. Pospíšil, M., Hofer, M., Znojil, V., Vácha, J., Netíková, J., Holá, J. (1995) Synergistic effect of granulocyte colony-stimulating factor and drugs elevating extracellular adenosine on neutrophil production in mice. *Blood* **86**, 3692-3697.
30. Pospíšil, M., Netíková, J., Kozubík, A. (1986) Enhancement of hemopoietic recovery by indomethacin after sublethal whole-body gamma irradiation. *Acta Radiol. Oncol.* **25**, 195-198.
31. Pospíšil, M., Netíková, J., Kozubík, A.., Pipalová, I.(1989a) Effect of indomethacin, diclofenac sodium and sodium salicylate on peripheral blood cell counts in sublethally gamma-irradiated mice. *Strahlenther. Onkol.* **165**, 627-631.
32. Pospíšil, M., Netíková, J., Pipalová, I., Kozubík, A. (1989) Enhancement of radioprotective effectiveness of adenosine monophosphate by dipyridamole. *Stud. Biophys.* **132**, 203-208.
33. Rathbone, M.P., Christjanson, L., DeForge, S., DeLuca, B., Gysbers, J.W., Hindley, S., Jovetich, M., Middlemiss, P., Takhal, S. (1992) Extracellular purine nucleosides stimulate cell division and morphogenesis: pathological and physiological implications. *Med. Hypothes.* **37**, 333-340.
34. Rathbone, M.P., Middlemiss, P.J., Gysbers, J.W., DeForge, S., Costello, P., DelMaestro, R.F. (1992) Purine nucleosides and nucleotides stimulate proliferation of a wide range of cell types. *In Vitro Cell. Developm. Biol.* **28A**, 529-536.
35. Serushago, B.A., Tanaka, K., Koga, Y., Taniguchi, K., Nomoto, K. (1987) Positive effects of indomethacin and interleukin-2 on murine hemopoietic stem cells. *Immunopharmacol.* **14**, 21-26.
36. Wallace, J.L., Reuter, B., Cicala, C., McKnight, W., Grisham, M.B., Cirino, G. (1994) Novel non steroidal anti-inflammatory drug derivatives with markedly reduced ulcerogenic properties in the rat . *Gastroenterol.* **107**, 173-179.

EFFECT OF NORADRENALINE ON RADIOPROTECTIVE EFFECT OF CALCIUM CHANNEL BLOCKERS

J. VÁVROVÁ, P. PETÝREK

Purkyne Military Medical Academy, Department of Radiobiology
500 01 Hradec Králové, Czech Republic

Abstract

Calcium channel blockers can be classified into three main subgroups on the basis of their pharmacologic features. In earlier work the radioprotective effect of dilthiazem and nifedipine was proved, when these antagonists were given to mice s.c. or i.p. In this work mechanism of radioprotective activity was studied.

For indirect investigation of the cardiovascular effects leading to hypoxia in mice, measurements of rectal temperature were used with assumption that hypothermic response reflects hypometabolic effects which are considered as compensatory reaction of the organism to hypoxia. I.p. administration of Diacordin (diltiazem hydrochlorid, Léciva, Praha) in the doses of 25-100 mg/kg to mice led to a significant decrease rectal temperature at the interval of 15 minutes to 2 hours after injection. This decrease was influenced by Noradrenaline administration. The dose of 2.8 mg/kg was optimal for significant inhibition of the hypothermic response first hour after administration of Diacordin. Administration of Noradrenaline before Diacordin reduced radioprotective effect of Diacordin. We studied radioprotective effect on:

- survival of lethal irradiated mice
- hematological indicators (number of CFU-S, CFU-GM in the bone marrow and the spleen).

Our experiment suported hypothesis that radioprotective activity of Diacordin is induced by at least two mechanisms. These two mechanisms are:

- protection by means of hypoxia as a result of the effect of treatment on cardiovascular system
- inhibition of the influx of Ca^{2+} through specialized channels into cells.

Abbreviations: CFU-GM - colony-forming unit-granulocyte/macrophage, CFU-S - colony-forming unit-spleen, E-CFU-S - endogenous colony-forming unit-spleen, Di - Diacordin, Na - Noradrenaline, Di + Na Diacordin and Noradrenaline, C- nonirradiated group

C. Baumstark-Khan et al. (eds.),
Fundamentals for the Assessment of Risks from Environmental Radiation, 427-432.

1. Introduction

Calcium antagonist, dilthiazem, with a benzothiazepene structure, appears to be a good radioprotective agent in whole-body irradiated mice. Earlier, we observed that dilthiazem alone (100 mg/kg) or in the combination with the immunomodulation agent Ribomunyl enhanced survival rate in the mice after whole body irradiation [1]. Ionizing radiation leads to an increased production of free radicals and peroxides with the result of cell membranes damage. The formation of reactive oxygen species can be inhibited by the induction of hypoxia, which is a widespread protective mechanism. A decrease of calcium entry into a cell may be another protective mechanism against ionizing radiation.

Radioprotective effect of Diacordin was in the case of whole-body irradiation significant mainly in the area of the doses when mice are dying of bone-marrow syndrome of radiation illness [1]. In this study, we wanted to prove how large participation in radioprotective effect of Diacordin hypoxia has. For indirect investigation of the cardiovascular effects leading to hypoxia in mice, measurements of rectal temperature were used with the assumption that hypothermic response reflects hypometabolic effects which are considered compensatory reaction of the organism to hypoxia [2].

To eliminate cardiovascular effects of Diacordin, the ability of Noradrenaline to induce vasoconstriction and increase of the blood pressure was used. Nonadrenaline was applied 2 minutes before Diacordin and rectal temperature was monitored. Radioprotective effect was observed according to lethality of whole-body irradiated mice and some hematological indicators.

2. Materials and Methods

2.1. EXPERIMENTAL ANIMALS AND IRRADIATION

Female ICR and (C57BL x DBA/2)F1 mice (Charles River Deutschland-Sulzfeld) were used in the experiments. They were 2 months old and weighed between 18 and 20 g. The mice received whole-body irradiation from a 60 Co unit Chisotron Chirana, target to subject distance 1 m, dose rate 0.3 Gy/min, doses of 8 -11 Gy.

2.2. DIACORDIN AND NORADRENALINE

Diacordin (dilthiazem hydrochloride, Léciva Praha) was dissolved in distilled water and administered in the doses 25-100 mg/kg i.p. 15 min before irradiation. Noradrenaline inj. 1mg/1ml (Léciva Praha) was diluted by saline (1 ml Noradrenaline + 1.5 ml saline) and administered in the dose 2.8 mg/kg s.c. 2 minutes before Diacordin injection.

2.3. DETERMINATION OF ENDOGENOUS AND EXOGENOUS CFU-S NUMBER

Number of CFU-S was determinated by method of Till and McCulloch [3, 4]. The number of macroscopically visible modules (colonies) was counted on pariental side of the spleen.

2.4. DETERMINATION OF CFU-GM

Hemopoietic progenitor cells committed to granulocyte and or macrophage development (CFU-GM) were assayed using technique as described by Testa and Molineux [5]. The cell suspensions used for this assay represented the pool of bone marrow from 6-8 iradiated, or treated and irradiated mice at each time. The cells were cultured in Iscove°s modified Dulbecco°s medium (ÚHKT Praha) supplemented with 20% fetal calf serum, 1% bovine serum albumin, 0.33 % agar, and a source of CSF (WEHI-3B conditioned medium served as the source of colony-stimulating activity). After an incubation period of 12 days in a fully humidified incubator in an atmosphere of 5% CO_2 at 37°C, colonies containing 30 cells or more were scored using a dissection microscope at 32 times magnification.

2.5. MEASUREMENTS OF RECTAL TEMPERATURE

Rectal temperature was measurement in conscious mice at room temperature with a thermometer.

2.6. STATISTICS

The results were statistically evaluated with the Student's t -test. Mean SD of sets of experiments at 95% confidence levels are shown on the figures. Statistical significance of the differences between compared sets is described for $P<0.05$.

3. Results and Discussion

An indirect prove for induction of hypoxia in mice without anesthesia can be the decrease of body temperature [2]. It is known that the decrease of body temperature is an response of organism to hypoxia by increasing the metabolism as compensatory reaction of organism. The average rectal temperature of ICR mice (without anesthesia) was determined in the group of 25 mice 37.8 C. Fig. 1 shows that Diacordin alone applied s.c. in the dose of 100 mg/kg led to a significant decrease of temperature from 15 minutes to 4 hours after application with a follow-up normalization. S.c. application led to the smaller decrease the first hour after application than i.p. application. In spite of the differences in the degree of induced hypotermia during i.p and s.c. application, we did not noticed differences in the surviving rate of lethally irradiated mice to which Diacordin was administered s.c. or i.p. [6].

Noradrenaline is a substance inducing vasoconstriction. Noradrenaline application in the dose of 2.8 mg/kg s.c. led to a significant increase of rectal temperature 15-60 minutes after application in comparison with the initial values. Since the 2nd hour, there was a significant decrease of temperature lasting up to 24 hours after application. In the case of Noradrenaline application in the dose of 2.8 mg/kg of mice weight 2 minutes before s.c. administration of Diacordin in the dose of 100 mg/kg, it prevents the decrease of rectal temperature the first hour after application of Diacordin. Temperature was at the level of the initial values up to 1 hour after Diacordin application. In two

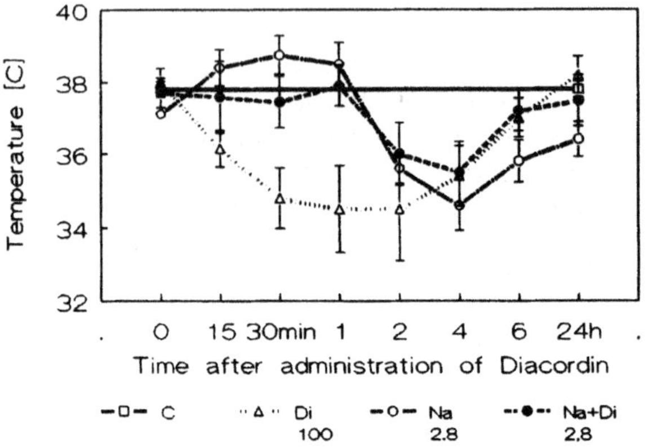

Figure 1. Effect of Diacordin and Noradrenalin on body temperature

hours, the decrease of rectal temperature in Na + Di group comparable only with the decrease after Diacordin (Di) was seen (Fig. 1).

Figure 2 shows the changes of rectal temperature in mice irradiated with a lethal dose of 10 Gy. Diacordin (100 mg/kg) was applied s.c. 15 minutes before irradiation and Noradrenaline (s.c. 2.8 mg/kg) 2 minutes before Diacordin. It is seen in Fig. 2 that there was a significant decrease of rectal temperature in all the observed groups already 15 minutes after irradiation. This decrease lasted all the observed interval (24 hours). So even the irradiation with the dose of 10 Gy led to the decrease of rectal temperature. The decrease was the largest 15 minutes after irradiation and Diacordin application before irradiation. Noradrenaline led to a significant increase of rectal temperature in the group with Diacordin 15 and 30 minutes after irradiation. 1 hour after irradiation the decrease of rectal temperature was same in Di and NA+Di groups. Since the 4th hour after irradiation, there was seen a trend towards normalization. Since this interval, there were no significant differences among the groups irradiated with the dose of 10 Gy. On the basis of these results, it is possible to assume that in the course of irradiation, in the case of Noradrenaline application before Diacordin injection, the hypoxic effect of Diacordin was eliminated.

Fig. 3 shows the surviving rate of mice after irradiation with a lethal dose of 11 Gy (10 animals in the group). It is seen in this figure that in the case of application of saline or Noradrenaline no animals survived the dose of 11 Gy. In the case of Diacordin application (s.c. 100 mg/kg) all the animals survived. Noradrenaline application 2 minutes before Diacordin application worsened the survival of mice, only 50% of animals survived. The results prove that hypoxic effect of Diacordin has a 50% participation in radioprotective effect and radioprotective effect is the result of effect of more mechanisms that only tissue hypoxia.

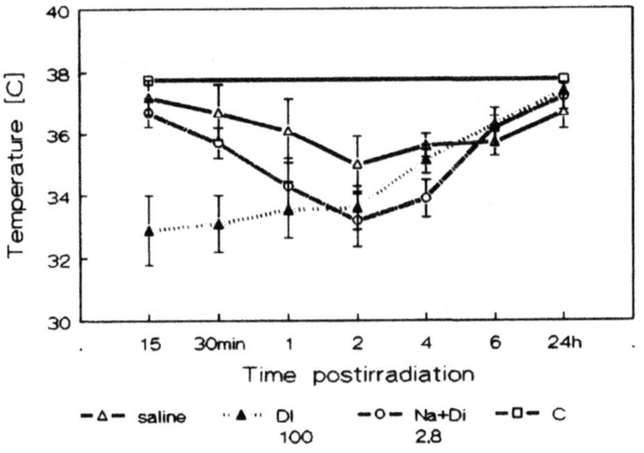

Figure 2 Effect of Diacordin and Noradrenalin on body temperature after 10 Gy X-irradiation

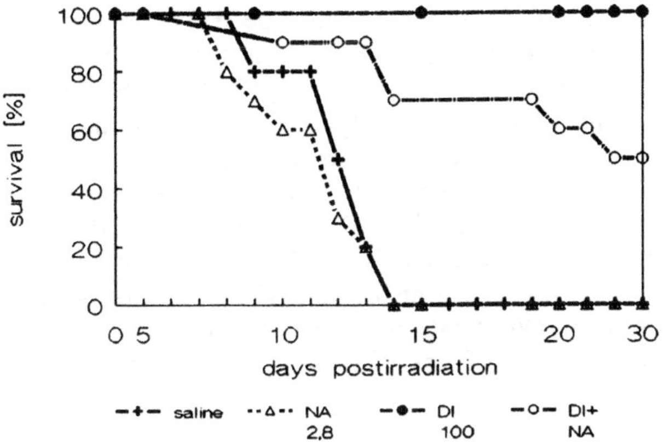

Figure 3. Effect of Diacordin and Noradrenalin on survival of 11 Gy X-irradiated mice

Table 1 shows the changes of hematological indicators after the application of Noradrenaline and Diacordin combination on day 8 after the irradiation of (C57BLx DBA/2)F1 mice with the dose of 8 Gy (sublethal dose). On day 8 after the irradiation with the dose of 8 Gy, the number of CFU-S in the femur of mice, to which Noradrenaline was administered, was very small. Diacordin application before irradiation increased the number of CFU-S statistically significantly. In the case of Noradrenaline application before Diacordin injection, this increase was not seen and the numbers were comparable with the group to which only Noradrenaline was administered. Practically the same trend was seen in the case of observation of the number of CFU-GM in bone

TABLE 1: The changes of hematological indicators on day 8 after 8 Gy X-irradiation of mice

Group	CFU-S/femur	CFU-GM/femur	E-CFU-S
C	35106 (34953 –35259)	17606 (14703-20509)	-
Na	94.2 (64.8-123.6)	148.5 (102.2 -194.8)	0.9 (0-1.6)
Di	187 (147-227)	1076 (845-1307)	2 (1.1-2.9)
Di + Na	85.5 (37.7-133.3)	226.8 (100.4 -353.2)	2.6 (1.4 -3.8)

marrow of the femur. A significant increase in comparison with Na group was only seen in the group protected by Diacordin. It is seen in the table that in the case of Noradrenaline application the number of endogenous colonies on the spleens is very small. Diacordin application led to the increase which was not influenced by Noradrenaline application before Diacordin application.

Interestingly, although there was a significant reduction in the quickness of regeneration of the number of CFU-S and CFU-GM in the bone marrow induced by Noradrenaline, the survival rate of mice in the area of lethal doses was influenced substantially less. The results of Pospíšil et al. [2] concerning the influencing of radioprotective effect of Dipyridamole and AMP by Noradrenaline had a different character. Noradrenaline did not practically influenced the quickness of hemopoietic regeneration but it reduced significantly the survival of irradiated mice. One of the possible hypothesis is that the damage caused by whole-body irradiation has a complex character where a number of systems is damaged, which means above all gastrointestinal system where different mechanisms of radioprotective effect of Diacordin can act. These mechanisms are significant from the point of view of survival of irradiated mice.

4. References

1. Vávrová, J. (1995) Stimulated hemopoiesis and enhanced survival following Diltiazem and Ribomunyl treatment in lethally irradiated mice. In: S.Kozubek and G.Horneck (eds.) *Proceedings of the Symposium Radiation Biology and its Application in Space Research at Brno, Czech Republic*, Kiramo pp. 188-193.
2. Pospíšil, M., Hofer, M., Vacek A., Netíková, J., Pipalová,I., Viklická, Š. (1993) Noradrenaline reduces cardiovascular effects of the combined Dipyridamole and AMP administration but preserves radioprotective effects of the drugs on hematopoiesis in mice. *Physiol.Res.* **42**, 333-340.
3. Till, J. E., McCulloch, E. A. (1961) A direct measurement of the radiation sensitivity of normal mouse bone marrow cells. *Radiat. Res.* **14**, 213-222.
4. Till, J. E., McCulloch, E. A. (1963) Early repair processes in marrow cells irradiation and proliferating in vivo. *Radiat.Res.* **18**, 96-105.
5. Testa, N. G, Molineux, G. (1993) *Haematopoiesis. A practical Approach,* Oxford University Press, New York.
6. Vávrová, J., Petýrek, P., Rešl, M. (1995) Radioprotective effect of calcium chanel blocker in combinatiom with cytokines. *Záverecná výzkumná zpráva . Hradec Králové, VLA JEP 93 p.*

THE RADIOPROTECTIVE EFFECTS OF WR-2721 IN MICE EXPOSED TO SUBLETHAL OR LETHAL FRACTIONATED DOSIS OF GAMMA-RADIATION

P. PETÝREK, J. ÖSTERREICHER AND J. VÁVROVÁ

Purkyne Military Medical Academy, Department of Radiobiology
50001 Hradec Králové, Czech Republic

Abstract

The radioprotective agents with sulphur have a sulphydryl groups that are able cath a free radicals when are present in organism in irradiation time. The optimal radioprotection is dependent on dose and way of radioprotective agent administration. The majority of radioprotective agents have a maximum radioprotective effect close to dose when express side toxical effects. WR-2721 have a small toxicity close to high dose reduction factor (to DRF=2.7). The radioprotective agent WR-2721 (Ethyol) is used in clinical medicine for radioprotection of nontumorous cells against effects of ionising radiation and cytostatic drugs in present time.

In our experiment, the mice were exposed to sublethal or lethal fractionated doses (5x 1 Gy to 5 Gy) of gamma-radiation aplicated in course of 5 days. The radioprotective drug WR-2721 in doses 50 mg/kg i.p. or 300 mg/kg i.m. 20 min before irradiation was administered. The radioprotective effect accorging modifications of haemopoietic stem cell function indicators, peripheral blood cells changes and survival of experimental animals, was investigated. Our findings indicate that radioprotective effectivity of WR-2721 is dependent partly on doses of fractionated irradiation partly in dose of radioprotective drug.

Abbreviations: CFU-S - colony-forming units-spleen, WBC -white blood cells, BM - bone marrow, DRF - dose reduction factor

1. Introduction

WR-2721 (amifostine; S-2-(3-aminopropylamino)-ethylphosphorothioic acid) was developed as a radioprotective agent. In animals, it protects normal tissues from the damaging effects of irradiation. In animals models, WR-2721 is selectively dephosphorylated to WR-1065 which is absorbed by normal tissue cells. Injection of WR-2721 protect against both early and late irradiation-induced injuries in a number of animal models. Maximum radioprotection results when WR-2721 is administered 30 to 60 minutes before irradiation. Degree of protection varies for different tissues, and in many

C. Baumstark-Khan et al. (eds.),
Fundamentals for the Assessment of Risks from Environmental Radiation, 433-436.

tissues is reduced when a lower dose of irradiation or fractionated irradiation is administered. Mortality following irradiation was also reduced by WR-2721. The severity and frequency of haematological toxicity produced by irradiation was also decreased by WR-2721. The faster return to baseline white blood cell counts in WR-2721 recipients suggested that the drug may improve progenitor cell survival after irradiation [1].

This report concerns a series of fractionated experiment to determine the degree of protection by WR-2721 in mouse haematopoiesis.

2. Materials and Methods

2.1. EXPERIMENTAL ANIMALS AND IRRADIATION

Female ICR mice (Charles River Deutschland-Sulzfeld) were used in the experiments. They were 2 monts old and weighed between 18 and 20 g. The mice received whole-body irradiation from a 60 Co unit Chisotron Chirana, target to subject distance 1 m, dose rate 0.3 Gy/min. The doses fractionated irradiation were 5x 2 Gy to 5x 4 Gy in 5 days with 24 h intervals.

2.2. DRUG

WR-2721 was dissolved in distilled water. Drug doses of 50 mg/kg i.p. and of 300 mg/kg i.m. were tested. The control groups were tested with saline. WR-2721 and saline were administered 20 minutes before irradiation.

2.3. DETERMINATION OF ENDOGENOUS CFU-S NUMBER

Number of CFU-S was determinated by method of Till and McCulloch [2, 3]. The number of macroscopically visible colonies was counted on pariental side of the spleen.

2.4. DETERMANATION OF HAEMATOLOGICAL INDICATORS

Bone marrow cells and leukocyte counts were perforformed by manual methods.

2.5. IMAGE ANALYSIS

The histologic sections were analysed by Image Analysis LUCIA M (Laboratory Imaging, Praha) with a microscope BX 40 (Olympus). The fields of the histologic section were grabbed and translated into gray scale. The nucleus of the nucleated cells was defined as 0-18 of gray scale (0-36). Five fields (240457 μm^5) were counted in each histologic section.

2.6. STATISTICS

The results were statistically evaluated with the Student`s t-test. Means SD of sets of experiements at 95% confidence levels are shown on the figures. Statistical significance of the differences between compared sets is described for P<0.05 (*), P<0.01(**) and P<0.001 (***).

3. Results and Discussion

In Fig. 1a, curves of the spleen weights after fractionated irradiation are described in a time interval of 30 days after irradiation. The decrease of the spleen weight of all experimental groups is statistically significant on day 4 and day 7. The decrease of the spleen weight is not statistically significant at all the other intervals. In Fig. 1b, we can also see the equal decrease thymus weights. This decrease is not statistically significant only on day 30. In Fig. 1c, we can see a statistically significant decrease of the bone marrow cells (femur) numbers in all experimental intervals. In Fig. 1d, we can also see equal decrease of white blood cells in all experimental intervals.
Fig. 2a-d show the values of some haematological indicators on day 10 after the end of fractionated irradiation. In Fig. 2a, we can see the changes of the spleen weights. The values of this indicator are not significantly different between the group 1 (5x2 Gy and saline) and the group 2 (5x2 Gy and WR-2721). On the contrary, values of this indicators are statistically significant between the group 3 (5x3 Gy and saline) and the group 4 (x3 Gy and WR-2721). In Fig. 2b-d, we can see the changes of the CFU-S, BM-cells (femur), and BM-cells (histologic section). The changes of all the indicators are equal as spleen. Values of the indicators are not statistically significant between the group 1 and the group 2. They are statistically significant between the group 3 and the group 4.

We have results from our earlier experiments with a long time administration of the WR-2721. The administrations of this drug in a dose of 50 mg/kg i.p. or of 300 mg/kg

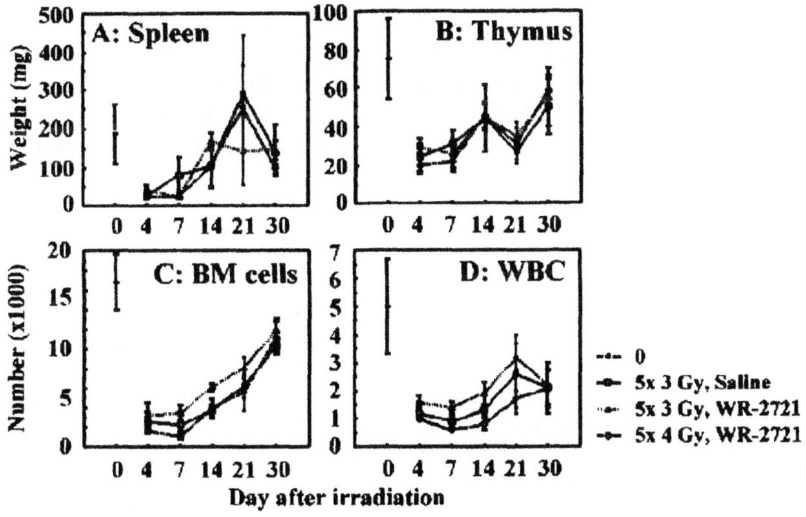

Figure 1. Decrease in organ weights after irradiation

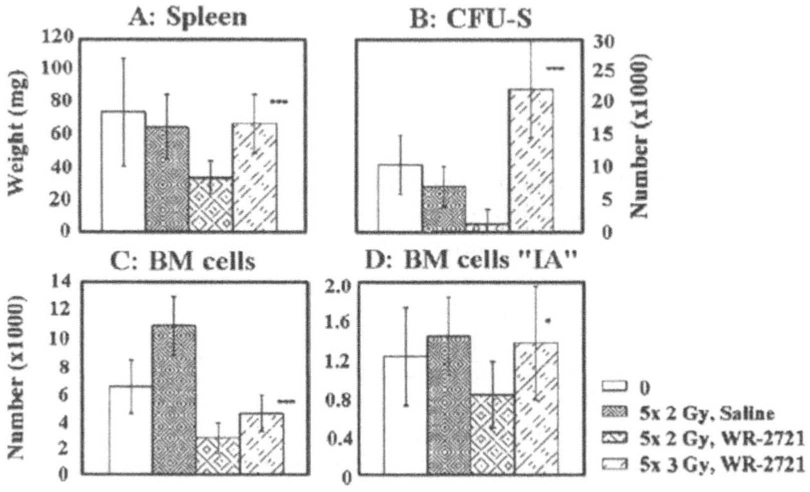

Figure 2. Some haematological indicators on day 10 after the end of fractionated irradiation

i.m. did not cause changes of the haematopoiesis or the survival of the non-irradiated experimental animal (mice). WR-2721 in a dose of 300 mg/kg i.m. protects experimental animals (mice) against high doses of irradiation (up to 13 Gy). The literature data show that the small dose fractionated irradiations are probable in accidents. The protective effect is decreased (DRF=1.1 to 1.2) at fractionated irradiation [4].

Our experimental results are in accordance with this reality. The results suggest that the protective effects of the WR-2721 administration are hardly verified in a small dose of fractionated irradiation. The differences in the survival of the experimental animals and changes of the haematological indicators are not usually significantly different. The results of the high doses of fractionated irradiations are in accordance with the results of a single dose of irradiation. In these experiments, the difference between protected and non-protected groups is statistically significant. The future clinical and experimental experience will solve the problem with the use of the WR-2721 in accidental small dose fractionated irradiation.

4. References

1. Spencer,C.M.,Goa,K.L.(1995) Amifostine A Review of its Pharmacodynamic and Pharmacokinetic Properties, and Therapeutic Potential as a Radioprotector and Cytotoxic Chemoprotector. *Drug*, **50**, 1001-1031.
2. Till,J.E.,McCulloch,E.A.(1961) A direct measurement of the radiation sensitivity of normal mouse bone marrow cells. *Radiat. Res.* **14** ,213-222.
3. Till, J.E.,McCulloch,E.A. (1963) Early repair processes in marrow cells irradiation and proliferating in vivo. *Radiat. Res.* **18**, 96-105.
4. Travis, E.L.,Thames,H.D. Jr.,Tucker, S.L.,Watkins, T.L. and Kiss, I. (1986) Protection of mouse jejunal crypt by WR-2721 after small doses of radiation. *Int. J. Radiat. Oncol. Biol. Phys.* **12**, 807-814.

TISSUE AND ORGANISM RESPONSES.

Recommendations for Research Needed

M. WLASCHEK [1], I. AHLERS [2] AND
E. AHLERSOVÁ [2], Y.E. DUBROVA[3], J. FÜLLER [4], P. KNUSCHKE [5],
E. PAULÍKOVÁ [2], P. PETÝREK [6], J. VÁVROVÁ [6]

[1] *Department of Dermatology, University of Cologne, Joseph-Stelzmann Str. 9, D-50924 Cologne Germany*
[2] *Institute of Animal Physiology, Faculty of Science, P.J.Šafárik University, Moyzesova 11, 041 67 Košice, Slovak Republic*
[3] *Vavilov Institute of General Genetics, Moscow B-333, Russia*
[4] *Department of Radiation Oncology, University Hospital, Friedrich-Schiller-University Jena, Bachstr. 18, D-07740 Jena, Germany*
[5] *Department. of Dermatology, Dresden University of Technology, Fetscherstr. 74, D-01307 Dresden, Germany*
[6] *Purkyne Military Medical Academy, Department of Radiobiology, 50001 Hradec Králové, Czech Republic*

1. General Introduction

Environmental radiation affects organisms. The physical nature of UV and ionizing radiation is different. In respect to environmental stress the organism will respose in the stepwise cascade of change, damage, harm and detriment. The aim of risk assessment must be quantitation of risks to the organism based on proper risk definition. For this purpose, the identification of the most important effects will be used to define research priorities. Dealing with organisms is hampered by different complicated interactions. Therefore, approximations to the response have to be used. Both, the genetic and the epigenetic reactions have to be taken into consideration. For an understanding of the mechanisms of radiation effects detailed knowledge on the physiology, biochemistry and genetics of an organism is needed. For research, standardization and improvements of methods are necessary. This includes radiation measurement and criteria of experienced pathological, physiological, biochemical and genetical examination. For different organs, the response to radiation is different due to distinct tasks of these organs in the organism. In this regard, different organ responses have to be integrated into the complex organism response.

C. Baumstark-Khan et al. (eds.),
Fundamentals for the Assessment of Risks from Environmental Radiation, 437-438.

2. State of the Art and Unsolved Problems

Due to the complex nature of the organism response to radiation, in this chapter only very few selected topics will be mentioned. Nevertheless, for risk assessment, new directions of organism related research will arise in future out of molecular research. These new research directions have carefully to be adapted to and validated for their relevance in the organism response.

2.1 SELECTED TOPICS

Radiation will suppress the immune system of the organism. Even local radiation will decrease the immune capacity of the organism limiting its defense capabilities against other external stress factors. The inhibition of the immun defense has been demonstrated for circulating blood cells and is postulated to play an important role in UV-related melanoma developing at non-exposed body sites. The stress response to radiation is organ specific indicating the need for tissue weighting factors of radiation.

Radiation caused changes in lipids, carbohydrates and hormone metabolism contribute to the acceleration of atherosclerosis, diabetes and aging. The direct correlation between lipid changes and these degenerative diseases has been confirmed in epidemiological studies. Therefore radiation not only is responsible for cellular transformation leading to tumorigenesis but also contributes to many other degenerative diseases. Estimating the effect of radiation exposure on somatic cells and on germline is needed.

Biological dosimetry in organisms is difficult. The propagation of germline mutation is a hazard to the offspring. The application of new methods will extend the available appropriate monitoring systems. To improve the radioresistance of an organism is a fundamental task in radiobiology. However, in contrast to the positive effects of chemical and biological radioprotective agents the mechanism of action, efficacy, practical utilization are only poorly understood.

3. Recommended Approach

Different research goals have to be addressed depending on the mentioned topics. In general, improvement in methodology by standardization in ionizing and non-ionizing radiation measurement is one important goal. This has to be done in combination with adequate measurement of the effect by pathological and biochemical methods. The improvement of models is required. The role of immunosuppression, of adaptive mechanisms and of the susceptibility of individual organisms has to be addressed. The characterization of the underlying cellular and molecular mechanisms will allow the identification of relevant biomarkers of organism response required for the quantitation of the effects.

SKIN CANCER PROTECTION CONCEPTS.

Worldwide Experiences With Their Realization in Special Consideration of The German Concepts

E.W. BREITBART AND M. BREITBART

Dermatologisches Zentrum Buxtehude
D-21614 Buxtehude, Germany

1. Introduction

The consensus meeting on "educational needs for primary and secondary prevention of melanoma in Europe" by the EORTC Melanoma group, published in 1991 [1] as well as a consensus conference on "early melanoma" by the National Institutes of Health, USA, in 1992 and the consensus meeting "How to decrease morbidity and mortality of skin cancer" by the "Commission of Early Detection and Prevention of Skin Cancer" of the Society of Dermatological Prevention (ADP e.V.) in Hamburg, 1994 [2] emphazised the importance of "public education" to achieve the reduction of morbidity and mortality of melanoma and nonmelanoma skin cancer.

In the 1960s Australia (particulary Queensland) started the first education campaign on secondary prevention of skin cancer and on hazards of ultraviolet rays (UV). In the 1980s many countries initiated similar campaigns in Europe (i.e. Austria, Belgium, Denmark, England, France, Germany, north-Italy, Netherlands, Scotland, Sweden and Switzerland). Overseas public skin cancer education campaigns were started i.e. in Canada, Hawai, New Zealand and in the United States of America.

To implement far-reaching educational strategies the cooperation of different social and scientific groups is required. For this reason in many countries these campaigns were organized in collaboration of i.e. different medical/scientific societies, state-run institutions (public health services) and private enterprises [3, 4].

2. Strategies in Sreening and Education Programs

Mainly marketing stratgies were used; the population was repeatedly instructed i.e. by information leaflets, posters, newspapers, radio and television; sometimes advertisements and television spots were developed. Special educational programs i.e. for schools and kindergartens were elaborated [5-7]. Some countries initiated in addition professional education campaigns (i.e. Australia, Scottland, Germany, Austria). Some intervention strategies were run nationwide (i.e.United States of America, Swiss, Germany and Australia), some were run locally (i.e. north-Italy, Austria and UK).

C. Baumstark-Khan et al. (eds.),
Fundamentals for the Assessment of Risks from Environmental Radiation, 439-450.

In different countries miscellaneous types of "screening" were realized:

- "mass screening":
  - Screening offers in docter`s offices;
  - Screening offers of public institutions (public health services)
  - Screening days/weeks on beaches, in towns etc.
- screening of "risk groups":
  - regular check-up of (relatives of) melanoma patients
  - regular check-up of individuals with multiple and/or "atypical" moles
- self-screening or screening of relatives:
  - Instructions to keep a check on oneself or on relatives were given during most
  - educational campaigns)

Prevention education programs and screening for cancer are usually undertaken in the hope that cancer will either be prevented primarily, or a malignancy will be discovered sufficiently early to cure patients. The desired end points are a reduction of cancer-associated morbidity and mortality. In the short term it is necessary to try to change beliefes, attitudes and habits of cancer risk groups and of the general population. For public health efforts to be effective in changing attitudes and habits regarding UV-protection, and to decrease mortality and morbidity of skin cancer, it is essential to understand the responses of the public to skin cancer detection efforts and educational programs. Different short- and long-term measures were evaluated to investigate the effectiveness of intervention strategies.

3. Evaluation of Effectiveness of Intervention Strategies

Already 1982 Putnam and Yanagisako evaluated a "skin cancer comic book" in Hawai and reported positive changes of sun-behavior as a direct result of readership [8]. In Australia it was shown that there was "an enormous increase in public knowledge" and a change in the "social style of life" [9-11]. In 1991 Borland et al. reported that the "cover yourself against skin cancer" campaign in Australia led to a significant improvement in overall protection in outdoor workers. In Sweden and England interventions in schools resulted in improved knowledge about "sun and skin cancer" [6, 12]. Hill et al. [13] stated that in Australia the crude proportion of sunburn dropped from 11% to 10% to 7% over 3 years. Hat wearing increased significantly each year (19%, 26%, 29%), as did sunscreen use (12%, 18%, 21%). Bourke et al. [14] reported the result of a survey to assess the level of awareness of melanoma, and the current sunbathing practices subsequent to an intervention campaign from 1987 to 1989 in Leicester (England). Although many people knew about melanoma they continued to put themselves at risk by sunbathing. This was especially true for teenagers, young adults and males.

4. Outcomes of International Screenings

Two research groups demonstrated that persons attending the melanoma/skin cancer screening programs were likely to be at increased risk of skin cancer compared to the general population, and appropriately selected themselves to be screened [15, 16] but most were women, well educated (with college or advanced degrees), and white [16]. In contrast to these findings Rampen et al. [17] described a lack of selective attendance of participants at skin cancer/melanoma screening clinics. Public education campaigns resulted in considerable increases in the workload of physicians in the UK [18, 19] and in Germany [3]. In the Netherlands extra workload has been negligible [20].

Significant numbers of melanomas were detected on screening events in the USA [21-23], in Germany [24], in the Netherlands [25, 26] and in the UK [18]. In Australia four invasive malignant melanomas and two in situ melanomas were detected in 3397 individuals screened [27]. In England 2150 employees of the Head Office of a large UK retailer were screened and 4 melanomas were discovered [28]. Recent evaluations of nationwide screenings (United States) indicated that most screen detected cutaneous melanomas are low risk-melanomas [23].

A melanoma-screening campaign was organized the Netherlands, in 1993. The results were compared with similar campaigns in the Netherlands, in 1990 which addressed skin cancer in general. Public education focusing on melanoma yields more selective attendance than focusing on skin cancer in general. The authors stated that these findings may have important implications with regard to the cost-effectiveness of skin cancer screening efforts [29].

In a large scale community survey in Australia of the prevalence of screening, self-screening or screening by another person was reported by 48% of the sample. The authors conclude that current health education programmes have been effective [30]. Koh et al. [23] outlined that only a minority of screenes are knowledgeable of self-screening steps and a low percentage performs it monthly.

In their review on "screening and case-finding of high risk persons" Koh et al. [16] described the surveillance-effects of family members of melanoma kindreds and patients with multiple and/or "atypical" nevi. In the Netherlands, in Scotland and in the USA increased numbers of melanomas were found; the tumor thickness was reduced.

4.1 TUMOR THICKNESS

A reduction of tumor thickness of melanoma was achieved in Austria [31], in Scottland [18, 32] in New Zealand [33], in Australia [11] and the United States of America [23].

4.2 MORTALITY

Recently, data have shown a slight decrease in mortality rates in women in Scotland, stabilization of the overall mortality rate in Australia and among younger birth cohorts and women in the United States, and a decling mortality rate among women in an intervention region in Italy compared with control towns [18, 34-37]. The disproportionately increasing incidence of thin melanoma, the divergence between incidence and mortality trends and the recent sharp increases in incidence in some populations suggest that earlier diagnosis or greater detection of less aggressive melanomas may have contributed to the incidence trends [38]. Streetly and Markowe [39] reported that in England & Wales in the 15-34 age group mortality rates have declined in women since the late 1970s. Giles et al [40]. concluded that Melanoma mortality in Australia peaked in about 1985 and has now plateaued. On the basis of trends in cohorts it can be expected to fall in coming years.

4.3 COST-EFFECTIVENESS

Two studies evaluated cost-effectiveness of public education and screening [41, 42]. The outcome of these studies was a recommendation for the use of such programs in other countries and regions.

5. Skin Cancer Prevention Concepts in Germany

In 1989, nationwide public education campaigns concerning primary and secondary prevention of skin cancer were started in Germany. The short term aims of the campaigns were:

1. Primary prevention:
   - Information of the public and health professionals about benefits and hazards of ultraviolet (UV) radiation.
   - Change in "social style of life", i.e. to deal more critically and consciously with UV radiation
2. Secondary prevention:
   - Improvement of the knowledge of the public and medical professionals about means of early detection of skin cancer
   - self-screening or screening of "risk groups", e.g. individuals with multiple and/or "atypical" moles or relatives of melanoma patients

The first campaign (1989) was named "How to protect yourself against malignant melanoma", which focused on signs and early symptoms of CMM and described risk factors for the development of CMM. In 1990-1992 public education was continued with the campaign "Beware of the sun" and finally in 1993/1994 informations about "Child and sun" were started. In the public education campaigns marketing techniques were used to promote sun protection behavior, and to inform about risk groups and early signs of skin cancer. To assess the impact of the intervention strategies a representative population sample was interviewed by random telephone calls in May 1989 (to) before the intervention program was started in June 1989. The second evaluation (t1) was done 1½ years (December 1990 and January 1991) after the initiation of the public education campaign; the third data collection (t2) was done in November and December 1993. In the following the main results are reported.

6. Realization of the Campaign

The preparation of the information material (posters, information booklets, advertisements) was done in cooperation with an advertising agency. The campaign used marketing strategies. The German population was informed repeatedly on prevention and early detection of skin cancer by the media, including articles, interviews in newspapers, interviews and reports in television, and in the radio. Free skin cancer screening-days were organized.

When the campaign was initiated, all German dermatologists, general practioners, internists, urologists, gynecologists, pediatricians, and body-caring (beauty culture) professions (e.g. hairdressers, chiropodists, attendants-lifeguards, masseurs) were provided with information booklets, and posters about the increasing incidence of melanoma, about melanoma and nonmelanoma skin cancer risk groups and the clinical appearence of pigmented lesions and early melanoma, respectively. A letter announcing the public education campaign was enclosed.

In order to prepare and to accomplish the nationwide intervention strategies the involvement and support of the following organizations was indispensable:

- German Pediatric Society
- German Gynecologic Society
- General Medical Council
- German Ministry of Health (BMG)
- German Ministry of Environmental Protection (Ministry of Environment, BMU)
- Institute for Radiation Hygiene (ISH) of the Federal Office for Radiation Protection (BfS), Germany
- Commission on Radiological Protection (SSK), Germany
- Health insurance companies
- German Cancer Society
- Public Health Services

6.1 POPULATION BASED SURVEY

A representative population sample was interviewed by random phone calls by trained interviewers using a standardized questionnaire. The content of the questionnaire is given in Table 1.

The unit of observation consisted of the residential population of German adults, ≥18 years of age, (unweightened: t_0 n=1773, t_1 n=1805, t_2 n=1800) in 3 geographic regions:

- the city of Hamburg
- the city of Munich
- the city of Würzburg and the rural districts Würzburg and Kitzingen

6.2 STATISTICAL EVALUATION

Comparisons among rates of answers (proportions) were done by z-test (z-distribution, precondition: $n_1 \geq 50$ and $n_2 \geq 50$) at a 0.025 level of significance. To compare two means, t-test for different variances was used.

All 3 surveys (1989, 1990/91, 1993) were weightened to the age and sex distribution of the year 1991.

In the cross-sectional analysis the following target variables were chosen:

1.) Image of brown skin being healthy skin
2.) Behavior to acquire sun tan:
    - Purposeful sun exposure
    - Use of sunscreen and protection factor
    - Visits in tanning parlors
3.) Assessment of the dangers of skin cancer
4.) Estimation of the possibilities to cure skin cancer

The target variables were analyzed concerning the following groups:

1.) Whole population (stratified: male, female)

2.) Stratification regarding level of education
3.) <30 years of age and >60 years of age of the whole population
4.) Population of each region

TABLE 1. Content of the standardized questionaire

Age	18-24, 25-29, 30-34, 35-39, 40-44, 45-49, 50-59, 60-69, ≥70
Gender	Male, female
Personal status	single, living alone; single with permanent partner; married, living with partner; married, not living together; divorced; widowed; no statement
School education	elementary school leaving certificate, first public examination in secondary school, university entrance qualification
Health insurance	public or private health insurance, health insurance of special companies, none, don`t know, no statement
hair colour	red, light, dark
skin type	Fitzpatrick I-IV
skin problems within the last 3 years	Yes, no
Interests concerning medical themes	very strong, strong, less, none
Attitude towards "brown skin",	healthy skin: minimal brown, brown
Opinion regarding the dangers of (skin) cancer	(main cause of death (cancer?), most fear of (cancer?), skin cancer is extreme / less dangerous)]
Knowledge about clinical signs of skin cancer	none, non-healing wounds, skin tumors, changes in skin tumors
Knowledge about possibilities to cure skin cancer	good, intermediate, bad
Knowledge about types of cancer looked for in the state-run cancer medical check-up (KFU)	skin cancer mentioned
Status of knowledge (connoisseur) about the skin cancer education program	
Amount of staying outdoors in summertime	5 hours or more
Frequency of visits in tanning salons	seldom, often, > 20 times/year
Number of days sunbathing	never, mean of days
Sunscreen use during sunbathing	always, nearly never
Protection factor	<5, 5-9, ≥10
Family history of melanoma	

7. Results

7.1 HEALTH-BELIEF

The percentage of individuals supposing "brown skin to be healthy skin" decreased dramatically from 1989 (35.5%) via 1990/91 (19.8%) to 1993 (21.7%). This significant cut-back was found also in all subgroups (women, men: <29 years of age, >60 years of age, lower level and higher level of education; in Munich, in Hamburg, and in Würzburg). Interestingly this drastic change happened in the beginning of the intervention strategies and remained constantly until 1993.

7.2 REPORTED CHANGE OF "LIFE-STYLE"

7.2.1 No sun exposition to acquire a sun tan

The proportion of persons, who *never* expose themselves to sun with the intention of acquiring a sun tan, rose significantly from 49.8% (t_0) to 56.4% (t_1, t_2) regarding the total population of the three areas. This significant trend was seen in men as well as in men with a higher education level. In men and women with a higher education level this effect was not registered till 4 years after the start of the public education campaigns. Concerning the group of "women" there was a significant upward movement from 52.5% (t_0) to 59.4% (t_1); in 1993 (t_2) this proportion decreased (57.6%) to a nonsignificant value. But significantly more "young women" (<29 years of age) avoided the sun in 1993 (31.7%, t_2) compared to 1989 (22.6%, t_0); nevertheless in this target population the percentage (31.7%) lied clearly below the value (56.4%) of the total population.

7.2.2 Sun exposition more than 6 weeks / year to acquire a sun tan

The proportion of people, who intentionally exposed themselves to sun more than 6 weeks / year decreased signifcantly from 8.6% in 1989 via 5.1% in 1990/91 to 5.6% in 1993. Especially young men and women (<29 years of age) changed their behavior. In 1989 14.5 % of young men exposed themselves to the sun; in 1990/91 only 4.9%, and in 1993 7.4% reported this risk behaviour. A similar significant trend was seen concerning "young women" (t_0 16.0%, t_1 6.0%, and t_2 4.4%). No change was registered in people with a lower education level. Concerning the different geographic regions, this trend was only seen in Hamburg (t_0 8.8%, t_1 4.6%, and t_2 4.9%).

7.2.3 Regular sunscreen use during sunbathing

An important aim of the public education was the change of the live-style with regard to the application of sunscreens. This effect was not registered till 4 years after the start of the public education campaigns. Regarding the total population of all three geographic areas the reported regular application of sunscreens increased significantly between 1989 (36.9%) and 1993 (49.9%). With the exception of "old males" (>60 years of age) and "males in Munich" we found in all subgroups a significant increase in 1993 vs 1989. Corresponding to these results the proportion of the total population, who never used sunscreens during sunbathing decreased significantly from 27.1% in 1989 via 26.9% in 1990/91 to 20.6% in 1993.

7.2.4 Regular application of protection factor ≥ 10

The trend recognized concerning regular application of sunscreens in the total population was confirmed. Again we did not see any change during 1989 (5.9%) and 1990/91 but in 1993 (18.9%) vs t_0 and t_1 The use of sunscreens with a high protection factor has been even tripled. With the exception of "old men" (>60 years of age), and "men with a low level of education" we found in all subgroups significant increases in 1993 vs 1989.

7.2.5 Visits in tanning parlors

The percentage of persons, who visit tanning parlors *more than 20 times / year* did not change during the three evaluation waves. Unlike this finding, the proportion of persons who go *sometimes* into a tanning saloon showed an inconsistent course. During the second evaluation wave a significant decrease (10.6%) of visits was recognized in the total population compared to the baseline evaluation (t_0 14.1%) in 1989. But in 1993 the frequency of visits increased again (17.1%). Similar trends were seen in young men, women and in the three geographic regions. Especially young women (42.4%) visit tanning parlors much more frequently than the average of the population (17.1%) in 1993.

7.3 KNOWLEDGE ABOUT SKIN CANCER

7.3.1 "Knowledge about the possibilities to cure skin cancer":

Significant rises of the knowledge, that skin cancer has good chances to be cured if detected early, were observed in 1993 vs 1989. This increase of knowledge was registered with reference to the total population (t_2 59.7%; t_0 52.2%), to the group of "young men" (t_2 68.2%; t_0 55.4%), to "men with lower levels of education" (t_2 62.8%; t_0 51.4%), to the group of "women" (t_2 56.9%; t_0 48.0%), to "old women" (t_2 56.5%; t_0 46.4%), to Hamburg (t_2 62.3%; t_0 52.0%), to "women in Hamburg" (t_2 58.4%; t_0 45.2%), and with reference to Würzburg (t_2 57.0%; t_0 50.6%). This trend was already seen in t_1 in seven subgroups.

7.3.2 "Knowledge of skin cancer signs"

Although the knowledge about cancer signs in general did not change in the total population (t_0 30.6% t_1 27.9% t_2 28.9%), the knowledge about indicators of possible skin cancer ("changes concerning moles and warts") raised significantly regarding the total population from 1989 (19.4%) via 1990/91 (25.4%) to 1993 (28.3%). Similar changes were seen in men (t_2 29.1%; t_0 17.0%), in men with a lower education level (t_2 26.4%; t_0 14.8%), in women (t_2 27.6%; t_0 21.5%), and in young women (t_2 43.9%; t_0 27.2%). Comparing the different regions it can be stated, that in Würzburg there was no change, and in Hamburg was the biggest change (t_2 30.5%; t_0 17.6%).

7.4 GENERAL EFFECTS OF THE NATIONWIDE INTERVENTION STRATEGIES

7.4.1 "Is skin cancer espcially dangerous?"

The participants of the study were asked, whether all types of cancer are comparably dangerous. Those who stated, that there are differences, should give an example. We did not recognize any increase concerning the statement "skin cancer is particulary dangerous" (t_0 6.6%; t_1 6.7%; t_2 5.2%) between 1989 and 1993 in the whole population and in all subgroups.

7.4.2 Knowledge about the skin cancer intervention strategies
In 1990/91, 17.6% of the total population had heard about the skin cancer education campaign. The percentage of connoisseurs of the skin cancer intervention raised significantly in 1993 to 28.7%. This increase was registered in nearly all subgroups with the exception of old men, men with a lower level of education, young women and women with a university entrance qualification.

Eleven percent of the total population could name concrete messages of the skin cancer intervention campaign in 1990/91 and 20.2% could name concrete messages in 1993. Thus we registered significant more people with a concrete rememberence of the education campaign in 1993 compared with 1990/91. Again similar changes were seen in all subgroups with the exception of the above mentioned groups.

8. Discussion

Worldwide experience with the realization of educational campaignes show that despite considerable success so far, still a long to be gone. Koh et al. [43] summarized the current state of melanoma and skin cancer control: Concerning "primary prevention" knowledge about ultraviolet radiation increased but there was only some change of behavior. Knowledge and prevention practices were significantly worse for men than for women and for those with a lower level of education. Most public education campaigns did not consider that people pass through discrete stages on their way to adopting healthy behaviors. Many authors agree that there is as yet little evidence of change in behaviour with regard to sun exposure and that especially teenagers, adolescents, men and low educated people are not appropriately targeted by public skin cancer campaigns [14, 44-47]. Especially in Australia there have been significant increases in knowledge and sun protective behavior reflected in the reduced proportions of people sunburnt at the weekend in summer in recent years [10, 11].

Concerning "screening" Koh et al. [43] stated that high-risk persons approprately select themselves. It was especially proposed to screen high risk groups with "atypical" nevi and or a family history of melanoma. Brandt [48] summarizes that many questions are still unresolved, but skin cancer screening by general internists, family practitioners, dermatologists, and other health professionals could prove cost-effective and reliable. However,in many countries it is difficult to monitor tumor thickness and melanoma epidemiology because there are no reliable population based incidence and mortality rates available.

To change behaviors requires shifts of attitudes; in Germany such a shift could be achieved. The "ideal of beauty" has been changed. Significant less people associated "brown skin to be healthy skin". In 1989, 35.5% of the participants stated "brown skin being a sign of health". In 1993 there was a cut-back to 21.7%. This significant cut-back was found in all subgroups regardless of gender, level of education, and age.

Concerning the *reported change of "life-style"* and *knowledge about skin cancer* we found also impressive significant changes in concordance with the study aims. The proportion of persons, who *never* expose themselves to sun with the intention of acquiring a sun tan, rose significantly from 49.8% (1989) to 56.4% (1993). In concordance with this finding the proportion of people, who intentionally exposed themselves to sun more than 6 weeks / year decreased signifcantly from 8.6% in 1989 to 5.6% in 1993. Especially young men and women (<29 years of age) changed their behavior. This cut-back was especially high in the group of young women (<29 years of age). In 1989, 16% exposed themselves intentionally to sun for more than 6 weeks/year, in 1993 only 4.4% of the young women reported this behaviour. This result is especially important, because one of our main target groups were young people. The differences between the geographic regions may be explained by the fact, that the "Commission of Early Detection and Prevention of Skin Cancer" of the ADP e.V. was located in Hamburg. Thus, the public relations work was particulary intensive there. As expected, the regular application of sunscreens increased significantly in 1993 (49.9%) vs 1989 (36.9%), but his effect was not registered till 4 years after the start of the public education campaigns. Corresponding to this finding the proportion of the total population, who *never* used sunscreens during sunbathing decreased significantly from 27.1% in 1989 via 26.9% in 1990/91 to 20.6% in 1993. Similar trends were registerd concerning "regular application of protection factor $\geq$ 10 in nearly all groups. Only old men (>60 years of age) did not change their behaviour. It was not possible to place the message that visits in tanning parlors may be harmful. People who visit tanning parlors frequently (>20 times/year) did not change their behavour at all. The same is true for people who sometimes use sunbeds. On the contrary, there seemed to be a nonsignificant positive trend towards tanning saloons. Possibly the sunbed-industry managed to promote the sunbed tanning to be a healthy tanning in cotrast to the sun. We conclude that in Germany less people expose themselves to the sun with the intention to acquire suntan; more people use sunscreens. This is especially true for high risk groups like young people.

In our study significant rises of the estimation, that skin cancer has good chances to be cured, were observed in 1993 (59.7%) vs 1989 (52.2%). In addition the knowledge about *indicators of possible skin cancer* raised significantly regarding the total population from 1989 (19.4%) to 1993 (28.3%). This message did not reach men with a university entrance qualification, old women and inhabitants of Wuerzburg. We registered this trend, although the knowledge about *cancer signs in general* did not change in the total population. But the proportion of people who knew skin cancer signs is still small (about 30%) and further detailed education is needed.

Regarding the total population the percentage of connoisseurs of the skin cancer campaigns raised significantly between 1990/91 and 1993 from 11.1% to 20.2%. This increase was registered in nearly all subgroups. We did not recognize any differences of proportions in the whole population concerning the statement "skin cancer is particulary dangerous". About 6% of the interviewed persons considered skin cancer to be particulary dangerous, while about 18% considered skin cancer to be less dangerous than other types of cancer. There was no change of these proportions over time. We conclude, that the public skin cancer education campaigns did not induce widespread fear of skin cancer.

In summary our study clearly shows that nationwide public skin cancer education campaigns significantly influence attitudes and habits in accordance with the study aims.

9. References

1. MacKie R; Østerlind A; Ruiter D; Fritsch P; Aapro M; Cesarini J; Vanwijck R; Drzewiecki K; Kölmel K; Muir C. (1991). Report on consensus meeting of the EORTC melanoma group on educational needs for primary and secondary prevention of melanoma in Europe. *Eur J Cancer* **27**, 1317-1323
2. Bernhardt J; Boldeman C; Breitbart EW; Breitbart M; Christophers et al. (1996). "How to decrease morbidity and mortality of skin cancer". Primary prevention of skin cancer. Screening of skin cancer. Consensus statement. *Europ J Cancer Prev*; **accepted**
3. Breitbart EW; Ebert A; Flatten G; Heidbreder G; Jung EG; Kimmig W; Kirschner W; Kölmel K; Landthaler M; Paul E; Roser M; Tilgen W; Verres R; Weichenthal M; Wiskemann A; Christophers E. (1992). Ziele und Ergebnisse der Hautkrebskampagne 1989/90. *Deutsches Ärzteblatt - "Ärztliche Mitteilungen"* **14**, 1199-1202
4. Melia J; Ellman R; Chamberlain J. (1994). Meeting The Health of the Nation target for skin cancer: problems with tackling prevention and monitoring trends. *J Public Health Med* **16**, 225-32
5. Girgis A; Sanson-Fisher RW; Tripodi DA; Golding T. (1993). Evaluation of interventions to improve solar protection in primary schools. *Health Educ Q* **20**, 275-87
6. Hughes BR; Altman DG; Newton JA. (1993). Melanoma and skin cancer: evaluation of a health education programme for secondary schools. *Br J Dermatol* **128**, 412-7 .
7. Kölmel FK; Cleffmann U; Roser M; Breitbart EW. (1993). Sonne, Kind und Melanome-Melanomprävention im Kindesalter. *Der Kinderarzt* **4**, 470-481
8. Putnam GL; Yanagisako KL. (1982). Skin cancer comic book: Evaluation of a public educational vehicle. *Cancer Detect Prev* **5**, 349-356
9. Chapman S; Marks R; King M. (1992). Trends in tans and skin protection in Australian fashion magazines, 1982 through 1991. *Am J Public Health* **82**, 1677-1680
10. Marks R. (1994). Melanoma prevention: Is it possible to change a population's behavior in the sun? *Pigment Cell Res* **7**, 104-106
11. Marks R. (1995). Skin Cancer Control in Australia. The balance between primary prevention and early detection. *Arch Dermatol* **131**, 474-479
12. Boldeman C; Jansson B; Holm LE. (1991). Primary prevention of malignant melanoma in a Swedish urban preschool sector. *J Cancer Educ* **6**, 247-53
13. Hill D; White V; Marks R; Borland R. (1993). Changes in sun-related attitudes and behaviours, and reduced sunburn prevalence in a population at high risk of melanoma. *Eur J Cancer Prev* **2**, 447-56
14. Bourke JF; Healsmith MF; Graham-Brown RA. (1995). Melanoma awareness and sun exposure in Leicester. *Br J Dermatol* **132**, 251-6
15. McGee R; Elwood M; Williams S; Lowry F. (1994). Who comes to skin checks? *N Z Med J* **107**, 58-60
16. Koh HK; Geller AC; Miller DR; Lew RA. (1995). The early detection of and screening for melanoma. International status. *Cancer* **75**, 674-83
17. Rampen FH; Berretty PJ; Van Huystee BE; Kiemeney LA; Nijs CH. (1993). Lack of selective attendance of participants at skin cancer/melanoma screening clinics. *J Am Acad Dermatol* **29**, 423-7
18. Doherty VR; MacKie RM. (1988). Experience of a public education programme on early detection of cutaneous malignant melanoma. *BMJ* **297**, 388-391
19. MacKie RM; Hole D. (1992). Audit of public education campaign to encourage earlier detection of malignant melanoma. *BMJ* **304**, 1012-1015
20. Rampen FH; Berretty PJ; van Huystee BE; Kiemeney LA; Nijs CH. (1993). General practitioners' workload after skin cancer/melanoma screening clinics in The Netherlands. *Dermatology* **186**, 258-60
21. Field SI. (1987). Melanoma skin cancer screening in Michigan. *J Am Acad Dermatol* **16**, 578-83
22. Koh HK, Caruso A, Gage I, Geller AC, Prout MN, White H, O'Connor K, Balash EM, Blumental G, Rex IH Jr, et al. (1990). Evaluation of melanoma/skin cancer screening in Massachusetts. Preliminary results. *Cancer* **65**, 375-9

23. Koh HK; Geller AC; Miller DR; Caruso A; Gage I; Lew RA. (1991). Who is being screened for melanoma/skin cancer? Characteristics of persons screened in Massachusetts. *J Am Acad Dermatol* **24**, 271-7
24. Hoffmann K; Matthes U; Stucker M; Segerling M; Altmeyer P. (1990). 1989 Bochum health education program for malignant melanoma *Offentl Gesundheitswes* **52**, 9-13
25. Krol S; Keijser LM; van der Rhee HJ; Welvaart K. (1991). Screening for skin cancer in The Netherlands. *Acta Derm Venereol* **71**, 317-21
26. Rampen FH; van Huystee BE; Kiemeney LA. (1991). Melanoma/skin cancer screening clinics: experiences in The Netherlands. *J Am Acad Dermatol* **25**, 776-7
27. Katris P; Crock JG; Gray BN. (1996). Research note: the Lions Cancer Institute and the Western Australian Society of Plastic Surgeons skin cancer screening programme. *Aust N Z J Surg* **66**, 101-4
28. Curley RK; Taylor FG; Marsden RA; Cox J; McLaughlin CA. (1993). Screening for skin cancer: experience of an occupational health screening programme. *Occup-Med-Oxf* **43**, 207-10
29. de Rooij MJ; Rampen FH; Schouten LJ; Neumann HA. (1995). Skin cancer screening focusing on melanoma yields more selective attendance. *Arch Dermatol* **131**, 422-5
30. Girgis A; Campbell EM; Redman S; Sanson-Fisher RW. (1991). Screening for melanoma: a community survey of prevalence and predictors. *Med J Aust* **154**, 338-43
31. Pehamberger H; Binder M; Knollmayer S; Wolff K. (1993). Immediate effects of a public education campaign on prognostic features of melanoma. *J Am Acad Dermatol* **29**, 106-108
32. Herd RM; Cooper EJ; Hunter JA; McLaren K; Chetty U; Watson AC; Gollock J. (1995). Cutaneous malignant melanoma. Publicity, screening clinics and survival. The Edinburgh experience 1982-90. *Br J Dermatol* **132**, 563-70
33. Tonks T; Duncan G; Crichton S. (1995). Public education and depth of primary malignant melanoma, Hutt Hospital, 1985-92. *N Z Med J* **108**, 344-6
34. Cristofolini M; Bianchi R; Boi S; DeCarli A; Micciolo R; Cristofolini P; Zumiani G. (1993). Effectiveness of the health campaign for the early diagnosis of cutaneous melanoma in Trentino, Italy. *J Dermatol Surg Oncol* **19**, 117-20
35. Marks R. (1991). Early detection of melanoma. Improving the outlook for all members of the population at risk [editorial] *Med J Aust.* **154**, 574-5
36. Scotto J; Pitcher H; Lee JAH. (1991). Indications of future decreasing trends in skin-melanoma mortality among whites in the United States. *Int J Cancer* **49**, 490-97
37. Elwood JM; Koh HK. (1994). Etiology, epidemiology, risk factors, and public health issues of melanoma. *Curr Opin Oncol* **6**, 179-87
38. Armstrong BK; Kricker A. (1994). Cutaneous melanoma. *Cancer Surv* **19-20**, 219-40
39. Streetly A; Markowe H. (1995). Changing trends in the epidemiology of malignant melanoma: gender differences and their implications for public health. *Int J Epidemiol* **24**, 897-907
40. Giles GG; Armstrong BK; Burton RC; Staples MP; Thursfield VJ. (1996). Has mortality from melanoma stopped rising in Australia? Analysis of trends between 1931 and 1994. *BMJ* **312**, 1121-5
41. Cristofolini M; Bianchi R; Boi S; Decarli A; Hanau C; Micciolo R; Zumiani G. (1993). Analysis of the cost-effectiveness ratio of the health campaign for the early diagnosis of cutaneous melanoma in Trentino, Italy. *Cancer* **71**, 370-4
42. Limpert GH. (1995). Skin-cancer screening: a three-year experience that paid for itself. *J Fam Pract* **40**, 471-5
43. Koh HK; Geller AC; Miller DR; Grossbart TA; Lew RA. (1996). Prevention and early detection strategies for melanoma and skin cancer. Current status [editorial]. *Arch-Dermatol.* **132**, 436-43
44. Rossi JS; Blais LM; Redding CA; Weinstock MA. (1995). Preventing skin cancer through behaviour change: implications for interventions. *Dermatol Clin* **13**, 613-622
45. MacKie RM. (1995). Melanoma prevention and early detection. *Br Med Bull* **51**, 570-83
46. Campbell HS; Birdsell JM. (1994). Knowledge, beliefs, and sun protection behaviors of Alberta adults. *Prev Med* **23**, 160-6
47. Austoker J. (1994). Melanoma: prevention and early diagnosis. *BMJ* **308**, 1682-6
48. Brandt TP. (1996). Skin cancer screening. *Med Clin North Am* **80**, 99-114

BIOLOGICAL UV DOSIMETRY AS TOOL FOR ASSESSING THE RISKS FROM AN INCREASED ENVIRONMENTAL UV-B RADIATION

G. HORNECK, P. RETTBERG AND R. FACIUS

Deutsches Zentrum für Luft- und Raumfahrt, Institut für Luft- und Raumfahrtmedizin, Abteilung Strahlenbiologie
D-51170 Köln, Germany

Abstract

Global scale decreases in total ozone have resulted in an increase in UV-B radiation at ground level. To assess the risks to human health and ecosystems from an enhanced UV-B radiation, an accurate and reliable UV monitoring system is required that weights the spectral irradiance according to the biological responses under consideration. Biological dosimetry meets these requirements by directly weighting the incident UV components of sunlight in relation to the biological effectiveness of the different wavelengths and to potential interactions between them. Bacteria, viruses and biomolecules have been developed as biological dosimeters. Their responses to environmental UV radiation, indicated as inactivation, mutagenesis or photochemical injury, reflect the UV sensitivity of DNA. Biological dosimeters are complementary to physical measurements as field dosimeters in terrestrial and aquatic ecosystems for long-term monitoring of environmental UV radiation and as personal dosimeters.

1. Introduction

The stratospheric ozone layer constitutes a protective atmospheric filter against harmful solar UV radiation. Satellite based ozone measurements, conducted since 1979, have revealed statistically significant downward trends in the ozone column on a global scale, except in the tropics [1]. A decrease in total column ozone shifts the edge of the solar spectrum that reaches the surface of the Earth towards shorter wavelengths. It is this highly wavelength specific absorption characteristic of atmospheric ozone and the wavelength specificity of the biological action spectra in the UV-B range that give rise to the global concern on the impact of a depletion of the stratospheric ozone layer and thereby an increase in UV-B upon the biosphere. The shape of the action spectrum of the biological phenomenon under consideration determines whether an incremental change in ozone results in significant changes in the biological effectiveness of solar UV radiation. This phenomenon stresses the need for a biological weighting of solar UV irradiance for assessing its effects on biological systems.

C. Baumstark-Khan et al. (eds.),
Fundamentals for the Assessment of Risks from Environmental Radiation, 451-456.

2. Biological Weighting of Solar UV Radiation

Generally, biologically weighted radiometric quantities are derived from the spectral solar irradiance data by multiplication with an action spectrum of a relevant photobiological reaction, e.g. DNA damage, erythema formation, skin cancer, reduced productivity of terrestrial plants or UV sensitivity of aquatic ecosystems (Figure 1). The biologically effective irradiance E_{eff} in $(W/m^2)_{eff}$ is then determined as follows:

$$E_{eff} = \int E_{\lambda}(\lambda) \cdot S_{\lambda}(\lambda)\, d\lambda \tag{1}$$

where $E_{\lambda}(\lambda)$ is the solar spectrum (W/m^2 nm), $S_{\lambda}(\lambda)$ is the action spectrum (relative units) and λ the wavelength (nm). Integration of the biologically effective irradiance E_{eff} over time, e.g. a full day, gives the biologically effective dose (BED) H_{eff} in $(J/m^2)_{eff}$. Figure 1 shows an example where the spectral response function of the biological UV dosimeter biofilm [2] which is indicative for the DNA damaging potential of solar UV is taken as biological weighting function. The solar biological effectiveness spectra, which are obtained by multiplication of the solar spectrum with the action spectrum, exhibit peaks in the range of 305 to 315 nm with a shift towards shorter wavelengths with decreasing solar zenith angle (sza). Besides this strong dependence on the sza, the weighted irradiance is influenced by clouds, aerosols and changes in the vertical ozone profiles. With decreasing ozone concentrations the peak of the biological effectiveness spectra is shifted towards shorter wavelengths.

However, even assuming highly accurate spectroradiometric data the calculated E_{eff} values may be erroneous or may lead to misinterpretations due to potential pitfalls, as outlined in the following.

2.1. DEMAND FOR HIGHLY ACCURATE BIOLOGICAL WEIGHTING FUNCTIONS

In the past, most action spectra have been produced for the purpose of identifying the UV sensitive target, e.g. the DNA. In this case, a qualitative or semiquantitative approach was mostly sufficient. When comparing the action spectra determined in different laboratories, discrepancies up to an order of magnitude can be observed, even for the same biological system as compiled e.g. by Setlow [3] and Tyrrell and Pidoux [4]. However, in biologically weighted UV dosimetry, small deviations in the biological weighting function are amplified resulting in large differences in the calculated E_{eff} values.

2.2. DEFICIENCY OF ACTION SPECTRA IN THE UV-A REGION

Because the spectrum of solar radiation at the Earth's surface increases by orders of magnitude from the UV-B to the UV-A range, the tails of action spectra in the UV-A region become considerably significant, although the biological effectiveness of UV-A radiation might be lower by orders of magnitude compared to the peak response. However, several action spectra from the literature do not extend to this region or show large

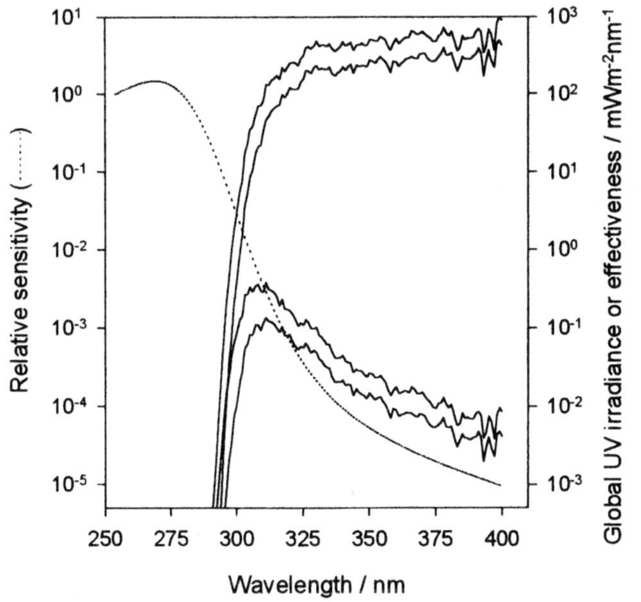

Figure. 1. Solar irradiance spectra, action spectrum of the biofilm biological UV dosimeter and calculated biological effectiveness spectra for August 7, 1996, at 10:00 (right curve) and 13:00 (left curve) MEST at DLR, Köln (50.52°N, 7.7°E, altitude 92 HNN).

error bars, mainly due to the inaccessibility of powerful UV-A sources. Including the tails of the action spectra in the calculations significantly affects the resulting E_{eff} values.

2.3. MONOCHROMATIC VERSUS POLYCHROMATIC UV RADIATION

Most action spectra have been determined by exposing organisms to monochromatic radiation. Sunlight however is a broad-band source where radiation at different wavelengths may interact. Apparent interactions of radiation at different wavelengths of the solar spectrum have been reported for various biological endpoints, as reviewed by Jagger [5]. Such interactions are indicative of the involvement of more than one photoprocess in the observable overall response of the system. In order to prove the potential involvement of interactive effects of solar radiation, monochromatic action spectra have to be complemented by polychromatic action spectra.

2.4. ABSOLUTE VERSUS RELATIVE ACTION SPECTRA.

In the calculations of biologically effective irradiance, action spectra in relative units are generally used with a reference wavelength (mostly the peak wavelength of the response function) at which the biological response is taken equal to one. Examples are 298 or 300 nm for skin damage, 280 nm for plant damage, and 254 nm for DNA injury.

In this approach, the calculated E_{eff} values represent only the relative biological responses at different wavelengths and not the absolute dimensions of the biological effects. Therefore, the BED calculated according to Equation 1 for two different photobiological processes cannot be compared numerically, only with respect to relative changes, e.g. of location, time or ozone concentration.

3. Biological UV Dosimetry

Biological dosimeters automatically weight the incident UV components of sunlight in relation to their biological effectiveness and to interactions between them in absolute terms. Ideally, the spectral response of the biological dosimeter is identical to that of the action spectrum of the photobiological effect under consideration. In this case, H_{eff} is given by the following relation:

$$H_{eff} = F \tag{2}$$

where F is the equivalent dose of a standard radiation in J/m^2 producing the same response of the detector as the radiation under consideration. As reviewed by Rontó et al. [6] and Horneck [7, 8], in most cases simple test systems have been used as biological dosimeters, such as uracil, DNA, bacteriophages, and bacteria. They directly measure the DNA damaging capacity of solar UV radiation as the main initiating event in a variety of harmful UV effects to human health and the whole biosphere. Induction rates for lethality, mutagenesis and dimerization have been used to determine directly the BED. For ecological questions, more complex dosimetric systems, such as flagellates or *Daphnia* have been applied as biological indicators.

We have developed a biofilm as biological UV dosimeter which is based on dry spores of *B. subtilis* immobilized on a polyester sheet. After exposure to sunlight and calibration at 254 nm, the biofilm is incubated in nutrient medium and the biomass formed is stained. The inhibition of biological activity is determined as relative optical density (OD) [2, 9]. The BED in $(J/m^2)_{eff}$ is given as equivalent dose to that at 254 nm which according to Equation 2 produces the same effect as the environmental radiation. The action spectrum deviates from the standard erythema spectrum by less than a factor of 2 over a wide range from UV-B to UV-A (Figure 2). The response of the biofilm to UV is additive, follows the reciprocity law, and it is independent of temperature between -20°C and 70°C, humidity in the range of 20 % to 80 % r. h. and storage up to 1 a [2]. Hence, the biofilm combines both, the direct biological weighting of a biological dosimeter and the robustness and simplicity of a chemical film dosimeter.

4. Applications of Biological Dosimetry

Biological dosimeters integrate over the exposure time, thereby providing an accurate record of the total biologically effective dose over an designated period regardless of changes in weather conditions. Even low levels of UV, as they occur e.g. in-water at deeper layers or after the polar night, can be recorded with high sensitivity by prolonged exposure of biological dosimeters. This makes them especially applicable for long-term

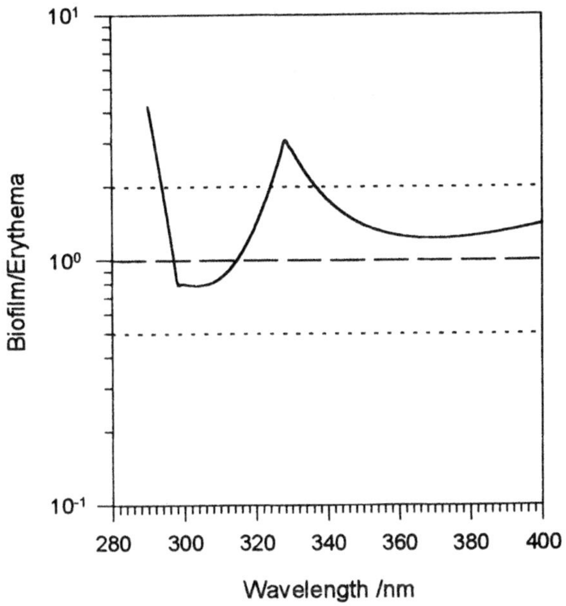

Figure. 2. Ratio between the spectral response of the biological UV dosimeter biofilm and the standard erythema curve (modified from [7]). The dotted lines indicate a deviation by a factor of ≤2.

monitoring of environmental UV radiation, even in remote locations, for scanning the distribution of BED in an ecosystem and for monitoring the personal UV exposure.

Acknowledgement. This study was supported by the Commission of the European Union through the projects no. EV-CT93-0342 and ENV4-CT95-0044.

5. References

1. Frederick, J.E. (1993) Ultraviolet sunlight reaching the Earth's surface: a review of recent research, *Photochem. Photobiol.* **57**, 175-178.
2. Quintern, L.E., Horneck, G., Eschweiler, U., Bücker, H. (1992) A biofilm used as ultraviolet dosimeter, *Photochem. Photobiol.* **55**, 389-395.
3. Setlow, R.B. (1974) The wavelengths in sunlight effective in producing skin cancer: a theoretical analysis, *Proc. Natl. Acad. Sci. USA* **71**, 3363-3366.
4. Tyrrell, R.M., Pidoux, M. (1987) Action spectra for human skin cells: estimates of the relative cytotoxicity of the middle ultraviolet, near ultraviolet, and violet regions of sunlight on epidermal keratinocytes, *Cancer Res.* **47**, 1825-1829.
5. Jagger, J., (1985). *Solar Actions on Living Cells*, Praeger, New York
6. Rontó, G., Gáspár, S., Gróf, P., Bérces, A., Gugolya, Z. (1994) Ultraviolet dosimetry in outdoor measurements based on bacteriophage T7 as a biosensor, *Photochem. Photobiol.* **59**, 209-214.
7. Horneck, G. (1995) Quantification of the biological effectiveness of environmental UV radiation, *J. Photochem. Photobiol. B: Biol.* **31**, 43-49.

8. Horneck, G. (1997) Biological UV dosimetry. In: D.-P. Häder (ed.). The Effects of Ozone Depletion on Aquatic Ecosystems, Academic, Landes, Austin, pp. 119-142.
9. Horneck, G., Rettberg, P., Rabbow, E., Strauch, W., Seckmeyer, G., Facius, R., Reitz, G., Strauch, K., Schott, J.-U. (1996) Biological dosimetry of solar radiation for different simulated ozone column thickness, *J. Photochem. Photobiol. B: Biol.*, **32**, 189-196.

EFFECTS OF SOLAR UV RADIATION ON AQUATIC ECOSYSTEMS

D.-P. HÄDER

Friedrich-Alexander-Universität, Institut für Botanik und Pharmazeutische Biologie
Staudtstr. 5, D-91058 Erlangen, Germany

Abstract

The marine ecosystem represents the single most important ecosystem on our planet. While the standing crop is comparatively small, it has been estimated to produce more biomass than all terrestrial ecosystems taken together. Thus, it constitutes also an important sink for atmospheric carbon dioxide. Earlier studies have shown that many phytoplankton organisms are under UV-B stress even at ambient levels. Phytoplankton is not equally distributed in the water column, but rather uses highly specific orientation strategies to move to and stay in zones of suitable light intensity for growth and survival. In some ecologically significant groups these zones undergo pronounced vertical daily migrations. However, these mechanisms are impaired by UV-B radiation. When these orientation strategies are impaired by short wavelength solar radiation (UV-B, 280-315 nm) the organisms cannot adapt to the constantly changing conditions in their environment. As a consequence, the photosynthetic pigments as well as the structural proteins of the photosynthetic apparatus of micro- and macroalgae are affected by solar radiation as shown by FPLC and gel electrophoresis as well as spectroscopic methods and pulse amplitude fluorescence. Also chromoproteins of the photoreceptor apparatus responsible for phototaxis have been found to be affected. Several targets of solar ultraviolet radiation have been identified. Any increase in solar UV-B radiation due to partial stratospheric ozone depletion caused by the anthropogenic emission of gaseous pollutants bears the risk of affecting the phytoplankton ecosystems. This situation may lead to a decrease in biomass production which in turn affects the whole intricate marine food web. Also the CO_2 sink capacity will be reduced; a hypothetical loss of 10 % phytoplankton productivity will result in 5 gigatons of carbon (in the form of CO_2) not removed from the atmosphere.

1. Introduction

Aquatic ecosystems equal terrestrial ecosystems in biomass production and are regarded as large sinks of atmospheric carbon estimated between 90 and 100 gigatons annually [1]. Since only 0.5 % of the water surface is freshwater, marine systems are by far the

C. Baumstark-Khan et al. (eds.),
Fundamentals for the Assessment of Risks from Environmental Radiation, 457-461.

most important. However, lakes are excellent model systems, and many of these systems are locally important. Most macroalgae are restricted to coastal areas, phytoplankton contributes the largest share in biomass production. Phytoplankton are the basis for the intricate food web in the oceans. Previous investigations indicate that many aquatic ecosystems are under considerable stress even at ambient UV-B levels [2, 3]. Since they depend on solar energy for photosynthesis, phytoplankton are restricted to the upper water column where they are simultaneously exposed to high levels of short wavelength radiation [4-9].

A number of basic questions need to be answered to determine the effects of solar UV-B radiation on aquatic ecosystems:

- What is the present and future spectral distribution of solar radiation at the surface of the water?
- What is the penetration of solar UV-B in different water types?
- What is the vertical distribution of the aquatic organisms in the water column?
- What is the sensitivity of the organisms affected by solar UV-B?
- How effective are repair and adaptation mechanisms?

2. Primary Production

Solar UV-B radiation affects DNA, impairs photosynthesis, enzyme activity and nitrogen incorporation, bleaches cellular pigments and inhibits motility and orientation [10, 11]. DNA is one of the UV-B targets, but many other chromophores and proteins are affected. Therefore many different action spectra have been measured in phytoplankton which are further complicated by antagonistic and repair processes stimulated by UV-A and visible radiation.

The highest concentrations of phytoplankton are found in the high latitude regions while the tropics and subtropics show far lower concentrations. In temperate oceans phytoplankton blooms occur in spring and disappear during summer. This has been taken as an indication for the sensitivity of phytoplankton to solar UV, and significant increases in irradiation are expected to have detrimental effects on phytoplankton productivity. Since depletions in stratospheric ozone are expected to continue well into the next century the effects of resulting UV-B increases on marine primary productivity may also be relevant in polar regions as well as at mid and low latitudes.

In the past, the role of nano- and picoplankton has been underestimated because of their small size; recently their contribution to the total biomass has been estimated to be at least 40 %. Bacterioplankton which are responsible for degradation and cycling of organic matter in the sea has a similarly important role. UV-B has been found to strongly affect nano- and bacterioplankton [12]. Several researchers have pointed out that small organisms are likely to be more susceptible to UV than larger ones [8]. Therefore, UV-B may affect species composition and lower food web processes. This effect will also affect the community structure within the primary producers.

3. Penetration of Solar Radiation onto the Water Column

The transparency of the water strongly depends on the water type [13]: in coastal waters with high concentrations of particular substances and gelbstoff UV-B may penetrate less than 1 m to the 1% level; on the other hand, in clear oceanic waters UV-B has been shown to penetrate to several tens of meters [15]. Phytoplankton productivity is limited to the euphotic zone, the top layer of the water column. The position of the organisms in the euphotic zone is governed by wind and waves as well as active movements. Solar UV-B irradiation affects motility and orientation in phytoplankton [2, 4].

Recent work indicated that solar UV-B affects nitrogenase activity and carbon dioxide uptake in rice paddy cyanobacteria [15, 16]. Some cyanobacteria, however, seem to be better adapted to high solar radiation than closely related forms.

4. Macroalgae and Seagrasses

Interest in UV-B effects on macroalgae and seagrasses has increased during the last few years [17]. Most macrophytes are attached to their growing site and are therefore restricted to certain depth zones above, below or within the tidal zone. If the UV-B/PAR ratio increases, the algae will be exposed to enhanced short wavelength radiation to which they may not be adapted. Using PAM (pulse amplitude modulation) fluorescence measurements, deep-water benthic algae were found to be most sensitive while intertidal algae were least sensitive [18-20].

5. Screening Pigments

Screening pigments which protect the organism from solar UV radiation have been found both in marine and freshwater organisms [7]. Scytonemin has been described in cyanobacteria where it is induced by UV [21]. In addition, cyanobacteria as well as many eukaryotic phytoplankton and macroalgae use other water soluble substances, mycosporines, as screening pigments which are passed on through the food web to the primary and secondary consumers. Other organisms use carotenoids to dissipate the excess radiation energy from the photosynthetic pigments.

6. Consequences

Though there is increasing evidence that enhanced UV-B is harmful to aquatic ecosystems, quantitative estimates are preliminary at this stage. In order to evaluate the current productivity in the oceans and a possible decrease in the future, increased efforts are necessary. A major loss in primary biomass productivity may have significant consequences for the intricate food web in aquatic ecosystems and even affect the harvest of fish and crustaceans for human consumption. Consequences of increased solar UV-B levels may be further complicated by feedback loops not yet considered and understood.

Biological effects of small changes in UV-B may be difficult to determine because the biological uncertainties and variations are large. The second major consequence of decreased phytoplankton productivity may be a reduced sink capacity for atmospheric carbon dioxide which may result in an augmented greenhouse effect and global climate change.

The predicted changes in the ozone layer may be too fast to be countered by genetic adaptation to higher UV-B levels. Different species differ in their sensitivity toward solar short wavelength radiation; therefore shifts in species composition may occur. UV seems to affect smaller phytoplankton more than larger organisms. Since primary feeders prey by size and not by species preference, changes in the food web may be a consequence.

References

1. Siegenthaler, U., and Sarmiento, J.L. (1993) Atmospheric carbon dioxide and the ocean, *Nature* **365**, 119-125.
2. Häder, D.-P. (1993) Effects of enhanced solar ultraviolet radiation on aquatic ecosystems, in M. Tevini, (ed.), *UV-B radiation and ozone depletion. Effects on Humans, Animals, Plants, Microorganisms, and Materials*, Lewis Publ., Boca Raton, Ann Arbor, London, Tokyo, pp. 155-192.
3. Smith, R.C., Prézelin, B.B., Baker, K.S., Bidigare, R.R., Boucher, N.P., Coley, T., Karentz, D., MacIntyre, S.,. Matlick, H.A, Menzies, D., Ondrusek, M., Wan, Z., and Waters, K.J. (1992) Ozone depletion: Ultraviolet radiation and phytoplankton biology in Antarctic waters, *Science* **255**, 952-959.
4. Häder, D.-P. (1993b) Risks of enhanced solar ultraviolet radiation for aquatic ecosystems, *Progress in Phycological Research* **9**, 1-45.
5. Weiler, C.S., and Penhale, P.A. *Ultraviolet Radiation in Antarctica: Measurements and Biological Effects*, American Geophysical Union, Washington, D.C., pp. 1-62.
6. Williamson, C.E., and Zagarese, H.E. (1994) *Impact of UV-B Radiation on Pelagic Freshwater Ecosystems, Arch. Hydrobiol.* Beiheft 43 (Special issue).
7. Karentz, D., Cleaver, J.E., and Mitchell, D.L. (1991) Cell survival characteristics and molecular responses of Antarctic phytoplankton to ultraviolet-B radiation, *J. Phycol.* **27**, 326-341.
8. Karentz, D., Bothwell, M.L., Coffin, R.B., Hanson, A., Herndl, G.J., Kilham, S.S., Lesser, M.P., Lindell, M., Moeller, R.E., Morris, D.P., Neale, P.J., Sanders, R.W., Weiler, C.S., and Wetzel, R.G. (1994) Report of working group on bacteria and phytoplankton. in C.E. Williamson and H.E. Zagarese (eds.), *Impact of UV-B Radiation on Pelagic Freshwater Ecosystems, Arch. Hydrobiol.* Beiheft 43 (Special issue), pp. 31-69.
9. Smith, R.C., and Cullen, J.J. (1995) *Implications of increased solar UVB for aquatic ecosystems, U.S. National Report to the IUGG (1991-1994)*, American Geophysical Union.
10. Döhler, G., Hagmeier, E., Grigoleit, E., and Krause, K.-D. (1991) Impact of solar UV radiation on uptake of ^{15}N-ammonia and ^{15}N-nitrate by marine diatoms and natural phytoplankton, *Biol. Phys. Pfl.* **187**, 293-303.
11. Häder, D.-P., Worrest, R.C., Kumar, H.D., and Smith, R.C. (1995) Effects of increased solar ultraviolet radiation on aquatic ecosystems, *AMBIO* **24**, 174-180, Reprint of *UNEP* document.
12. Herndl, G.J., Müller-Niklas, G., and Frick, J. (1993) Major role of ultraviolet-B in controlling bacterioplankton growth in the surface layer of the ocean, *Nature* **361**, 717-719.
13. Piazena, H., and Häder, D.-P. (1994) Penetration of solar UV irradiation in coastal lagoons of the Southern Baltic Sea and its effect on phytoplankton communities, *Photochem. Photobiol.* **60**, in press.
14. Smith, R.C., and Baker, K.S. (1979) Penetration of UV-B and biologically effective dose-rates in natural waters, *Photochem. Photobiol.* **29**, 311-323.

15. Tyagi, R., Kumar, H.D., Vyas, D., and Kumar, A. (1991) Effects of ultraviolet-B radiation on growth, pigmentation, $NaH^{14}CO_3$ uptake and nitrogen metabolism in *Nostoc muscorum*, in Y.P. Abrol, P.N. Wattal, A. Gnanam, Govindjee, D.R. Ort and A.H. Teramura (eds.) *Impact of Global Climatic Changes on Photosynthesis and Plant Productivity*, Proceedings of the Indo-US Workshop held on January 8-12,1991 at New Delhi, India, Oxford and IBH Publishing Co. New Delhi, Bombay, Calcutta, pp. 109-124.
16. Tyagi, R., Srinivas, G., Vyas, D., Kumar, A., and Kumar, H.D. (1992) Differential effect of ultraviolet-B radiation on certain metabolic processes in a chromatically adapting *Nostoc*, *Photochem. Photobiol.* **55**, 401-407.
17. Häder, D.-P., and Figueroa, F.L. (1997) Photoecophysiology of marine macroalgae, *Photochem. Photobiol.* **66**, 1-14.
18. Häder, D.-P., Worrest, R.C., Kumar, H.D., and Smith, R.C. (1994) Effects of increased solar ultraviolet radiation on aquatic ecosystems, *UNEP Environmental Effects Panel Report*, 65-77.
19. Häder, D.-P. (1994) UV-B effects on aquatic systems, in R.H Biggs and M.E.B. Joyner (eds.), *Stratospheric Ozone Depletion/UV-B Radiation in the Biosphere, NATO ASI Series, Vol.* I18, Springer, Berlin, Heidelberg, pp. 155-160.
20. Larkum, A.W.D., and Wood, W.F. (1993) The effect of UV-B radiation on photosynthesis and respiration of phytoplankton, benthic macroalgae and seagrasses, *Photosynth. Res.* **36**, 17-23.
21. Garcia-Pichel, F., and Castenholz, R.W. (1991) Characterization and biological implications of scytonemin, a cyanobacterial sheath pigment, *J. Phycol.* **27**, 395-409.

EFFECTS OF OZONE DEPLETION AND INCREASED ULTRAVIOLET-B RADIATION ON TERRESTRIAL PLANTS

L.O. BJÖRN

Plant Physiology, Lund University
Box 117, S-221 00 Lund, Sweden

Abstract

The reasons for the present worry about ozone depletion and its effects on terrestrial plants are explained. Only by a combination of fortunate circumstances has a catastrophe been avoided. For several reasons our group has focused its research on the high latitude terrestrial ecosystems, and is investigating several of them in northern Sweden and on the island of Spitzbergen. One reason is that the radiation amplification factor, i.e. the percent increase in biologically effective radiation per percent ozone depletion is higher in plants with low amounts of protective pigments, as is the case with subarctic and arctic plants. Another is the slow repair of radiation damage at low temperature. Some effects of simulated ozone depletion on subarctic vegetation are described.

1. Introduction: Why Worry About the Ozone Layer?

Thanks to the sun the temperature of our planet is just what we need. The sun is also powering life on earth by providing free energy, directly or indirectly by the process of photosynthesis. These effects are due almost entirely to the solar radiation of wavelength above 315 nm, i.e. UV-A, visible light (photosynthetically active radiation) and infrared radiation. But the sun also emits electromagnetic radiation of shorter wavelength, which has destructive effects on all cells. Various gases of the atmosphere protect us from electromagnetic radiation of wavelength below 280 nm: γ radiation, X-rays and UV-C radiation. But in the window 280-315 nm, protection of the biosphere from lethal solar radiation depends on a single constituent of the atmosphere: ozone. That ozone can fulfil this role is almost incredible in view of the small amount present: if all the ozone in the atmosphere could be condensed to a liquid and smeared over the surface of the earth, the film would be only about 4 µm thick. One reason that the tiny amount can have this important role is that its absorption spectrum coincides with that of DNA, a substance that is essential to all living beings but can be easily damaged by ultraviolet radiation. And it may be sufficient to damage a single DNA molecule in a cell to cause the death of the cell, or to transform it into a malignant cell, which can kill a man.

C. Baumstark-Khan et al. (eds.),
Fundamentals for the Assessment of Risks from Environmental Radiation, 463-470.

During the 1970s and 1980s about a million tons of synthetic chlorofluorocarbons, organic compounds containing chlorine and fluorine, were released *each year* into the atmosphere. This is a small amount compared to that of some other substances added to the atmosphere as a consequence of human activities, and these substances were first considered to be quite harmless. But then it was found that they act as transport vehicles for the chlorine, which is released as free chlorine atoms when the substances eventually reach the upper atmosphere and are hit by the unattenuated radiation there. And it was found that a single chlorine atom, by catalytic action, causes the destruction of, on average, about 100,000 ozone molecules. Bromine compounds, as well as nitrogen oxides, also released by human activities into the atmosphere, have similar catalytic properties.

We are very fortunate that, thanks to the Nobel laureates Rowland, Molina and Crutzen, and others, and also by the dramatic warning given by the sudden appearance of the antarctic ozone hole in the part of the earth of least biological importance, the ozone destroying power of these substances was discovered at such an early date. We are fortunate that the politicians could be convinced to impose restrictions on the manufacture and use of ozone destroying substances at such an early date, starting with the Montreal protocol in 1987, and followed up by several later, and more stringent agreements. Finally, we are very fortunate that the substances chosen for refrigeration and other large-scale use were chlorofluorocarbons and not bromofluorocarbons. Compared to chlorine, the bromine atoms are many times more potent in destroying ozone. We have indeed been very careless in tinkering with the atmosphere. The good side of this is that it may be a warning for the future.

2. Present Ozone Depletion, and its Effect on Biologically Active UV-B radiation

The ozone hole in Antarctica is rather well known, and I will here focus on high northern latitudes, where the decrease in annual average ozone is presently 6% per decade. Fortunately (again) the largest ozone depletion occurs early in the year when the sun is low in the sky, and the filtering effect even of a thinned ozone layer is large due to the long path of the solar rays through the ozone layer.

Our special interest in high northern latitudes merits some explanation, because the UV-B radiation is much less intense there than at lower latitudes. One reason is that near the equator there is no significant depletion of stratospheric ozone. Another reason is a little more subtle, and has to do with the concept of radiation amplification factor and its relation to internal protective filters, which is explained in a following section. A third reason is the general fragility of arctic and subarctic ecosystems, and a fourth is the relation between damage and repair, which will also be explained below.

3. Spectral Weighting of Radiation

All UV-B radiation does not have the same destructive effect per photon. In general, the shorter wavelength bands, closer to the UV-C region, have a greater effect per photon, and even per energy unit. In other words, the action spectrum for damage rises from long wavelengths toward short wavelengths. On the other hand, the spectral irradiance of daylight has an opposite slope: it rises from short wavelengths towards longer wavelengths. Because of the steepness of the ozone absorption spectrum, this rise is very steep.

Instead of quantifying UV-B in a purely physical manner, saying, for instance, that the irradiance (the amount of radiation falling on a horizontal surface) is at a certain instant 1.3 W m^{-2}, it makes more sense to quote *weighted* radiation values. Weighting implies multiplying the spectral irradiance (or even more properly the spectral fluence rate, although this is seldom known) by the action spectrum for damage (the weighting function) for each wavelength band, and then integrating (summing) over all relevant wavelengths. A convention has been introduced that the weighting function should be normalized to have the value 1 at 300 nm. The procedure is explained in Figure 1.

A main difficulty with this is, that there exists a great deal of uncertainty concerning proper weighting functions. One function which has been very much used for terrestrial plants is the "generalized plant action spectrum" [1]. In reality the true functions vary not only between plant species, but also depending on the state of the particular plant, and among different kinds of radiation effect. It would be impossible to determine and keep track of all true weighting functions, and some simplifications are necessary. Still, as we shall show, the concept of weighting function is useful, and can help in reaching some important insights.

4. Radiation Amplification Factors

When ozone is depleted, the daylight spectrum changes. Generally speaking, the short wavelength cutoff is displaced towards shorter wavelengths. This causes an increase in biologically effective radiation, which can be calculated using the weighting function.

It is then found that the percent increase in biologically effective radiation $\Delta\%R$ is usually not the same as the percent decrease in ozone $-\Delta\%Oz$. The ratio $-\Delta\%R/\Delta\%Oz$ is called the *radiation amplification factor*, and it may be smaller or larger than unity (this explanation is slightly simplified, and applies exactly only for small changes).

The effect of ozone depletion depends not only on the sensitivity of the process under consideration to ultraviolet radiation in general, but also on the radiation amplification factor. A high factor means high sensitivity to ozone depletion.

5. DNA Damage under Various Conditions

It must be emphasized, that DNA is not the only molecule that can be damaged by ultraviolet radiation, but we shall continue our considerations using DNA as an example, partly because DNA is one important target, perhaps the most important, partly because the radiation effects on DNA are relatively well understood.

Ultraviolet radiation may cause different kinds of lesion in DNA. Induction of the two most frequent types (not necessarily the most important types under all conditions) have action spectra that for "naked" DNA closely follow the well known absorption spectrum for DNA. The lesions in question are cyclobutane type pyrimidine dimers and (6-4) photoproducts. For DNA in living plants on the other hand, the action spectra are

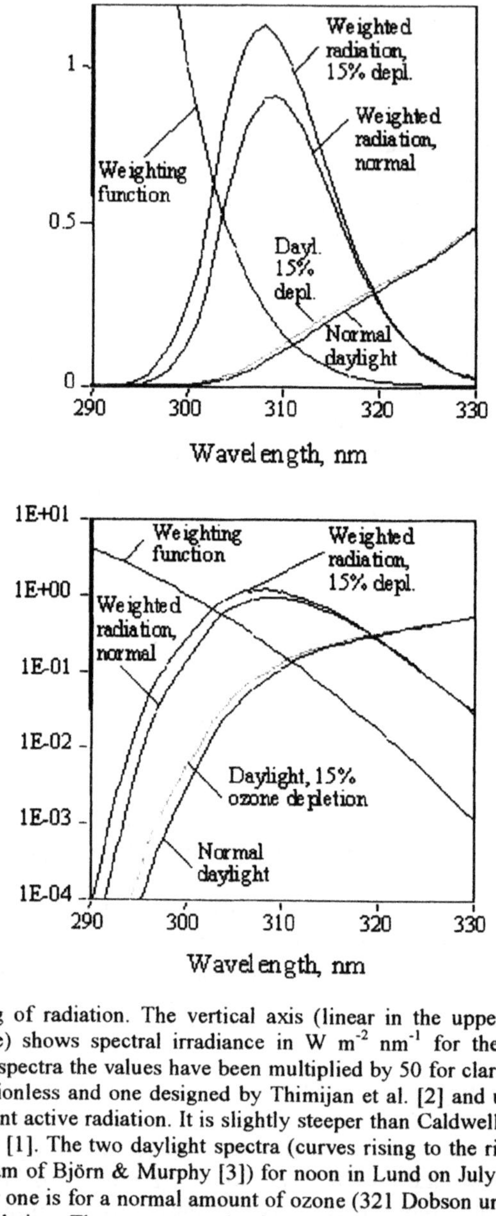

Figure 1. Weighting of radiation. The vertical axis (linear in the upper panel, logarithmic in the lower one) shows spectral irradiance in W m^{-2} nm^{-1} for the daylight spectra. For the weighted spectra the values have been multiplied by 50 for clarity. The weighting function is dimensionless and one designed by Thimijan et al. [2] and used in our group for computing plant active radiation. It is slightly steeper than Caldwell's "generalized plant action spectrum" [1]. The two daylight spectra (curves rising to the right) were computed (using the program of Björn & Murphy [3]) for noon in Lund on July 10 (unweighted radiation). The lower one is for a normal amount of ozone (321 Dobson units), the upper one for 15% ozone depletion. The two curves having maxima show weighted spectral irradiance (enlarged 50 times to show up more clearly) and are obtained by multiplication of the unweighted radiation values by the weighting function. The two latter curves (the upper one for 15% ozone depletion) differ considerably, although the difference between the unweighted spectral irradiance values appears very small. The weighted radiation curve for depleted ozone encloses an area above the abscissa that is 24.2% larger than the corresponding curve for normal ozone, implying a radiation amplification factor of 24.2/15=1.62. In this example the calculations were carried out for a certain point in time only, while in practice one would integrate over whole days.

modified by the absorption of radiation in protecting pigments present in the outer layers of the plant. The action spectrum for cyclobutane dimer production has been measured in alfalfa leaves by Quaite et al. [4]. The corresponding radiation amplification factors for "naked" [5] and "inside plant" DNA [4] under conditions similar to those of Figure 1 are 2.1 [6] and 0.5, respectively.

The alfalfa action spectrum was obtained with plants that are native to low latitudes. Such plants contain considerable amounts of UV-B absorbing, protective pigments (mainly isoflavonoids in this case), even if they are not subjected to UV-B during growth. It is the presence of these pigments that modifies the action spectrum and decreases the radiation amplification factor. Plants native to high latitudes (subarctic to arctic regions), generally contain much lower amounts of protective pigments [7]. Consequently it can be anticipated that the action spectrum for dimer formation in these plants more closely resembles that of "naked" DNA, and hence that higher radiation amplification factors are valid for these plants. This is the background for an important reason we have for focusing our research on plants growing at high latitudes.

In general, action spectra sloping more steeply towards shorter wavelength correspond to higher radiation amplification factors than "flatter" spectra, and appreciable tails extending into the UV-A region decrease the radiation amplification factors. No known process has a higher radiation amplification factor than those for which DNA is the radiation absorber (for some DNA lesions, DNA is not the main absorber, but the damage to DNA is inflicted by free radicals produced in photochemical reactions with other radiation targets). One "non-DNA" reaction (or group of reactions), which has often been discussed in connection with UV-B effects on plants is the inactivation of photosystem II, an important part of the photosynthetic system. The action spectrum for this process, however, is rising towards shorter wavelength with a very moderate slope, corresponding to a radiation amplification factor of only 0.1 to 0.2. Therefore, even if photosystem II is easily inactivated by UV-B (or UV-A), it is not likely that ozone depletion produces an important change in this inactivation rate, and plants are presumably equipped to deal with it. Photosynthesis can be experimentally inhibited by UV-B also in other ways, but there is no hard evidence that inhibition of photosynthesis plays any important role in the changes in growth observed in terrestrial plants as a consequence of increased UV-B.

6. Methods for Investigating Effects of Increased Environmental UV-B

In early experiments plants growing in growth chambers or greenhouses were exposed to UV-B from fluorescent lamps. The results were very alarming. But soon it was found that the impact of UV-B was exaggerated by the artificial conditions under which the plants were growing. In particular, it was found that the weak (as compared to natural daylight) visible light in the growth chambers made the plants more sensitive to UV-B. Nowadays experiments aimed at predicting the impact of ozone depletion are usually carried out in the field, sometimes in specially designed, high quality phytotrons.

Field experiments can be carried out as exclusion experiments, where the treatment consists in excluding the short wavelength part of the daylight spectrum with filters, and the control has a filter transmitting this radiation. This type of experiment is relatively cheap and easily set up. Alternatively experiments are carried out as enhancement experiments, in which the treatment consists of irradiation of plants with UV emitting fluorescent lamps. A small emission of UV-C radiation is filtered away using cellulose acetate film, and the control has similar lamps with filters removing both UV-C and UV-B.

In a simpler form of enhancement experiments, lamps are switched on and off by clocks. The times are chosen to make the biologically active radiation exposure per day correspond to that which would ensue from a certain ozone depletion, the effects of which one wishes to simulate. However, it can be suspected that the sudden switching on of the full irradiance can result in special shock effects in the plant, and it is also difficult to take the cloud effect into account with such a system. When the sky is cloud-covered, the amount of UV-enhancement required to simulate a certain amount of ozone depletion is lower than if the sky is free of clouds.

In a more sophisticated type of experiment, the lamps are regulated by the output of ultraviolet sensors, which sense both the ambient UV-B, and the sum of ambient and lamp UV-B. An electronic circuit keeps the ratio between the two at a value which closely corresponds to a certain percent ozone depletion, irrespective of solar elevation, clouds and other weather factors, and the actual amount of ozone.

7. Experiments of Our Own Group

7.1. GENERAL

We have started enhancement experiments both of the simpler and the more sophisticated type within projects supported by the Swedish Environmental Protection Agency and the European Commission [8-11, literature cited therein, and unpublished work). For the work outlined here, the Swedish group, in addition to the present author, has consisted mainly of the following scientists and students, in alphabetical order: M. Bertgren, T.V. Callaghan, C. Gehrke, U. Johanson, M. Sonesson. We have cooperated closely with foreign, mainly British colleagues, and the following are worth special mention: D. Gwynn-Jones, J.A. Lee, and C.H. Robinson. F. Bowles, Woods Hole, USA, has constructed the modulation system for administration of radiation. Most experiments performed are of an intermediate type, in which the radiation from the lamps is switched on and off in steps, to decrease the shock effect of going immediately from no enhancement to the full enhancement level. For some of the experiments we have results, in most cases of a preliminary nature, since the experiments continue.

7.2. EXPERIMENTAL SITES AND TYPES OF VEGETATION INVESTIGATED

Our field experiments are carried out at Abisko Scientific Research Station in northern Sweden (68.4°N, 18.8°E, 360 m above sea level and close to Lake Torne) and also on the arctic island of Spitzbergen (78°N, in Adventdalen valley near the town of Longyearbyen). At each of these locations, we have selected two different natural ecosystems for investigation. In Abisko a mountain heath with ericaceous dwarf shrub (and moss) vegetation, and an ombotrophic bog with two species of moss. On Spitzbergen the two ecosystems are a grassy plain and a sloping heath dominated by *Cassiope tetragona* .

7.3. SOME RESULTS OBTAINED SO FAR

Experimental results obtained so far support our suspicion, that even in the Arctic and Subarctic moderate ozone depletion can result in measurable effects on vegetation.

In the dwarf shrub heath 15-20% simulated ozone depletion resulted in reduced stem extension growth in all four species investigated. In the deciduous species (*Vaccinium myrtillus* and *V. uliginosum* it amounted to only about 10% in each of two consecutive years of irradiation [8], but in the evergreen species *V. vitis-idaea* and *Empetrum hermaphroditum* it was about 30% during the second year (for *V. vitis-idaea* there was no inhibition at all in the first year). The increasing inhibition in the evergreen species may mean that damage is accumulating in the leaves from year to year.

Also with respect to leaf growth the different species of dwarf shrubs react differently. In *V. uliginosum* dry weight per leaf as well as leaf area are increased by UV-B enhancement, while leaf thickness is decreased. In *V. vitis-idaea,* on the other hand, leaf thickness is increased by UV-B enhancement.

Over 4 years no changes in percent ground cover, due to UV-B enhancement, by the various species of dwarf shrub were detected, and the same holds for the moss layer. In all tested species of dwarf shrub (*V. vitis-idaea, V. myrtillus* and *Cassiope tetragona*) the amounts of soluble UV-B absorbing pigments were increased as a consequence of UV-B enhancement [9, 10]. Extension growth of the bog mosses *Sphagnum fuscum* and *Dicranum elongatum* was inhibited by UV-B enhancement when provided by a switched (on/off) lamp system, but the inhibition is not significant using a "modulated" lamp system, i.e. one that is continuously regulated in step with ambient radiation.

UV-B enhancement decreases the rate of plant litter breakdown in two ways [12]: by direct inhibiting action on the decomposing fungi and by changing the chemical composition of the plant litter in such a way as making it more resistant to decomposition. Although an effect on plant nutrient mobilization has not been demonstrated, it may be assumed that such an effect exists, and that this may indirectly affect plant growth.

Acknowledgements: Thanks are due to the Swedish Environmental Protection Agency, the European Commission, the Swedish Council for Planning and Coordination of Research (FRN) and Astra Draco AB for providing funding over the years, to the Royal Swedish Academy of Sciences and the personnel of Abisko Scientific Research Station and of the Department of Plant Physiology at Lund University (especially Mr. Per Westergren) for support in various ways, and to the Governor of Svalbard, the University of Svalbard Director Jarle Nygaard and Mr Sigmund Spjelkavik for facilitating the experimentation in Svalbard.

References

1. Caldwell, M.M. (1971). Solar ultraviolet radiation and the growth and development of higher plants. In: A.C. Giese, A.C. (ed.), *Photophysiology Vol. 6*, Academic Press, New York. pp 131-177.
2. Thimijan, R.W., Carns, H.R. & Campbell, L.E. (1978). Final report (EPA-IAG- D6-0168): Radiation sources and relative environmental control for biological and climatic effects of UV research (BACER). Environmental Protection Agency, Washington, D.C.

3. Björn, L.O. & Murphy, T.M. (1985). Computer calculation of solar ultraviolet radiation at ground level. *Physiol. Vég.* **23**, 555-561.
4. Quaite, F.E., Sutherland, B.M. & Sutherland, J.C. (1992). Action spectrum for DNA damage in alfalfa lowers predicted impact of ozone depletion. *Nature* **358**, 576-578.
5. Chan, G.L., Peak, M.J., Peak, J.G. and Haseltine, W.A. (1986). Action spectrum for the formation of endonuclease-sensitive sites and (6-4) photoproducts induced in a DNA fragment by ultraviolet radiation. *Int. J. Radiat. Biol.* **50**, 641-648.
6. Madronich, S., McKenzie, R.L., Caldwell, M.M. & Björn, L.O. (1995). Changes in ultraviolet radiation reaching the Earth's surface. *Ambio* **24**, 143-152
7. Barnes, P.W., Flint, S.D. & Caldwell, M.M. (1987). Photosynthesis damage and protective pigments in plants from a latitudinal arctic/alpine gradient exposed to supplemental UVB-radiation in the field. *Arct. Alp. Res.* **19**, 21-27.
8. Johanson, U., Gehrke, C., Björn, L.O. & Callaghan, T.V. (1995). The effects of enhanced UV-B radiation on the growth of dwarf shrubs in a subarctic heathland. *Functional Ecology* **9**, 713-719.
9. Johanson, U. (1997). Future stratospheric ozone depletion will affect a subarctic dwarf shrub ecosystem. Diss. Lund University. ISBN 91-628-2404-X.
10. Björn, L.O., Callaghan, T., Gehrke, C., Gunnarsson, T., Holmgren, B., Johanson, U., Snogerup, S., Sonesson, M., Sterner, O. & Yu, S.-G. (1997). Effects on subarctic vegetation of enhanced UV-B radiation. In: Lumsden, P. (ed.) *Plant and UV-B: Responses to Environmental Change,* Cambridge University Press. pp. 233-246.
11. Björn, L.O., Callaghan, T.V., Johnsen, I., Lee, J.A., Manetas, Y., Paul, N.D., Sonesson, M., Wellburn, H.A.R., Coop, D., Heide-Jørgensen, H.S., Gehrke, C., Gwynn-Jones, D., Johanson, U., Kyparissis, A., Levizou, E., Nikolopoulos, D., Petropoulou, Y. & Stephanou, M. (1997). The effects of UV-B radiation on European heathland species. *Plant Ecol.* **128**, 252-264.
12. Gehrke, C., Johanson, U., Callaghan, T.V., Chadwick, D. & Robinson, C.H. (1995). The impact of enhanced ultraviolet-B radiation on litter quality and decomposition processes. In *Vaccinium* leaves from the Subarctic. *Oikos* **72**, 213-222.

BIOLOGICAL EFFECTS OF LOW DOSES OF RADIATION.

A Review of Cancer Mortality Data in High Natural Background Radiation Areas in Asia and Relevant Experimental Data

H. TANOOKA

Genetics Division, National Cancer Center Research Institute
5-1-1 Tsukiji, Chuo-ku, Tokyo 104, Japan

Abstract

To study effects of low level ionizing radiation on human health, epidemiological data on high natural background radiation areas are very important. Such data obtained in China, India, and Japan are reviewed here. No elevated radiation risk with regard to cancer mortality was shown. Humans seem to tolerate a 10-fold increase of environmental external radiation. A virtually safe dose to human health is considered to be 0.1 Gy per year at conservative estimation. Experimental data on biological response specifically to low level ionizing radiation are reviewed.

1. Introduction

In principle, life produces replication errors in its own genes. Life is also affected by environmental hazards. Oxidative damage is unavoidable in life. The chemical stucture of the damage is equivalent to that of ionizing radiation-induced DNA damage. This damage is eliminated by cellular repair capacity, including not only DNA repair activity but also apoptotic function and immunological activity in the broad sense, and only a small fraction, one out of a thousand or less, is converted to mutation. Recent findings indicate that a defect in the repair system elevates cancer risk. If life can repair or tolerate replication errors and oxidative damage, it must also tolerate radiation damage of the same magnitude. In reality, the amount of DNA damage caused by natural environmental radiation is much less than that of replication errors and oxidative DNA damage. The low dose radiation problem should be extended to this far point, and the magnitude of naturally occurring replication errors and oxidative damage should be taken as a base line for assessment of radiation risk to human health.

When addressing the problem of effects on health stemming from environmental ionizing radiation, it is important to obtain data on the health effects of continuous exposure to radiation at low dose rates. Such effects are to be distinguished from those observed for single exposure at a high dose rate, such as in the case of A-bomb survi-

C. Baumstark-Khan et al. (eds.),
Fundamentals for the Assessment of Risks from Environmental Radiation, 471-478.

vors. Substantial data on the low dose-rate effects are now being obtained by epidemiological studies on residents living in a high natural background radiation area (HBRA). An important point in these studies is that accurate dosimetry is repeatable. Such evidence has been reviewed previously by Kondo [1]. In this article, three HBRA studies conducted in China, India, and Japan, and relevant experimental data are reviewed.

2. Review of Epidemiological Data on High Natural Background Radiation Areas in Asia

2.1. YANGJIANG, GUANDONG PROVINCE, CHINA

The location of the Yangjiang district in China is shown in Figure 1. Data are quoted from literature published by High Background Research Group, China [2-4]. The external γ-radiation level is 2.1 mGy/year, which is about 3 times higher than the average in the control area. The study population was 78,614 in the HBRA and 27,903 in the control area, in 1987. Importantly, a large number of people who have lived in the same place for many generations can be studied in both the HBRA and the control area, and accurate dosimetry is possible. Cytogenetic study showed an age- and dose-dependent increase of chromosome aberrations in lymphocytes of peripheral blood of HBRA residents [2, 4]. However, the epidemiological data from 1970 to 1986 showed that the cancer mortality in the HBRA was slightly lower than in the control area (relative risk: 0.93, P=0.65, for males and 0.96, P=0.35, for females)[2]. Most recent results for 1987-90 [3] are plotted in Figure 2, where mortalities for all cancers, except nasopharynx cancer, were again lower in the HBRA. This study is still under way.

2.2. KARUNAGAPPALLY, KERALA, INDIA

The γ-radiation background level in Karunagappally is 0.50-110.9 mGy per year; the average is 7-10 fold higher than that of the control area. The population was 79,450 in the HBRA and 100,213 in the control area, in 1992. The epidemiological study for 1990-94 data showed no difference in mortalities for all causes of deaths and all cancers between the HBRA and the control area (Figure 2). Breast and thyroid cancer mortalities in females were even lower in the HBRA [5, 6].

2.3. MISASA, TOTTORI PREFECTURE, JAPAN

Misasa has been known for a radon-containing hot spa (radioactivity of spa water: 17-9,400 Bq/l). The town is old and the population (3,381 in 1985) is relatively stable compared with cities in Japan. It should be noted that people had used radon-containing water from wells (70-100 Bq/l) as drinking water until 1953, when the central water supply system (radon content, 30-40 Bq/l) was built. Epidemiological study on cancer mortality showed lower mortalities due to all cancers, stomach cancer, lung cancer, and colon cancer in the Misasa radon spa area than those in the control area. However, only all cancer and stomach cancer mortality data were statistically significant [7] (Fig. 2). The data were reanalyzed by defining the radon area according to measured radon levels. Results showed that lung cancer mortality became higher than in the previous study and reached the same level as that in the control area. However, stomach cancer mortality was again lower [8].

In a hot spa area without radon, Beppu, Japan, cancer mortalities were at the same levels as those in the control areas with no hot spa [9]. Therefore, the effect of radon hot spa could not be explained by the hot spa effect.

2.4. SUMMARY AND DISCUSSION ON HBRA AND RADON SPA DATA

None of the above data showed an increase in cancer mortality in the HBRA or the radon spa area. It seems that human cancer mortality does not increase up to a 10-fold background radiation level. It is interesting to see a lower stomach cancer mortality in both the Chinese HBRA and the Japanese radon spa area, indicating a "hormetic" effect of radiation [10]. The study by Cohen [11] showed a decrease of lung cancers with an increase of indoor radon concentration. It should be noted that the dose-response for chromosome aberration observed in human peripheral blood does not necessarily correlate with that for human cancer mortality. The difference of the target cell explains this situation.

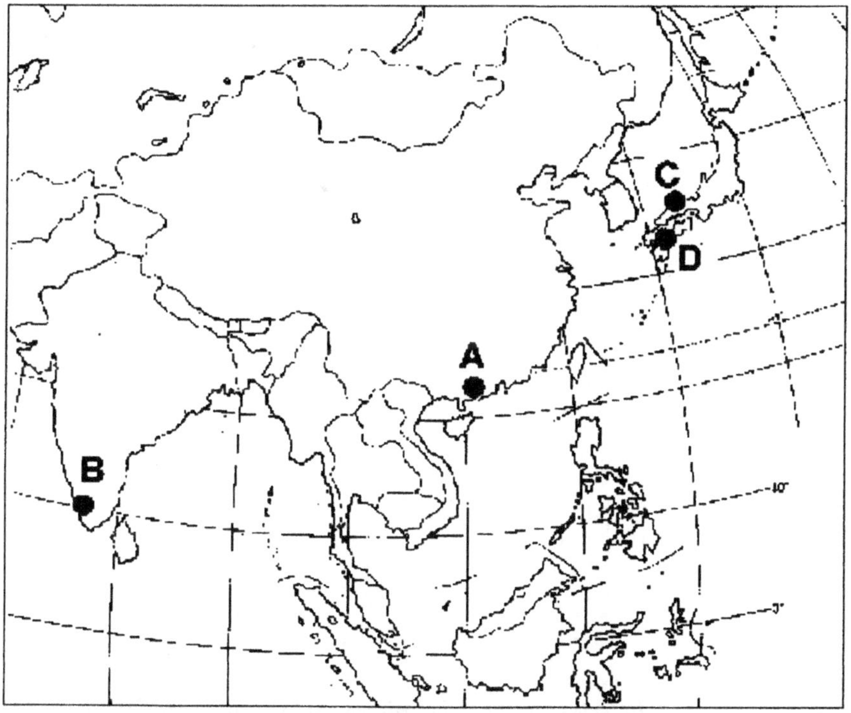

Figure 1. Geographical map showing high natural background radiation areas: (A) Yangjiang, Guangdong Province, China; (B) Karunagappally, Kerala, India; (C) Misasa radon spa area, Tottori Prefecture and (D) Beppu non-radon spa area, Oita Prefecture, Japan.

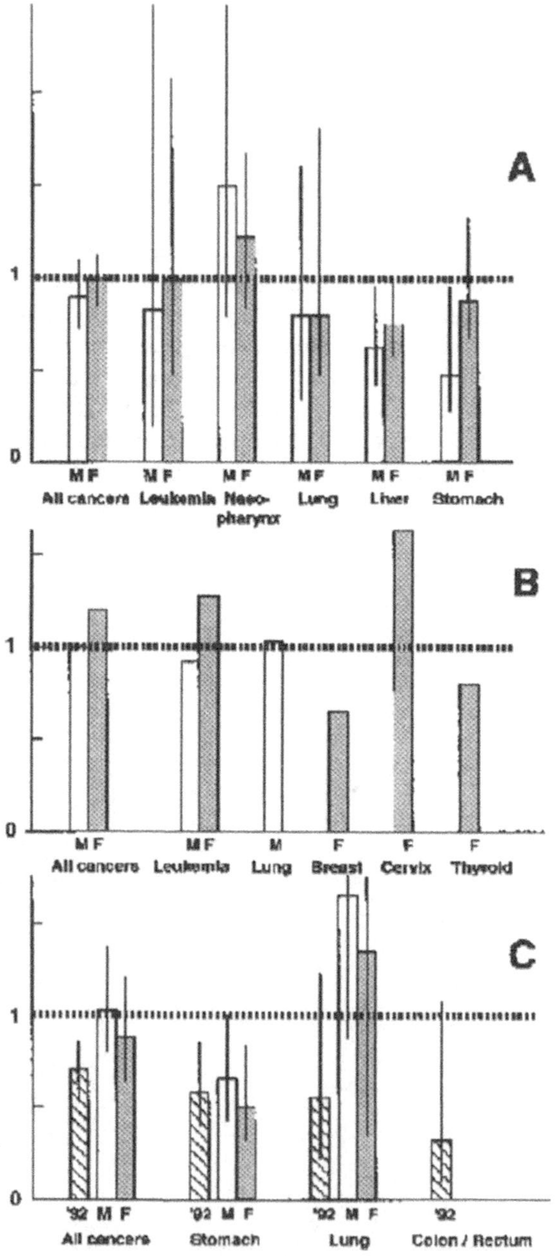

Figure 2. Relative risks for cancer mortality in high natural background radiation areas. (A) Yangjiang, China, data from Ref. 3. (B) Karunagappally, India, data from ref. 5. (C) Misasa radon spa area, Japan, data from ref. 7 (marked '92) and ref. 8. M, males. F, females. (Data quoted with permission).

As an example of continuous human exposure to radiation, the case of radium painters should be mentioned, where a sharp threshold exists at 10 Gy of a cumulative absorbed dose for bone tumor induction [12]. This very high tolerant dose may represent an extreme case of human tolerance to continuous radiation with a low dose rate. Such an effect of continuous exposure should be clearly distinguished from a case of single acute exposure such as the A-bomb survivor case [13]. Even in the A-bomb case, the increase of cancer risk is not recognized below 0.1 Gy [1], and a longer life span is observed in both males and females who received 0.1-1 Gy of A-bomb (γ) radiation in Nagasaki as compared with controls [14].

3. Experimental Evidence for Low Dose- or Low Dose Rate-Specific Biological Effects

3.1. THRESHOLD DOSE RESPONSE FOR MOUSE TUMOR INDUCTION

We have observed tumors induced on the backs of mice during repeated radiation of ^{90}Sr-^{90}Y β-rays with a frequency of 3 times a week throughout their life span. The radiation dose given per week ranged from 1.5 to 35.4 Gy. Tumor incidence plotted as a function of dose, summarized from 1988-93 data [15], showed a non-linear threshold type (Figure 3). Tumors produced with supra-threshold doses were squamous cell carcinomas, basal cell carcinomas, fibrosarcomas of the skin and osteosarcomas. The tumors had a mutation in the *p53* gene with a considerably high frequency of 20/65 [16]. However, with a sub-threshold dose of 1.5 Gy per week, no tumor was present for the life span of mice [17].

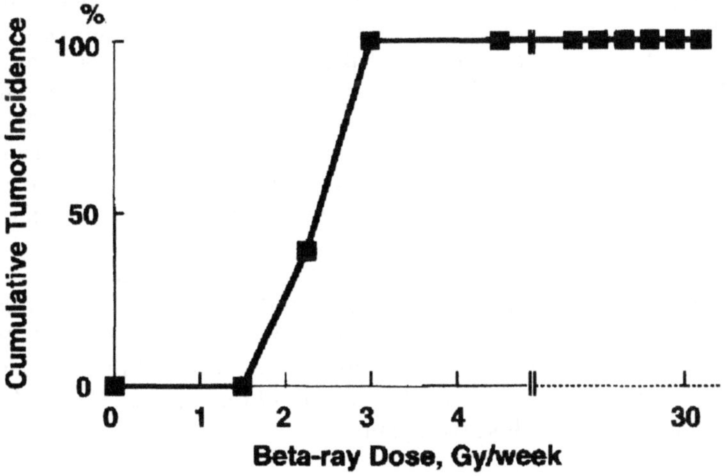

Figure 3. Threshold dose response of mouse tumor induction by beta irradiation repeated 3 times weekly on the backs of mice. (Redrawn from data in refs. 15 and 17).

The lower leukemia incidence with the lower dose rate of radiation was recognized in whole body-irradiated mice already in early experiments by Upton *et al.* [18], and confirmed to a further extent of lower dose rates, showing reduction of mouse leukemia by a factor of 20 (Nomura, personal communication). Moreover, the dose rate effect was observed even with high LET radiation for lung tumor incidence in radon-inhaling rats [19]. Therefore, high LET radiations exert the dose rate effect in the whole body system.

It should be noted that the threshold dose response known for malformation of mice irradiated *in uetro* is explained by the apoptotic activity of the *p53* gene efficiently functioning at low doses [20].

3.2 LOW DOSE-SPECIFIC BIOLOGICAL RESPONSE

Many biological effects of ionizing radiation were observed specifically at sublethal low doses. Expression of the gene of thioredoxin, an essential protein in cell cycle regulation as well as being a radio-protective protein, is induced with an optimum dose of 0.25 Gy in human lymphocytes [21]. An adaptive response to ionizing radiation was first demonstrated on chromosome aberration in cultured human lymphocytes [22]. Recently, such an adaptive response was found in mouse survival; mice which received whole-body preirradiation with an optimum dose of 0.05 Gy showed a higher survival (70%) as compared with 30% for mice without preirradiation after 7.4 Gy whole body irradiation [23]. Furthermore, the aggressive behavior of male mice is suppressed by whole-body irradiation with an optimum dose of 0.05Gy [24]. Together with many other observations, the biological response specifically to a low level of radiation seems to occur at a sublethal dose range of 0.05-0.5 Gy.

4. Radiation Protection Concept: Virtually Safe Dose

It is considered to be reasonable to set a "virtually safe dose" of continuous radiation for human health. People are exposed to natural external radiation, about 0.8 mGy per year in the whole body. No evidence is reported for an increase in cancer mortality with radiation doses of 10-100 fold of this level even for acute exposure. The virtually safe dose is thus considered by conservative estimation to be 0.1 Gy per year for continuous radiation. The definition of the unit Sv should be revised in the future, because the dose rate effect is now known to exist for continuous exposure to high LET radiation and the presently adopted RBE values for high LET radiation might not be applicable.

Acknowledgements. I thank Dr. L. Wei, China, and Dr. M. K. Nair, India, for providing me with permission to quote their epidemiological data, Dr. S. Kondo, Professor Emeritus, Osaka University, for valuable discussions, Dr. S. Hattori, Central Research Institute for Electric Power Industry, for his efforts for support of the low dose and hormesis study in Japan, and Dr. M. Terada, Director of the National Cancer Center Research Institute, for support in this investigation.

5. References

1. Kondo, S. (1994) Health Effects of Low Level Radiation, Kinki University Press, Osaka, and Medical Physics Publishing, Madison.
2. Wei, L., Zha, Y., Tao, Z., He, W., Chen, D., and Yuan, Y. (1990) Epidemiological investigation of radiological effects in high background radiation areas of Yangjiang, China. *J. Radiat. Res.*, **31**, 119-136.
3. Tao, Z., Kato, H., Zha, Y., Akiba, S., Sun, Q., He, W., Lin, Z., Zou, J., Zhang, S., Liu, Y., Sugahara, T., and Wei, L. (1997) Study on cancer mortality among the residents in high background radiation area of Yangjiang, China, in L. Wei, T. Sugahara, and Tao, Z. (eds.) *High Levels of Natural Radiation 1996 Radiation Dose and Health Effects,*. Elsevier, Amsterdam-Lausanne-New York-Shannon-Tokyo, pp. 249-254.
4. Jiang, T., Wang, C., Chen, D., Yuan, Y., Wei, L., Hayata, I., Morishima, H.,Nakai, S., and Sugahara, T. (1997) Preliminary report on quantitative study of chromosome aberrations following life time exposure to high background radiation in China, in L. Wei, T. Sugahara, and Tao, Z. (eds.) *High Levels of Natural Radiation 1996 Radiation Dose and Health Effects,*. Elsevier, Amsterdam-Lausanne-New York-Shannon-Tokyo, pp. 301-306.
5. Nair, M. K. ed. (1996) Natural Background Radiation Cancer Registry. A Project of the Regional Cancer Centre, Trivandram, India, Technical Report 1.
6. Nair, M. K., Amma, N. S., Gangadharan, P., Padmanabhan, V., Jayalekshmi, P., Jayadevan, S., and Mani, K. S. (1997) Epidemiological study of cancer in the high background radiation area in Kerala, in L. Wei, T. Sugahara, and Tao, Z. (eds.) *High Levels of Natural Radiation 1996*, Elsevier, Amsterdam-Lausanne-New York-Shannon-Tokyo, pp. 271-276.
7. Mifune, M., Sobue, T., Arimoto, H., Komoto, Y., Kondo, S., and Tanooka, H. (1992) Cancer mortality survey in a spa area (Misasa, Japan) with a high radon background, *Jpn. J. Cancer Res.* **83**, 1-5.
8. Ye, W., Sobue, T., Lee, V.S., Tanooka, H., Mifune, M., Suyama, A., Koga, T., Morishima, H., and Kondo, S. (1998) Mortality and cancer incidence in Misasa, a spa area in Japan, a spa area with elevated radon levels. *Jpn. J. Cancer Res.* **89**, 789-796
9. Suzuki, Y., Honjo, S., Kawamura, H., Koishi, F., Suzuki, T. (1994) Cancer mortality in low radon spa area, *Jpn. J. Cancer Res.* **85**, 1063-1066.
10. Luckey, T. D. (1982) Physiological benefits from low levels of ionizing radiation, *Health Phys.* **43**, 771.
11. Cohen, B. L. (1990) Expected indoor 222Rn levels in counties with very high and very low lung cancer rates. *Health Phys.* **75**, 897-907.
12. Rowland, R. E. (1995) Dose-response relationships for female radium dial workers: A new look, in G. van Kaick, A. Karaoglou, and A. M. Kellerer (eds.) *Health Effects of Internally Deposited Radionuclides: Emphasis on Radium and Thorium*, World Scientific, Singapore-New Jersey-London-Hong Kong, pp. 135-143.
13. Shimizu, Y., Kato, H., Schull, W. J., Preston, D. L., Fujita, S., and Pierce, D. A. (1989) Studies of the mortalityof A-bomb survivors based on the DS86 and T65DR shielded kerma and organ doses, *Radiat. Res.* **118**, 502-524.
14. Mine, M., Okumura, Y., Ichimaru, M., Nakamura, T., and Kondo, S. (1990) Apparently beneficial effect of low to intermediate doses of A-bomb radiation on human life span. *Intern. J. Radiat. Biol.* **58**, 1035-1043.
15. Tanooka, H. and Ootsuyama, A. (1993) Threshold-like dose response of mouse skin cancer induction by repeated beta irradiation and its relevance to radiation-induced human skin cancer, *Recent Results in Cancer Research*, Springer-Verlag, Berlin-Heidelberg, 128, pp. 231-241.
16. Ootsuyama, Y., Makino, H., Nagao, M., Ochiai, A., Yamauchi, Y., and Tanooka, H. (1994) Frequent p53 mutation in mouse tumors induced by repeated beta-irradiation, *Mol. Carcinog.* **11**, 236-242.

17. Ootsuyama, A. and Tanooka, H. (1993) Zero tumor incidence in mice after repeated lifetime exposure to 0.5 Gy of beta radiation, *Radiat. Res.* **134**, 244-246.
18. Upton, A., C. (1964) Comparative aspects of carcinogenesis by ionizing radiation, *Nat. Cancer Inst. Monogr.* **14**, 221-242.
19. Morlier, J. P., Morin, M., Monchaux, G., Fritsch, P., Pineau, J. F., Chameaud, J., Lafuma, J., and Masse, R. (1994) Lung cancer incidence after exposure of rats to low doses of radon: Influence of dose rate, *Radiation Protection Dosimetry*, **56**, 93-97.
20. Norimura, T., Nomoto, S., Katsuki, M., Gondo, Y., and Kondo, S. (1996) p53-Dependent apoptosis suppresses radiation-induced teratogenesis, *Nature Med.* **2**, 577-580.
21. Hoshi, Y., Tanooka, H., Miyazaki, K., and Wakasugi, H. (1997) Induction of thioredoxin in human lymphocytes with low-dose ionizing radiation. *Biochim. Biophys. Acta*, **1359**, 65-70.
22. Olivieri, G., Bodycote, J., and Wolf, S. (1984) Adaptive response of human lymphocytes to low concentrations of radioactive thymidine, *Science* **223**, 594-597.
23. Yonezawa, M., Misonoh, J., and Hosokawa, Y. (1996) Two types of X-ray-induced radioresistance: Presence of 4 dose ranges with distinct biological effects, *Mutation Res.* **358**, 237-244.
24. Miyachi, H. and Yamada, T. (1994) Changes of aggressive behavior and brain serotonin turnover after very low dose X-irradiation in mice. *Neuro Sci. Lett.* **175**, 92-94.

RISKS FROM RADON INHALATION

L. TOMÁŠEK

National Radiation Protection Institute
Šrobárova 48, Prague, Czech Republic

Abstract

Radon is estimated to represent about half the average exposure to ionizing radiation of the population. Alpha emissions from inhaled radon progeny can result in lung cancer. The first studies of miners demonstrated a linear relationship between cumulative exposure and relative risk, further ones a decreasing effect of age at exposure, time since exposure, and exposure rate. Recent Czech results confirmed these findings and demonstrated an increased risk already at low exposures -15WLM. The association of lung cancer and residential radon has not been clearly shown, partly because of much lower exposures and large uncertainties related to radon exposure in homes. In spite of this, a meta-analysis of indoor studies found a significant trend in the relative risk by exposure, which was not different from results of a large Czech cohort indoor study.

1. Introduction

The radioactive gas radon is an immediate decay product of radium, which is a radioactive product of uranium. Both uranium and radium are present in most soils and rocks and therefore radon is ubiquitous in air. Radon decays into a series of solid short-lived radioisotopes - called radon daughters (^{218}Po, ^{214}Pb, ^{214}Bi, and ^{214}Po). Two of these (^{218}Po and ^{214}Po) emit α-particles. The α-emissions can cause damages of the DNA of epithelial cells in the respiratory tract, which may result in lung cancer [1].

For centuries, miners in the Erz mountains were known by an unusually high mortality from lung diseases (Agricola, 1556, Paracelsus, 1567). The disease was eventually identified as lung cancer by Haerting and Hesse in 1879. The increased mortality from lung cancer was first ascribed to ore dust, particularly arsenic. Radon was identified as a possible cause by the end of the 1930s, after large measurements in the mines of Schneeberg. The role of inhaled short-lived decay products was discovered later by Bale (1951) and Harley (1953).

2. Risk of Lung Cancer

The first epidemiological evidence relating excess of lung cancer to different levels of potential alpha energy concentration was reported in the Colorado miners by Lundin et

C. Baumstark-Khan et al. (eds.),
Fundamentals for the Assessment of Risks from Environmental Radiation, 479-488.

al (1971) and in the Czech uranium miners in Jáchymov by Ševc et al (1972). Both epidemiological cohort studies concluded that the risk of lung cancer increased monotonically with cumulative exposure to radon progeny [1]. Several other cohorts of miners have been studied since. In 1988, four such studies (Colorado, Ontario, Saskatchewan, and Sweden - 360cases) were combined to obtain a projection model for the relative risk of lung cancer [2]. Six years after, a larger set of 11 studies (2736 cases) was analyzed by Lubin et al [3]. Basic information of cohorts that were used in the last joint analysis [4] is given in Table 1.

The most recent joint analysis confirmed the linear relationship between cumulative 5-year lagged exposure and relative risk of lung cancer and general pattern of declining excess relative risk (ERR) per unit exposure with attained age, time since exposure, and exposure rate [4].

The Czech studies involve three cohorts (Table 2). The oldest cohort (S) comprises 4320 uranium miners at Jáchymov and Horní Slavkov firstly exposed in the period 1948-59, 5624 miners of the second cohort (N) entered the Príbram uranium mines in 1968-74, and the third cohort (L) includes 914 burnt-clay miners in the Rakovník district employed underground in the period 1960-80. The follow-up of the studies relied on the population registry, the pensions office, and local enquiries [5]. Exposure estimates in the S study were derived from large numbers of measurements of radon. Mean

TABLE 1. Basic data of miners studies used in joint analysis

Cohort	Size	Follow-up	kPY	O	O/E	Exposed	WLM [1]
Czech (S)	4320	1952-90	107	705	4.96	1948-75	189
Colorado	3346	1950-90	89	377	4.53	1936-68	606
New Mexico	3286	1957-85	56	57	3.04	1953-85	87
Ontario	21346	1955-86	380	291	1.68	1937-85	24
Newfoundland	2092	1950-90	58	147	4.35	1936-77	252
Saskatchewan	8486	1950-80	113	65	1.71	1950-80	12
NWT Canada	2103	1950-80	52	57	2.23	1942-60	138
Sweden	1294	1951-91	33	79	2.36	1908-77	77
France	1785	1948-85	44	45	2.20	1946-85	53
Australia	2516	1948-87	52	54	1.46	1948-61	4
China	17120	1976-87	187	975		1922-76	249

kPY=1000person-years, O=observed cases, E=expected cases by national rates

1 One Working Level equals any combination of radon progeny in one liter of air which results in the ultimate emission of 1.3 □105 megaelectronvolts (MeV) of energy from alpha particles. WLM is time-integrated exposure measure, i.e. the product of time in working months (170 hours) and working levels (1WLM=3.54 mJh m-3).

TABLE 2. Czech cohorts status by 1995

Cohort	Since	Size	Cases	O/E	Died	WLM	Duration
S	1952	4320	771	4.67	66 %	155	9
L	1960	914	67	2.23	45 %	18	14
N	1969	5624	50	1.38	11 %	6	5

number per year and shaft was 223 in the period 1949-60. Each man's annual exposures to radon progeny were estimated combining measurement data with the men's registered employment details (time spent in mine-shafts and job category). In the whole N study and in the L study since 1978, the exposure estimates were based on personal dosimetric records. Before 1978, the exposures in the L study were estimated by extrapolation using ventilation data.

By 1995, a total of 888 cases of lung cancer were observed in the three Czech cohorts, most of them in the old and high exposed cohort (Tab.2).

The statistical analyses were based on relative risk models in the general form

$$O = cE(1 + ERR(W, Z)) \quad (1)$$

where O denotes the observed number of cases, E is the number expected from national mortality data, ERR is the excess relative risk function depending on exposure W and modifying variables Z, and c is an intercept term that allows the mortality rate for the 'unexposed' cohort to differ from that in the general population. The most simple model of relative risk assumes a linear dependence on total cumulative exposure W lagged by 5 years. The coefficient of proportionality b (=ERR/WLM - Excess Relative Risk per WLM) in this model is constant: ERR(W,Z) = b W.

The simple model of relative risk gives the ERR/WLM estimate of 0.014 (95% CI: 0.010-0.018; CI=confidence interval). for the whole range of cumulative exposures 1-860 WLM (Fig. 1). The intercept estimate in this model is 1.58 (95% CI 1.32-1.89). The relative risks relevant to low exposure categories suggest a somewhat steeper slope.

3. Risk at Low Exposure Rates

The present situation in mines, which is greatly influenced by radiation protection measures, is characteristic by lower exposures and exposure rates than in the past (1945-65). In almost all studies of miners, the relative risk coefficients (ERR/WLM) corresponding to very high concentrations of radon were found lower in comparison to lower concentrations. The explanation of this inverse exposure rate effect is not clear, one might be nonlinearity between exposure and dose to target cells or cell killing at high exposure rates [6].

When exposures are restricted to <100 WLM (Fig.2), the ERR/WLM estimate for the Czech cohorts is 0.023 (95%CI: 0.004-0.043) and the intercept estimate of 1.25 (95%CI: 1.14-1.37) corresponds better to relative mortality (O/E) in the cohorts for very low exposure categories. It is worth noticing that the relative risk in the exposure category 10-20 WLM already shows a significant difference from the baseline rate.

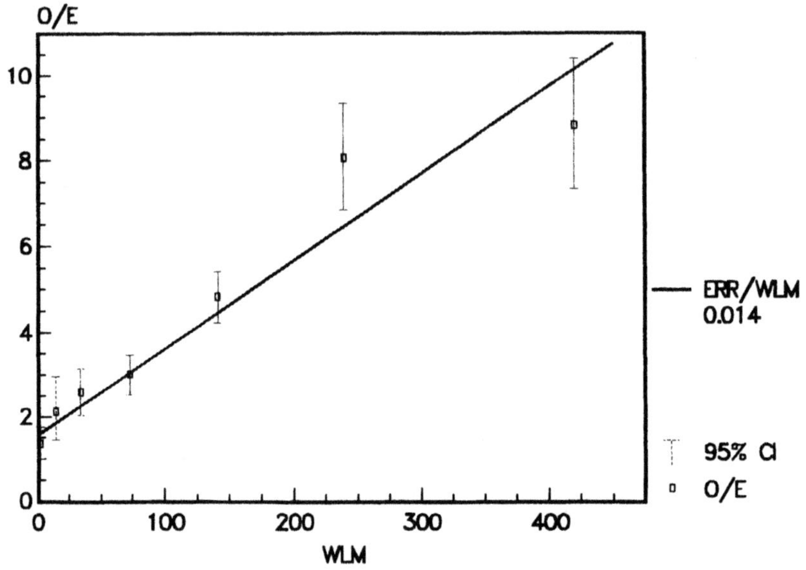

Figure 1. Relative risk by cumulative exposure in the joint Czech cohorts

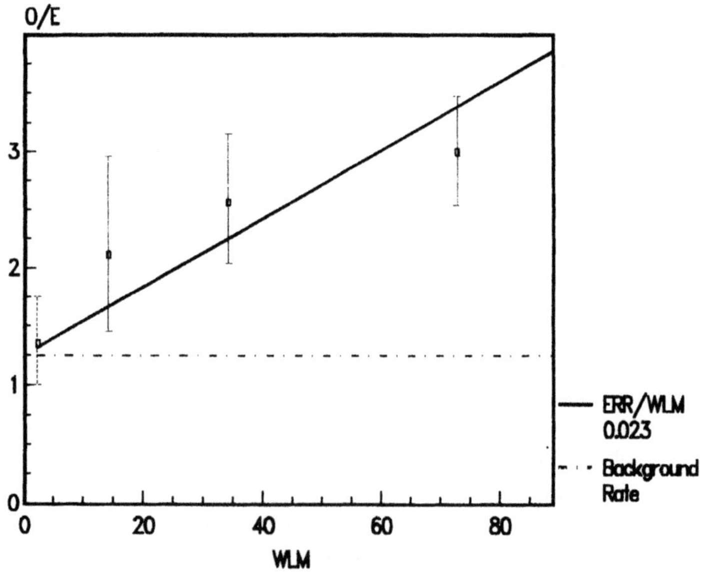

Figure 2. Relative risk by cumulative exposure below 100 WLM in joint Czech cohorts

4. Temporal Modifiers of Exposure-Response Relationship

The temporal changes in the relative risk assessment were firstly analyzed in the BEIR IV [2]. It was found that exposures received in more than 15 years ago had half the effect of exposures in the period of 5-14 years. The patterns of decline in the Czech study by 1985, that was not included in the joint BEIR IV analyses, were similar [7].

The effect of time and age related modifying factors is studied by the method of exposure windows which consists in separating the individual total past exposure into segments corresponding to different time periods and age categories. For instance, to analyze the effect of age at exposure (-29, 30-39, 40-) and time since exposure (5-14, 15-24, 25-39), the total W divides into 9 parts. In addition, the exposure rate factor can be analyzed in the same way, i.e. by windows, in which exposures from lower exposure rates (<8WL) are separated from exposures from higher exposure rates. In this way, the total cumulative exposure splits up to 18 components. Simultaneous marginal effects of each factor is estimated by maximum likelihood method (Tab.3).

The parameters of the complex model translated into age-at-exposure and time-since-exposure specific ERR/WLM coefficients illustrate temporal and age related changes in the magnitude of risk coefficients (Tab.4).

TABLE 3. Linear relative risk model modified by time-since-exposure, age at exposure, and exposure rate in the Czech joint cohort (S+N+L)

		95% CI
ERR/WLM *)	0.123	0.085 - 0.181
Time Since Exposure		
5-14	1	
15-24	0.25	0.16 - 0.32
25-	0.06	0.03 - 0.09
Age at Exposure		
-29	1	
30-39	0.55	0.33 - 0.80
40-	0.25	0.16 - 0.33
Exposure Rate		
<8 WL	1	
>8 WL	0.35	0.03 - 0.69

*) relative to TSE: 5-14 and AE: -29

TABLE 4. TSE & AE specific ERR/WLM for exposure rate <8 WL.

	Age at Exposure		
TSE	-29	30-39	40-
5-14	0.123	0.068	0.031
15-24	0.031	0.017	0.008
25-	0.007	0.004	0.002

TABLE 5. Relative risk from radon and smoking

Smoking	Cases	O/E	ERR/WLM	95% CI
Low (1)	11	0.43	.054	-0.027 - 0.136
High (2)	82	3.04	.060	0.006 - 0.115

5. Interaction With Smoking

The above models of lung cancer risk were developed without considering any influence of smoking. As the S cohort was set up in 1970, when most of the members were not available for collecting smoking data, the interaction of smoking and radon exposure was studied only in the N and L cohorts, where smoking habits were available for 83% of miners. The earlier results of these studies found a higher (not significant) relative risk for non-smokers [6, 8]. In such analyses, the crucial point is a sufficient number of cases among non-smokers. By 1995, only 3 cases of lung cancer were observed in the non-smoker group. In order to analyze the effect of smoking, the miners were classified into two groups: (1) non-smokers, ex-smokers, and regular smokers of less than 10 cigarettes per day, and (2) regular smokers of at least 10 cigarettes per day.

The estimated risk coefficients in the two groups show very little differences (Tab.5). In the joint analysis of 6 cohorts of miners [3], where data on smoking were available (64 cases among non-smokers), no significant differences were found as for the smoking status, though the risk coefficient among non-smokers was found about three times higher. So far, the present results from the Czech cohorts suggest that the linear relative risk model is justifiable in any group of workers with any smoking prevalence. A conclusive inference can be made when larger numbers are available.

6. Risk of Other Cancers

While the effect of radon exposure to lung cancer is known and well described for decades, the associations to cancers other than lung cancer are much less convincing. As observed numbers for separate cancer sites are usually much lower, a joint analysis in this respect was essential. The pooling included 11 studies of miners, practically all (10) that were used for analyses of lung cancer risk [9]. In 10 studies where external mortality rates were available (China study excluded), a total of 1179 non-lung cancers were observed (O/E=1.01). Relative risk of this group of diseases was lower in the first decade since first exposure (O/E=0.87) and higher later (O/E=1.03), as would be expected because of selection of healthy men for employment. For 28 individual cancer sites, significantly higher overall risks were found for stomach (O/E=1.33) and primary liver (O/E=1.73) and for leukaemia (O/E=1.93) in the first decade since start. For none of these diseases was mortality significantly related to cumulative exposure, except for pancreatic cancer. The increase for leukaemia in the early period, though not related to radon, might be influenced by γ-radiation in uranium mines. The joint study provided considerable evidence that increased exposure to radon do not cause a risk of mortality from cancers other than lung cancer [9]. The Czech results by 1990 [10] did not find any significant differences from the above joint findings, except a weak positive relationship of cumulative radon exposure to mortality from malignant melanoma in the S cohort. Similar results in incidence were observed by Ševcová and Ševc [11].

7. Indoor Exposure to Radon

The levels of radon concentration in dwellings are generally much lower than in mines, even under present mining conditions. The risk from residential radon was not considered to constitute a public hazard until 1980s, when it was realized that concentrations of radon could vary in magnitude of two orders. Therefore in some proportion of population, the levels can be comparable to those in mines. For instance, based on miner studies, some 10% of lung cancers are estimated in the USA each year to occur from residential radon [4].

In order to evaluate directly the effect of radon in houses, many studies were launched. Results have not been consistent. Some case-control studies reported positive relationship, while other did not found any increase in risk. Eight such studies of at least 200 cases were combined by Lubin and Boice [12] in a meta-analysis representing 4263 lung cancers. The combined trend in the relative risk was significantly different from zero and the relative risk of 1.14 (CI: 1.0-1.3) at 150 Bq/m^3 and 25-year residence was estimated, which corresponds to the ERR of 0.04 (CI: 0.00-0.08) per unit exposure (1 kBq m^{-3}.a corresponds to exposure at 1000 Bq/ m^3 for 1 year; kBq m^{-3}.a). However, there were significant differences in the study specific estimates of the exposure-response relationship.

Radon levels in Czech dwellings are estimated to some 120 Bq/m^3, highest reported value was 20 kBq/m^3. In 1990, a large cohort study [13] started in an area of elevated concentrations (mean 650 Bq/ m^3). The study population includes nearly 12 000 inhabitants, who have resided in the specific area (80 villages - 2500 houses) for at least 3 years. During follow-up period 1960-90, a total of 142 lung cancers were observed (O/E=1.08; CI: 0.91-1.27). Although not significant in the total, separate figures in exposure categories suggest a significant linear trend with cumulative exposure. The estimated coefficient of ERR per 1 kBq.m^{-3} a 0.009 (CI: 0.003-0.015) falls in the range estimated by Lubin and Boice [12].

8. Uncertainty of Estimates

The models for lung cancer risk are potentially limited by sources related both to response and exposure. The response errors include losses to follow-up, misspecification of causes of death, or selection of control in case-control studies. The disease outcome may be subject to error in classification, depending sometimes on age or place of death. Studies of the accuracy of death certificates for lung cancer suggest that this misclassication does not represent a significant bias, but the degree of bias might be different in the past. This type of errors can be reduced by including an intercept term or by the so-called internal comparison.

The errors in exposure estimates is unavoidable. The concentration of radon and its decay products vary substantially in time and location. In most studies, recent exposures were measured with higher accuracy than those in the past. Early measurements were in terms of radon concentrations and exposure to radon progeny was derived using assumptions about equilibrium of radon and radon progeny in air. The determining role in lung cancer risk consists in radiation dose of alpha energy to target cells in the respiratory tract, which can be estimated only indirectly from exposure to radon progeny. The relationship between the exposure and dose is extremely complex and is dependent on a series of biological and physical factors. The size of the radioactive particles determines their penetration through the upper respiratory system and the pattern of deposition within the tracheobronchial region. Short-lived radon decay products can be deposited directly in the respiratory tract (unattached fraction) or through the surface of aerosol particles (attached fraction). Biological factors include tidal volume and respiratory frequency, proportion of nasal breathing, bronchial morphometry, mucociliary clearence rate, and thickness of mucus [1].

All the above factors are addressed by lung dosimetric models in order to estimate the dose from the exposure to radon progeny, but the range of published dose-conversion coefficients is relatively wide (0.3-0.003 Sv/WLM) [15]. The conversion can be also obtained by a direct comparison of the detriment associated with a unit effective dose and a unit radon progeny exposure. The conversion based on lifetime risk coefficients recommended by the ICRP [15] is 0.005 Sv/WLM for workers and 0.004 for the general public. The comparison of time-since-exposure (TSE) relative risk models in BEIR IV [2] and BEIR V [16] (Tab. 6) suggests a mean conversion factor of 0.017 Sv/WLM for workers.

The differences between occupational and residential exposure estimates do not only consist in physical conditions of these two environments or biological differences

TABLE 6. BEIR IV and BEIR V models of relative risk for lung cancer

	BEIR IV	BEIR V	Conversion factor
TSE	ERR/WLM	ERR/Sv	Sv/WLM
10y	0.025	1.72	0.015
20y	0.0125	0.64	0.020

between heavy working miners and the general population, but also in different approaches to estimate the duration of stay in given places. The mine records and conditions are usually more concise than data on residence occupancy. This might be one of the reasons why indoor studies have not clearly demonstrated the effect of radon to lung cancer risk in general population. Sample size for case-control studies under realistic scenarios of measurement errors and population mobility are estimated to several thousands [17]. So far, the numbers of cases in reported studies are one order lower.

9. Radon Programme

Czech policy on indoor radon is outlined from 1990, when the Radon Committee was established in order to (1) survey houses with elevated radon concentrations, (2) propose preventive and remedial strategies, (3) organize guidance and training in radon measurement and mitigation measures, (4) supervise certification in the field, and (4) evaluate public remediation projects. A decree issued in 1991 set up action levels for radon in existing and new dwellings, radon in soil and water, and radium in building materials. Intervention levels are given for the equivalent equilibrium concentration of radon in air in existing houses - 200 Bq/m^3 and in new houses - 100 Bq/m^3. Houses with levels exceeding the action level are recommended by radiation protection inspectors for appropriate countermeasures. The expenses are partly (10-100 kCZK) paid for by the state according to radon concentration, soil gas risk, and mitigation techniques.

The radon program includes a country wide survey in order to identify the houses with the highest radon concentrations. This is based on geological mapping, and large scale radon measurements organized by the National Radiation Protection Institute and regional radon inspectors of the State Office for Nuclear Safety. Two open detectors are generally used for 1-year measurement in the dwelling. By 1996, more than 100 000 houses have been measured. It is estimated that more than 60 000 houses (2-3 %) exceed the action level of 200 Bq/m.

10. Conclusions

Epidemiological studies of workers have provided a great part of knowledge on carcinogenic effects of ionizing radiation. Historical studies of uranium miners, who experienced in the past relatively high exposures, have clearly demonstrated the linear effect of cumulative exposure. As occupational exposures today are generally much lower, studies at such conditions have to involve large populations to show an effect. This is important for indoor exposures in dwellings, where clear and consistent risk estimation has not been made. So far, the risk from lower exposures is assumed to follow the same linear non-threshold model. This assumption can be verified by a further research in low exposed populations.

11. References

1. Samet J.M. (1991) Diseases of uranium miners and other underground miners exposed to radon. *Occupational Medicine* **6**: 629-639.

2. National Research Council, Committee on Biological Effects of Ionizing Radiation (1988) *Health risk of radon and other internally deposited alpha-emitters (BEIR IV)*. National Academy Press, Washington DC.
3. Lubin J.H., Boice J.D., Hornung R.W., Edling C., Howe G.R., Kunz E., Kusiak R.A., Morrison H.I., Radford E.P., Samet J.M., Tirmarche M., Woodward A,. Yao S.X., Pierce D.A. (1994) *Radon and lung cancer risk: a joint analysis of 11 underground miners studies*. National Institutes of Health, NIH Publ 94-3644.
4. Lubin J.H., Tomášek L., Edling C., Hornung R.W., Howe G.R., Kunz E., Kusiak R.A., Morrison H.I., Radford E.P., Samet J.M., Tirmarche M., Woodward A., Yao S.X. (1997) Estimating lung cancer risk from residential radon using low exposure miner data. *Radiation Research* **147**: 126-134.
5. Tomášek L., Darby S.C., Fearn T., Swerdlow A.J., Placek V., Kunz E. (1994) Pattern of lung cancer mortality among uranium miners in West Bohemia with varying rates of exposure to radon and its progeny. *Radiation Research* **137**: 251-261.
6. Ševc J., Kunz E., Tomášek L., Placek V., Horácek J. (1988) Cancer in man after exposure to Rn daughters. *Health Physics* **54**: 27-46.
7. Ševc J., Tomášek L., Placek V. (1991) Risk of malignant lung tumours as a result of inhalation of radon daughter products. *Èeskoslovenská hygiena* **36**: 3-13, in Czech.
8. Placek V., Tomášek L., Heribanová A., Kunz E. (1997) Lung cancer and exposure to radon under present mining conditions. *Pracovní lékaøstvi* **49**: 14-20, in Czech.
9. Darby S.C., Whitley E., Howe G.R., Hutchings S.J., Kusiak R.A., Lubin J.H., Morrison H.I., Tirmarche M., Tomášek L., Radford E.P., Roscoe R.J., Samet J.M., Yao S.X. (1995) Radon and cancers other than lung cancer in underground miners: a collaborative analysis of 11 studies. *J. National Cancer Institute* **87**: 378-384.
10. Tomášek L., Darby S.C., Swerdlow A.J., Placek V., Kunz E. (1993) Radon exposure and cancers other than lung cancer among uranium miners in West Bohemia. *Lancet* **341**: 919-923.
11. Ševcová M., Ševc J. and Thomas J. (1978) Alpha irradiation of the skin and the possibility of late effects. *Health Physics* **35**: 803-806.
12. Lubin J.H., Boice J.D. (1997) Lung cancer risk from residential radon: meta-analysis of eight epidemiologic studies. *J. National Cancer Institute* **89**:49-57.
13. Kunz E., Tomášek L., Müller T., Burian I., Placek V., Heribanová A., Matzner J. (1996) Indoor exposure to radon - cohort study. Proc. 9th International Congress IRPA, Vienna, Vol 2:284-286.
14. National Research Council (1991) *Comparative dosimetry of radon in mines and homes*. National Academy Press, Washington DC.
15. ICRP (1993) *Protection against Radon-222 at home and at work*. ICRP Publication 65. Annals of the ICRP 23 (2).
16. National Research Council, Committee on Biological Effects of Ionizing Radiation (1990) *Health effects of exposure to low levels of ionizing radiation (BEIR V)*. National Academy Press, Washington DC.
17. Lubin J.H., Samet J.M., Weinberg C. (1990) Design issues in epidemiologic studies of indoor exposure to Rn and risk of lung cancer. *Health Physics* **59**: 807-817.

RADIATION EXPOSURE IN CIVIL AIRFLIGHT

G. REITZ

Deutsches Zentrum für Luft- und Raumfahrt, Institut für Luft- und Raumfahrtmedizin, Abteilung Strahlenbiologie
D-51170 Köln, Germany

Abstract

The International Commission on Radiological Protection recommended in ICRP Publication 60 that exposures to natural sources should be included as part of occupational exposure for the operation of jet aircraft. As a consequence the revision of the Euratom Basic Safety Standards Directive (BSS) include operation of jet aircrafts and exposures by natural sources which are considerable elevated against the public exposure as occupational exposure for which necessary arrangements for the control of the exposures has to be established by the member states. The national authorities are required to develop an appropriate protection philosophy and radiation protection guidelines for flight crews and member of the public, especially frequent flyers and pregnant women. The results of measurements given in this paper reflects the data used as base for the recommendations of assessing the radiation exposures and the risk involved of the working groups of SSK (Strahlenschutzkommission) in Germany and the European Radiation Dosimetry (EURADOS) Group. The excess cancer risk due to solar and galactic cosmic radiation is discussed and future activities in this area to reduce uncertainties in dose and risk assessment in this flight altitude are described.

1. Introduction

Since the beginning of its existence mankind is exposed to the terrestrial and cosmic radiation. The terrestrial radiation has its origin in the presence of radioactive material in the earth crust. Mainly, potassium and the nuclides of the uranium-radium chain and the thorium chain which create as intermediate product radon, are of importance for the radiation exposure. The incident cosmic radiation, mostly high energetic protons, is modulated by the earth magnetic field and by interaction processes in the atmosphere in such a way that it contributes only to a small amount to the total radiation exposure on the earth surface which accounts to about 2.4 mSv annually. With increasing altitude from ground the protection of the earth atmosphere decreases which results in an increase of the cosmic radiation exposure to several times as much as the exposure on ground. Depending on flight routes the annual additional exposure amounts to between

C. Baumstark-Khan et al. (eds.),
Fundamentals for the Assessment of Risks from Environmental Radiation, 489-496.

2 and 9 mSv. Therefore the air crew belongs to the highly exposed groups of workers, for whom a permanent monitoring of the radiation exposure is indispensable. Little controversial discussed is the fact, that the number of radiation induced damages certainly increases with increasing radiation exposure and the same is true with the probability of cancer incidence. Some studies of cancer incidence and mortality for air crews point towards an increased risk, but from none of the studies it can be concluded, that this is due to the increased radiation exposure.

2. Composition of the Radiation Field in the Atmosphere

2.1. GALACTIC COSMIC RADIATION

With the exception of very rare solar flares, the composition of the radiation field inside the atmosphere is determined by the galactic cosmic radiation. Galactic cosmic radiation originate outside the solar system and impinge isotropically on earth. There is no conclusive proof of the mechanisms accelerating them and of the astrophysical sites where matter becomes cosmic particle radiation. There is no information about the direction of their sources since these particles are scrambled by irregular interstellar magnetic fields on their way towards the earth. Because of their high energies up to 10^{20} eV they most probably originate from supernova explosions, neutron stars, pulsars or other sources where high energetic phenomena are involved. Detected particles consist of 98% baryons and 2% electrons. The baryonic component is composed of 85% protons (hydrogen nuclei), with the remainder being alpha particles (14%) and heavier nuclei (about 1%).

Three factors determine the extent of the radiation exposure in the atmosphere, the solar magnetic field, the earth magnetic field and the absorption depth in the atmosphere. The influence of all factors increases with decreasing energy of the incident primary particles.

During the solar cycle the sun moves through a phase of maximum activity associated with an increase of the intensity of the emitted plasma (solar wind) and an increase of the solar magnetic field. This increase causes a modulation of the galactic cosmic radiation in the energy range below 2 GeV, e.g. particles with lower energies are not capable to enter the interplanetary space; the total amount of particles available decreases. With increasing solar activity the intensity of particles impinging on earth decreases.

A cosmic ray particle has to penetrate the Earth's magnetic field in order to enter the atmosphere. This penetrating ability is related to the ions magnetic rigidity which is given by its momentum divided by its charge. All particles with the same rigidity follow a track with the same curvature in a given magnetic field. For each point inside the magnetosphere and each direction from that point there exists a rigidity threshold below which the cosmic rays are not able to reach this point. This rigidity is called the geomagnetic cut-off rigidity. It is proportional to the magnetic field component perpendicular to the direction of particle motion, i.e. for a particle moving towards the center of Earth the cut-off rigidity has a maximum value at the equator, since the particle moves perpendicular to the field lines and the cut-off rigidity vanishes at the pole, since the particle moves in the direction of the field lines. Therefore geomagnetic shielding is less effective for high inclination orbits than for low inclination orbits. This means that in low inclination orbits only particles of high energy have access. Towards higher inclinations additionally particles of lower energies are observed. Due to the latitude dependent shielding the number of particles increases from lower inclinations towards higher inclinations, and at the poles particles of all energies can enter in the direction of the magnetic field lines. This effect is reduced at civil aviation altitudes by absorption of low energy particles in the higher atmosphere, with the result that at geomagnetic latitudes greater than about 50° no further increase occurs and a plateau in the particle fluence is reached.

A particle which penetrates into the atmosphere loses most of its energy by ionisation of the atoms and molecules. At high energies also nuclear interactions take place. Through fragmentation the number of particles with high atomic number decreases strongly with increasing atmospheric depth. This particles are therefore of minor importance in civil airflight altitudes. Mainly protons are responsible for the production of secondaries in this altitudes. The protons and secondary nucleons create in interactions a million or more secondary particles. Figure 1 shows a schematic representation of the development of such a cascade which consists of three components, the electron-photon, the hadronic and the muonic component [1]. The principal reaction that takes place is the production of secondary neutrons and protons and the pion triplet by an interaction of a primary proton or a neutron with air. Through the production of secondary particles the radiation field builds up to a maximum at about 20 km, at lower altitudes the absorption becomes predominant. With increasing number of particles their mean energy decreases until the energy of the produced particles is below the ionizing energy, the cascade dies out.

2.2. SOLAR COSMIC RAYS

The solar cosmic radiation is mainly produced by solar particle events, sudden sporadic eruptions of the chromosphere of the sun. They develop in minutes, and last for hours up to several days. In the years of maximum solar activity up to 10 events occur, during the years of minimum solar activity only one event can be observed on average only. The largest events, in the average 1-3 per cycle, take place at the end of the period of maximum solar activity. Charged particles (mostly protons, some helium and heavier ions) with energies up to several GeV are emitted. The slopes of the spectra as well as the proton fluences differ considerable from event to event. Not all events produce particles that will reach the Earth. Only high energy particles in the GeV range may contribute to the radiation exposure in the atmosphere following the same mechanism as galactic cosmic rays do.

3. Measurements of Radiation Exposure by Galactic Cosmic Rays

Exposure rates as function of altitude latitude and were measured in a series of flights (e.g. Foelsche [2], Hewitt [3], Schaefer [4]). A compilation of data is given in Reitz et al.[5]. The secondary particle contribution to the equivalent dose rate at solar minimum conditions and 55° geomagnetic latitude is shown in figure 2 [6]. Between 9 and 20 km flight altitude electrons and neutrons deliver the major part to the dose.

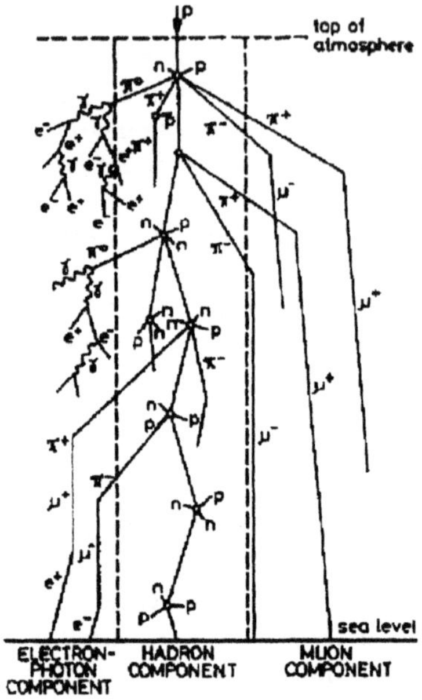

Figure 1: Schematic Representation of the particle production in the atmosphere [2]

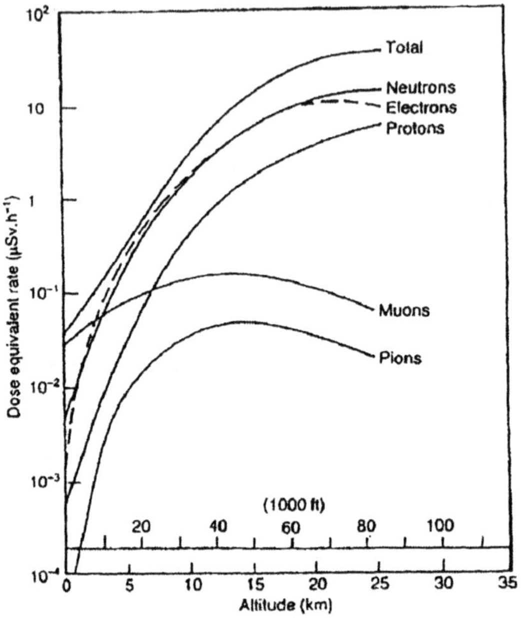

Figure 2: Cosmic ray dose equivalent rates as a function of altitude for different particles at 55° geomagnetic latitude during solar minimum conditions [6]

Based on a survey of experimental and theoretical data the Advisory Committee for Radiation Biology Aspects of the SST derived best estimates on dose equivalent index rates of galactic radiation in dependence of altitude and geomagnetic latitude (Figure 3)[7]. From this figure one may estimate a 65% decrease in equivalent dose rate from solar minimum to solar maximum conditions. At the equator the exposure is lower up to a factor of two to three compared to high latitudes. In dependence of altitude the exposure remains constant above latitudes of 55° to 65°.

In 1990 three facts call for additional measurements: All solar cycles are different, with the exception that they last all approximately 11 years; the strength of the earth magnetic field decreases with time; availability of improved instruments; the difficulty to calculate effective doses or equivalent doses accurately from earlier measurements. Such measurements are performed with several instruments, including tissue equivalent proportional counters (TEPC), Reuter-Stokes counters, Rem-counters, an AEG monitor and passive detector systems. Figure 4 shows recent results for latitudes above 50° in dependence of altitude [8]. The calculation of the dose equivalent uses quality factors according ICRP26 [9], applying quality factors defined in ICRP60 [10] these values has to be multiplied by a factor of 1.1. The measurements are performed at half way from solar maximum condition to minimum conditions. For the maximum exposure during

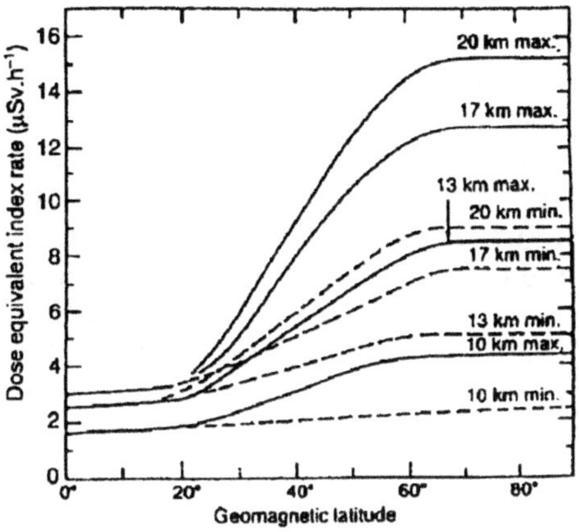

Figure 3: Best values for maximum and minimum galactic dose equivalent rates as function of the geomagnetic latitude for different altitudes. Note: minimum indicates solar maximum conditions, and reverse [7]

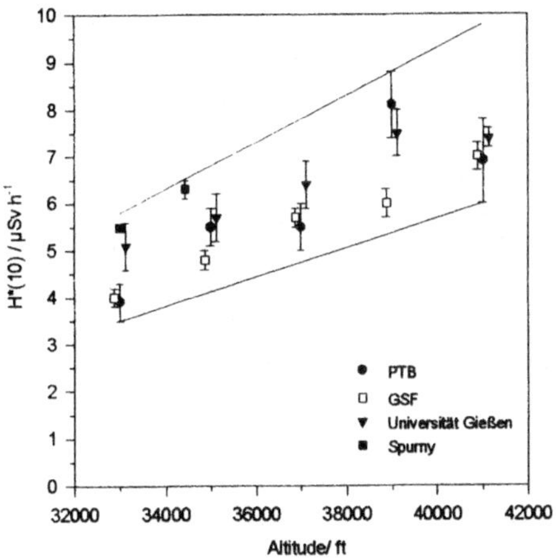

Figure 4: Measurement of ambient dose equivalent rates above a geomagnetic latitude of 50° in dependance of altitudes using quality factors of ICRP26. Error bars are statistical errors only (one standard deviation) [8]

solar minimum, the values has to be multiplied by a factor of 1.3. Actual annually flight hours above 30 000 ft are 920 (maximum allowable duty time of 1000 hours corrected by ground-wait time and ascent and descent times). In the GSF/BG/DLH measuring program it was shown, that 33% of the flight hours were below 33000 ft, 53% between 35000 and 37000 ft, 12.5% between 38000 and 39000 ft and 2.5% up to 41000 ft. Based on this altitude distribution and using the values of the upper limit line in Figure 4 and the above outlined assumptions the maximum exposure during solar minimum arrives at 8.7 mSv per annum. A small group of air crew performs flights in latitudes above 50°, only.

In the flight altitude of the Concorde at about 18 km the shielding of the atmosphere is much more smaller, resulting in higher exposure rates for both, the galactic cosmic radiation and the solar radiation. The Concorde is equipped with a radiation monitoring device. During British Concorde flights maximum dose rates of 45 µSv/h were measured, the mean dose rate is about 12 µSv/h. With a maximum of 547 flight hours the annual exposure arrives at 6.7 mSv. On the French Concorde the measured annual mean exposure is 13 µSv/h, with only 257 flight hours the exposure is 3.3 mSv. The exposure of the Concorde air crew is therefore comparable to that of the subsonic air crew.

4. Measurements of Exposure to Solar Cosmic Rays

There are no direct measurements of exposures to solar cosmic rays in airflight altitudes. Data about energy and fluence spectra are based on measurements of earth bound neutron monitors or by satellites which allow to perform a trustworthy estimate of upper limits of the event. Reports are available from Foelsche for the solar event of February 1956 [3]. The upper limit of the equivalent dose rate was about 30 mSv at 20 km altitude, and still 10 mSv at 10 km. Since this time only events were observed for which the dose rates are very much smaller. This is true also for the large flares of August 1972 and August 1989. During the U.S. Shuttle mission STS-28 (altitude 300 km, inclination 57°) doses between 0.012 and 1.1 mGy were observed at various Orbiter locations during the event of August 1989 [11]. Onboard the Russian space station MIR (altitude 379 km to 410 km, 51.6° inclination) an additional dose of 3.1 mSv was measured during the event of September 1989 [12]. The Space Environment Service Center (SESC) calculated a equivalent dose of 1.1 mSv h^{-1} at an altitude of 19.8 km over the North Pole for the September 1989 event [13].

5. Estimation of Cancer Risks

The annual exposure of air crew is between 2 mSv and 9mSv dependent on the flight route. After an occupational period of 35 years the exposure arrives at a value between 70 and 315 mSv (this is very conservative assumption since at solar maximum times the exposure is lower by a factor of 1.3). This values may be only slightly increased by a potential contribution of the solar cosmic rays to the total exposure. The annual limits recommended by ICRP will be not exceeded even if a combined exposure occurs of the galactic and a solar component similar to the event of February 1956.

The risk of cancer death is assumed to be $4x10^{-5}$ per mSv. A simple calculation indicates an excess risk for the air crew between 0.28% and 1.26%. A solar event like in February 1956 would add a risk of 0.04%. These excess risks to the air crew should be judged by comparing it with the spontaneous risk of cancer death for the public in Germany of 20.5± 0.2%, or in Schleswig-Holstein, a federal state, of 19.8±0.9%. The above given percentages for the air crew excess risks are of the same order as the variations of the spontaneous risks in the public.

There is one problem remaining namely the question whether the biological effectiveness of neutrons is judged correctly or is underestimated. An underestimation would lead strictly to a higher dose equivalent. Two methods are available for investigation this potential underestimate, the biological dosimetry and epidemiological studies.

Biological dosimetry is used for the retrospective determination of exposures. Peripheral lymphocytes in venous blood are analysed for chromosomal aberrations, such as dicentric chromosomes, centric ring chromosomes and translocations. First investigations on pilots point towards an increase of aberrations. The low number of subjects and the limited knowledge on exposure history makes additional studies necessary. In Germany a study on aberrations for female air crew is in progress. The increase of the number of aberrations allows no direct conclusion on the added cancer risk. With a calibration curve the dose has to be estimated from which excess risks can be calculated. Biological dosimetry allows the assessment of doses above 20 mSv.

Epidemiological studies are the base for the estimate of the consequences of the radiation exposure. At low doses such studies alone are not able to deliver useful data for the assessment of the radiation risk, extrapolations to low doses has to be verified by experimental data.

Studies in populations with increased natural radiation exposure in comparison with the residual public (factor 2-3) as example in Kerala, India, and the Gandong region, China, show no definite higher cancer mortality rates in this regions. No increase in leukemia and tumors is observed among nuclear power plant workers. For the Sellafield power plant, UK, this is true up to exposures of 400 mSv and even higher. The number of some tumors increased significantly, but they are too low to make a conclusive judgment.

From 1990 to 1996 five studies are performed on cancer mortality among pilots and four studies on cancer incidence, three among pilots and one among female crew attendants. In most cases their findings remained non-conclusive, due to the limited size of the study population. In none of these studies the exposure history of the individuals are known. Four of the mortality studies showed a reduced risk, in one study this result was significant. One study showed an significant increased risk. The cancer incidence studies among pilots is inconsistent in the type and amount of observed tumors. The study for female flight attendance showed an increased risk for breast cancer, but on very small case numbers. In addition the study population was too young. The estimated mean dose (15-20 mSv) was judged too small by the authors to entirely explain the access risk observed. In summary, the number of observed cancer cases are lower mostly compared to that of the control group, in some studies the contrary is the case. From these studies it is not appropriate to claim an increase of cancer risk among air crew with increased radiation exposure. Other factors has to be taken into account, such as electromagnetic radiation, increased ozone concentration, jet fumes, changed dining behavior, circadian rhythm disturbances and tobacco smoke, to explain the observed results. Lifestyle differences of air crew as compared with the general population have furthermore been recognized. Flight crews are extremely healthy, pilots are subject of at least one medical examination per year. Special care has to be given to the selection of the control groups.

6. Final Conclusions

Although there is no clear answer that air crew members carry an excess risk of cancer induced by cosmic radiation exposure, radiation protection measures are required absolutely, since they are receive indeed a remarkable higher radiation exposure than the public. The contribution to the radiation exposure due to solar activities is certainly low. Since in an individual case a solar flare may cause an exposure of some mSv, pregnant women shall be informed about the increased risk of an exposure especially at times of high solar activity. For the fetus the radiation limit is set to 1 mSv. The working conditions for women should be such that the exposure of the fetus will not exceed 1 mSv following the declaration of pregnancy. By restriction of flight hours on high latitude routes a reduction of the radiation exposure of the crew may be achieved. The residual duty time has to be performed to locations with lower radiation exposure.

The SSK (Strahlenschutzkommission) in Germany and the European Radiation Dosimetry (EURADOS) Group established each a Working group to consider the extend of the exposure to cosmic radiation of the air crew and the scientific and technical problems associated with radiation protection dosimetry for this group. The procedures of estimating the radiation exposure were reviewed. The SSK working group concludes that dose estimation by an appropriate computer program would give adequately accurate values of ambient dose for each individual. Variations of the exposure due to the modulation of the cosmic radiation by the solar cycle has to be determined in appropriate time intervals by additional measurements [8]. The EURADOS working group classifies the air crew as category B workers and concludes that there is a need for record and control in jet aircrafts in order to produce a figure for effective dose. Route doses based on measurements or calculations from flight parameters or in future sub-sonic aircrafts on-board monitoring are considered as appropriate methods for dose estimates [14].

7. References

1. Allkofer, O. C. and Grieder, P. K. F. *Cosmic Rays on Earth*. Physics Data 25-1, ISSN 0344-8401, 1984
2. Foelsche, T., Mendell, R. B., Wilson, J. W., and Adams, R. R. *Measured and calculated neutron spectra and dose equivalent rates at high altitudes; relevance to SST operations and space research*. NASA TN D-7715, October 1974
3. Hewitt, J. E., Hughes, L., McCaslin, J. B., Smith, A. R., Stephens, L. D., Syvertson, C. A., Thomas, R. H., and Tucker, A. B. *Exposure to cosmic-ray neutrons at commerical jet aircraft altitude.* Natural Radiation Environment III, CONF-780422 (Vol.2), Technical Information Center/U.S. Dept. of Energy, 855-881, 1980
4. Schaefer, H. J. Public health aspects of galactic radiation exposure at supersonic transport altitudes. *Aerospace Med.* **39**, 1298-1303 (1968)
5. Reitz, G., Schnuer K., and Shaw, K. (Eds.) *Radiation Exposure of Civil Air Crew. Rad Prot. Dosim.* **48** (1993)
6. Baily, P. E. Neutron radiation dosimetry in high altitude flight personnel. *Aviat. Space Environ. Med.* **53(8)** 808-817 (1982)
7. Advisory Committee for Radiation Biology Aspects of the SST, Final Report. Cosmic radiation exposure in supersonic and subsonic flight. *Aviat. Space Environ. Med.* **46(9)** 1170-1185 (1975)
8. SSK: Die Ermittlung der durch kosmische Strahlung verursachten Strahlenexposition des fliegenden Personals, Stellungnahme der Strahlenschutzkommission (SSK), Bundesministerim für Umwelt, Naturschutz und Reaktorsicherheit. Stuttgart: Gustav Fischer Verlag, Heft 1 (1995)
9. International Commission on Radiological Protection, Recommendations of the International Commission on Radiological Protection, ICRP Publication 26, ICRP (1977)
10. International Commission on Radiological Protection, Recommendations of the International Commission on Radiological Protection, ICRP Publication 60, ICRP (1991)
11. Golightly, M. J., Hardy, A. C., Atwell, W., and Hardy, K. Description, analysis and impact of major solar activity during recent U.S. shuttle missions. *Adv. Space Res.* **12(2-3)** (2)335-(2)338 (1992)
12. Dachev, Ts. P., Matviichuk, Yu. N., Bankov, N. G., Semkova, J. V., Koleva, R. T., Ivanov, Ya. J., Tomov, B. T., Petrov, V. M., Shurshakov, V. A., Bengin, V. V., Machmutov, V. S., Panova, N. A., Kostereva, T. A., Temny, V. V., Ponomarev, Yu., N., and Tykva, R. "MIR" radiation dosimetry results during the solar proton events in September - October 1989. *Adv. Space Res.* **12(2-3)** (2)321-(2)324 (1992)
13. Barish, R. J. Health physics concerns in commercial aviation. *Health Phys.* **59(2)** 199-204 (1990)
14. European Commission, Radiation Protection 85, Exposure of Air Crew to Cosmic Radiation, EURADOS Report 1996-1, 1996

MICRODOSIMETRY OF ENVIRONMENTAL RADIATION FIELDS

F. SPURNY [1] AND J.-F. BOTTOLLIER-DEPOIS [2]

[1] *Department of Radiation Dosimetry, NPI AS CR*
Na Truhlarce 39/64, 18086 Praha 8, Czech Republic
[2] *Institut de Protection et de Sûreté Nucléaire, Département de Protection de la Santé de l'Homme et de Dosimétrie, Service de Dosimétrie, BP n°6*
92265 Fontenay-aux-Roses Cedex, France

1. Introduction

Risks from environmental radiation depend both on the radiation quantity (absorbed dose) and the radiation quality (space and time distributions of dose, energy deposition distributions on the microscopic level). The contribution discusses the possibilities to characterise experimentally the energy deposition distributions on the microscopic level. Three experimental microdosimetry methods are presented: a) tissue equivalent proportional counters (TEPC) [1]; b) chemically etched polyallyldiglycolcarbonate solid state nuclear track detector (SSNTD) [2]; and c) bubble detectors (BD) [3].

Their principle are reviewed, the spectra of microdosimetric quantities in some radiation fields are shown. The microdosimetric spectra determined by means of these methods in some environmental radiation fields are presented, analysed and discussed. The measurements have been performed on board of both subsonic and supersonic aircraft; and behind the shielding of high energy particle accelerators. The spectra obtained are compared, some conclusions drawn from such comparison are outlined.

2. Experimental

2.1. PRINCIPLES OF METHODS

2.1.1. Tissue Equivalent Proportional Counter (TEPC)

The use of TEPC for experimental microdosimetry is based on the simulation of small tissue element by the low pressure gas filling [1]. Several TEPC were constructed, some of them are commercially available. We used NAUSICAA equipment developed in France [4]. Its sensor is a cylindrical low gas pressure TEPC. Its sensitive volume is filled with a propane based tissue equivalent gas mixture at the pressure corresponding to a tissue target with a diameter of 3 μm. The counter wall thickness is equivalent to 10

C. Baumstark-Khan et al. (eds.),
Fundamentals for the Assessment of Risks from Environmental Radiation, 497-502.

mm of tissue. The NAUSICAA equipment measures directly the lineal energy between 0.15 and 1500 keV/µm. Microdosimetric spectra obtained with this equipment are shown in the Figure 1. The instrument is calibrated with a ^{60}Co photon source for low linear energy transfer (LET) region (below ~10 keV/µm), and in the field of AmBe neutron source for high LET region. The actual performance is regularly checked by means of an internal ^{244}Cm alpha particle source. It can be used in the fields of radiation with the equivalent dose rates between ~1 µSv/h and several mSv/h.

2.1.2. LET Spectrometer Based on a SSNTD

SSNTD's are also devices whose response, i.e. the revealing of a particle track through a chemical etching, is related to the energy deposition density, the dimensions of critical volume are estimated to be several nm [5]. The dependence of track parameters on the type and the LET of a particle permits both the particle identification and the spectrometry. We have developed a LET spectrometer based on the chemically etched SSNTD, a polyallyldiglycolcarbonate (PADC), including the automatic evaluation of particle track parameters by means of an optical image analyser based on a Leitz microscope and specially developed software [2, 5]. The etch rate ratio $V = V_T/V_B$ (where V_T is the track etch rate and V_B is the bulk etch rate) is established for each developed track and the spectra of V are converted to LET spectra on the basis of the calibration performed by means of heavy charged particles. Actually, we are able to determine the LET spectra of particles in the region between 12 and 700 keV/µm in tissue. There are practically no limitations on the use of this spectrometer from the point of view of the equivalent dose rate; the background and the superposition of tracks limits their use on the region between ~1 mSv and 100 mSv. Microdosimetric spectra obtained with this PADC LET spectrometer are shown in Figure 2.

2.1.3. Microdosimetry with Bubble Detectors

Bubble detectors [3, 6] can be used as neutron dosemeters. The samples with different neutron energy thresholds are proposed. However, these thresholds are a little margined. Actually, BD's are the equipment for which the effect, the bubble appearance, is related

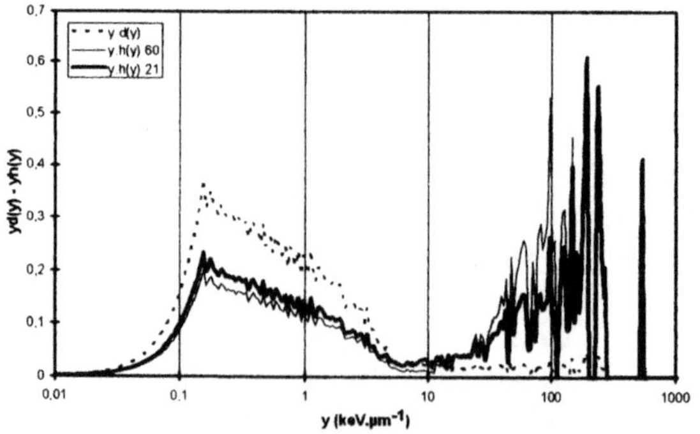

Figure 1. Dose and equivalent dose distributions in lineal energy established with NAUSICAA equipment on board the supersonic aircraft.

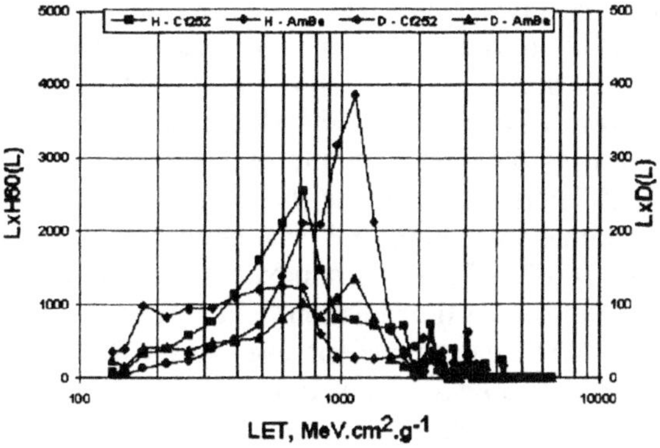

Figure 2. Dose and equivalent dose distributions in LET established with PADC spectrometer in the field of AmBe and ^{252}Cf neutrons.

to the local energy deposition. It is estimated that the dimensions of critical volume of energy deposition in them are of the order of tens of nm [6]. The neutron energy thresholds are deduced from the changes of the LET spectra of secondary particles created through the neutron interactions in the detector or its surroundings. The use of bubble detectors with the different neutron energy thresholds would therefore permit to obtain the integral spectra of the local energy deposition density on the tens of nm scale. Region of LET values covered would be above few tens of keV/μm. The limit of use differs for two types of BD's. There is no influence of the equivalent dose rate on the reading of bubble damage track detectors [6], the maximum equivalent dose depends on the evaluation mode. With eye counting it is of the order of tens of μSv. For the superheated drop detectors and the acoustic mode of the reading [3], a saturation is observed when the equivalent dose rate approaches to 1 mSv/h.

2.2. ENVIRONMENTAL RADIATION FIELDS STUDIED

Environmental radiation represents risk for both external and internal exposure. Only the external exposure is treated in this contribution. In the most of cases, environmental external radiation exposure is caused by a radiation with low LET. Only TEPC can give an estimation of microdosimetric characteristics in low LET region, the threshold for other methods mentioned is above 10 keV/μm. However, there are some environmental radiation fields in which the high LET component of an external exposure is also important. On the Earth's surface, such conditions are typical for the occupational exposure behind the shielding of high energy particle accelerators. Similar fields exists already at the flight altitude of subsonic civil transport, the collective effective dose from global air travel is estimated to about 10,000 manSv for the year [7]. The importance of the dosimetry and the microdosimetry of such environmental radiation fields has been further stressed through the ICRP 60 publication [8]. It recommends to consider the exposure to radiation in jet aircraft as occupational exposure to enhanced natural radiation in the cases when the annual equivalent doses are liable to exceed 1 mSv. We have therefore decided to test the methods mentioned in following radiation fields:

The reference fields behind the shielding of high energy particle accelerators. Such fields have been recently formed behind the shielding of the SPS facility at CERN [9],and/or the phasotron of the Joint Institute of Nuclear Research [10]. Particularly interesting are the fields behind the concrete shielding at CERN and «the hard» field at Dubna. In both cases the neutron (high LET) component contains important fraction of high energy neutrons(~100 MeV). It is assumed that their composition resembles the neutron (high LET) component of the radiation field created by cosmic rays at commercial flight altitudes [11]. The mean equivalent dose rates reach up to 1 mSv/h.

The radiation fields on board of subsonic and/or supersonic aircraft. Many studies of the exposure on board aircraft have been realised last years, also in our two laboratories [12, 13]. Many different dosimetry systems were tested. The results obtained have been compared, analysed and discussed in many publications and some review articles [11, 13]. It should be mentioned that the values of the equivalent dose rate are below 10 μSv/h for subsonic transport, approach to 15 μSv/h in the case of supersonic aircraft.

3. Results and Discussion

3.1. MICRODOSIMETRIC SPECTRA

Some of microdosimetric spectra are presented in Figures 3 and 4. The spectra obtained with NAUSICAA TEPC and with PADC LET spectrometer agree well, also their form is similar in an on-Earth's high energy reference field and on board aircraft.

The first remark is important because no attempt has been done to normalise calibration procedures for both methods. The importance of the second remark consist in an additional prove that the high energy reference fields are suitable to calibrate the detectors to be used on board aircraft.

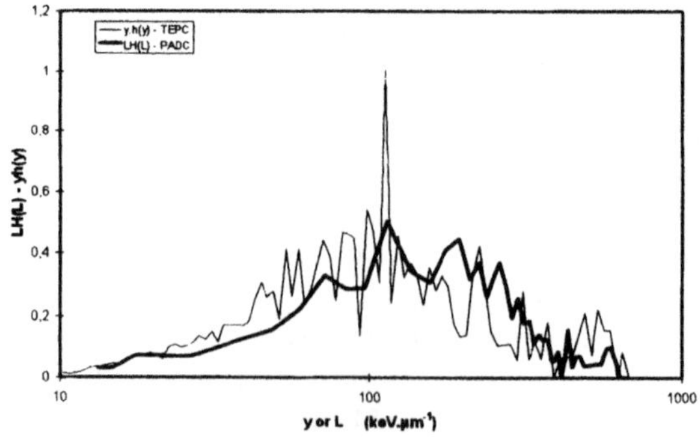

Figure 3. Microdosimetric spectra obtained with TEPC and with PADC LET spectrometer in CERN high energy reference field (top concrete)

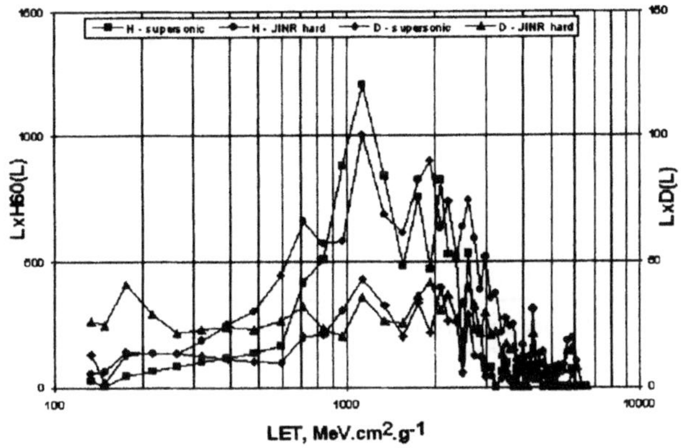

Figure 4. Dose and equivalent dose distributions established with PADC LET spectrometer in Dubna hard field and on supersonic aircraft board

TABLE 1. Total dosimetric characteristics obtained from microdosimetric spectra

Radiation field	ICRP	H - TEPC, μSv[1]	H - PADC, μSv[1]	QF - TEPC	QF - PADC
Supersonic	21	4.6±0.7	6.6±1.0	7.9±0.5	9.9±1.2
Aircraft board	60	6.3±0.9	7.7±1.2	10.9±0.7	12.7±1.6
CERN - top	21	328±45	300±40	8.7±0.7	8.0±1.1
Concrete	60	378±50	330±45	9.0±0.7	9.1±1.2

1 per hour at aircraft; per 106 counts of PIC monitor at CERN

3.2. COMPARISON OF TOTAL DOSIMETRIC AND MICRODOSIMETRIC DATA

Total dosimetric and microdosimetric characteristics are calculated from the spectra established by means of NAUSICAA TEPC and PADC LET spectrometer. The values obtained for some radiation fields are presented in Table 1. Both approaches give comparable results. Besides these results agree also reasonably with reference values, equal for CERN-top concrete field and the equivalent dose to (260±31) (ICRP 21), resp. (300±30) pSv (ICRP 60) [13].

For bubble detectors, we tested the superheated drop detectors in the CERN and subsonic board aircraft fields. The results are presented in the Table 2.

One can see there again the resemblance of both types of radiation fields tested. When one compare these relative responses with the microdosimetric spectra established with PADC LET spectrometer, one can estimate the LET threshold for SDD studied. Analysis has given the preliminary values about 80, 60, resp. 40 keV/μm for SDD 6000, 1000, resp. 100.

TABLE 2. Relative responses of bubble detectors with different thresholds in some radiation fields

Detector	EN(thresh), keV	Relative response [1]	
		CERN - top concrete	subsonic aircraft board
SDD 100	100	0.56±0.08	0.50±0.10
SDD 1000	1000	0.80±0.11	0.70±0.12
SDD 6000	6000	1.11±0.14	1.12±0.15

[1] in terms of H*(10) of AmBe neutrons, relatively to TEPC reference values

4. Conclusions

The results obtained have confirmed the possibilities of all three methods to characterise both dosimetric and microdosimetric characteristics of environmental radiation fields. Further studies are needed to get these possibilities more quantitative. It is particularly important for bubble detectors where the studies of detectors with more thresholds are necessary. The studies should be also complemented by a theoretical analysis of the dependence of local energy deposition density distributions on the critical volume size.

Acknowledgements. The studies were supported through the grants of the AS CR Nos.335402 and 3048606

5. References

1. Rossi H.H.:(1968), in *Radiation Dosimetry* vol. I, eds.: Attix F.H., Roesch W.C., Academic Press, p. 43-93
2. Spurny, F Bednar, J. Johansson, L.and Satherberg, A. (1996), LET spectra of secondary particles in CR 39 track etch detector, *Radiation Measurements* **26**, 645-649
3. Apfel, R.E. (1979), The superheated drop detector, *Nucl.Instrum.Meth.* **162**, 603-608
4. Bouisset, P. et al. (1992), Description de l'instrument NAUSICAA conçu pour effectuer des mesures instantanées de H,D, et du spéctre TLE en champs complexes, *In: IRPA 8, Montréal, mai 1992*, p;463-466
5. Charvat, J. (1986), *LET spectrometry with polymer SSNTD's* . PhD Thesis, Univ. of Prague,
6. Ing, H. (1996), Bubble detection and the assessment of biological risk from space radiations, *Rad.Prot.Dos.* **65**, 421-425
7. *Sources and Effects of Ionizing Radiation* (1993), UNSCEAR Report ; ISBN 92-1-142200-0, Annex A, p.38
8. ICRP Publication 60 (1991), Recommendations; *Annals of ICRP No. 1-3,*
9. Hôfert, M. and Stevenson, G.R. (1994),The CERN-CEC high energy reference field facility, *Report CERN/TIS-RP/94-02/CF*, Geneve, January 1994
10. Aleinikov, E.A. *etal.* (1994), Reference fields for metrology of radiation monitoring, *Rad.Prot.Dos.* **54**, 57-60
11. *Exposure of Air Crew to Cosmic Radiation.* (1996), eds.: I.R. McAulay et al.; EURADOS report 1996-01,
12. Spurny, F. (1997). Experimental approach to the exposure of aircrew to cosmic rays, *Rad. Prot. Dos.* **70**, 409-412
13. Bottollier-Depois, J.-F.(1997), Evaluation de l'exposition au rayonnement cosmique à bord d'avion long courrier, *Rapport SDOS 1997-01*, Fontenay-aux-Roses, février 1997

SPACE RADIATION.

Effects and Risks

R.J.M. FRY

Life Science Division, Oak Ridge National Laboratory
Oak Ridge, TN 37830

1. Introduction

The day of the International Space Station is drawing near. It will not be long before the cargo carried by the shuttle will include the pieces that will take shape as the International Space Station is constructed. When complete the number of crew members will swell the relatively small cohort that have spent months on Mir. Radiation will remain a concern, especially for those on missions with sojourns lasting many months.

Not far behind is the engineers' need for more precise information about the radiation that must be taken into account in designing the shielding of the spacecraft destined for deep space missions. This means that equivalent doses in relevant tissues and the potential risks that such doses may pose must be estimated with considerable confidence.

This review will discuss briefly the information that has, perhaps, improved the assessment of the radiation risks but will concentrate on the research needs.

2. Factors Determining Risks

Shuttle flights have become numerous, with a schedule that is being met with an impressive record. Mir has had an extended life, occupied for months by increasing numbers of cosmonauts and astronauts. These activities and the development of the approaches to dosimetry have increased the understanding of the radiation environments.

There are two fundamental aspects of the radiation environments in space that influence the levels of the potential effects, namely, dose rate and radiation quality; in addition, the duration of the mission dictates the total dose. The shielding may influence the dose rate and the radiation quality at the level of the tissues at risk. The design and management of the missions can influence the total exposure.

2.1. DOSE AND FLUENCE RATE

In the absence of a solar flare in low-earth orbit the spacecraft experiences two distinct radiation sources: the trapped protons from the so-called inner radiation belt and the

C. Baumstark-Khan et al. (eds.),
Fundamentals for the Assessment of Risks from Environmental Radiation, 503-512.

galactic cosmic rays. Exposures to the trapped photons occur with the traversal of the South Atlantic Anomaly (SAA), the region of the inner belt where the protons spiral down closer to earth than in other regions. The duration of the traversals of the SAA depends on altitude and orbital inclination. Measurements made [1] on STS-40, for which the altitude ranged from 278 km to 296 km at an orbital inclination of 39°, indicated the duration of the traversals averaged 19 minutes and ranged from 8-31 minutes. The dose rate rises to about 0.002 mGy min^{-1} in the SAA. The dose integrated over a day is about 0.096 mGy and the dose equivalent is 0.023 mSv d^{-1}. The important point from the biological point of view is that at such dose rates for both the stochastic and deterministic effects there will be a maximum reduction in the effect and will be characteristic of low dose rates. The influence of altitude is well illustrated by the dose rate of 1.64 mGy d^{-1} measured on STS-31, which orbited at about 617 km and at an inclination of 28.5°.

The radiation belts are not static; the configuration of the proton inner belt is affected by changes in magnetic field caused by solar magnetic activity. Golightly et al. [2] noted the westward drift of SAA to be about 0.49 degrees y^{-1} at 300 km and a northward of the centre of the SAA by about 0.12 degrees y^{-1}. Badhwar et al [3] also measured the drifts and found that they were comparable with the rates of drift of the geomagnetic field.

The intensity of the protons of the inner belt varies with the solar cycle; it decreases as the solar activity increases. The mechanism of this modulation is the thermospheric heating of the upper atmosphere by the increasing solar ultraviolet radiation. Studies of the dose contribution of the belt protons on Mir, the space shuttles and the CRRES satellite are resulting in improvements in the previous models such as the AP-8 model [4].

Somewhere outside of our solar system the composite of high-energy charged particles originate and fill the interstellar space. These particles are called galactic cosmic rays (GCR) and consist predominately of protons but also of an important component, the heavier ions with Zs ranging up to 92 (uranium). The energy spectrum of the GCR outside the earth's magnetic field is from some tens of MeV per nucleon to greater than 10^{20} eV. The highest fluence is for the particles in the 100 MeV per nucleon to 10 GeV per nucleon range and the composition is 87 percent protons, 12 percent helium, and one percent heavier ions.

The flux of GCR is affected by the solar activity and is maximum during the period of minimum solar activity. With increasing solar activity, the emitted hot, highly ionized gas from the sun increases. This gas carries magnetic fields away from the sun; the stronger solar magnetic field leads to a reduction of the intensity of the GCR. The variations in the intensity follows the 11-year cycle of the sun's activity.

The difference in fluence, about a factor of 2-3, between the maximum and minimum phases of the solar cycle is sufficient to be of possible importance to the planners of missions to realms beyond the magnetosphere. The maximum GCR fluence in free space is about 4 particles cm^{-2} s^{-1}. In low-earth orbit the GCR fluxes are lower because of the earth's magnetic field, and it is only in the polar regions that the entry of GCRs is less restricted. The contribution of GCR to the total equivalent dose in low-earth orbit varies with the orbital inclination and the solar activity. The LET spectra of the primary and secondary radiations must be known for the calculation of the effective dose, and the necessary measurements are now becoming available.

2.2. THE IMPORTANCE OF DOSE RATE

An important aspect of the interpretation of the biological effects that is known is the dose rates of the trapped proton radiation and the galactic cosmic rays. What is not known is the relationship of LET and the track structure of the heavy ions and their biological effects and therefore the effective dose rates. The understanding of the dose rates of the radiations encountered in space and the relationship of dose rate to the level of effect are important in the estimates of: 1) the risk of cancer, 2) of the hazards that very large solar particle events might pose and 3) of the risk of sterility.

While information about the carcinogenic effect of protons is sparse, it is reasonable to assume that at low dose rates the risk of cancer will be reduced. The question is by how much and what is the effect not only of dose rate but of protraction. The UNSCEAR 1993 report [5] suggested that 0.1 mGy min^{-1} was a suitable definition of low dose rate, at which level both stochastic and deterministic effects would be at a minimum. The problem with such a definition is that there is no indication of over what period of protraction would this definition hold. Consider that 0.1 mGy min^{-1} is about 53 Gy y^{-1} and there is little reason to believe that such a total dose, even at such a low dose rate, would not be disastrous for spermatogenesis. Life shortening in mice becomes dose-rate independent and total-dose dependent at about 0.1 mGy min^{-1} [6] and may be considered to be support for the UNSCEAR definition. The question is by how much is the risk of cancer reduced by lowering the dose rate.

The International Commission on Radiological Protection (ICRP) has adopted a dose-rate reduction factor of 2 which has some theoretical support. There is both experimental [7] and epidemiological [8] evidence that the effect of lowering the dose rate and fractionation is tissue-dependent. The study of the women examined frequently with fluoroscopy, who incurred relatively high total doses but in very small fractions, showed an increase in breast cancer but no increase of lung cancer [8]. These and other studies indicate that a better understanding of dose-rate and fractionation effects would improve the risk estimates for low-LET radiation in general, and in the case of space radiation, protons.

2.3. DOSE RATE AND SOLAR PARTICLE EVENTS

In the last few years there has been important new information about the dose rate of the radiation in large solar particle events. In the past, the interpretation of the effect of solar particle events has led to dire predictions about the severity of the effects, even suggesting that they would be life threatening and have been described in the vernacular as “show stoppers.“

There appeared to be an assumption that the dose rate of the proton radiation in the large solar particle events would be high enough to cause deterministic effects in a manner characteristic of high-dose-rate exposures. It can be seen from fig. 1 that equivalent dose rates during the large solar particle event in October, 1989, fluctuated over the 10 or so days, appearing as if there were multiple events [9]. It is estimated that

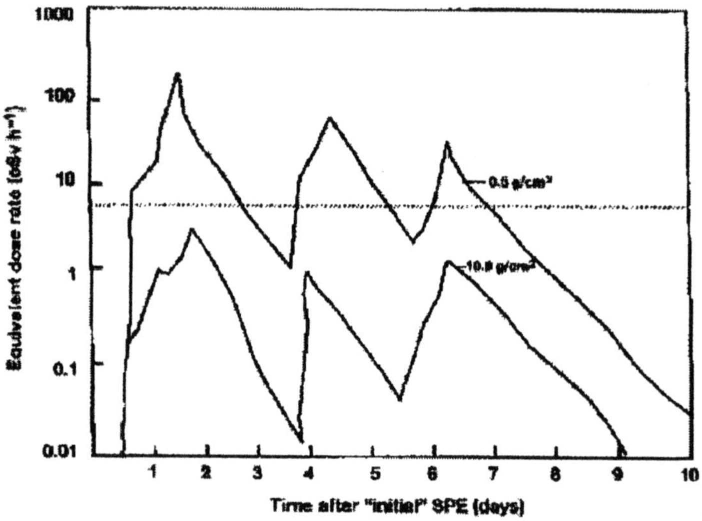

Figure 1. The estimated equivalent dose rate to the skin during the solar particle event in October 1989 behind two levels of shielding (from [9])

behind 10 g/cm^2 shielding the equivalent dose rate at the level of the skin might have reached about 40 mSv h^{-1} but only for a fraction of the total period of the event. At the peak dose rate the effect on the gut would be considered a low-dose-rate effect. It appears that shielding that is feasible will mitigate the acute effects by reducing the dose rate of the protons. However, predictions of the equivalent doses involved in large solar particle events are somewhat compromised by new evidence that there is a higher contribution of heavy charged particles in the solar emission than previously thought.

A thorough analysis of the equivalent dose rates in bone marrow, gut and skin that could occur in a very large solar particle event should be carried out. This information could ensure that the designers of spacecraft for deep space missions have the necessary information to provide adequate shielding.

3. Missions Beyond the Earth's Magnetic Field

Galactic cosmic rays are the predominant source of radiation exposure; solar radiation makes only a small contribution to the radiation environment during quiet times on the surface of the sun. The unpredictable intermittent release of high fluxes of energetic ions can occur in extraordinary amounts, even greater than 10^{32} ergs in a fairly small volume of the solar atmosphere. The terminology describing the events has evolved along with a greater understanding of the Sun and the remarkable eruptions on its surface. Solar flare was used to describe events that now have been distinguished more clearly [10], but as yet there is no single theory of solar flares. The releases of energy that are of most concern in the assessment of risk are now often referred to as solar-proton events instead of the well established term, solar particle event. Fortunately, the very large events are rare, and although short periods of penetrating energies at dose rates that could cause high-dose-rate effects in some tissues can occur, as discussed above, adequate shielding should protect crew members from debilitating acute effects. Within the magnetosphere the geomagnetic shielding is very effective. For example, the large event in September, 1989, increased the radiation on the Mir space station by 0.5 mGy. The October, 1989, event occurred at the time of a large geomagnetic storm that reduced the shielding by the radiation belts and increased the radiation by 0.15 Gy [11]. Solar particle events are not only important because of a potential to cause acute damage under special circumstances but they can contribute also to the total dose incurred by crew members on deep space missions. Because of the long duration of such missions it is essential to restrict the absorbed doses to levels that will not increase the excess cancer mortality later in the lives of the members of the crews to an unacceptable degree. To do so, the risks must be estimated and to do that it must be possible to assess the carcinogenic effects of the heavy-ion component of the galactic cosmic rays as well as the protons. Since there are no human data for the risk of cancer induction by radiation of these qualities, experimental data have to be used. Currently, the approach is to construct relationships between the linear energy transfer (LET) of various radiations and their quality factor (Q). The derivation of the values of Q depends on the determination of the relative biological effectiveness (RBE) of relevant radiations. Based on relative abundances an average Q can then be derived. The dose multiplied by the average Q provides an average equivalent dose. Then it is a simple matter to derive risk estimates based on the risk estimates for low-LET radiation [12].

In 1991 ICRP [12] modified its recommendation about the relationship between quality factor and unrestricted linear energy transfer (L) because of the increased RBE_M values and the reduced effectiveness of heavy ions with LET values greater than 100 keV μm^{-1} reported for cell killing and mutation. The recommendation for the Q-L relationships are shown in table 1.

This results in the change in the shape of the curve for Q as a function of LET shown in fig. 2. The difference between the curves is the precipitous drop in Q at LETs greater than 100 keV μm^{-1} seen in ICRP's curve [12]. This change has little or no impact on the radiation protection standards for earthly beings but it does affect the average Q for galactic cosmic rays to some degree. The effect will largely be due to the

TABLE 1. Q-L Relationships

L in water keV μm^{-1}	Q
< 10	1
10 - 100	0.32 L - 2.2
> 100	$300/\sqrt{L}$

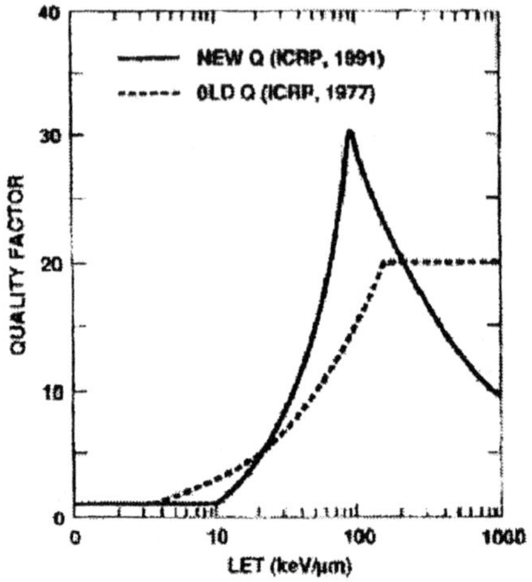

Figure 2. The quality factor as a function of LET (keV µm^{-1}) suggested by ICRP [14]; - - - -, and ICRP [12]; ———

change in the Q for iron, the most abundant of the high-Z particles. Whether the modified relationship is correct for cancer induction remains to be seen. Currently, there are data only for the Harderian gland over the range of pertinent LETs. The relationship of RBE as a function of LET (fig. 3) or prevalences as a function of fluence of particles ranging from 0.4 keV µm^{-1} to 953 keV µm^{-1} do not suggest a reduction in Q at LETs considerably in excess of 100 keV µm^{-1} (fig. 4). Until data for other tumor systems become available this question must remain subjudice. To obtain the required Q entails determination of RBE_Ms for the relevant heavy charged particles and therefore the initial slopes of the dose responses for the induction of cancer by ions of LETs ranging from about 10 keV µm^{-1} to about 400 keV µm^{-1}. The RBE-LET relationship for a more restricted range of LETs, for example, ions with LETs from 80 keV µm^{-1} to 250 keV µm^{-1}, for a number of representative tissues would be an acceptable compromise. The determination of the initial slopes of the responses of the reference radiation, almost certainly γ-radiation, which are required for the calculation of the RBE_Ms is no trivial task. Unfortunately, the same type of information is required for the alternative approach that has been suggested. Curtis et al. [13] have proposed the use of risk cross section. This is defined as the probability per unit fluence, averaged over the organ of interest, of the induction of cancer in that specific organ for a particular type of particle of a given energy and LET. In other words, the cross section is the response for cancer induction per unit fluence. Obviously, similar data are required for both the risk cross-section method and the approach deriving equivalent doses using the Q values obtained experimentally.

The outlook of carrying out the requisite experiments before the engineers begin designing the spacecraft seems dim. Some Q-LET relationship for which there is some

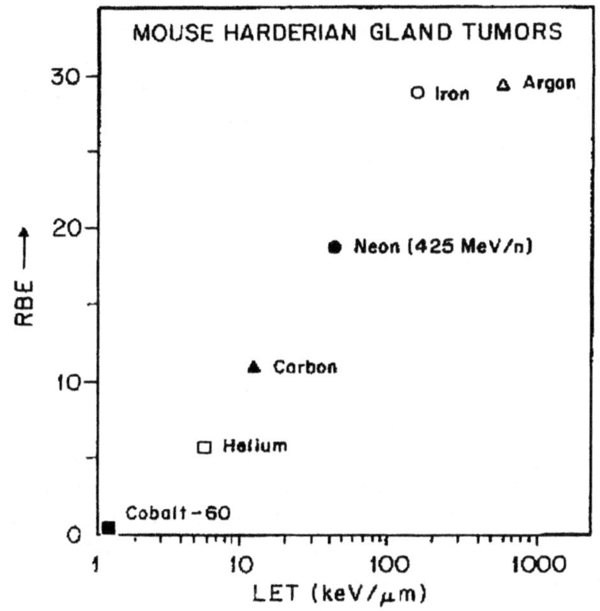

Figure 3. The RBE for the induction of tumors in the Harderian gland of the mouse as a function of LET, keV μm^{-1}. (from [15])

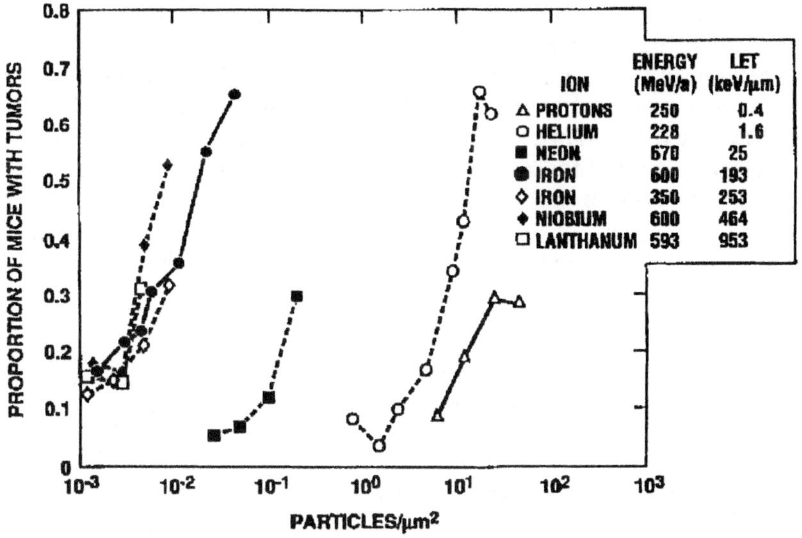

Figure 4. The prevalence of Harderian gland tumors as a function of fluence, particles/μm^{-2} (from [16])

consensus, and the risk estimates from the atomic bomb survivors seem the likely approach that will have to be taken. Hopefully, much better estimates of the absorbed doses will be available when the selection of the shielding material is made, the fragmentation characteristics are known, and a more complete picture of the production of secondaries is available. At least this information would reduce the uncertainties in the estimates of the doses and the risk estimates for low-LET radiation do continue to improve.

4. Non Cancer Effects of Galactic Cosmic Rays

There is a further concern about the risks posed by exposure to galactic cosmic rays that remains unresolved. The nature of the disposition of energy with traversal of the ions with high Zs, such as iron, is so radically different from that of low-LET radiation or even neutrons that our concepts and dogmas of radiobiology of these radiation qualities are not necessarily applicable. The fact that the tracks of the more heavily charged particles may traverse many cells and also deposit energy from delta rays in cells neighboring the particle track raised the question many years ago that lesions unlike any seen with low-LET radiation might occur. Such lesions have been referred to as microlesions, a term that seems singularly inept considering the extent of the deposition of energy. The potential for serious effects is still a matter of concern. It is of course the possibility of damage to contiguous neurons in the CNS that might have consequences that cannot be predicted without more experimental evidence.

It is not that there is solid evidence that exposure over a period of say three years in deep space will result in a loss neurons that will present as a clinically important deficit in the years long after the mission. The concern is that the possibility of an undesirable late effect in CNS cannot be unequivocally discounted. It is essential that an accurate estimate of such a risk is at hand long before long-duration missions beyond the earth's magnetic field are undertaken.

The neurons of the central nervous system (CNS) are considered to be relatively radioresistant, which no doubt is related to their lack of mitotic activity. The same cannot be said of the important support tissues such as neuroglia or the vasculature. Nevertheless, the tolerance dose of the brain for fractionated exposures of low-LET radiation has been estimated in terms of thousands of grays [17].

Curtis et al [18] have estimated the frequency of traversals of cells and nuclei in specific sites in the brain and of retinal cells. As can be seen from tables 2 and 3, the

TABLE 2. Probabilities of Traversals of Cells of the Retina and Brain On Three-Year Deep Space Mission [18]

Target Cells	$Z \geq 15$	Iron Ions ($Z = 26$)
Retinal Cells (180 μm^2)	0.2 - 0.24	0.05 – 0.07
Brain 60 μm^2	0.06 - 0.08	0.01 – 0.02
471 μm^2	0.4 - 0.5	0.1 – 0.13

TABLE 3. Percentage of Nuclei Traversed on a Three-Year Deep Space Mission [18]

	Nuclei Traversed (%)
Ions: $Z > 15$	6.0 - 12.0
Ions: $Z = 26$	1.6 - 3.0

probability of cells in nuclei being hit varies with the size of the cells and nuclei and the Z of the radiation.

These estimates do not take the delta rays into account. There are still no quantitative experimental data on early and late loss of neurons in relation to dose and LET. An important question is whether neurons that have been traversed by high-Z particles, and which survive, develop changes much later in life. Lett and his co-workers [19-20] found that the initial damage, DNA strand breaks induced by either low-LET radiation or high-Z particles, was repaired, but a subsequent breakdown of DNA with age occurred in the retinal cells exposed to high-Z particles, in particular ^{56}Fe ions. These findings are consistent [21] with the finding that the type of DNA damage caused by iron ions is poorly repaired and it is possible that a late breakdown of DNA could occur. But the late changes in DNA irradiated with heavy charged particles require further study. There are also data on the behavioral changes induced by high-Z particles [22-23] that suggest that relatively low doses of iron ions can cause effects. While no reliable estimate of risk of clinically important damage to the CNS can yet be given, the accumulated evidence of DNA damage, loss of neurons, altered motor function and behavior is sufficient to require a careful assessment.

5. References

1. Badhwar G.D.; Konradi A.; Braby L.A.; Atwell W.; Cucinotta F.A., (1994). Measurements of trapped protons and cosmic rays from recent Shuttle flights. *Adv. Space Res.* **14(10)**, 67-72
2. Golightly M.J., Hardy A.C., Hardy K. (1994). Results of time-resolved radiation exposure measurements made during U.S. shuttle missions with a tissue -equivalent proportional counter *Adv. Space Res.* **14(10)**, 923-926
3. Badhwar G.D., Golightly M.J., Konradi A., Atwell W., Kern J.W., Cash B., Benton E.V., Frank A.L., Sanner D., Keegan R.P., Frigo L.A., Petrov V.M., Tchernykh I.V., Akatov Y.A., Shurshakov V.A., Arkhangelsky V.V.,. Kushin V.V, Klyachin N.A., Vana N. and Schoner W. (1995). In-Flight Radiation Measurements on STS-60, *Radiat. Meas.* **26**, 17-34
4. Sawyer D.M. and Vette, J.I. (1976). AP-8 trapped proton environment for solar maximum and solar minimum National Science Data Center, Goddard Space Flight Center, NSSDC/WDC-A-R&S 76-06
5. UNSCEAR United Nations Scientific Committee on the Effects of Atomic Radiation. (1993). UNSCEAR 1993 Report to the General Assembly with Scientific Annexes Annex F, „Influence of Dose and Dose Rate on Stochastic Effects of Radiation“ United Nations New York
6. Sacher G.A., (1964). Dose, dose rate, radiation quality and host factors for radiation-induced life shortening, in: Aging, Carcinogenesis and Radiation Biology, K.C. Smith ed., pp 493-517, Plenum Press, New York
7. Ullrich R.L. and Storer J.B., (1979). Influence of γ-Irradiation on the development of neoplastic disease in mice III. Dose-rate effects, *Radiat. Res* **80**, 325-342

8. Howe G.R. (1995). Lung cancer mortality between 1950 and 1987 after exposure to fractionated moderate-dose-rate ionizing radiation in the Canadian fluoroscopy cohort study and a comparison with lung cancer mortality in the atomic bomb survivors study, *Radiat. Res* **142**, 295-304
9. Simonsen L., Cucinotta F.A., Atwell W. and Nealy J.E. (1991). Temporal analysis of the October 1989 proton flare using computerized anatomical models, *Radiat. Res.* **133**, 1-11
10. Smart D.F and. Shea M.A., (1996). Solar Radiation, in: *Encyclopedia of Applied Physics, Vol 18*, pp. 393-429, VCH Publishers, New York
11. Smart D.F.; Shea M.A.; Dachev T.P.; Bankov N.G.; Petrov V.M.; Bengin V.V. (1994). The dose rate observed on 19-21 October 1989 and its modulation by geophysical effects. *Adv Space Res.* **14(10)**, 651-654.
12. ICRP. (1991). International Commission on Radiological Protection Recommendations of the International Commission on radiological protection, ICRP Publication 60, Annals of the ICRP 21, Pergamon Press, Elmsford, New York
13. Curtis S.B., Nealy J.E., andWilson J.W. (1995). Risk cross sections and their application to risk estimation in the galactic cosmic ray enrironment, *Radiat. Res.* **114**, 57-65
14. ICRP. (1977). International Commission on Radiological Protection Recommendations of the International Commission on Radiological Protection, ICRP Publication 26, Annals of the ICRP (3), Pergamon Press, Oxford
15. Fry R.J.M., Powers-Risius P., Alpen E.L. and Ainsworth E.J. (1989). High-LET radiation carcinogenesis, *Radiat. Res.* **104**, *Suppl.* **8**, S188-S195
16. Alpen E.L., Powers-Risius P., Curtis S.B., Deguzman R. and Fry R.J.M. (1994). Fluence-based relative biological effectiveness for charged particle carcinogenesis in mouse Harderian gland, *Adv. Space Res.* **14(10)**, 573-581
17. Kramer S., Southard M.E., and Mansfield C.M. (1972). Radiation effects and tolerance of the central nervous system, *Front. Radiat. Ther. Oncol* **6**, 332-345
18. Curtis S.B.,. Vasquez M.E, Wilson J., Atwell W., Kim M., and Capala J. (1997). Cosmic Ray hit frequencies in critical sites in the central nervous system, *Adv. Space Res*, in press
19. Lett J.T., Keng P.C., Bergtold D.S. and Howard J. (1987). Effects of heavy ions on rabbit tissues: Induction of DNA strand breaks in retinal photoreceptor cells by high doses of radiation, *Radiat. Environ. Biophys.* **26**, 23-26
20. Williams G.R. and Lett J.T. (1996). Damage to the photoreceptor cells of the rabbit retina from ^{56}Fe ions: Effect of age at exposure, *Adv. Space Res.* **18**, 55-58
21. Rydberg B., Lobrich M., and Cooper P.K. (1994). DNA double-strand breaks induced by high energy neon and iron ions in human fibroblasts. 1. Pulsed field gel electrophoresis method, *Radiat. Res* **139**, 133-141
22. Hunt W.A., Joseph J.A. and Rabin B.M. (1989). Behavioral and neurochemical abnormalities after exposure to low doses of high-energy iron particles, *Adv. Space Res.* **9**, 333-336
23. Rabin B.M., Joseph J. A., Hunt W.A., Dalton T.B., Kardasamy S.B., Harris S.B. and Ludewigt B. (1994). Behavioral endpoints for radiation injury. *Adv Space Res.* **14(10)**, 457-466

SPACE RADIATION ASSESSMENT OF RADIOSENSITIVE BODY ORGANS IN THE INTERNATIONAL SPACE STATION

W. ATWELL

Boeing North American, Space Systems Division
Houston, Texas 77058 USA

1. Introduction

The current NASA astronaut radiation exposure limits [1] for the skin, eye, and blood-forming organs (BFO) for 30 days are: 150, 100, and 25 cSv, respectively; for 1 year: 300, 100, and 50 cSv, respectively; and for an astronaut's career: 600, 400, and 100-400 cSv, respectively. The 100-400 cSv BFO career limit is both age- and gender-specific. Risk coefficients are also listed for other male and female radiosensitive body organs, i.e., lung, thyroid, esophagus, stomach, colon, liver, pancreas, kidney, bladder, and female breast.

In an earlier paper, Atwell [2] presented preliminary astronaut space radiation exposure data for several locations in the International Space Station Alpha (ISSA) using an interim shielding model of the ISSA [3] and the Computerized Anatomical Male (CAM) [4] and Computerized Anatomical Female (CAF) [5,6] models. Since that time, Atwell [7] has corrected and improved the CAM and CAF models. In addition, improvements have been made in the HZETRN code [8,9], which is used to compute space radiation exposures from galactic cosmic radiation (GCR).

In this paper, results of an exhaustive parametric human body organ space radiation exposure study are presented for locations in the ISSA U.S. Laboratory (LAB), U. S. Habitation (HAB), and the ESA Columbus modules from both the Van Allen Belt protons "trapped" in the South Atlantic Anomaly (SAA) and GCR for both solar minimum and maximum conditions. These exposure data should serve as a baseline reference for the radiosensitive body organ risk estimates identified by the National Council on Radiation Protection and Measurements (NCRP) [1].

2. Space Radiation Environment Models

2.1. GEOMAGNETICALLY TRAPPED RADIATION ENVIRONMENT

Models of the trapped proton [10] and electron [11] environments have been developed and can be obtained from the NASA Goddard Space Flight Center/National Space Science Data Center. Since electrons are easily shielded by most typical spacecraft shield-

C. Baumstark-Khan et al. (eds.),
Fundamentals for the Assessment of Risks from Environmental Radiation, 513-518.

ing, only the protons trapped in a region known as the SAA will be considered in this paper. The trapped proton populations vary by approximately a factor of two with solar maximum being about ½ solar minimum. We have assumed a 400 km x 51.6° orbit for the ISSA and computed the 30-day external trapped proton spectra using the AP-8MIN and AP-8MAX models [10]. These proton spectra are attenuated through the respective ISSA shielding distributions [3] computed for various locations in the HAB, LAB, and Columbus modules. These transmitted spectra are then attenuated through the CAM and CAF shielding models to determine the absorbed dose and dose equivalent [12] at various body organs from the SAA protons. The ISSA shielding model and the CAM and CAF models are discussed below.

2.2 GALACTIC COSMIC RADIATION (GCR) ENVIRONMENT

The free-space GCR model of Badhwar-O'Neill [13] attenuated by the geomagnetic field at the 400 km x 51.6° orbit was used to compute the GCR spectrum incident on the ISSA shielding model. This spectrum is further attenuated by the ISSA shielding and is the incident GCR spectrum inside the ISSA at the body organ location of interest. This spectrum is then attenuated by the CAM or CAF body self-shielding to compute the absorbed dose or dose equivalent at the desired location (organ site). This complex multi-attenuation/nuclear interaction process is accomplished with the HZETRN code [8,9] developed at NASA Langley Research Center.

2.3 TRAPPED ELECTRON AND SOLAR PARTICLE EVENT ENVIRONMENTS

Although not considered in this paper, it is worth mentioning that the construction of the ISSA will occur during the rise of the next solar cycle and will require over 1500 crew-hours of EVA [14]. Since the spacesuit is quite thin, trapped electrons in the SAA and outer belt "horns," which are located at high +/- latitudes, represent a significant radiation exposure concern to the EVA crewmen. In addition, solar protons emitted during periods of increased solar activity can possibly be an additional source of radiation exposure to the ISSA crews.

3. Preliminary ISSA Shielding Model

Colborn, et al [3] have developed an interim shielding model of the complete assembly ISSA. The ISSA coordinate system is a right-hand system with the Y-axis along the velocity vector, the X-axis is out the starboard side, and the Z-axis is vertical (up). Using this model and the CADrays program [3], which performs a ray-tracing technique to generate shielding material thickness distributions at selected locations of interest, we have produced ISSA shielding distributions for five (5) locations in the US HAB module, five (5) locations in the US LAB module, and three (3) locations in the ESA Columbus module. Table 1 shows the coordinates of the 13 ISSA locations selected in this study. Figure 1 shows the shielding distributions at location #1 in the US HAB module, location #16 in the US LAB module, and location #1 in the ESA module. The ray-tracing process involved the systematic selection of 720 rays having equal solid angle

TABLE 1. ISSA coordinates/dose point locations*

MODULE	PT. #	X	Y	Z	LOCATION
US HABITATION (HAB)	1	0	1816	-284	ZENITH - CENTER
	5	0	1816	-1240	NADIR - CENTER
	9	0	1816	-762	MIDPOINT - CENTER
	16	0	1816	-1032	NADIR END - CENTER
	23	0	1816	-499	ZENITH END - CENTER
US LABORATORY (LAB)	1	0	2031	-61	AFT - CENTER
	5	0	2987	-61	FORWARD - CENTER
	9	0	2246	-61	AFT END CENTER
	16	0	2509	-61	MIDPOINT - CENTER
	23	0	2779	-61	FWD. END - CENTER
ESA COLUMBUS	1	752	3324	-61	S'BOARD END - CTR
	2	539	3324	-61	MIDPOINT - CENTER
	3	326	3324	-61	PORT END - CENTER

* coordinate units: cm

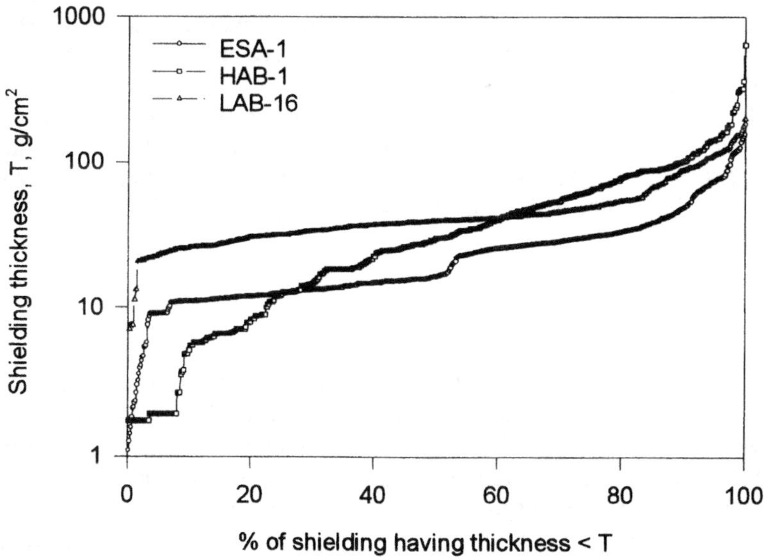

Figure 1. ISSA shielding distributions (720 rays) for the HAB, LAB, and ESA modules.

and to determine the thicknesses of ISSA shielding material intercepted by each ray to produce a 720-ray (thickness) shielding distribution.

4. Male and Female Anatomical Shielding Models

Very detailed human anatomical models have been developed for use in calculating space radiation exposures at various locations in the body. The Computerized Anatomical Male (CAM) [4] represents the 50th percentile US Air Force male. The CAM weighs 73.59 kg (161.9 lb.) and has a height of 175.51 cm (69.1 in.). Using the same type of ray-tracing technique described above, shielding distributions have been generated for numerous locations (organ sites) in the CAM model.

Similarly, the Computerized Anatomical Female (CAF) [5,6] model was developed and was derived by modifying the CAM model by deleting the male organs and adding the female organs: breasts, ovaries, and uterus. The CAF model was then scaled by a factor of 0.92 to represent the 50th percentile US Air Force female. Also, shielding distributions have been generated for numerous locations (organ sites) in the CAF model.

5. Results

Due to the page limitation of this paper, we show (see Table 2) only the absorbed dose and dose equivalent results for the CAM and CAF skin, eye, and BFO at solar minimum for 30 days and 1 year in the US HAB (Pt. #1), US LAB (Pt. #16), and the ESA module (Pt. #1). The crew exposures are given for the contributions from the SAA protons, the GCR environment, and the total (sum of the SAA + GCR). Also shown in comparison are the current NASA astronaut dose limits. The CAF results are only slightly higher (few %) than the CAM. The other four HAB, four LAB, and two ESA results can be obtained from the author; however, these results lie within the range of the HAB-01 and LAB-16 results shown in Table 2. Also, all values are approximately 60-65% less for solar maximum conditions.

The GCR contribution to the deeper-lying radiosensitive organs are comparable to the BFO, but the SAA proton contribution to these organs range from 8% (lung) to 30% (bladder) less than the SAA BFO exposure for both the CAM and CAF. Whereas, the CAF breast is about 18% and the CAM and CAF thyroid is 23% greater than the SAA BFO exposure.

6. Conclusions

Scientific Committee 75 [1] has stated "Risk estimates for radiation induced cancer in individual organs should be based on organ doses. Therefore, it is very important that organ doses be determined for the various radiation environments both within the space vehicle and for extravehicular activity." To this end this capability now exists. In addition, the expected ISSA crew space radiation exposures are well below the current limits and are comparable or less than the crew exposures on the *Mir* space station [15].

TABLE 2. CAM and CAF total space radiation exposures on ISSA (solar minimum)

30-day		HAB Pt. 1			LAB Pt. 16			ESA Pt. 1			Crew Limit	
MODEL	ORGAN	SAA	GCR	Total	SAA	GCR	Total	SAA	GCR	Total	30-day	
		(cSv)	(cSv)	(cSv)	(cSv)	(cSv)	(cSv)	(cSv)	(cSv)	(cSv)		(cSv)
CAM	SKIN	0.74	1.26	2.00	0.24	1.21	1.45	0.64	1.26	1.90	SKIN:	150
CAM	EYE	0.72	1.23	1.95	0.24	1.18	1.42	0.63	1.24	1.87	EYE	100
CAM	BFO	0.43	1.14	1.57	0.17	1.10	1.27	0.41	1.13	1.54	BFO	25
CAF	SKIN	0.76	1.27	2.03	0.25	1.22	1.47	0.65	1.27	1.92	SKIN	150
CAF	EYE	0.74	1.24	1.98	0.24	1.19	1.43	0.64	1.25	1.89	EYE	100
CAF	BFO	0.45	1.16	1.61	0.17	1.12	1.29	0.42	1.16	1.58	BFO	25
1 year		HAB Pt. 1			LAB Pt. 16			ESA Pt. 1			Crew Limit	
MODEL	ORGAN	SAA	GCR	Total	SAA	GCR	Total	SAA	GCR	Total	1-year	
		(cSv)	(cSv)	(cSv)	(cSv)	(cSv)	(cSv)	(cSv)	(cSv)	(cSv)		(cSv)
CAM	SKIN	9.00	15.33	24.33	2.92	14.72	17.64	7.79	15.33	23.12	SKIN	300
CAM	EYE	8.76	14.97	23.73	2.92	14.36	17.28	7.67	15.09	22.75	EYE	200
CAM	BFO	5.23	13.87	19.10	2.07	13.38	15.45	4.99	13.75	18.74	BFO	50
CAF	SKIN	9.19	15.45	24.64	2.99	14.84	17.83	7.87	15.45	23.32	SKIN	300
CAF	EYE	8.98	15.09	24.07	2.96	14.48	17.44	7.80	15.21	23.01	EYE	200
CAF	BFO	5.43	14.11	19.54	2.12	13.63	15.75	5.11	14.11	19.22	BFO	50

7. References

1. "Guidance on radiation received in space activities," (1989) NCRP Report No. 98, National Council on Radiation Protection and Measurements, Bethesda, MD 20814, USA.
2. Atwell, W. (1996) The space radiation environment: an overview for Mir and ISSA missions, AIAA Paper 96-0928, American Institute of Aeronautics and Astronautics, Washington, DC 20024, USA.
3. Colborn, B.L., Ringler, S.J., and Armstrong, T.W. (1995) CADrays 3-D mass model of international space station alpha: interim version through flight assembly stage 19A, Science Applications International Corporation Report No. SAIC-TN-9528, Prospect, TN 38477, USA.
4. Billings, M.P. and Yucker, W.R. (1973) The computerized anatomical man (CAM) model, NASA CR-134043.
5. Yucker, W.R. and Huston, S.L. (1990) Computerized anatomical female. Final report, McDonnell Douglas Corporation Report MDC H 6107, Huntington Beach, CA 92647, USA.
6. Yucker, W.R. (1992) Computerized anatomical female body self-shielding distributions, McDonnell Douglas Space Systems Company Report MDC 92H0749, Huntington Beach, CA 92647, USA.
7. Wu, H., Atwell, W., Cucinotta, F.A., and Yang, C. (1996) Estimate of space radiation-induced cancer risks for international space station orbits, NASA Technical Memorandum 104818.

8. Wilson, J.W., Chun, S.Y., Badavi, F.F., Townsend, L.W., and Lamkin, S.L. (1991) HZETRN: a heavy ion/nucleon transport code for space radiations, NASA Technical Paper 3146, NASA Langley Research Center, Hampton, VA 23665, USA.
9. Private communication, Simonsen, L.C., NASA Langley Research Center, Hampton, VA 23665, USA.
10. Sawyer, D.M. and Vette, J.I. (1976) AP-8 trapped proton environment for solar maximum and solar minimum, Report NSSDC/WDC-A-R&S 76-06, NASA Goddard Space Flight Center, Greenbelt, MD 20771, USA.
11. Vette, J.I. (1991) The AE-8 trapped electron model environment, Report NSSDC/WDC-A-R&S 91-24, NASA Goddard Space Flight Center, Greenbelt, MD 20771, USA.
12. ICRP (1990) *Recommendations of the International Commission on Radiological Protection.* ICRP Publication 60, Pergamon Press, Oxford.
13. Badhwar, G.D. and O'Neill, P.M. (1996) Galactic cosmic radiation model and its application, *Adv. Space Res.* **17**, 7-17.
14. Private communication, Bingham, J., NASA Headquarters, Washington, DC, USA.
15. Badhwar, G.D., Atwell, W., Cash, B.L., Petrov, V.M., Akarov, Yu.A., Tchernykh, I.V., Shurshakov, V.A., and Arkhangelsky, V.A. (1996) Radiation environment on the *Mir* orbital station during solar minimum, presented at Committee on Space Research 31st Scientific Assembly, The University of Birmingham, England (to be published in *Adv. In Space Res.*).

RADIATION PROTECTION CONCEPTS.

Recommendations for Research Needed

D. HARDER [1], V. KLENER [2] AND
W. ATWELL [3], M. BENO [4], E.W. BREITBART [5], C. CREMER [6],
R.J.M. FRY [7], R. GREINERT [5], D.-P. HÄDER [8], P. KNUSCHKE [9],
G. OBE [10], G. REITZ [11], P. RETTBERG [11], F. SPURNY [12],
H. TANOOKA [13], L. TOMÁŠEK [14], B. VOLKMER [5]

[1] *Institute of Medical Physics and Biophysics, Göttingen, Germany*
[2] *State Office for Nuclear Safety, Praha, Czech Republic*
[3] *Boeing North American, Space Systems Division, Houston, USA*
[4] *Institute of Preventive and Clinical Medicine, Bratislava, Slovak Republic*
[5] *Dermatologisches Zentrum Buxtehude, Buxtehude, Germany*
[6] *Applied Optics and Information Processing, Institute of Applied Physics, Heidelberg, Germany*
[7] *Life Science Division, Oak Ridge National Laboratory, Oak Ridge, USA*
[8] *Friedrich-Alexander-Universität, Institut für Botanik und Pharmazeutische Biologie, Erlangen, Germany*
[9] *Dept. of Dermatology, University of Technology, Dresden, Germany*
[10] *Dept. of Genetics, University of Essen, Essen, Germany*
[11] *Deutsches Zentrum für Luft- und Raumfahrt, Institut für Luft- und Raumfahrtmedizin, Abteilung Strahlenbiologie, Germany*
[12] *Department of Radiation Dosimetry, NPI AS CR, Praha, Czech Republic*
[13] *Genetics Division, National Cancer Center Research Institute, Tokyo, Japan*
[14] *National Radiation Protection Institute, Prague, Czech Republic*

1. Introduction

Environmental radiation exposure is here understood to be the exposure to natural or man-made sources of ionizing or non-ionizing radiation due to private or professional activities, including radiation exposure incurred by medical examinations and treatments. Ultraviolet light (UV) will be treated here as the main source of risk for non-ionizing radiation. *Radiation protection concepts* are the main strategies devised to prevent or duly restrict adverse health effects of the irradiation of persons or populations. For instance, in the case of solar UV radiation, the exposure to UV can be largely

C. Baumstark-Khan et al. (eds.),
Fundamentals for the Assessment of Risks from Environmental Radiation, 519-526.

reduced by a change in life style. *Needs for future research*, from which radiation protection will benefit, will be the main subject of the following considerations.

2. Brief statement of knowledge

2.1. IONIZING RADIATION

Contemporary formulations of the conceptual framework for the protection of man from adverse health effects of ionizing radiation have been given in the „1990 Recommendations of the International Commission on Radiological Protection" and in the „International Basic Safety Standards for Protection Against Ionizing Radiation and for the Safety of Radiation Sources". In short, the strategy of radiation protection aims at providing a practical and pragmatic system of safety for man (and indirectly for the protection of the environment) without unduly limiting the beneficial practices of medicine and industry giving rise to radiation exposure. The philosophy of radiation protection is based on social as well as scientific judgements, first of all on the knowledge of the biological effects of ionizing radiation. Radiation protection intends to prevent the occurrence of deterministic effects by keeping doses below the relevant thresholds and to promote all reasonable steps to reduce the induction of stochastic effects.

Everyone in the world is exposed to radiation from natural and artificial sources. In order to achieve its above aim, the system of radiation protection has to identify particular exposure circumstances and conditions requiring and allowing appropriate regulatory actions. The procedures controlling radiation exposures could be viewed differently for two rather distinctive categories of exposure. Some human activities increase the overall exposure to radiation, either introducing new radiation sources, operational programs, pathways, and individuals, or modifying the existing pathways from existing sources to man and thus increasing the exposure of individuals or the number of individuals exposed. This sort of human activity, related to sources being under control, is described as *„practices"*. Other human activities can decrease the already existing overall exposure by removing sources, modifing pathways or reducing the number of exposed individuals. These activities are described as *„interventions"* and include, e.g., remedial actions after radiological emergencies.

In case of the practices, radiation protection is guided by three principles: (1) *The justification of a practice*: No practice involving exposures to radiation should be adopted unless it produces sufficient benefit to the exposed individual or to society to offset the radiation detriment it causes. (2) *The optimization of radiation protection*: In relation to any particular source within a practice, the magnitude of individual doses, the number of people exposed, and the likelihood of incurring exposures where these are not certain to be received should all be kept as low as reasonably achievable, economic and social factors being taken into account (ALARA principle). (3) *The limitation of individual doses and risks*: The exposure of individuals resulting from the combination of all the relevant practices should be subject to dose limits, or to some control of risk in the case of potential exposures. The dose limits are set higher for radiation workers than for members of the public, as the former accept the working and exposure conditions willingly, are trained in avoiding dangers of their workplaces, and are regularly checked for the health status and monitored for doses received.

In the case of interventions the remedial actions should do more good than harm, i.e. the reduction in detriment resulting from the reduction in dose should be sufficient to justify the harm and the costs of the intervention. The form, scale, and duration of intervention should be optimized so that the net benefit of the reduction of dose should be maximised.

2.2. ULTRAVIOLET RADIATION

During the last decade skin cancer incidence has increased worldwide, with the steepest gradient compared to other cancers under investigation in the Caucasian population. The United Nations Environment Programme (UNEP) has estimated that over 2 million nonmelanoma skin cancers and 200,000 malignant melanomas (MM) occur globally each year. In the case of nonmelanoma skin cancers (basal cell carcinoma, BCC, and squamous cell carcinoma, SCC) this substantial rise in skin cancer has been mainly attributed to earlier changes in lifestyle of large sections of the population (e. g. spending more leisure time and holidays in sunny environments and using artificial UV sources like sun-beds), resulting in an increased exposure to both UVB and UVA. While an enhanced cumulative UV dose increases the risk for SCC and to a lesser extent for BCC, MM induction turns out to be more dependent on sunburns in childhood and intermittent UV exposure. WHO has estimated that up to 20 % of the cataracts resulting in blindness of 12-15 million people could be due to UV-exposure. In addition to skin cancer and eye diseases UV exposure causes suppression of the immune system in all populations over the world. This might lead to adverse consequences, e. g., to a reduction in the success of infectious disease immunization programs, particularly in areas where UV intensities are high.

Beyond this, all other factors unchanged, the depletion in stratospheric ozone and a corresponding increase at ground level of highly genotoxic UVB radiation has to be considered as a serious risk factor for higher incidences of skin cancer in the future. There are indications that in the northern hemisphere the thickness of the stratospheric ozone layer is decreasing by several percent per decade. In the event of a 10 % decrease in stratospheric ozone, other conditions and behaviour remaining unchanged, UNEP expects an additional 300,000 non-melanoma and 4,500 melanoma skin cancers globally each year. However, these numbers correspond to present estimates and have to be further corroborated by a) highly standardized measurements of ozone concentration and UVB irradiation at ground level and b) the establishment of more accurate dose-effect relationships. From epidemiological studies several risk factors especially for the induction of skin cancer by UV-radiation are known, including skin type (for BCC and SCC) and lentigines, multiple melanocytic nevi, atypical and congenital nevi as well as familial malignant melamoma (for MM).

The consensus meeting on "educational needs for primary and secondary prevention of melanoma in Europe" by the EORTC Melanoma Group, published in 1991, as well as a consensus conference on "early melanoma" by the National Institutes of Health, USA, in 1992, and the consensus meeting "How to decrease morbidity and mortality of skin cancer" by the "Commission of Early Detection and Prevention of Skin Cancer" of the Society of Dermatological Prevention (ADP e.V.) in Hamburg, 1994, emphazised the importance of "public education" to achieve the reduction of morbidity and mortality of melanoma and nonmelanoma skin cancer. In the 1960s Australia, particularly Queensland, started the first education campaign on secondary prevention of skin cancer and on hazards of ultraviolet radiation. In the 1980s many European countries (Austria, Belgium, Denmark, England, France, Germany, North Italy, Netherlands, Scotland, Sweden and Switzerland) initiated similar campaigns. Overseas public skin cancer education campaigns were started in Canada, Hawai, New Zealand and in the United States of America. In order to implement far-reaching educational strategies, the cooperation of different social and scientific groups such as medical and scientific societies, state-run institutions, public health services and private enterprises is required. Experience has shown that „marketing“ strategies and methods such as by information leaflets, posters, newspapers, radio and television, advertisements and television spots, are useful instruments for public information and education. Special educational programs, i.e., for schools and kindergartens, were elaborated. In addition, some countries initiated professional education campaigns, some intervention strategies were run nationwide, and some were promoted locally. In different countries miscellaneous types of "screening" were realized, such as „mass screening“ in doctors' offices, by public health services, on beaches or in towns, as well as screening or self-screening of "risk groups" such as melanoma patients and their relatives, and individuals with multiple and/or "atypical" moles. Instructions for self-screening were given during most educational campaigns.

Due to very successful results, the use of therapeutical UV radiation in the treatment of different diseases, especially chronic inflammatory skin diseases, has increased within the last few decades. Taken together, the risk of skin cancer development by phototherapy seems to be small if all quality standards concerning artificial UV sources are applied.

2.3. ECOSYSTEMS

Terrestrial and aquatic ecosystems are under considerable stress from solar short wavelength UV irradiation, and the increasing depletion of stratospheric ozone may lead to detrimental effects. Decreases in biomass productivity may effect both crop plants and aquatic ecosystems, leading to decreased food production in agriculture and the oceans. Because of different sensitivity and adaptability of species, changes in species composition within the ecosystem can be predicted. The biosphere cycles a considerable amount of atmospheric carbon (200 gt), and aquatic ecosystems are believed to remove up to 4 gt of carbon annually from the atmosphere by sedimenting organic and inorganic carbon into the deep sea (biological pump). Any reduction in carbon uptake by aquatic ecosystems will augment the greenhouse effect and global climate change.

3. Unsolved Problems

3.1. IONIZING RADIATION:

3.1.1. Radiobiology

More data are needed for the amendment of the *accuracy of risk assessment* for both deterministic and stochastic effects. Mechanisms of radiation oncogenesis should be elucidated on the basis of recent advances in biophysics and molecular biology. Loss of function of tumor suppressor genes due to changes in chromosomal structure is considered a major factor in oncogenesis. Evidence of these genes being targets of radiation action comes from studies of germ-line mutations that predispose to cancer. *Epidemiological studies* on exposed human populations are expected to bring more precise quantitative data, enabling the amendment of the values of risk coefficients and the clarification in more detail of the influence of dose and dose rate on stochastic effects. The reliability of epidemiological results will greatly benefit from increased accuracy in the assessment of exposure, and insufficient, out-of-date estimates of exposure will have to be replaced by realistic estimates of doses due to external and internal radiation exposures. In the presentation of results, epidemiologists will put more effort into clarifying the limits of uncertainty of their results, especially the possible magnitude of false positive results and the possible blurring by confounders such as pre-medications with immunosuppressors or viral contributions to cancer induction. The search for more sensitive systems of *biological monitoring* will continue, utilizing among others the methods of cytology and molecular biology. Efforts should also be directed to the attempt to apply cytogenetic methods in cases of local irradiation of target tissues (e.g., skin in the case of accidents and bronchial epithelium with uranium miners). *Adaptive response to radiation, induced radiorestistance or radiosensitivity* in cells and organisms appear to be now reliably documented as existing facts, but still many unsolved questions in this field remain, especially the quantification of these effects in humans and their conceivable relevance for setting or correcting standards in radiation protection. The same holds for *individual variations of radiosensitivity* in certain subgroups of the population. For example, an area of further studies could be the specific features of radiation hazard in the *developing human organism* in utero and in children from the aspect of both deterministic (e.g., brain development) and stochastic effects. Studies of the *radioprotective effect* of selected pharmaceuticals and biological factors could improve the understanding of defending and repairing mechanisms and moreover contribute to the application of these factors in radiotherapy. The *microdosimetry of nanometer regions* is presently being developed into a valuable instrument with the potential to clarify radiobiological mechanisms in various branches of radiation protection research. For instance, the proportionality of the yield of locally clustered DNA damage with dose-mean restricted LET - contrary to the absence of any strong LET dependence of the yield of locally uncorrelated double-strand breaks - together with the same proportionality observed with exchange-type chromosomal aberrations and in-vitro cell transformations in the low-dose region, has been a strong argument to associate the latter processes with clustered DNA damage. Further development of computer-based as well as nanotechnological experimental microdosimetry is therefore regarded as a promising aspect for the advancement of radiation protection research.

3.1.2. Monitoring and Dose Assessment:

Organizational and technical standards of *personal dosimetry* will increase. Further progress in the dosimetry of heavy and high energetic particles with the perspective of applications in jet-aircraft flights, space missions and upon workers around accelerators is necessary. Also, the methods of the monitoring of *internal contaminations* of individuals should be amended and broadly implemented in operational radiation protection.

The lowering of dose limits by ICRP may require enhanced accuracy of individual dose assessments. The release of radionuclides into the environment and the difficult problem of the depositories of radioactive waste need continuing studies of the fields of *migration of pollutants* within the environment, developing and testing models aiming at the evaluation of dose assessment for the members of the public, and establishing satisfactory monitoring programs. With regard to the significance of potential (emergency) exposures an appropriate *warning system* should be introduced both in the personal and environmental monitoring areas.

3.1.3. Optimized procedures of radiation protection:
Regulatory measures should be reconsidered in detail for some specific categories of exposure. Efforts should be directed at decreasing the probability of *potential exposures* from non-nuclear sources – such as industrial irradiators and particle accelerators – which have been the cause of deaths and severe injuries in recent years. Another category is *chronic exposures,* including the exposure of the public from *land contaminated in the past.* There are particular difficulties, especially in appropriating funds and attributing the costs of intervention and amelioration to those who have caused large-area contaminations. Another problem is to define a well-balanced approach to the protection of *jet-aircraft crews, frequent flyers and the participants in space missions.* Their exposure could be envisaged as a case of enhanced natural background and distinguished from those situations where the natural background is excluded from the radiation protection considerations as an uncontrollable source. Another example of enhanced natural exposure is the *inhalation exposure to radon daughters in dwellings and at underground work.* This issue needs also continuing attention, and the rational attitude to the extent and means of the protection of the public is still questioned. In operational radiation protection it is important to continue developing pragmatic tools facilitating application of *risk/benefit analyses* and other decision-aiding techniques. A concept of monetary equivalent of man-Sv has been recommended for this purpose, but it has not proven to be applicable in its present form in the international framework. Instead, the optimization of practices and interventions in the form of internationally recommended *reference values and reference procedures,* for instance in medical diagnostic uses of radiation, appears to be realistic and to find broad acceptance. Improved methods of assessing the *probability of causation* in cases of compensation, especially improvements in retrospective dosimetry, and improved practices of converting the probability of causation into the actual extent of a compensation, also present a desire for achievements in the future.

3.1.4. Radiation protection education and information:
In the past it has been shown that public information about basic facts of radiobiology and the education of the public to understand basic approaches of establishing radiation safety in peaceful applications of radiation sources have largely been the matter of non-specialists and have therefore sometimes suffered from diffuse presentation of facts and from considerable over- or underestimates of risks. It is now clear that ways of improving this situation are not only urgently needed, but form a task that needs scientific ideas and views, not the least because the freedom of opinion should not be entangled. Therefore, a most appropriate target of safety research has to be the sum of approaches

leading to objective information and education of the public in matters of radiation protection.

3.2. ULTRAVIOLET RADIATION

3.2.1. Photobiology:
A variety of problems have to be solved in order to understand the biological mechanisms of skin cancer induction. These include: a) accurate measurement of the action spectra and repair of UV induced DNA damages in different cell types of human skin, b) investigation of the mechanisms of damage to DNA occuring at longer wavelength, particulary by the more penetrating UVA, c) screening for mutations in oncogenes and tumor suppressor genes which are associated with skin cancer, d) understanding of the influences of human genetic predispositions, particulary in the case of malignant melanoma, e) proper description of the role of UV radiation in skin cancer promotion and its influence on intra- and extracellular signalling events, and search for UV-induced epigenetic changes which may be relevant for phenotypic change, f) identification of molecular processes which lead to UV-induced immunosuppression, and g) development of molecular biological methods to carry out molecular epidemiology in the case of nonmelanotic skin cancers as well as malignant melanoma.

3.2.2. Monitoring and Dose Assessment:
Up to now there are not enough data available for the assessment of the individual UV exposure in epidemiological investigations of selected groups. Physical measurements with stationary instruments in combination with retrospective questionnaires are not applicable due to the strong dependency of the individual UV dose on the behavior, the continuous changes in the orientation of the subject to the sun and the uncertainties in remembering and protocolling the periods of outdoor activities. Integrating chemical and recently-developed biological UV dosimeters have to be tested and judged carefully with regard to their dosimetric properties, especially for the application with UVA radiation as well as with regard to their applicability as personal dosimeters. Personal UV dosimetry systems have to be improved into certified methods, especially by alining their response characteristics with standardized action spectra as defined for UV erythema by CIE in 1987 and for UV hazard by ACIGH in 1995.

3.2.3. Radiation protection education and information:
Prevention education programs and screening for cancer are usually undertaken in the hope that cancer will either be prevented primarily or a malignancy will be discovered sufficiently early to cure patients. The desired endpoints are a reduction of cancer-associated morbidity and mortality. In the short term it is necessary to try to change beliefs, attitudes and habits of cancer risk groups and of the general population. For public health efforts to be effective in changing attitudes and habits regarding UV protection, and to decrease mortality and morbidity of skin cancer, it is essential to understand the responses of the public to skin cancer detection efforts and educational programs. The above-mentioned intervention programs carried out in different countries have to be standardized to an internationally-accepted level and should be applied taking into account the national requirements and specifities. Epidemiological studies

related to personal UV-exposure levels, parameters of individual behavior and objective skin properties will be very useful for supporting public education campaigns.

Different short- and long-term criteria have to be applied in order to investigate and evaluate the effectiveness of intervention strategies. Among these criteria, the knowledge and change of behaviour, the outcomes of screenings, tumor thicknesses of malignant melanoma, mortality, cost effectiveness and other measures will have to be applied.

3.3. ECOSYSTEMS

Global long-term monitoring of biologically relevant radiation is needed to predict significant increases in UVB radiation. Understanding the effects on molecular processes in photosynthesis may permit to derive quantitative predictions on damage and crop losses. Investigation of the protective measures (induction of screening substances and repair mechanisms) on the cellular, organismal and ecosystem level are important to evaluate adaptability. Molecular biological countermeasures can be taken to enhance UVB resistance in crop plants. However, this approach is not practicable in wild plant populations and aquatic organisms.

4. Justification for Further Research Needed and Recommended Approach.

The review of unsolved problems in (3) shows topics relevant for further progress in radiation protection. Without clarification of yet unresolved issues, radiation protection would stagnate at today's standard. Urgently-needed new data could in the future be collected from the results of research institutions in many countries and from the documents published from the reputed scientific bodies, such as the United Nations Scientific Committee on the Effects of Atomic Radiations, the International Commission on Radiological Protection, etc. The involvement of NATO in supporting world-wide radiation protection research would form a most welcome avenue for its peaceful mission. The activities needed to fulfill the requirements of radiation protection interventions are rather costly, and therefore it might be economical to invest money in radiation protection research with the aim to rationalize operational measures, to withdraw activities considered as non-effective and to introduce new ones justified by the results of research. Not the least of the desires to be noted at this occasion is that even better cooperation and avoidance of contradictions between international and governmental commissions and agencies be growing in the field of radiation protection. The participants of the NATO ARW on the assessment of risk from environmental radiation would greatly appreciate future contributions of NATO to support research aimed at the establishment of optimized procedures in global radiation protection.

Author Index

Subject Index

A

B

C

D

S

T

GPSR Compliance

The European Union's (EU) General Product Safety Regulation (GPSR) is a set of rules that requires consumer products to be safe and our obligations to ensure this.

If you have any concerns about our products, you can contact us on ProductSafety@springernature.com

In case Publisher is established outside the EU, the EU authorized representative is:

Springer Nature Customer Service Center GmbH
Europaplatz 3
69115 Heidelberg, Germany

Batch number: 10824484

Printed by Printforce, the Netherlands